Bohmian Mechanics

Detlef Dürr · Stefan Teufel

Bohmian Mechanics

The Physics and Mathematics of Quantum Theory

Prof. Detlef Dürr
Universität München
Fak. Mathematik
Theresienstr. 39
80333 München
Germany
detlef.duerr@lmu.de

Prof. Stefan Teufel
Mathematisches Institut
Auf der Morgenstelle 10
72076 Tübingen
Germany
stefan.teufel@uni-tuebingen.de

ISBN 978-3-540-89343-1 e-ISBN 978-3-540-89344-8
DOI 10.1007/978-3-540-89344-8
Springer Dordrecht Heidelberg London New York

Library of Congress Control Number: 2009922149

Cover design: deblik, Berlin

Printed on acid-free paper

Springer is part of Springer Science+Business Media (www.springer.com)

Preface

This book is about Bohmian mechanics, a non-relativistic quantum theory based on a particle ontology. As such it is a consistent theory of quantum phenomena, i.e., none of the mysteries and paradoxes which plague the usual descriptions of quantum mechanics arise. The most important message of this book is that quantum mechanics, as defined by its most general mathematical formalism, finds its explication in the statistical analysis of Bohmian mechanics following Boltzmann's ideas.

The book connects the physics with the abstract mathematical formalism and prepares all that is needed to achieve a commonsense understanding of the non-relativistic quantum world. Therefore this book may be of interest to both physicists and mathematicians. The latter, who usually aim at unerring precision, are often put off by the mystical-sounding phrases surrounding the abstract mathematics of quantum mechanics. In this book we aim at a precision which will also be acceptable to mathematicians.

Bohmian mechanics, named after its inventor David Bohm,[1] is about the motion of particles. The positions of particles constitute the primitive variables, the primary ontology. For a quantum physicist the easiest way to grasp Bohmian mechanics and to write down its defining equations is to apply the dictum: Whenever you say particle, mean it!

The key insight for analyzing Bohmian mechanics lies within the foundations of statistical physics. The reader will find it worth the trouble to work through Chap. 2 on classical physics and Chap. 4 on chance, which are aimed at the understanding of the statistical analysis of a physical theory as it was developed by the great physicist Ludwig Boltzmann. Typicality is the ticket to get to the statistical import of Bohmian mechanics, which is succinctly captured by $\rho = |\psi|^2$. The justification for

[1] The equations were in fact already written down by the mathematical physicist Erwin Madelung, even before the famous physicist Louis de Broglie suggested the equations at the famous 1927 Solvay conference. But these are unimportant historical details, which have *no significance* for the understanding of the theory. David Bohm was not aware of these early attempts, and moreover he presented the full implications of the theory for the quantum formalism. The theory is also called the pilot wave theory or the de Broglie–Bohm theory, but Bohm himself called it the causal interpretation of quantum mechanics.

this and the analysis of its consequences are two central points in the book. One major consequence is the emergence of the abstract mathematical structure of quantum mechanics, observables as operators on Hilbert space, POVMS, and Heisenberg's uncertainty relation. All this and more follows from the theory of particles in motion.

But this is not the only reason for the inclusion of some purely mathematical chapters. Schrödinger's cat story is world famous. A common argument to diminish the measurement problem is that the quantum mechanical description of a cat in a box is so complicated, the mathematics so extraordinarily involved, that nobody has done that yet. But, so the argument continues, it could in principle be done, and if all the mathematics is done properly and if one introduces all the observables with their domains of definitions in a proper manner, in short, if one does everything in a mathematically correct way, then there is no measurement problem. This answer also appears in disguise as the claim that decoherence solves the measurement problem. But this is false! It is precisely because one can in principle describe a cat in a box quantum mechanically that the problem is there and embarrassingly plain to see. We have included all the mathematics required to ensure that no student of the subject can be tricked into believing that everything in quantum physics would be alright if only the mathematics were done properly.

Bohmian mechanics has been around since 1952. It was promoted by John Bell in the second half of the last century. In particular, it was the manifestly nonlocal structure of Bohmian mechanics that led Bell to his celebrated inequalities, which allow us to check experimentally whether nature is local. Experiments have proved that nature is nonlocal, just as Bohmian mechanics predicted. Nevertheless there was once a time when physicists said that Bell's inequalities proved that Bohmian mechanics was impossible. In fact, all kinds of criticisms have been raised against Bohmian mechanics. Since Bohmian mechanics is so simple and straightforward, only one criticism remains: there must be something wrong with Bohmian mechanics, otherwise it would be taught. And as a consequence, Bohmian mechanics is not taught because there must be something wrong with it, otherwise it would be taught. We try in this book to show how Bohmian mechanics could be taught.

Any physicist who is ready to quantize everything under his pen should know what quantization means in the simplest and established frame of non-relativistic physics, and learn what conclusions should be drawn from that. The one conclusion which cannot be drawn is that nothing exists, or more precisely, that what exists cannot be named within a mathematically consistent theory! For indeed it can! The lesson here is that one should never give up ontology! If someone says: "I do not know what it means to exist," then that is fine. That person can view the theory of Bohmian mechanics as a precise and coherent mathematical theory, in which all that needs to be said is written in the equations, ready for analysis.

Our guideline for writing the book was the focus on the *genesis* of the ideas and concepts, to be clear about *what it is* that we are talking about, and hence to pave the way for the hard technical work of learning *how it is done*. In short, we have tried not to leave out the letter 'h' (see the Melville quote on p. 4).

References

The references we give are neither complete nor balanced. Naturally the references reflect our point of view. They are nevertheless chosen in view of their basic character, sometimes containing many further references which the reader may follow up to achieve a more complete picture.

Acknowledgements

The present book is a translation and revision of the book *Bohmsche Mechanik als Grundlage der Quantenmechanik* by one of the present authors (D.D.), published by Springer in 2001. The preface for this book read as follows: "In this book I wrote down what I learned from my friends and colleagues Sheldon Goldstein and Nino Zanghì. With their cooperation everything in this book would have been said correctly – but only after an unforeseeably long time (extrapolating the times it takes us to write a normal article). For that, I lack the patience (see the Epilogue) and I wanted to say things in my own style. Because of that, and only because of that, things may be said wrongly." This paragraph is still relevant, of course, but we feel the need to express our thanks to these truly great physicists and friends even more emphatically. Without them this book would never have been written (but unfortunately with them perhaps also not). We have achieved our goal if they would think of this as a good book. With them the (unwritten) book would have been excellent. Today we should like to add one more name: Roderich Tumulka, a former student, now a distinguished researcher in the field of foundations who has since helped us to gain further deep insights.

The long chapter on chance resulted from many discussions with Reinhard Lang which clarified many issues. It was read and corrected several times by our student Christian Beck. Selected chapters were read, reread, and greatly improved by our students and coworkers Dirk Deckert, Robert Grummt, Florian Hoffmann, Tilo Moser, Peter Pickl, Sarah Römer, and Georg Volkert. The whole text was carefully read and critically discussed by the students of the reading class on Bohmian mechanics, and they are all thanked for their commitment. In particular, we wish to thank Dustin Lazarovici, Niklas Boers, and Florian Rieger for extensive and thoughtful corrections. We are especially grateful to Sören Petrat, who did a very detailed and expert correction of the whole text.

The foundations of quantum mechanics are built on slippery grounds, and one depends on friends and colleagues who encourage and support (with good criticism) this research, which is clearly not mainstream physics. Two important friends in this undertaking are Giancarlo Ghirardi and Herbert Spohn. But there are more friends we wish to acknowledge, who have, perhaps unknowingly, had a strong impact on the writing of the book: Angelo Bassi, Gernot Bauer, Jürg Fröhlich, Rodolfo Figari, Martin Kolb, Tim Maudlin, Sandro Teta, and Tim Storck.

We would also like to thank Wolf Beiglböck of Springer-Verlag, who encouraged and supported the English version of this book. Every author knows that a book can be substantially improved by the editorial work done on it. We were extremely fortunate that our editor was Angela Lahee at Springer, who astutely provided us with the best copy-editor we could have wished for, namely Stephen Lyle, who brilliantly transformed our Germanic English constructions into proper English without jeopardizing our attempt to incite the reader to active contemplation throughout. We are very grateful to both Angela and Stephen for helping us with our endeavour to present Bohmian mechanics as a respectable part of physics.

Landsberg and Tübingen,
October 2008

Detlef Dürr
Stefan Teufel

Contents

Chapter 1
Introduction

1.1 Ontology: What There Is

1.1.1 Extracts

Sometimes quantization is seen as the procedure that puts hats on classical observables to turn them into quantum observables.

Lewis Carroll on Hatters and Cats

Lewis Carroll (alias Charles Lutwidge Dodgson 1832–1898) was professor of mathematics at Oxford (where Schrödinger wrote his famous cat article):

> The Cat only grinned when it saw Alice. It looked good-natured, she thought: still it had VERY long claws and a great many teeth, so she felt that it ought to be treated with respect. "Cheshire Puss," she began, rather timidly, as she did not at all know whether it would like the name: however, it only grinned a little wider. "Come, it's pleased so far," thought Alice, and she went on. "Would you tell me, please, which way I ought to go from here?" "That depends a good deal on where you want to get to," said the Cat. "I don't much care where –" said Alice. "Then it doesn't matter which way you go," said the Cat. "– so long as I get SOMEWHERE," Alice added as an explanation. "Oh, you're sure to do that," said the Cat, "if you only walk long enough." Alice felt that this could not be denied, so she tried another question. "What sort of people live about here?" "In THAT direction," the Cat said, waving its right paw round, "lives a Hatter: and in THAT direction," waving the other paw, "lives a March Hare. Visit either you like: they're both mad." "But I don't want to go among mad people," Alice remarked. "Oh, you can't help that," said the Cat: "we're all mad here. I'm mad. You're mad." "How do you know I'm mad?" said Alice. "You must be," said the Cat, "or you wouldn't have come here." Alice didn't think that proved it at all; however, she went on "And how do you know that you're mad?" "To begin with," said the Cat, "a dog's not mad. You grant that?" "I suppose so," said Alice. "Well, then," the Cat went on, "you see, a dog growls when it's angry, and wags its tail when it's pleased. Now I growl when I'm pleased, and wag my tail when I'm angry. Therefore I'm mad." "I call it purring, not growling," said Alice. "Call it what you like," said the Cat. "Do you play croquet with the

D. Dürr, S. Teufel, *Bohmian Mechanics*, DOI 10.1007/978-3-540-89344-8_1,

> Queen today?" "I should like it very much," said Alice, "but I haven't been invited yet." "You'll see me there," said the Cat, and vanished.
>
> Alice was not much surprised at this, she was getting so used to queer things happening. While she was looking at the place where it had been, it suddenly appeared again.

Alice's Adventures in Wonderland (1865) [1].

Parmenides on What There Is

The Greek philosopher Parmenides of Elea wrote as follows in the sixth century BC:

> Come now, I will tell thee – and do thou hearken to my saying and carry it away – the only two ways of search that can be thought of. The first, namely, that It is, and that it is impossible for it not to be, is the way of belief, for truth is its companion. The other, namely, that It is not, and that it must needs not be, – that, I tell thee, is a path that none can learn of at all. For thou canst not know what is not – that is impossible – nor utter it; for it is the same thing that can be thought and that can be.
>
> It needs must be that what can be spoken and thought is; for it is possible for it to be, and it is not possible for what is nothing to be. This is what I bid thee ponder. I hold thee back from this first way of inquiry, and from this other also, upon which mortals knowing naught wander two-faced; for helplessness guides the wandering thought in their breasts, so that they are borne along stupefied like men deaf and blind. Undiscerning crowds, who hold that it is and is not the same and not the same, and all things travel in opposite directions!
>
> For this shall never be proved, that the things that are not are; and do thou restrain thy thought from this way of inquiry.

The Way of Truth [2].

It is sometimes said that, in quantum mechanics, the observer calls things into being by the act of observation.

Schrödinger on Quantum Mechanics

Erwin Schrödinger wrote in 1935:

> One can even set up quite ridiculous cases. A cat is penned up in a steel chamber, along with the following device (which must be secured against direct interference by the cat): in a Geiger counter there is a tiny bit of radioactive substance, so small, that perhaps in the course of the hour one of the atoms decays, but also, with equal probability, perhaps none; if it happens, the counter tube discharges and through a relay releases a hammer which shatters a small flask of hydrocyanic acid. If one has left this entire system to itself for an hour, one would say that the cat still lives if meanwhile no atom has decayed. The psi-function of the entire system would express this by having in it the living and dead cat (pardon the expression) mixed or smeared out in equal parts.
>
> It is typical of these cases that an indeterminacy originally restricted to the atomic domain becomes transformed into macroscopic indeterminacy, which can then be resolved by direct observation. That prevents us from so naively accepting as valid a "blurred model" for

> representing reality. In itself it would not embody anything unclear or contradictory. There is a difference between a shaky or out-of-focus photograph and a snapshot of clouds and fog banks.

Die gegenwärtige Situation in der Quantenmechanik [3].
Translated by J.D. Trimmer in [4].

The difference between a shaky photograph and a snapshot of clouds and fog banks is the difference between Bohmian mechanics and quantum mechanics. In itself a "blurred model" representing reality would not embody anything unclear, whereas resolving the indeterminacy by direct observation does. As if observation were not part of physics.

Feynman on Quantum Mechanics

Feynman said the following:

> Does this mean that my observations become real only when I observe an observer observing something as it happens? This is a horrible viewpoint. Do you seriously entertain the thought that without observer there is no reality? Which observer? Any observer? Is a fly an observer? Is a star an observer? Was there no reality before 10^9 B.C. before life began? Or are you the observer? Then there is no reality to the world after you are dead? I know a number of otherwise respectable physicists who have bought life insurance. By what philosophy will the universe without man be understood?

Lecture Notes on Gravitation [5].

Bell on Quantum Mechanics

According to John S. Bell:

> It would seem that the theory is exclusively concerned about "results of measurement", and has nothing to say about anything else. What exactly qualifies some physical systems to play the role of "measurer"? Was the wavefunction of the world waiting to jump for thousands of years until a single-celled living creature appeared? Or did it have to wait a little longer, for some better qualified system [...] with a Ph.D.? If the theory is to apply to anything but highly idealized laboratory operations, are we not obliged to admit that more or less "measurement-like" processes are going on more or less all the time, more or less everywhere? Do we not have jumping then all the time?

Against "measurement" [6].

Einstein on Measurements

> But it is in principle quite false to base a theory solely on observable quantities. Since, in fact, it is the other way around. It is the theory that decides what we can observe.

Albert Einstein, cited by Werner Heisenberg[1] in [7].

[1] Quoted from *Die Quantenmechanik und ein Gespräch mit Einstein*. Translation by Detlef Dürr.

Melville on Omissions

> While you take in hand to school others, and to teach them by what name a whale-fish is to be called in our tongue leaving out, through ignorance, the letter H, which almost alone maketh the signification of the word, you deliver that which is not true.

Melville (1851), Moby Dick; or, The Whale, Etymology [8].

1.1.2 In Brief: The Problem of Quantum Mechanics

Schrödinger remarks in his laconic way that there is a difference between a shaky or out-of-focus photograph and a snapshot of clouds and fog banks.

The first problem with quantum mechanics is that it is not about what there is. It is said to be about the microscopic world of atoms, but it does not spell out which physical quantities in the theory describe the microscopic world. Which variables specify what is microscopically there? Quantum mechanics is about the wave function. The wave function lives on configuration space, which is the coordinate space of all the particles participating in the physical process of interest. That the wave function lives on configuration space has been called by Schrödinger *entanglement* of the wave function. The wave function obeys a linear equation – the Schrödinger equation. The linearity of the Schrödinger equation prevents the wave function from representing reality.[2] We shall see that in a moment. The second problem of quantum mechanics is that the first problem provokes many rich answers, which to the untrained ear appear to be of philosophical nature, but which leave the problem unanswered: What is it that quantum mechanics is about?

It is said for example that the virtue of quantum mechanics is that it is *only* about what can be measured. Moreover, that what can be measured is *defined* through the measurement. Without measurement there is nothing there. But a measurement belongs to the macroscopic world (which undeniably exists), and its macroscopic constituents like the measurement apparatus are made out of atoms, which quantum mechanics is supposed to describe, so this entails that the apparatus itself is to be described quantum mechanically. This circularity lies at the basis of the measurement problem of quantum mechanics, which is often phrased as showing incompleteness of quantum mechanics. Some things, namely the objects the theory is about, have been left out of the description, or the description – the Schrödinger equation – is not right. Mathematically the measurement problem may be presented as follows. Suppose a system is described by linear combinations of wave functions φ_1 and φ_2. Suppose there exists a piece of apparatus which, when brought into interaction with the system, measures whether the system has wave function φ_1 or φ_2. Measuring

---

[2] Even if the equation were nonlinear, as is the case in reduction models, the wave function living on configuration space could not by itself represent reality in physical space. There must still be some "beable" (related to the wave function) in the sense of Bell [9] representing physical reality. But we ignore this fine point here.

means that, next to the 0 pointer position, the apparatus has two pointer positions 1 and 2 "described" by wave functions Ψ_0, Ψ_1, and Ψ_2, for which

$$\varphi_i \Psi_0 \xrightarrow{\text{Schrödinger evolution}} \varphi_i \Psi_i , \qquad i = 1,2 . \tag{1.1}$$

When we say that the pointer positions are "described" by wave functions, we mean that in a loose sense. The wave function has a support in configuration space which corresponds classically to a set of coordinates of particles which would form a pointer.

The Schrödinger equation is linear, so for the superposition, (1.1) yields

$$\varphi = c_1\varphi_1 + c_2\varphi_2 , \qquad c_1, c_2 \in \mathbb{C} , \qquad |c_1|^2 + |c_2|^2 = 1 ,$$

$$\varphi\Psi_0 = (c_1\varphi_1 + c_2\varphi_2)\Psi_0 \xrightarrow{\text{Schrödinger evolution}} c_1\varphi_1\Psi_1 + c_2\varphi_2\Psi_2 . \tag{1.2}$$

The outcome on the right does not concur with experience. It shows rather a "macroscopic indeterminacy". In the words of Schrödinger, observation then resolves this macroscopic indeterminacy, since one only observes either 1 (with probability $|c_1|^2$) or 2 (with probability $|c_2|^2$), i.e., observation resolves the blurred description of reality into one where the wave function is either $\varphi_1\Psi_1$ or $\varphi_2\Psi_2$. In Schrödinger's cat thought experiment $\varphi_{1,2}$ are the wave functions of the non-decayed and decayed atom and $\Psi_{0,1}$ are the wave functions of the live cat and Ψ_2 is the wave function of the dead cat. Schrödinger says that this is unacceptable. But why? Is the apparatus not supposed to be the observer? What qualifies us better than the apparatus, which we designed in such a way that it gives a definite outcome and not a blurred one?

The question is: what in the theory describes the actual facts? Either those variables which describe the actual state of affairs have been left out of the description (Bohmian mechanics makes amends for that) or the Schrödinger equation which yields the unrealistic result (1.2) from (1.1) is false. (GRW theories, or more generally, dynamical reduction models, follow the latter way of describing nature. The wave function collapses by virtue of the dynamical law [10].)

The evolution in (1.2) is an instance of the so-called decoherence. The apparatus decoheres the superposition $c_1\varphi_1 + c_2\varphi_2$ of the system wave function. Decoherence means that it is in a practical sense impossible to get the two wave packets $\varphi_1\Psi_1$ and $\varphi_2\Psi_2$ superposed in $c_1\varphi_1\Psi_1 + c_2\varphi_2\Psi_2$ to interfere. It is sometimes said that, taking decoherence into account, there would not be any measurement problem. Decoherence is this practical impossibility, which Bell referred to as fapp-impossibility (where fapp means for all practical purposes), of the interference of the pointer wave functions. It is often dressed up, for better looks, in terms of density matrices. In Dirac's notation, the density matrix is

$$\rho_G = |c_1|^2 |\varphi_1\rangle|\Psi_1\rangle\langle\Psi_1|\langle\varphi_1| + |c_2|^2 |\varphi_2\rangle|\Psi_2\rangle\langle\Psi_2|\langle\varphi_2| .$$

This can be interpreted as describing a statistical mixture of the two states $|\varphi_1\rangle|\Psi_1\rangle$ and $|\varphi_2\rangle|\Psi_2\rangle$, and one can say that in a fapp sense the pure state given by the right-hand side of (1.2), namely

$$\rho = |c_1\varphi_1\Psi_1 + c_2\varphi_2\Psi_2\rangle\langle c_1\Psi_1\varphi_1 + c_2\Psi_2\varphi_2| ,$$

is close to ρ_G. They are fapp-close because the off-diagonal element

$$c_1c_2^*|\Psi_1\varphi_1\rangle\langle\Psi_2\varphi_2|$$

where the asterisk denotes complex conjugation, would only be observable if the wave parts could be brought into interference, which is fapp-impossible. The argument then concludes that the meaning of the right-hand side of (1.2) is fapp-close to the meaning of the statistical mixture. To emphasise the fact that such arguments miss the point of the exercise, Schrödinger made his remark that there is a difference between a shaky or out-of-focus photograph and a snapshot of clouds and fog banks.

To sum up then, decoherence does not create the facts of our world, but rather produces a sequence of fapp-redundancies, which physically increase or stabilize decoherence. So the cat decoheres the atom, the observer decoheres the cat that decoheres the atom, the environment of the observer decoheres the observer that decoheres the cat that decoheres the atom and so on. In short, what needs to be described by the physical theory is the behavior of real objects, located in physical space, which account for the facts.

1.1.3 In Brief: Bohmian Mechanics

Bohmian mechanics is about point particles in motion. The theory was invented in 1952 by David Bohm (1917–1992) [11] (there is a bit more on the history in Chap. 7). In a Bohmian universe everything is made out of particles. Their motion is guided by the wave function. That is why the wave function is there. That is its role. The physical theory is formulated with the variables $\mathbf{q}_i \in \mathbb{R}^3$, $i = 1,2,\ldots,N$, the positions of the N particles which make up the system, and the wave function $\psi(\mathbf{q}_1,\ldots,\mathbf{q}_N)$ on the configuration space of the system. If the wave function consists of two parts which have disjoint supports in configuration space, then the system configuration is in one or the other support. In the measurement example, the pointer configuration is either in the support $\operatorname{supp}\Psi_1$ of Ψ_1 (pointing out 1) or in the support $\operatorname{supp}\Psi_2$ of Ψ_2 (pointing out 2).

Bohmian mechanics happens to be deterministic. A substantial success of Bohmian mechanics is the explanation of quantum randomness or Born's statistical law, on the basis of Boltzmann's principles of statistical mechanics, i.e., Born's law is not an axiom but a theorem in Bohmian mechanics. Born's statistical law concerning $\rho = |\psi|^2$ says that, if the wave function is ψ, the particle configuration is $|\psi|^2$-distributed. Applying this to (1.2) implies that the result i comes with probability

$|c_i|^2$. Suppose the enlarged system composed of system plus apparatus is described by the configuration coordinates $\mathbf{q} = (\mathbf{x}, \mathbf{y})$, where $\mathbf{x} \in \mathbb{R}^m$ is the system configuration and $\mathbf{y} \in \mathbb{R}^n$ the pointer configuration. We compute the probability for the result 1 by Born's law. This means we compute the probability that the true (or actual) pointer configuration Y lies in the support $\operatorname{supp} \Psi_1$. In the following computation, we assume that the wave functions are all normalized to unity. Then

$$\mathbb{P}(\text{pointer on } 1) = \int_{\operatorname{supp} \Psi_1} |c_1 \varphi_1 \Psi_1 + c_2 \varphi_2 \Psi_2|^2 \mathrm{d}^m x \mathrm{d}^n y \tag{1.3a}$$

$$= |c_1|^2 \int_{\operatorname{supp} \Psi_1} |\varphi_1 \Psi_1|^2 \mathrm{d}^m x \mathrm{d}^n y + |c_2|^2 \int_{\operatorname{supp} \Psi_1} |\varphi_2 \Psi_2|^2 \mathrm{d}^m x \mathrm{d}^n y$$

$$+ 2\Re \left[c_1^* c_2 \int_{\operatorname{supp} \Psi_1} (\varphi_1 \Psi_1)^* \varphi_2 \Psi_2 \mathrm{d}^m x \mathrm{d}^n y \right] \tag{1.3b}$$

$$= |c_1|^2 \int |\varphi_1 \Psi_1|^2 \mathrm{d}^m x \mathrm{d}^n y = |c_1|^2 , \tag{1.3c}$$

where $\Re$ in (1.3b) denotes the real part of a complex quantity. The terms involving Ψ_2 yield zero because of the disjointness[3] of the supports of the pointer wave functions Ψ_1, Ψ_2. This result holds fapp-forever, since it is fapp-impossible for the wave function parts Ψ_1 and Ψ_2 to interfere in the future, especially when the results have been written down or recorded in any way – the ever growing decoherence.

Suppose one removes the positions of the particles from the theory, as for example Heisenberg, Bohr, and von Neumann did. Then to be able to conclude from the fapp-impossibility of interference that only one of the wave functions remains, one needs to add an "observer" who by the act of observation collapses the wave function with probability $|c_i|^2$ to the Ψ_i part, thereby "creating" the result i. Once again, we may ask what qualifies an observer as better than a piece of apparatus or a cat. Debate of this kind has been going on since the works of Heisenberg and Schrödinger in 1926. The debate, apart from producing all kinds of philosophical treatises, revolves around the collapse of the wave function. But when does the collapse happen and who is entitled to collapse the wave function? Does the collapse happen at all? It is, to put it mildly, a bit puzzling that such an obvious shortcoming of the theory has led to such a confused and unfocussed debate. Indeed, it has shown physics in a bad light. In Bohmian mechanics, the collapse does not happen at all, although there is a fapp-collapse, which one may introduce when one analyzes the theory. In collapse theories (like GRW), the collapse does in fact happen. Bohmian mechanics and collapse theories differ in that way. They make different predictions for macroscopic interference experiments, which may, however, be very difficult to perform, if not fapp-impossible.

We did not say how the particles are guided by the wave function. One gets that by simply taking language seriously. Whenever you say "particle" in quantum mechanics, mean it! That is Bohmian mechanics. All problems evaporate on the

[3] The wave functions will in reality overlap, but the overlap is negligible. Therefore the wave functions are well approximated by wave functions with disjoint supports.

spot. In a nutshell, the quantity $|\psi_t|^2$, with ψ_t a solution of Schrödinger's equation, satisfies a continuity equation, the so-called quantum flux equation. The particles in Bohmian mechanics move along the flow lines of the quantum flux. In other words the quantum flux equation is the continuity equation for transport of probability along the Bohmian trajectories. Still not satisfied? Is that too cheap? Why the wave function? Why $|\psi_t|^2$? We shall address all these questions and more in the chapters to follow.

Bohmian mechanics is defined by two equations: one is the Schrödinger equation for the guiding field ψ_t and one is the equation for the positions of the particles. The latter equation reads

$$\dot{\mathbf{Q}} = \mu \Im \frac{\nabla \psi_t(\mathbf{Q})}{\psi_t(\mathbf{Q})} , \tag{1.4}$$

where $\Im$ denotes the imaginary part, and μ is an appropriate dimension factor. The quantum formalism in its most general form, including all rules and axioms, follows from this by *analysis* of the theory. In particular, Heisenberg's uncertainty relation for position and momentum, from which it is often concluded that particle trajectories are in conflict with quantum mechanics, follows directly from Bohmian mechanics.

Bohmian mechanics is nonlocal in the sense of Bell's inequalities and therefore, according to the experimental tests of Bell's inequalities, concurs with the basic requirements that any correct theory of nature must fulfill.

A Red Herring: The Double Slit Experiment

This is a quantum mechanical experiment which is often cited as conflicting with the idea that there can be particles with trajectories. One sends a particle (i.e., a wave packet ψ) through a double slit. Behind the slit at some distance is a photographic plate. When the particle arrives at the plate it leaves a black spot at its place of arrival. Nothing yet speaks against the idea that the particle moves on a trajectory. But now repeat the experiment. The next particle marks a different spot of the photographic plate. Repeating this a great many times the spots begin to show a pattern. They trace out the points of constructive interference of the wave packet ψ which, when passing the two slits, shows the typical Huygens interference of two spherical waves emerging from each slit. Suppose the wave packet reaches the photographic plate after a time T. Then the spots show the $|\psi(T)|^2$ distribution,[4] in the sense that this is their empirical distribution. Analyzing this using Bohmian mechanics, i.e., analyzing Schrödinger's equation and the guiding equation (1.4), one immediately understands why the experiment produces the result it does. It is clear that in each run the particle goes either through the upper or through the lower slit. The wave function goes through both slits and forms after the slits a wave function with an

[4] In fact, it is the quantum flux across the surface of the photographic plate, integrated over time (see Chap. 16).

interference pattern. Finally the repetition of the experiment produces an ensemble which checks Born's statistical law for that wave function. That is the straightforward physical explanation.

So where is the argument which reveals a conflict with the notion of particle trajectories? Here it is:

Close slit 1 and open slit 2. (1.5a)
The particle goes through slit 2. (1.5b)
It arrives at $\mathbf{x}$ on the plate with probability $|\psi_2(\mathbf{x})|^2$, (1.5c)

where ψ_2 is the wave function which passed through slit 2. Next

close slit 2 and open slit 1. (1.6a)
The particle goes through slit 1. (1.6b)
It arrives at $\mathbf{x}$ on the plate with probability $|\psi_1(\mathbf{x})|^2$, (1.6c)

where ψ_1 is the wave function which passed through slit 1. Now open both slits.

Both slits are open. (1.7a)
The particle goes through slit 1 or slit 2. (1.7b)
It arrives at $\mathbf{x}$ with probability $|\psi_1(\mathbf{x}) + \psi_2(\mathbf{x})|^2$. (1.7c)

Now observe that in general

$$|\psi_1(\mathbf{x}) + \psi_2(\mathbf{x})|^2 = |\psi_1(\mathbf{x})|^2 + |\psi_2(\mathbf{x})|^2 + 2\Re\psi_1^*(\mathbf{x})\psi_2(\mathbf{x}) \neq |\psi_1(\mathbf{x})|^2 + |\psi_2(\mathbf{x})|^2 .$$

The $\neq$ comes from interference of the wave packets ψ_1, ψ_2 which passed through slit 1 and slit 2. The argument now proceeds in the following way. Situations (1.5b) and (1.6b) are the exclusive alternatives entering (1.7b), so the probabilities (1.5c) and (1.6c) must add up. But they do not. So is logic false? Is the particle idea nonsense? No, the argument is a red herring, since (1.5a), (1.6a), and (1.7a) are *physically* distinct.

1.2 Determinism and Realism

It is often said that the aim of Bohmian mechanics is to restore determinism in the quantum world. That is false. Determinism has nothing to do with ontology. What is "out there" could just as well be governed by stochastic laws, as is the case in GRW or dynamical reduction models with, e.g., flash ontology [12, 13]. A realistic quantum theory is a quantum theory which spells out what it is about. Bohmian mechanics is a realistic quantum theory. It happens to be deterministic, which is fine, but not an ontological necessity. The merit of Bohmian mechanics is not determinism, but the refutation of all claims that quantum mechanics cannot be reconciled with a realistic description of reality. In physics, one needs to know what is going on. Bohmian mechanics tells us what is going on and it does so in the most straightforward way imaginable. It is therefore the fundamental description of Galilean physics.

The following passage taken from a letter from Pauli to Born, concerning Einstein's view on determinism, is in many ways reminiscent of the present situation in Bohmian mechanics:

> Einstein gave me your manuscript to read; he was not at all annoyed with you, but only said that you were a person who will not listen. This agrees with the impression I have formed myself insofar as I was unable to recognise Einstein whenever you talked about him in either your letter or your manuscript. It seemed to me as if you had erected some dummy Einstein for yourself, which you then knocked down with great pomp. In particular, Einstein does not consider the concept of "determinism" to be as fundamental as it is frequently held to be (as he told me emphatically many times), and he denied energetically that he had ever put up a postulate such as (your letter, para. 3): "the sequence of such conditions must also be objective and real, that is, automatic, machine-like, deterministic." In the same way, he disputes that he uses as a criterion for the admissibility of a theory the question: "Is it rigorously deterministic?" Einstein's point of departure is "realistic" rather than "deterministic".

Wolfgang Pauli, in [14], p. 221.

References

1. L. Carroll: *The Complete, Fully Illustrated Works* (Gramercy Books, New York, 1995)
2. J. Burnet: *Early Greek Philosophy* (A. and C. Black, London and Edinburgh, 1930)
3. E. Schrödinger: Naturwissenschaften **23**, 807 (1935)
4. J.A. Wheeler, W.H. Zurek (eds.): *Quantum Theory and Measurement.* Princeton Series in Physics (Princeton University Press, Princeton, NJ, 1983)
5. R.P. Feynman, F.B. Morinigo, W.G. Wagner: *Feynman Lectures on Gravitation* (Addison-Wesley Publishing Company, 1959). Edited by Brian Hatfield
6. M. Bell, K. Gottfried and M. Veltman (eds.): *John S. Bell on the Foundations of Quantum Mechanics* (World Scientific Publishing Co. Inc., River Edge, NJ, 2001)
7. W. Heisenberg: *Quantentheorie und Philosophie, Universal-Bibliothek [Universal Library]*, Vol. 9948 (Reclam-Verlag, Stuttgart, 1987). Vorlesungen und Aufsätze. [Lectures and essays], With an afterword by Jürgen Busche
8. H. Melville: *Moby Dick; or, The Whale* (William Benton, Encyclopaedia Britanica, Inc., 1952). Greatest Books of the Western World
9. J.S. Bell: *Speakable and Unspeakable in Quantum Mechanics* (Cambridge University Press, Cambridge, 1987)
10. A. Bassi, G. Ghirardi: Phys. Rep. **379** (5–6), 257 (2003)
11. D. Bohm: Phys. Rev. **85**, 166 (1952)
12. J.S. Bell: Beables for quantum fields. In: *Speakable and Unspeakable in Quantum Mechanics* (Cambridge University Press, 1987)
13. R. Tumulka: J. Stat. Phys. **125** (4), 825 (2006)
14. I. Born: *The Born–Einstein Letters* (Walker and Company, New York, 1971)

Chapter 2
Classical Physics

What is classical physics? In fact it has become the name for non-quantum physics. This begs the question: What is quantum physics in contrast to classical physics? One readily finds the statement that in classical physics the world is described by classical notions, like particles moving around in space, while in modern physics, i.e., quantum mechanics, the classical notions are no longer adequate, so there is no longer such a "naive" description of what is going on. In a more sophisticated version, quantum mechanics is physics in which the position and momentum of a particle are operators. But such a statement as it stands is meaningless. One also reads that the difference between quantum physics and classical physics is that the former has a smallest quantum of action, viz., Planck's constant $\hbar$, and that classical physics applies whenever the action is large compared to $\hbar$, and in many circumstances, this is a true statement.

But our own viewpoint is better expressed as follows. Classical physics is the description of the world when the interference effects of the Schrödinger wave, evolving according to Schrödinger's equation, can be neglected. This is the case for a tremendously wide range of scales from microscopic gases to stellar matter. In particular it includes the scale of direct human perception, and this explains why classical physics was found before quantum mechanics. Still, the viewpoint just expressed should seem puzzling. For how can classical motion of particles emerge from a wave equation like the Schrödinger equation? This is something we shall explain. It is easy to understand, once one writes down the equations of motion of Bohmian mechanics. But first let us discuss the theory which governs the behavior of matter across the enormous range of classical physics, namely, Newtonian mechanics. In a letter to Hooke, Newton wrote: "If I have seen further it is by standing on the shoulders of giants."

D. Dürr, S. Teufel, *Bohmian Mechanics*, DOI 10.1007/978-3-540-89344-8_2,

2.1 Newtonian Mechanics

Newtonian mechanics is about point particles. What is a point particle? It is "stuff" or "matter" that occupies a point in space called its position, described mathematically by $\mathbf{q} \in \mathbb{R}^3$. The theory describes the motion of point particles in space. Mathematically, an N-particle system is described by the positions of the N particles:

$$\mathbf{q}_1, \dots, \mathbf{q}_N , \qquad \mathbf{q}_i \in \mathbb{R}^3 ,$$

which change with time, so that one has trajectories $\mathbf{q}_1(t), \dots, \mathbf{q}_N(t)$, where the parameter $t \in \mathbb{R}$ is the time.

Newtonian mechanics is given by equations – the physical law – which govern the trajectories, called the equations of motion. They can be formulated in many different (but more or less equivalent) ways, so that the physical law looks different for each formulation, but the trajectories remain the same. We shall soon look at an example. Which formulation one prefers will be mainly a matter of taste. One may find the arguments leading to a particular formulation more satisfactory or convincing than others.

To formulate the law of Newtonian mechanics one introduces positive parameters, called masses, viz., $m_1, \dots, m_N$, which represent "matter", and the law reads

$$m_i \ddot{\mathbf{q}}_i = \mathbf{F}_i(\mathbf{q}_1, \dots, \mathbf{q}_N) . \tag{2.1}$$

$\mathbf{F}_i$ is called the force. It is in general a function of all particle positions. Put another way, it is a function of the *configuration*, i.e., the family of all coordinates $(\mathbf{q}_1, \dots, \mathbf{q}_N) \in \mathbb{R}^{3N}$. The set of all such N-tuples is called *configuration space*. The quantity $\dot{\mathbf{q}}_i = \mathrm{d}\mathbf{q}_i/\mathrm{d}t = \mathbf{v}_i$ is the velocity of the ith particle, and its derivative $\ddot{\mathbf{q}}_i$ is called the acceleration.

Newtonian mechanics is romantic in a way. One way of talking about it is to say that particles accelerate each other, they interact through forces exerted upon each other, i.e., Newtonian mechanics is a theory of interaction. The fundamental interaction is gravitation or mass attraction given by

$$\mathbf{F}_i(\mathbf{q}_1, \dots, \mathbf{q}_N) = \sum_{j \neq i} G m_i m_j \frac{\mathbf{q}_j - \mathbf{q}_i}{\|\mathbf{q}_j - \mathbf{q}_i\|^3} , \tag{2.2}$$

with G the gravitational constant.

All point particles of the Newtonian universe interact according to (2.2). In effective descriptions of subsystems (when we actually use Newtonian mechanics in everyday life), other forces like harmonic forces of springs can appear on the right-hand side of (2.1). Such general forces need not (and in general will not) arise from gravitation alone. Electromagnetic forces will also play a role, i.e., one can sometimes describe electromagnetic interaction between electrically charged particles by the Coulomb force using Newtonian mechanics. The Coulomb force is similar to (2.2), but may have a different sign, and the masses m_i are replaced by the charges e_i which may be positive or negative.

One may wonder why Newtonian mechanics can be successfully applied to subsystems like the solar system, or even smaller systems like systems on earth. That is, why can one ignore all the rest of the universe? One can give various reasons. For example, distant matter which surrounds the earth in a homogeneous way produces a zero net field. The force (2.2) falls off with large distances and the gravitational constant is very small. In various general situations, and depending on the practical task in hand, such arguments allow a good effective description of the subsystem in which one ignores distant matter, or even not so distant matter.

Remark 2.1. Initial Value Problem

The equation (2.1) is a differential equation and thus poses an initial value problem, i.e., the trajectories $\mathbf{q}_i(t)$, $t \in \mathbb{R}$, which obey (2.1) are only determined once initial data of the form $\mathbf{q}_i(t_0)$, $\dot{\mathbf{q}}_i(t_0)$ are given, where t_0 is some time, called the initial time. This means that the future and past evolution of the trajectories is determined by the "present" state $\mathbf{q}_i(t_0)$, $\dot{\mathbf{q}}_i(t_0)$. Note that the position alone is not sufficient to determine the state of a Newtonian system.

It is well known that differential equations need not have unique and global solutions, i.e., solutions which exist for all times for all initial values. What does exist, however, is – at least in the case of gravitation – a local unique solution for a great many initial conditions, i.e., a solution which exists uniquely for some short period of time, if the initial values are reasonable. So (2.1) and (2.2) have no solution if, for example, two particles occupy the same position. Further, for the solution to exist, it must not happen that two or more particles collide, i.e., that they come together and occupy the same position. It is a famous problem in mathematical physics to establish what is called the existence of dynamics for a gravitating many-particle system, where one hopes to show that solutions fail to exist globally only for exceptional initial values. But what does "exceptional" mean? We shall answer this in a short while. ■

We wish to comment briefly on the manner of speaking about interacting particles, which gives a human touch to Newtonian mechanics. We say that the particles *attract each other*. Taking this notion to heart, one might be inclined to associate with the notion of particle more than just an object which has a position. But that might be misleading, since no matter how one justifies or speaks about Newtonian mechanics, when all is said and done, there remains a physical law about the motion of point particles, and that is a mathematical expression about changes of points in space with time. We shall explore one such prosaic description next.

2.2 Hamiltonian Mechanics

One can formulate the Newtonian law differently. Different formulations are based on different fundamental principles, like for example the principle of least action. But never mind such principles for the moment. We shall simply observe that it is

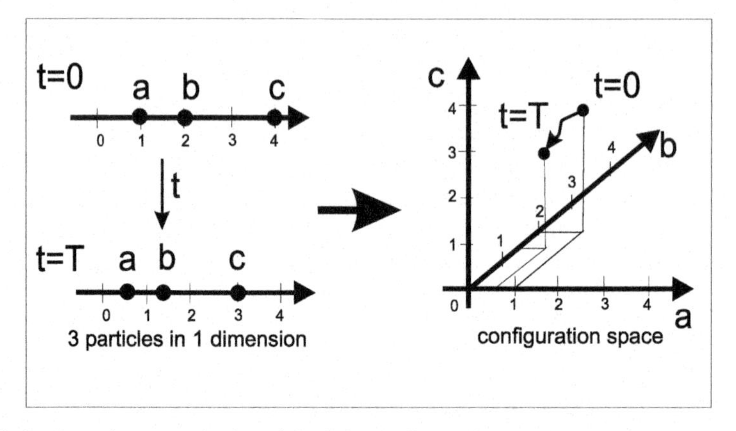

Fig. 2.1 Configuration space for 3 particles in a one-dimensional world

mathematically much nicer to rewrite everything in terms of configuration space variables:

$$\mathbf{q} = \begin{pmatrix} \mathbf{q}_1 \\ \vdots \\ \mathbf{q}_N \end{pmatrix} \in \mathbb{R}^{3N} ,$$

that is, we write the differential equation for all particles in a compact form as

$$m\ddot{\mathbf{q}} = \mathbf{F} , \tag{2.3}$$

with

$$\mathbf{F} = \begin{pmatrix} \mathbf{F}_1 \\ \vdots \\ \mathbf{F}_N \end{pmatrix} ,$$

and the mass matrix $m = (\delta_i^j m_j)_{i,j=1,\ldots,N}$.

Configuration space cannot be depicted (but see Fig. 2.1 for a very special situation), at least not for a system of more than one particle, because it is 6-dimensional for 2 particles in physical space. It is thus not so easy to think intuitively about things going on in configuration space. But one had better build up some intuition for configuration space, because it plays a fundamental role in quantum theory.

A differential equation is *by definition* a relation between the *flow* and the *vector field*. The flow is the mapping along the solution curves, which are integral curves along the vector field (the tangents of the solution curves). If a physical law is given by a differential equation, the vector field encodes the physical law. Let us see how this works.

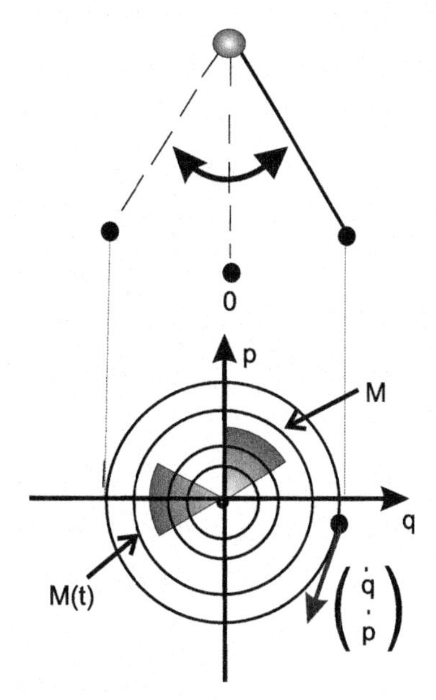

Fig. 2.2 Phase space description of the mathematically idealized harmonically swinging pendulum. The possible trajectories of the mathematically idealized pendulum swinging in a plane with frequency 1 are concentric circles in phase space. The sets M und $M(t)$ will be discussed later

The differential equation (2.3) is of second order and does not express the relation between the integral curves and the vector field in a transparent way. We need to change (2.3) into an equation of first order, so that the vector field becomes transparent. For this reason we consider the *phase space* variables

$$\begin{pmatrix} \mathbf{q} \\ \mathbf{p} \end{pmatrix} = \begin{pmatrix} \mathbf{q}_1 \\ \vdots \\ \mathbf{q}_N \\ \mathbf{p}_1 \\ \vdots \\ \mathbf{p}_N \end{pmatrix} \in \mathbb{R}^{3N} \times \mathbb{R}^{3N} = \Gamma ,$$

which were introduced by Boltzmann,[1] where we consider positions and velocities. However, for convenience of notation, the latter are replaced by momenta $\mathbf{p}_i = m_i \mathbf{v}_i$. One point in Γ represents the present state of the entire N-particle system. The phase space has twice the dimension of configuration space and can be depicted for one particle moving in one dimension, e.g., the pendulum (see Fig. 2.2).

Clearly, (2.3) becomes

[1] The notion of phase space was taken by Ludwig Boltzmann (1844–1906) as synonymous with the state space, the phase being the collection of variables which uniquely determine the physical state. The physical state is uniquely determined if its future and past evolution in time is uniquely determined by the physical law.

$$\begin{pmatrix} \dot{\mathbf{q}} \\ \dot{\mathbf{p}} \end{pmatrix} = \begin{pmatrix} m^{-1}\mathbf{p} \\ \mathbf{F}(\mathbf{q}) \end{pmatrix} . \tag{2.4}$$

The state of the N-particle system is completely determined by $\begin{pmatrix} \mathbf{q} \\ \mathbf{p} \end{pmatrix}$, because (2.4) and the initial values $\begin{pmatrix} \mathbf{q}(t_0) \\ \mathbf{p}(t_0) \end{pmatrix}$ uniquely determine the phase space trajectory (if the initial value problem allows for a solution).

For (2.2) and many other effective forces, there exists a function V on $\mathbb{R}^{3N}$, the so called potential energy function, with the property that

$$\mathbf{F} = -\mathrm{grad}_q V = -\frac{\partial V}{\partial \mathbf{q}} = -\nabla V .$$

Using this we may write (2.4) as

$$\begin{pmatrix} \dot{\mathbf{q}} \\ \dot{\mathbf{p}} \end{pmatrix} = \begin{pmatrix} \dfrac{\partial H}{\partial \mathbf{p}}(\mathbf{q},\mathbf{p}) \\ -\dfrac{\partial H}{\partial \mathbf{q}}(\mathbf{q},\mathbf{p}) \end{pmatrix} , \tag{2.5}$$

where

$$\begin{aligned} H(\mathbf{q},\mathbf{p}) &= \frac{1}{2}(\mathbf{p}\cdot m^{-1}\mathbf{p}) + V(\mathbf{q}) \\ &= \frac{1}{2}\sum_{i=1}^{N}\frac{\mathbf{p}_i^2}{m_i} + V(\mathbf{q}_1,\dots,\mathbf{q}_N) . \end{aligned} \tag{2.6}$$

Now we have the Newtonian law in the form of a transparent differential equation (2.5), expressing the relation between the integral curves (on the left-hand side, differentiated to yield tangent vectors) and the vector field on the right-hand side (which are the tangent vectors expressing the physics). The way we have written it, the vector field is actually generated by a function H (2.6) on phase space. This is called the Hamilton function, after its inventor William Rowan Hamilton (1805–1865), who in fact introduced the symbol H in honor of the physicist Christiaan Huygens (1629–1695). We shall see later what the "wave man" Huygens has to do with all this. The role of the Hamilton function $H(\mathbf{q},\mathbf{p})$ is to give the vector field

$$\mathbf{v}^H(\mathbf{q},\mathbf{p}) = \begin{pmatrix} \dfrac{\partial H}{\partial \mathbf{p}} \\ -\dfrac{\partial H}{\partial \mathbf{q}} \end{pmatrix} , \tag{2.7}$$

and the Hamiltonian dynamics is simply given by

$$\begin{pmatrix} \dot{\mathbf{q}} \\ \dot{\mathbf{p}} \end{pmatrix} = \mathbf{v}^H(\mathbf{q},\mathbf{p}) . \tag{2.8}$$

The function H allows us to focus on a particular structure of Newtonian mechanics, now rewritten in Hamiltonian terms. Almost all of this section depends solely on this structure, and we shall see some examples shortly. Equations (2.5) and (2.6) with the Hamilton function $H(\mathbf{q},\mathbf{p})$ define a Hamiltonian system.

The integral curves along this vector field (2.7) represent the possible system trajectories in phase space, i.e., they are solutions $\begin{pmatrix} \mathbf{q}\big(t,(\mathbf{q},\mathbf{p})\big) \\ \mathbf{p}\big(t,(\mathbf{q},\mathbf{p})\big) \end{pmatrix}$ of (2.8) for given initial values $\begin{pmatrix} \mathbf{q}\big(0,(\mathbf{q},\mathbf{p})\big) \\ \mathbf{p}\big(0,(\mathbf{q},\mathbf{p})\big) \end{pmatrix} = \begin{pmatrix} \mathbf{q} \\ \mathbf{p} \end{pmatrix}$. Note that this requires existence and uniqueness of solutions of the differential equations (2.8). One possible evolution of the entire system is represented by one curve in phase space (see Fig. 2.3), which is called a flow line, and one defines the Hamiltonian flow by the map $(\Phi_t^H)_{t\in\mathbb{R}}$ from phase space to phase space, given by the prescription that, for any t, a point in phase space is mapped to the point to which it moves in time t under the evolution (as long as that evolution is defined, see Remark 2.1):

$$\Phi_t^H\left(\begin{pmatrix} \mathbf{q} \\ \mathbf{p} \end{pmatrix}\right) = \begin{pmatrix} \mathbf{q}\big(t,(\mathbf{q},\mathbf{p})\big) \\ \mathbf{p}\big(t,(\mathbf{q},\mathbf{p})\big) \end{pmatrix} .$$

We shall say more about the flow map later on. The flow can be thought of pictorially as the flow of a material fluid in Γ, with the system trajectories as flow lines.

Hamiltonian mechanics is another way of talking about Newtonian mechanics. It is a prosaic way of talking about the motion of particles. The only romance left is the secret of how to write down the physically relevant H. Once that is done, the romance is over and what lies before one are the laws of mechanics written in mathematical language. So that is all that remains. The advantage of the Hamiltonian form is that it directly expresses the law as a differential equation (2.8). And it has the further advantage that it allows one to talk simultaneously about all possible trajectories of a system. This will be helpful when we need to define a *typical* trajectory of the system, which we must do later.

However, this does not by any means imply that we should forget the Newtonian approach altogether. To understand which path a system takes, it is good to know how the particles in the system interact with each other, and to have some intuition about that. Moreover, we should not lose sight of what we are interested in, namely, the behavior of the system in physical space. Although we have not elaborated on the issue at all, it is also important to understand the physical reasoning which leads to the mathematical law (for example, how Newton found the gravitational potential), as this may give us confidence in the correctness of the law. Of course we also achieve confidence by checking whether the theory correctly describes what we see, but since we can usually only see a tiny fraction of what a theory says, confidence is mainly grounded on theoretical insight.

The fundamental properties of the Hamiltonian flow are conservation of energy and conservation of volume. These properties depend only on the form of the equa-

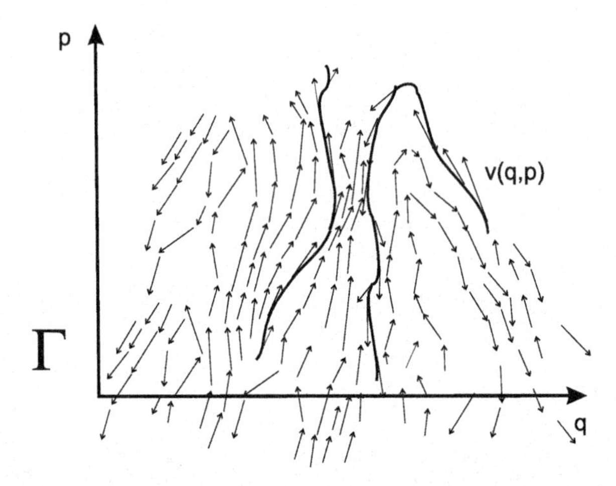

Fig. 2.3 The Hamilton function generates a vector field on the $6N$-dimensional phase space of an N-particle system in physical space. The integral curves are the possible trajectories of the entire system in phase space. Each point in phase space is the collection of all the positions and velocities of all the particles. One must always keep in mind that the trajectories in phase space are not trajectories in physical space. They can never cross each other because they are integral curves on a vector field, and a unique vector is attached to every point of phase space. Trajectories in phase space do not interact with each other! They are not the trajectories of particles

tions (2.8) with (2.7), i.e., $H(\mathbf{q},\mathbf{p})$ can be a completely general function of $(\mathbf{q},\mathbf{p})$ and need not be the function (2.6). When working with this generality, one calls $\mathbf{p}$ the canonical momentum, which is no longer simply velocity times mass. Now, conservation of energy means that the value of the Hamilton function does not change along trajectories. This is easy to see. Let $\big(\mathbf{q}(t),\mathbf{p}(t)\big)$, $t\in\mathbb{R}$, be a solution of (2.8). Then

$$\frac{\mathrm{d}}{\mathrm{d}t}H\big(\mathbf{q}(t),\mathbf{p}(t)\big)=\dot{\mathbf{q}}\frac{\partial H}{\partial\mathbf{q}}+\dot{\mathbf{p}}\frac{\partial H}{\partial\mathbf{p}}=\frac{\partial H}{\partial\mathbf{p}}\frac{\partial H}{\partial\mathbf{q}}-\frac{\partial H}{\partial\mathbf{q}}\frac{\partial H}{\partial\mathbf{p}}=0\,. \tag{2.9}$$

More generally, the time derivative along the trajectories of any function $f\big(\mathbf{q}(t),\mathbf{p}(t)\big)$ on phase space is

$$\frac{\mathrm{d}}{\mathrm{d}t}f\big(\mathbf{q}(t),\mathbf{p}(t)\big)=\dot{\mathbf{q}}\frac{\partial f}{\partial\mathbf{q}}+\dot{\mathbf{p}}\frac{\partial f}{\partial\mathbf{p}}=\frac{\partial H}{\partial\mathbf{p}}\frac{\partial f}{\partial\mathbf{q}}-\frac{\partial H}{\partial\mathbf{q}}\frac{\partial f}{\partial\mathbf{p}}=:\{f,H\}\,. \tag{2.10}$$

The term $\{f,H\}$ is called the Poisson bracket of f and H. It can also be defined in more general terms for any pair of functions f,g, viewing g as the Hamilton function and Φ_t^g the flow generated by g:

$$\{f,g\}=\frac{\mathrm{d}}{\mathrm{d}t}f\circ\Phi_t^g=\frac{\partial g}{\partial\mathbf{p}}\frac{\partial f}{\partial\mathbf{q}}-\frac{\partial g}{\partial\mathbf{q}}\frac{\partial f}{\partial\mathbf{p}}\,. \tag{2.11}$$

Note, that $\{f,H\} = 0$ means that f is a *constant of the motion*, i.e., the value of f remains unchanged along a trajectory ($\mathrm{d}f/\mathrm{d}t = 0$), with $f = H$ being the simplest example.

Now we come to the conservation of volume. Recall that the Hamiltonian flow $(\Phi_t^H)_{t\in\mathbb{R}}$ is best pictured as a fluid flow in Γ, with the system trajectories as flow lines. These are the integral curves along the Hamiltonian vector field $\mathbf{v}^H(\mathbf{q},\mathbf{p})$ (2.7). These flow lines have neither sources nor sinks, i.e., the vector field is divergence-free:

$$\operatorname{div}\mathbf{v}^H(\mathbf{q},\mathbf{p}) = \left(\frac{\partial}{\partial \mathbf{q}}, \frac{\partial}{\partial \mathbf{p}}\right)\begin{pmatrix} \dfrac{\partial H}{\partial \mathbf{p}} \\ -\dfrac{\partial H}{\partial \mathbf{q}} \end{pmatrix} = \frac{\partial^2 H}{\partial \mathbf{q}\partial \mathbf{p}} - \frac{\partial^2 H}{\partial \mathbf{p}\partial \mathbf{q}} = 0\,. \tag{2.12}$$

This important (though rather trivial) mathematical fact is known as Liouville's theorem for the Hamiltonian flow, after Joseph Liouville (1809–1882). (It has nothing to do with Liouville's theorem in complex analysis.) A fluid with a flow that is divergence-free is said to be incompressible, a behavior different from air in a pump, which gets very much compressed. Consequently, and as we shall show below, the "volume" of any subset in phase space which gets transported via the Hamiltonian flow remains unchanged. Before we express this in mathematical terms and give the proof, we shall consider the issue in more general terms.

Remark 2.2. On the Time Evolution of Measures.
The notion of volume deserves some elaboration. Clearly, since phase space is very high-dimensional, the notion of volume here is more abstract than the volume of a three-dimensional object. In fact, we shall later use a notion of volume which is not simply the trivial extension of three-dimensional volume. Volume here refers to a *measure*, the size or weight of sets, where one may in general want to consider a biased weight. The most famous measure, and in fact the mother of all measures, is the generalization of the volume of a cube to arbitrary subsets, known as the Lebesgue measure λ. We shall say more about this later.[2] If one feels intimidated by the name Lebesgue measure, then take $|A| = \int_A \mathrm{d}^n x$, the usual Riemann integral, as the (fapp-correct) Lebesgue measure of A. The measure may in a more general sense be thought of as some kind of weight distribution, where the Lebesgue measure gives equal (i.e., unbiased) weight to every point. For a continuum of points, this is a somewhat demanding notion, but one may nevertheless get a feeling for what is meant. For the time being we require that the measure be an additive nonnegative set function, i.e., a function which attributes positive or zero values to sets, and which is additive on disjoint sets: $\mu(A\cup B) = \mu(A) + \mu(B)$. The role of the measure will

[2] We need to deal with the curse of the continuum, which is that not all subsets of $\mathbb{R}^n$ actually have a volume, or as we now say, a measure. There are non-measurable sets within the enormous multitude of subsets. These non-measurable sets exist mathematically, but are not constructible in any practical way out of unions and intersections of simple sets, like cubes or balls. They are nothing we need to worry about in practical terms, but they are nevertheless there, and so must be dealt with properly. This we shall do in Sect. 4.3.1.

eventually be to tell us the size of a set, i.e., which sets are small and which are big. Big sets are important, while small ones are not.

It may be best to think for now of a measure in general as some abstract way of attributing "mass" to subsets. One is then led to ask how the measure (or the mass) changes with a flow. That question was first asked for the Hamiltonian flow, but it can be asked, and in fact has been asked, for flows of a general character. We shall do the same now. Let μ be a measure on the phase space Γ, which we take for simplicity as being $\mathbb{R}^n$. We consider now a general (not necessarily Hamiltonian) flow map on phase space, that is, a one-parameter family of maps $(\Phi_t)_{t\in\mathbb{R}}$, with parameter "time" t:

$$\big(\Phi_t(\mathbf{x})\big)_{t\in\mathbb{R}}\,,\ \mathbf{x}\in\mathbb{R}^n: \quad \Phi_t\circ\Phi_s(\mathbf{x})=\Phi_{t+s}(\mathbf{x})\,, \qquad \Phi_0(\mathbf{x})=\mathbf{x}\,. \tag{2.13}$$

In general, any flow Φ_t on $\mathbb{R}^n$ (or some general phase space) naturally defines the time evolution of the measure μ_t on $\mathbb{R}^n$:

$$\mu_t=\mu\circ\Phi_{-t}\,, \tag{2.14}$$

which means

$$\mu_t(A)=\mu(\Phi_{-t}A)\,, \qquad \text{or} \quad \mu_t(\Phi_t A)=\mu(A)\,, \tag{2.15}$$

for all (measurable) sets A and all $t\in\mathbb{R}$. Behind this definition is the simple logic that, if a measure μ is given at time $t=0$, then the measure μ_t of a set A is the measure μ of the set from which A originated by virtue of the flow. In other words the measure changes only because the set changes.

The notion of stationary measure is very important. This is a measure that does not change under the flow Φ_t, i.e.,

$$\mu_t(A)=\mu(A)\,,\ \forall t\in\mathbb{R}\,,\ \forall A\,. \tag{2.16}$$

The stationary measure plays a distinguished role in justifying probabilistic reasoning. Its importance was presumably first discovered by Boltzmann, and later on we shall spend some time considering Boltzmann's general ideas about statistical physics, which are valid for a whole range of theories.

The above-mentioned preservation of volume for Hamiltonian flows as a consequence of Liouville's theorem refers to phase space volume and is the assertion that

$$\lambda(\Phi_{-t}A)=\lambda(A)\,, \tag{2.17}$$

with λ the Lebesgue measure on phase space. This means that, under the Hamiltonian flow, sets change their shape but not their volume. This may also be referred to as Liouville's theorem, since it is a direct consequence of (2.12), as we shall show next. For the pendulum in Fig. 2.2, one sees this immediately, since the slices of the pie just rotate. The general situation is depicted in Fig. 2.4. ■

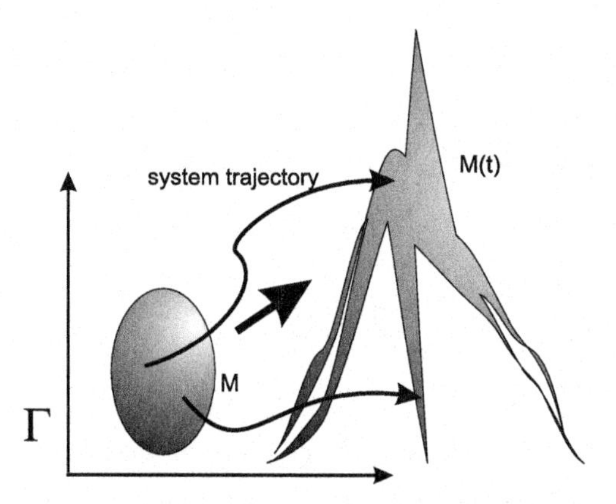

Fig. 2.4 The preservation of volume of the Hamiltonian flow. The set M in phase space changes under the flow but its volume remains the same

Remark 2.3. Continuity Equation
From the change in the measure (2.15) for a general flow generated by a vector field, one can derive a differential equation which governs the change in the *density* of the measure. That differential equation is called the continuity equation. If the measure has a density $\rho(\mathbf{x})$ (you may think of a mass density), then the change of measure with time defines a time-dependent density $\rho(\mathbf{x},t)$, and one has the logical relation[3]

$$\begin{aligned} \mu_t(A) &=: \int \chi_A(\mathbf{x})\rho(\mathbf{x},t)\mathrm{d}^n x \\ &\| \\ \mu(\Phi_{-t}A) &=: \int \chi_{\Phi_{-t}A}(\mathbf{x})\rho(\mathbf{x})\mathrm{d}^n x \\ &= \int \chi_A(\Phi_t(\mathbf{x}))\rho(\mathbf{x})\mathrm{d}^n x, \end{aligned} \tag{2.18}$$

where χ_A is the characteristic function of the set $A \subset \Gamma$, also called the indicator function of the set A, i.e., the function which is 1 on A and zero otherwise, and $\Phi_{-t}A = \{\mathbf{x} \in \Gamma \mid \Phi_t(\mathbf{x}) \in A\}$. Furthermore Φ_t is the solution flow map of some vector field $\mathbf{v}(\mathbf{x})$ on $\mathbb{R}^n$ (or some general phase space), i.e.,

$$\frac{\mathrm{d}}{\mathrm{d}t}\Phi_t(\mathbf{x}) = \mathbf{v}\big(\Phi_t(\mathbf{x})\big)\,. \tag{2.19}$$

We shall now show that the density $\rho(\mathbf{x},t)$ satisfies the continuity equation:

$$\frac{\partial}{\partial t}\rho(\mathbf{x},t) + \mathrm{div}\big[\mathbf{v}(\mathbf{x})\rho(\mathbf{x},t)\big] = 0\,. \tag{2.20}$$

---

[3] Note in passing that $\rho(\mathbf{x},t)$ can be computed from an obvious change of variables in the last integral, namely, $\rho(\mathbf{x},t) = \rho(\Phi_{-t}(\mathbf{x}))|\partial\Phi_{-t}(\mathbf{x})/\partial\mathbf{x}|$.

To see this, replace the indicator function in (2.18) by a smooth function f with compact support:

$$\int f\big(\Phi_t(\mathbf{x})\big)\rho(\mathbf{x})\mathrm{d}^n x = \int f(\mathbf{x})\rho(\mathbf{x},t)\mathrm{d}^n x \,. \tag{2.21}$$

Now differentiate (2.21) with respect to t to get

$$\int \frac{\mathrm{d}\Phi_t(\mathbf{x})}{\mathrm{d}t}\cdot\Big[\nabla f\big(\Phi_t(\mathbf{x})\big)\Big]\rho(\mathbf{x})\mathrm{d}^n x = \int f(\mathbf{x})\frac{\partial}{\partial t}\rho(\mathbf{x},t)\mathrm{d}^n x \,. \tag{2.22}$$

Replacing $\mathrm{d}\Phi_t(\mathbf{x})/\mathrm{d}t$ by the right-hand side of (2.19) and using (2.21) again with f replaced by $\mathbf{v}(\mathbf{x})\cdot\big[\nabla f(\mathbf{x})\big]$ (a wonderful trick), the left-hand side of (2.22) becomes

$$\int \mathbf{v}(\mathbf{x})\cdot\big[\nabla f(\mathbf{x})\big]\rho(\mathbf{x},t)\mathrm{d}^n x \,, \tag{2.23}$$

and after partial integration,

$$-\int f(\mathbf{x})\mathrm{div}\big[\mathbf{v}(\mathbf{x})\rho(\mathbf{x},t)\big]\mathrm{d}^n x \,. \tag{2.24}$$

Since this is equal to the right-hand side of (2.22) and since f is arbitrary, we conclude that (2.20) holds. ■

Now we ask whether there is a stationary measure (2.16). In terms of densities, the question is: Does there exist a stationary density, that is, a density independent of time, satisfying (2.20)? Since the time derivative part of (2.20) vanishes, i.e.,

$$\frac{\partial}{\partial t}\rho(\mathbf{x}) = 0 \,,$$

the density must satisfy the partial differential equation

$$\mathrm{div}\big[\mathbf{v}(\mathbf{x})\rho(\mathbf{x})\big] = 0 \,. \tag{2.25}$$

This is in general a very difficult question. In particular, it is almost impossible to find the solution for a general vector field. However, not so in the Hamiltonian case, where the answer turns out to be trivial! This is a consequence of (2.12). Setting

$$\mathbf{x} = \begin{pmatrix} \mathbf{q} \\ \mathbf{p} \end{pmatrix} ,$$

equation (2.12) reads $\mathrm{div}\,\mathbf{v}(\mathbf{x}) = 0$ for all $\mathbf{x}$. In this case, (2.20) becomes (after using the product rule)

$$\frac{\partial}{\partial t}\rho(\mathbf{x},t) + \mathbf{v}(\mathbf{x})\cdot\nabla\rho(\mathbf{x},t) = 0 \,. \tag{2.26}$$

Now $\rho(\mathbf{x},t) =$ constant, which we may choose to be unity, is obviously stationary. Putting $f = \chi_A$, $A \subset \Gamma$, and taking into account $\rho = 1$, (2.21) yields

$$\int \chi_A(\Phi_t(\mathbf{x}))\mathrm{d}^n x = \int \chi_{\Phi_{-t}A}(\mathbf{x})\mathrm{d}^n x = \int \chi_A(\mathbf{x})\mathrm{d}^n x = \lambda(A) . \tag{2.27}$$

In short, (2.27) says

$$\lambda(\Phi_{-t}A) = \lambda(A) . \tag{2.28}$$

We may as well put the future set $\Phi_t A$ into (2.28), instead of A, and use $\Phi_{-t}\Phi_t = \mathrm{id}$ (identity), whence

$$\lambda(A) = \lambda(\Phi_t A) . \tag{2.29}$$

In conclusion, *Liouville's theorem* implies that the Lebesgue measure ($\equiv$ volume) is stationary for Hamiltonian flows.

Remark 2.4. Time-Dependent Vector Fields
The continuity equation also holds for time-dependent vector fields $\mathbf{v}(\mathbf{x},t)$, in which case the flow map is a two parameter group $\Phi_{t,s}$ advancing points from time s to time t. All one needs to do is to replace the vector field in (2.20) by the time-dependent expression $\mathbf{v}(\mathbf{x},t)$, and the proof goes through verbatim. But now the notion of a stationary measure seems unsuitable, since the velocity field (representing the physical law) changes with time. But remarkably the notion still applies for Hamiltonian flows, i.e., even in the case where the Hamiltonian is time dependent (energy is not conserved), the volume (Lebesgue measure of a set) remains unchanged under the flow. ■

Remark 2.5. On the Initial Value Problem
The Lebesgue measure in phase space plays a distinguished role for the Hamiltonian flow. It is thus natural to weaken the problem of initial values in the sense of the measure, so that one is happy if it can be shown that the bad set of initial conditions for which no global solutions exist has Lebesgue measure zero. Be warned however, that a set which has measure zero may be small in the sense of the measure, but it is not necessarily small in the sense of cardinality (number of points in the set). The famous Cantor set, a subset of the interval $[0,1]$, has as many members as the reals, but has Lebesgue measure zero.

We close this section with what is a heretical thought for modern physicists, namely, a Newtonian universe. This is not physical (we know it ignores quantum mechanics and much more), but we can nevertheless conceive of it, and it is a good enough framework in which to ask a question which in an appropriate sense could be raised in any other theory: *Which initial values give rise to OUR Newtonian universe?* Put another way: *According to which criteria were the initial values of OUR universe chosen?* We do not ask *who* chose the initial conditions, but rather: Which physical law determines them? One possible answer to this could be that our universe is nothing special, that it could be *typical* in the sense that almost all initial conditions would give rise to a universe very much like ours (where "almost all" means that the Lebesgue measure of the set which does not give rise to a universe like ours is very small). It turns out that this is not the case, but we shall address this issue later. ■

2.3 Hamilton–Jacobi Formulation

The Hamiltonian structure and phase space are intimately connected with symplectic geometry. We shall say more about that at the end of the chapter. We wish to move on to another question. The Hamiltonian formulation of Newtonian mechanics is prosaic and brings out the particular structure shared by Newtonian mechanics and all Hamiltonian flows: conservation of energy (if H is time independent) and phase space volume. But that was not Hamilton's aim. He had a much deeper vision for mechanics. He was looking for an analogy between mechanics and wave optics, namely Huygens' principle and Fermat's extremal principle of geometric optics, according to which light rays take the path requiring the shortest time, and moreover follow the normals of wave fronts. Could mechanics be formulated by a similar guidance principle, where the mechanical paths are determined by the normal vectors of wave fronts? The extremal principle replacing Fermat's is the least action principle $\delta \int L \mathrm{d}t = 0$, where $L(\mathbf{q}, \dot{\mathbf{q}})$ is called the Lagrange function. The mechanical (Newtonian) trajectories between $t_0, \mathbf{q}_0$ and $t, \mathbf{q}$ (note that instead of initial position and initial velocity, we consider initial position and end position) are characterized as the extremals of

$$\int_{t_0}^{t} L\big(\mathbf{q}(t'), \dot{\mathbf{q}}(t')\big) \mathrm{d}t' .$$

We omit the derivation of the Euler–Lagrange equation, which is standard, but we recall that for Newtonian mechanics

$$L(\mathbf{q}, \dot{\mathbf{q}}) = \frac{1}{2} \dot{\mathbf{q}} \cdot m \dot{\mathbf{q}} - V(\mathbf{q}) . \tag{2.30}$$

For this Lagrange function, the Euler–Lagrange equations are the usual Newtonian equations. The Lagrange function and Hamilton function are Legendre transforms of one another.[4] (It is remarkable that almost all the great mathematicians of the 19th century have left some trace in theoretical mechanics.) Starting with H, we get L by changing from the variable $\mathbf{p}$ to $\dot{\mathbf{q}}$, so that $(\mathbf{q}, \mathbf{p})$ gets replaced by $(\mathbf{q}, \dot{\mathbf{q}})$, using the implicitly given function

$$\dot{\mathbf{q}} = \frac{\partial H(\mathbf{q}, \mathbf{p})}{\partial \mathbf{p}} ,$$

[4] Here is the definition. Let $f(x)$ be convex and let z be a given slope. Then look for the point $x(z)$ at which the tangent to the graph of f has slope z. You find $x(z)$ by minimizing $F(x,z) = f(x) - xz$ in x. By convexity of f, this is uniquely determined. The Legendre transform of f is $g(z) = F\big(x(z), z\big)$.

where the equation is solved by $\mathbf{p}$ as a function of $\dot{\mathbf{q}}$.

For "normal" Hamilton functions (quadratic in the momentum), that solution is immediate, and looking at (2.30), the Legendre transform pops right up:

$$L(\mathbf{q},\dot{\mathbf{q}}) = \mathbf{p}\cdot\dot{\mathbf{q}} - H(\mathbf{q},\mathbf{p}) . \tag{2.31}$$

Note in passing that if one starts with the least action principle as being fundamental, one can guess the form of the Lagrange function from basic principles such as symmetry, homogeneity, and simplicity. But that is not what we wish to discuss here.

We come now to Huygens' principle and the definition of waves $S_{\mathbf{q}_0,t_0}(\mathbf{q},t)$ which guide mechanical trajectories starting at $\mathbf{q}_0$ and moving along the vector field $\mathbf{p}(\mathbf{q},t) = \nabla S_{\mathbf{q}_0,t_0}(\mathbf{q},t)$. Hamilton suggested

$$S_{\mathbf{q}_0,t_0}(\mathbf{q},t) := \int_{t_0}^{t} L(\gamma,\dot{\gamma})\mathrm{d}t , \tag{2.32}$$

where $\gamma : \mathbf{q}_0, t_0 \longrightarrow \mathbf{q}, t$ is the extremum of the action principle, i.e., the Newtonian path. This function is often called the Hamilton–Jacobi function.

Unfortunately, this definition generally leads to a multivalued function. Take for example the harmonic oscillator with period T. There are many extremal trajectories for a harmonic oscillator with period T which go from $(0,T)$ to $(0,2T)$, so S is not uniquely defined. Or again, think of a ball which bounces off a wall. The position $\mathbf{q}$ in front of the wall can always be reached within a given time by two routes, one with and one without reflection from the wall.

However, the definition is good for short enough times. So never mind this difficulty, let us pursue the thought. Ignoring the dependence on $(\mathbf{q}_0,t_0)$ and considering[5]

$$\mathrm{d}S = \frac{\partial S}{\partial \mathbf{q}}\mathrm{d}\mathbf{q} + \frac{\partial S}{\partial t}\mathrm{d}t ,$$

and in view of (2.32) and (2.31), we can identify

$$\mathrm{d}S = \mathbf{p}\mathrm{d}\mathbf{q} - H(\mathbf{q},\mathbf{p})\mathrm{d}t .$$

Then by comparison

$$\mathbf{p}(\mathbf{q},t) = \frac{\partial S}{\partial \mathbf{q}}(\mathbf{q},t) , \tag{2.33}$$

whence

$$\frac{\partial S}{\partial t}(\mathbf{q},t) + H\left(\mathbf{q}, \frac{\partial S}{\partial \mathbf{q}}\right) = 0 . \tag{2.34}$$

[5] We may ignore that dependence because we assume uniqueness of the trajectory, and this implies that $S_{\mathbf{q}_0,t_0}(\mathbf{q},t) = S_{\mathbf{q}_1,t_1}(\mathbf{q},t) + S_{\mathbf{q}_0,t_0}(\mathbf{q}_1,t_1)$.

This is known as the Hamilton–Jacobi differential equation. For Newtonian mechanics, where $\dot{\mathbf{q}}_i = \mathbf{p}_i/m_i$, we then obtain the following picture. On the configuration space $\mathbb{R}^n$ for N particles ($n = 3N$), we have a function $S(\mathbf{q},t)$ ("unfortunately" multivalued), whose role it is to generate a vector field

$$\mathbf{v}(\mathbf{q},t) = m^{-1}\nabla S(\mathbf{q},t) \tag{2.35}$$

on configuration space. Integral curves $\mathbf{Q}(t)$ along the vector field are the possible trajectories of the N-particle system, i.e., they solve

$$\frac{\mathrm{d}\mathbf{Q}}{\mathrm{d}t} = \mathbf{v}\big(\mathbf{Q}(t),t\big) .$$

The function $S(\mathbf{q},t)$ is itself dynamical and solves the nonlinear partial differential equation

$$\frac{\partial S}{\partial t}(\mathbf{q},t) + \frac{1}{2}\sum_{i=1}^{N}\frac{1}{m_i}\left(\frac{\partial S}{\partial \mathbf{q}_i}\right)^2 + V(\mathbf{q}) = 0 . \tag{2.36}$$

This picture is, as we said, not quite right, because S is in general not well defined for mechanics. On the other hand, it is almost quantum mechanics. We shall soon understand which little quantum is missing to get the picture right.

2.4 Fields and Particles: Electromagnetism

Many hold the view that the particle is not a good concept for physics. They see it as a classical Newtonian concept which has been made obsolete by quantum mechanics. Fields on the other hand are generally well accepted, because relativity teaches us that the right physical theory will be a field theory, eventually quantized of course. To understand whether fields do work as well as one hopes, we shall have a quick look at electromagnetism, where dynamical fields come into play as fundamental objects needed to describe interactions between particles which carry a "charge". Electromagnetic fields act on a particle at position $\mathbf{q} \in \mathbb{R}^3$ via the Lorentz force:

$$m\ddot{\mathbf{q}} = e\left[\mathbf{E}(\mathbf{q},t) + \frac{\dot{\mathbf{q}}}{c}\times\mathbf{B}(\mathbf{q},t)\right] , \tag{2.37}$$

where $\mathbf{E}(\mathbf{q},t)$ und $\mathbf{B}(\mathbf{q},t)$ are the electric and magnetic fields, and c is the velocity of light. While the fields act on particles as described, in electromagnetism the fields are not independent agents living in a kingdom of their own, for they are themselves generated by particles. They are generated by particles and they act on particles, which is why one may say that they are there to represent the interaction between particles. But when the particles are point particles, which is the most natural relativistic possibility, this does not go down well with the fields. We shall explain this now. We shall also use the opportunity to introduce relativistic physics.

Albert Einstein (1879–1955) deduced from Maxwell's equations of electromagnetism that space and most importantly time change in a different way from Galilean physics when one changes between frames moving with respect to each other. The nature of this change is governed by the fact that the velocity of light does not change, when one moves from one frame to another. This led to the understanding, soon to be formalised in the four-dimensional description by Hermann Minkowski (1864–1909), that space and time are "of the same kind". That is, a particle needs for its specification not only a position in space, but also a location in time, implying that the coordinates of a particle in relativistic physics should be space and time coordinates. This is a revolution in Newtonian mechanics, where we are of course used to particles having different positions, but not different times. So in relativistic physics one must get used to the spacetime description of particles, with each particle having its own spacetime coordinates. In other words, the configuration space of classical mechanics, where we collect all positions of the particles of a system at the same time, no longer plays a fundamental role. Instead, Einstein showed that the overthrow of absolute time brings physics closer to a true description of nature. On this basis, he believed that physical theories must be local, in the sense that no physical effect can move faster than light. John Bell showed that this is wrong. We shall devote a whole chapter to this later, but now we must move on.

Minkowski introduced the so-called four-dimensional spacetime with a particular scalar product.[6] In spacetime, particles no longer move in a Newtonian way, but according to new dynamics. The particle position is now x^μ, $\mu = (0,1,2,3)$, where x^0 is selected as the time coordinate, since it is distinguished by the "signature" of the so-called Minkowski length[7]

$$\mathrm{d}s^2 = (\mathrm{d}x^0)^2 - \mathrm{d}\mathbf{x}^2 = (\mathrm{d}x^0)^2 - \sum_{i=1}^{3}(\mathrm{d}x^i)^2 .$$

In Newtonian mechanics, we are used to parameterizing paths by time, which is no longer natural. A natural parameter now is length – Minkowski length – normalized by $1/c$, i.e., on each trajectory we select a zero mark somewhere and go from there

[6] Minkowski suggested using imaginary numbers for the time coordinate in the spacetime coordinates $(x_0 = \mathrm{i}ct, \mathbf{x})$, because then the formal Euclidean scalar product yields the Minkowski metric $s^2 = (x^0)^2 + \mathbf{x}^2 = -ct^2 + \sum_{i=1}^{3}(x^i)^2$. The advantage is that all congruences, i.e., transformations leaving the scalar product invariant (which form the so-called Lorentz group) appear in Euclidean guise, and so can be viewed as four-dimensional rotations and reflections. This differs from the Galilean case, where the change to relatively moving frames (Galilean boosts) must be dealt with separately. In the Minkowski case, the corresponding Lorentz boost is simply a "rotation", but with an imaginary angle. When one considers the changes of x_0 and, say, x_1, such a rotation yields $x_0{}' = x_0 \cos\phi - x_1 \sin\phi$, $x_1{}' = x_0 \sin\phi + x_1 \cos\phi$, which requires in the Minkowski case $\phi = \mathrm{i}\psi$, an imaginary angle. Following the point $x_1 = 0$ in the primed frame (moving with relative velocity v), one then has $x_1{}'/ct' = v/c = \tanh\psi$, which yields the well-known formula for the Lorentz boost.

[7] The sign of $\mathrm{d}s^2$ is conventional: $\mathrm{d}s^2 < 0$ implies a spacelike distance, while $\mathrm{d}s^2 > 0$ implies a timelike distance.

with the length element of the ith particle

$$\mathrm{d}\tau_i = \frac{\mathrm{d}s_i}{c} = \frac{1}{c}\sqrt{\left(\frac{\mathrm{d}x_i^0}{\mathrm{d}\tau_i}\right)^2 - \left(\frac{\mathrm{d}\mathbf{x}_i}{\mathrm{d}\tau_i}\right)^2}\,\mathrm{d}\tau_i\,.$$

Thus

$$\dot{x}_i^2 = g_{\mu\nu}\dot{x}_i^\mu \dot{x}_i^\nu = c^2\,, \qquad g_{\mu\nu} = \begin{pmatrix} 1 & 0 \\ 0 & -\mathsf{E}_3 \end{pmatrix}, \tag{2.38}$$

where E_3 is the 3×3 identity matrix, and we use the Einstein summation convention according to which we sum automatically over those indices appearing more than once. The dot over x_i indicates the derivative with respect to Minkowski length τ_i, also called the proper time, of the ith particle (in the frame where the particle is at rest $x_i^0 = c\tau_i$). The metric tensor $g_{\mu\nu}$ which defines here the Minkowski scalar product can be used to lower indices:

$$x_\mu := g_{\mu\nu}x^\nu\,,$$

while the inverse of the metric tensor denoted by $g^{\mu\nu}$ is used to raise indices.

For us it is natural to parameterize the trajectory by the coordinate time $x^0 = ct$ of our rest frame where we see the particle move with velocity $\mathbf{v}$. We thus have a new function

$$\bar{x}^\mu(x^0) = \left(x^0, \mathbf{x}(x^0)\right) = x^\mu\left(\tau(x^0)\right),$$

for which we get by the chain rule

$$\frac{\mathrm{d}\bar{x}^\mu}{\mathrm{d}x^0} = \left(1, \frac{\mathbf{v}}{c}\right) = \dot{x}^\mu \frac{\mathrm{d}\tau}{\mathrm{d}x^0}\,,$$

or, taking the Minkowski square,

$$1 - \frac{\mathbf{v}^2}{c^2} = c^2\left(\frac{\mathrm{d}\tau}{\mathrm{d}x^0}\right)^2, \tag{2.39}$$

which allows us to switch between proper time and coordinate time.

The relativistic dynamics of a "free" particle may be defined by an extremal principle which determines the physical spacetime path from x to y as the path with the shortest Minkowski length. This means that the variation

$$\delta\int_x^y \mathrm{d}s = \delta\int_{\tau(x)}^{\tau(y)} \left(\dot{x}^\mu \dot{x}_\mu\right)^{1/2} \mathrm{d}\tau = 0\,.$$

If we wish to talk about relativistic mechanics in Newtonian terms, i.e., if we wish to use Newtonian concepts like energy, mass, and force and the like – and we might wish to do that because it may then be easier to arrive at Newtonian mechanics as a limiting description of relativistic mechanics – we can multiply the integral by dimensional constants to get an action integral, so that the terms in the Euler–Lagrange equation read analogously to the Newtonian terms. That is, writing $S = -mc\int ds$ and using (2.39) in the $x^0 = ct$ parametrization, the Lagrange function becomes

$$L(\dot{\mathbf{q}}) = -mc^2\sqrt{1-\frac{v^2}{c^2}} \,.$$

The Euler–Lagrange equations lead to the canonical momentum

$$\mathbf{p} = \frac{m}{\sqrt{1-v^2/c^2}}\mathbf{v} =: \tilde{m}\mathbf{v} \,,$$

from which we recognize m as the rest mass, because $\mathbf{p} \approx m\mathbf{v}$ when $v \ll c$.

The canonical momentum can be taken as the vector part of a canonical four-momentum

$$p^\mu = m\dot{x}^\mu \,,$$

the Minkowski length of which is [see (2.38)]

$$g_{\mu\nu}p^\mu p^\nu = p^\mu p_\mu = m^2\dot{x}^\mu\dot{x}_\mu = m^2c^2 \,. \tag{2.40}$$

Hence, parameterizing by $x^0 = ct$, we find

$$p^\mu = m\dot{x}^\mu = \frac{m}{\sqrt{1-v^2/c^2}}(c,\mathbf{v}) \,.$$

Observing further that

$$\frac{mc}{\sqrt{1-v^2/c^2}} \approx mc\left(1+\frac{1}{2}\frac{v^2}{c^2}+\dots\right) = mc + \frac{m}{2}\frac{v^2}{c} + \cdots \,,$$

we are led to set $E = p^0c$. With (2.40), we thus obtain the energy–momentum relation:

$$E = \sqrt{\mathbf{p}^2c^2 + m^2c^4} = \tilde{m}c^2 \,.$$

Now we have for N particles the spacetime trajectories $q_i = \left(q_i^\mu(\tau_i)\right)$, $\mu = 0,1,2,3$, $i = 1,\dots,N$. Let us introduce a force K^μ, which accelerates the particles. By virtue of (2.38), we have

$$\ddot{x}^\mu\dot{x}_\mu = 0 \,,$$

and this suggests

$$K^\mu\dot{x}_\mu = 0 \,,$$

that is, the force should be orthogonal to the velocity in the sense of the Minkowski metric. The simplest way to achieve this is to put

$$K^{\mu} = F^{\mu\nu}\dot{x}_{\nu} \, ,$$

with $F^{\mu\nu} = -F^{\nu\mu}$, an antisymmetric tensor of rank 2, i.e., an antisymmetric 4×4 matrix. One way to generate such a tensor is to use a four-potential A^{μ}:

$$F^{\mu\nu} = \frac{\partial}{\partial x_{\mu}}A^{\nu} - \frac{\partial}{\partial x_{\nu}}A^{\mu} \, . \tag{2.41}$$

The Maxwell–Lorentz theory of electromagnetic interaction has the force act on the particles through a law that involves not only masses as parameters, but also charges e_i:

$$m_i\ddot{q}_i^{\mu} = \frac{e_i}{c}F^{\mu}{}_{\nu}(q_i)\dot{q}_i^{\nu} \, . \tag{2.42}$$

In view of (2.37), one names the matrix elements as follows:

$$F^{\mu}{}_{\nu}(x) = \begin{pmatrix} 0 & E_1 & E_2 & E_3 \\ E_1 & 0 & B_3 & -B_2 \\ E_2 & -B_3 & 0 & B_1 \\ E_3 & B_2 & -B_1 & 0 \end{pmatrix} \, , \tag{2.43}$$

recalling that indices are lowered or raised by action of $g^{\mu\nu} = g_{\mu\nu}$, whence $F^{\mu}{}_{\nu} = F^{\mu\lambda}g_{\lambda\nu}$.

For the three-vector $\mathbf{q}_i(\tau_i)$, we then obtain

$$m_i\ddot{\mathbf{q}}_i = e_i\left(\mathbf{E} + \frac{\dot{\mathbf{q}}_i}{c} \times \mathbf{B}\right) \, ,$$

where dots over symbols still refer to derivatives with respect to τ_i. For small velocities (compared to the velocity of light), this is close to (2.37).

But the fields $F^{\mu}{}_{\nu}$ are themselves generated by the particles, and this is supposed to give the interaction between the charges. The equation which describes the generation of fields is not difficult to guess, by analogy with the gravitational potential which is given by the potential equation $\Delta V = \nabla \cdot \nabla V = \delta(x)$ for a point mass at the origin. (Note that the scalar product construction of the law is a good trick for making the law invariant under Euclidean congruences.) Taking the four-dimensional ∇ in the Minkowski sense suggests the four-dimensional potential equation (invariant under Minkowskian congruences)

$$\left[\left(\frac{\partial}{\partial x^0}\right)^2 - \left(\frac{\partial}{\partial \mathbf{x}}\right)^2\right]A^{\mu} = \Box A^{\mu} = \frac{4\pi}{c}j^{\mu} \, , \tag{2.44}$$

where j^μ is the current originating from moving particles carrying charge e_i. We discuss the current below.

Note in passing that (2.41) does not determine the vector potential uniquely, because a term of the form $\partial f/\partial x^\mu$ for some well-behaved f can always be added to A^μ without changing the forces. This is called gauge invariance. Equation (2.44) determines the potential in what is known as the Lorentz gauge, where

$$\frac{\partial}{\partial x^\mu}A^\mu = 0 .$$

The current is a distribution, because the charges are concentrated at the positions of the particles, which are points. One could think of smeared out charges, but a charge with a certain extension (a ball, for example) would not be a relativistic object because of Lorentz contraction, i.e., a ball would not remain a ball when moving.[8] Lorentz contraction is an immediate consequence of the loss of simultaneity, since the length of a rod is defined by the spatial distance of the end points of the rod when taken at the same time.

The current of a point charge is by itself unproblematic. It has the following frame-independent, i.e., relativistic, form:

$$j^\mu(x) = \sum_i e_i c \int_{-\infty}^{\infty} \delta^4\big(x - q_i(\tau_i)\big)\dot{q}_i^\mu(\tau_i)\mathrm{d}\tau_i , \tag{2.45}$$

with

$$\delta^4(x) = \prod_{\mu=0}^{3} \delta(x^\mu) ,$$

and we use x for the four vector $x = (x^0, \mathbf{x})$, since we have used the boldface notation for three-dimensional vectors. Better known is the form in a coordinate frame. With $x^0 = ct$, we obtain (writing the integral as a line integral along the trajectory in the second equality below)

[8] Suppose one did not care about relativistic invariance, and took a small ball in the rest frame of the electron (for instance, with radius 10^{-16} cm, which seems to be an upper bound from experimental data). Unfortunately, this yields an effective mass of the electron larger than the observed electron mass. The rough argument is that the Coulomb energy of a concentrated charge (infinite for a point charge) yields by the energy–mass relation a field mass which, since it moves with the electron must be accelerated as well, and is effectively part of the electron mass. Extended Lorentz invariant charge distributions entail dynamical problems as well, when strong accelerations occur [1–4].

$$\begin{aligned}
j^\mu(x) &= \sum_i e_i c \int_{-\infty}^{\infty} \delta^4\big(x - q_i(\tau_i)\big)\dot{q}_i^\mu(\tau_i)\mathrm{d}\tau_i \\
&= \sum_i e_i c \int_{\{q_i(\tau_i)|\tau_i \in \mathbb{R}\}} \delta^4(x-q_i)\mathrm{d}q_i^\mu \\
&= \sum_i e_i c \int_{-\infty}^{\infty} \delta\big(\mathbf{x} - \mathbf{q}_i(t_i)\big)\delta(ct - ct_i)\frac{\mathrm{d}q_i^\mu}{\mathrm{d}t_i}(t_i)\mathrm{d}t_i \\
&= \sum_i e_i \delta\big(\mathbf{x} - \mathbf{q}_i(t)\big)\frac{\mathrm{d}q_i^\mu}{\mathrm{d}t}(t) \,.
\end{aligned}$$

From (2.45) we easily obtain the continuity equation

$$\frac{\partial}{\partial x^\mu} j^\mu(x) = 0 \,.$$

Using the fact that the trajectories are timelike, we have

$$\begin{aligned}
\frac{\partial}{\partial x^\mu} j^\mu(x) &= \sum_i e_i c \int_{-\infty}^{\infty} \frac{\partial}{\partial x^\mu}\delta^4(x-q_i)\frac{\mathrm{d}q_i^\mu}{\mathrm{d}\tau_i}\mathrm{d}\tau_i \\
&= \sum_i e_i c \int_{\{q_i(\tau_i)|\tau_i \in \mathbb{R}\}} -\frac{\partial}{\partial q_i^\mu}\delta^4(x-q_i)\mathrm{d}q_i^\mu \\
&= \sum_i e_i c \left[\lim_{\tau_i \to \infty}\delta^4\big(x-q_i(\tau_i)\big) - \lim_{\tau_i \to -\infty}\delta^4\big(x-q_i(\tau_i)\big)\right] \\
&= 0\,, \qquad \forall x \in \mathbb{R}^4 \,.
\end{aligned}$$

The system of differential equations (2.41)–(2.44) defines the theory of charges interacting via fields. Now (2.42) is unproblematic if the field F^μ_ν is *given* as a well-behaved function. The linear partial differential equation (2.44) is likewise unproblematic if j^μ is *given*, even as a distribution, as in (2.45). Then one has a Cauchy problem, i.e., in order to solve (2.44), one needs initial data $A^\mu(0,x^1,x^2,x^3)$ and $(\partial A^\mu/\partial t)(0,x^1,x^2,x^3)$, and there is no obstacle to finding a solution.

However, we now have to solve (2.42)–(2.44) together, rather than separately, and this does not work. The system of differential equations is only formal, i.e., there are no functions $q^\mu(\tau)$ and $A^\mu(x)$, whose derivatives would satisfy the equations. It does not even matter whether we have more than one particle. Let us take one particle to see what goes wrong. First solve

$$A^\mu(x) = \left(\Box^{-1}\frac{4\pi}{c}j^\mu\right)(x) = \int \Box^{-1}_{x,x'}\frac{4\pi}{c}j^\mu(x')\mathrm{d}^4x' \,, \tag{2.46}$$

with a Green's function $\Box^{-1}_{x,x'}$ given by

$$\Box\Box^{-1}_{x,x'} = \delta^4(x-x') \,. \tag{2.47}$$

The Green's function is not unique. The different possible Green's functions differ by solutions of the homogeneous equation ($j = 0$), i.e., by functions in the kernel of $\Box$. A symmetric choice is[9]

$$4\pi\Box^{-1}_{x,x'} = \delta\big((x-x')^2\big) = \delta\big((x^\mu - x'^\mu)(x_\mu - x'_\mu)\big) \, .$$

Why is that symmetric? Using

$$\delta\big(f(x)\big) = \sum_k \frac{1}{|f'(x_k)|}\delta(x - x_k) \, , \tag{2.48}$$

where x_k are the single zeros of f we get

$$\delta(x^2 - y^2) = \frac{1}{2}\frac{\delta(x-y)}{y} + \frac{1}{2}\frac{\delta(x+y)}{y} \, ,$$

and thus

$$\begin{aligned} 4\pi\Box^{-1}_{x,y} &= \delta\left((x-y)^2\right) = \delta\left((x^0 - y^0)^2 - (\mathbf{x}-\mathbf{y})^2\right) \\ &= \frac{1}{2}\frac{\delta\big((x^0-y^0) - |\mathbf{x}-\mathbf{y}|\big)}{|\mathbf{x}-\mathbf{y}|} + \frac{1}{2}\frac{\delta\big((x^0-y^0) + |\mathbf{x}-\mathbf{y}|\big)}{|\mathbf{x}-\mathbf{y}|} \, , \end{aligned}$$

which is the sum of retarded and advanced Green's function, a notion which will become clear shortly. Any linear combination of these parts is a possible $\Box^{-1}_{x,y}$, and one commonly uses only the retarded part, using the argument that one experiences only radiation emitted in the past. We shall say more about that later. One may convince oneself by formal manipulations[10] that (2.47) holds for this or any other linear combination.

Now let us come to the end of the story. With (2.46) and (2.45), we get

$$A^\mu(x) = e\int \delta\left((x - q(\tau))^2\right)\dot{q}^\mu(\tau)\mathrm{d}\tau \, , \tag{2.49}$$

---

[9] It is quite natural for the Green's function to be like this. It is the most natural relativistic function one can write down. The points which have Minkowski distance zero from one another form the (backward and forward) light cones and they are special in a certain sense. So the function is not eccentric in any way.

[10] For example,

$$\begin{aligned} &\left[\left(\frac{\partial}{\partial x^0}\right)^2 - \Delta\right]\frac{\delta(x^0 - |\mathbf{x}|)}{|\mathbf{x}|} \\ &\quad = \frac{1}{|\mathbf{x}|}\delta'' - \Delta\left(\frac{1}{|\mathbf{x}|}\right)\delta(x^0 - |\mathbf{x}|) - \frac{1}{|\mathbf{x}|}\Delta\delta(x^0 - |\mathbf{x}|) - 2\nabla\left(\frac{1}{|\mathbf{x}|}\right)\cdot\nabla\delta(x^0 - |\mathbf{x}|) \\ &\quad = \frac{1}{4\pi}\delta(\mathbf{x})\delta(x^0 - |\mathbf{x}|) \, , \end{aligned}$$

using the chain rule on $\nabla\delta$.

and with (2.48), we find

$$A^\mu = e\frac{\dot{q}^\mu(\tau_a)}{2(x^\mu - q^\mu(\tau_a))\dot{q}_\mu(\tau_a)} + e\frac{\dot{q}^\mu(\tau_r)}{2(x^\mu - q^\mu(\tau_r))\dot{q}_\mu(\tau_r)} ,$$

where τ_a and τ_r are the solutions of

$$(x - q(\tau))^2 = 0 \quad \Longleftrightarrow \quad ct - c\tau = \mp|\mathbf{x} - \mathbf{q}(\tau)| .$$

We can now understand the terminology used. The retarded/advanced time is given by the intersection of the forward/backward light cone based at **x** with the trajectory of the particle (see Fig. 2.5, where **x** should be thought of as an arbitrary point).

We have thus derived a field (see for instance [5], if some manipulations seem unclear) A^μ which is well-behaved everywhere *except* at points x lying on the world-line q of the charge generating the field. For $x = q(\tau)$, we have $\tau_a = \tau_r = \tau$, and we see that the denominator is zero. But this is now the end of the story, since these are the x values which are needed in (2.42).

This problem is well known by the name of the electron self-interaction. The field which the electron generates acts back on the electron, and this back-reaction is mathematically ill-defined, since the electron is a point. Hence, the field idea for managing interactions between point charges does not work, unless one introduces formal manipulations like renormalization [6], or changes electromagnetism on small scales [7].

The Maxwell–Lorentz theory of electromagnetism works well (in the sense of describing physical phenomena correctly) when the fields are generated by smeared out charges (charge clouds), so one can describe the radiation from an antenna. It also works when the fields are given as "external" fields, which act on charges by the Lorentz force equation (see [3, 8, 9] for mathematical proofs concerning existence and uniqueness). In short, electromagnetism is fine for most non-academic life. One may ask why this is so. The reason may be that the Maxwell–Lorentz theory is the macroscopic description of the fundamental theory of electromagnetism we shall describe next. But that theory does not contain fields on the fundamental level.

2.5 No fields, Only Particles: Electromagnetism

What is bad about fields when it comes to describing interactions? The problem is that the field naturally acts everywhere, and thus also on the very particle which generates the field. But taking the meaning of relativistic interaction between particles seriously, why does one need fields at all? Why not get rid of the field and have the particles interact directly? In a sense one does this when solving (2.44) for A^μ and putting that into (2.42). Fokker thought that way [10], like many others before him, including Gauss [11], Schwarzschild [12], and Tetrode [13]. He wrote down the variational principle for a relativistic particle theory, which was later rediscov-

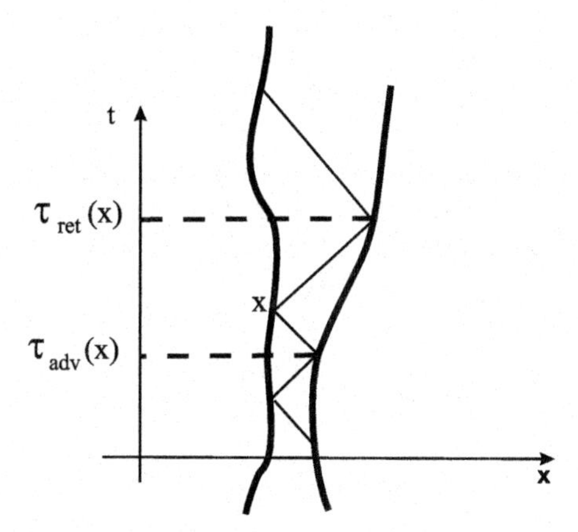

Fig. 2.5 In Feynman–Wheeler electromagnetism particles interact along backward and forward light cones. Here we set $c = 1$

ered by Wheeler and Feynman [14, 15], to explain retarded radiation. How can the particles interact directly in a relativistic way?

There is no natural notion of simultaneity whereby a force can act at the same time between particles, but we already know that the simplest choice is to take the Minkowski distance and to say that particles interact when there is distance zero from one to the other. Hence the particle at spacetime point q interacts with all other particles at spacetime points which are intersection points of light cones based at q with the other trajectories, that is, when

$$(q_i^\mu - q_j^\mu)(q_{i\mu} - q_{j\mu}) = (q_i - q_j)^2 = 0\,,$$

or put another way, when $\delta\big((q_i - q_j)^2\big)$ is not zero.

Note, that there are always two light cones based at one point, one directed towards the future and one directed towards the past (see Fig. 2.5), although of course the physical law is not concerned about such notions as past and future.

It is rather clear that dynamics which is defined by future times and past times can no longer be given by differential equations of the ordinary kind. But some differential equations can nevertheless be written down from a variational principle. The Fokker–Wheeler–Feynman action S is the simplest relativistic action one could think of describing interacting particles:

$$S = \sum_i \left[-m_i c \int \mathrm{d}s_i - \sum_{j>i} \frac{e_i e_j}{c} \int\!\!\int \delta\big((q_i - q_j)^2\big) \mathrm{d}q_i^\mu \mathrm{d}q_{j\mu} \right]. \tag{2.50}$$

Writing the trajectories $q_i^\mu(\lambda_i)$ with arbitrary parameters λ_i and using the notation $\dot{q}_i^\mu = \mathrm{d}q_i^\mu/\mathrm{d}\lambda_i$, we obtain

$$S=\sum_i\left[-m_ic\int\left(\dot{q}_i^{\mu}\dot{q}_{i\mu}\right)^{1/2}\mathrm{d}\lambda_i-\sum_{j>i}\frac{e_ie_j}{c}\int\int\delta\left((q_i-q_j)^2\right)\dot{q}_i^{\mu}\dot{q}_{j\mu}\mathrm{d}\lambda_i\mathrm{d}\lambda_j\right].$$

The most noteworthy feature is that there are no diagonal terms in the double sum. This is the major difference with the Maxwell–Lorentz theory, where the diagonal terms should normally be present. The contribution of the ith particle to the interaction reads

$$-\frac{e_i}{c}\int\mathrm{d}\lambda_i\dot{q}_i^{\mu}\int\sum_{j\neq i}e_j\delta\left((q_i-q_j)^2\right)\dot{q}_{j\mu}\mathrm{d}\lambda_j=-\frac{1}{c^2}\int j_i^{\mu}(x)A_{i\mu}(x)\mathrm{d}x\,,$$

with the "field"[11]

$$A_{i\mu}(x)=\sum_{j\neq i}e_j\int\delta\left((x-q_j)^2\right)\dot{q}_{j\mu}\mathrm{d}\lambda_j\,. \tag{2.51}$$

In the Wheeler–Feynman formulation, fields [like (2.51)] would only be introduced as a suitable macroscopic description, good for everyday applications of electromagnetism, like handling capacitors. On the fundamental level there is no radiation field and there is no radiation. Therefore, opposite charges may orbit each other without "losing energy" due to radiation. Famous solutions of that kind are known as Schild solutions [16].

Equation (2.49) shows that both the advanced and the retarded Green's functions appear in A_μ. But it is the "emission of radiation" which we typically see and which is solely described by the retarded Green's function. Wheeler–Feynman and also Maxwell–Lorentz electromagnetism are time reversible, i.e., the theory does not favor emission before absorption. The typicality of emission has become known as the problem of the electromagnetic arrow of time. The original motivation of Wheeler and Feynman was to reduce this arrow of time to the thermodynamic one, which had been so successfully explained by Boltzmann, by supposing a special initial distribution of the particles in the phase space of the universe. (We shall address this further in the chapter on probability.)

Wheeler and Feynman therefore considered the thermodynamic description of the particle system, i.e., they considered a distribution of charges throughout the universe which "absorb all radiation". In terms of the theory, this means that the sum of the differences of the retarded and advanced forces over all particles vanish. This is called the absorber condition. This macroscopic theory is still time-symmetric. But supposing further that at some early time the initial distribution of the particles was special (non-equilibrium in some sense), then time-directed radiation phenomena and in particular the observed radiation damping of an accelerated charge are reduced to Boltzmann's explanation of irreversibility (see Sect. 4.2).

[11] We computed this in (2.49), but it is important to understand that this is simply a mathematical expression, which plays no role unless x is a point on the worldlines of the other particles. There is no field in this theory.

We emphasize that the Wheeler–Feynman electromagnetic theory is a mathematically consistent relativistic theory with interaction. In fact it is the only such mathematically well defined and physically relevant theory existing so far, and it is about particles, not fields. The statistical mechanics of the theory leads to the well known description by electromagnetic fields and smeared out charges and is thus experimentally indistinguishable from Maxwell–Lorentz electromagnetism, whenever the latter is well defined, i.e., whenever no point charges are considered to generate fields.

The theory is, however, unfamiliar since its dynamical law is not of the familiar form. The intuition one has from solving differential equations with Cauchy data fails. So why is this? In fact, the Euler–Lagrange equations of (2.50) are not ordinary differential equations, since advanced and retarded times appear in them.[12] In contrast, the Maxwell–Lorentz theory is formally of the ordinary type, but with the serious drawback that the fields render the equations mathematically undefined when they are interpreted as fundamental, i.e., when point charges are considered.

Remark 2.6. On the Nature of Reality
Reality is a curious notion. Physics takes the view that something "out there" exists, and that the world is "made out of something". This is not curious at all. But it is not so easy to say what it is that the world is made of, since the only access we have to the world is through our senses and our thinking, and communication about the experience we have. What the world is made of is specified by our physical theory about the world, and it is only there where we can see what the world is made of. When our physical theory is about point particles and how they move, then there are point particles out there – if what the theory says about their motion is consistent with our experience, of course. The connection between the entities of the theory and our experience is often complicated, often not even spelt out in any detailed way. Nevertheless, one has some kind of feel for how it works, and a bit of pragmatism in this is alright.

When we wish to explain a physical phenomenon, we reduce it (in the ideal case) to the behavior of the ontological quantities the physical theory is about. In Maxwell–Lorentz electromagnetism, fields are ontological. Switch on your radio. What better explanation is there than to say that the fields are out there, and they get absorbed as radio waves by the radio antenna, and that the radio transforms them back into air waves? Music to your ears. But in Wheeler–Feynman electromagnetism, there are no fields and only particles. It explains the music as well. But the explanation is different [13, 14].

If the Maxwell–Lorentz theory (with point charges) were mathematically consistent, we could chose between fields and particles as being "real", or only particles as being "real". Since both would describe the macroscopic world as we see it, our choice would then have to be made on the grounds of simplicity and beauty of the theories. Perhaps in the future we shall find a simpler and nicer theory than the ones we have now, one which is solely about fields. Then only fields will be "real".

---

[12] It is an intriguing problem of mathematical physics to establish existence and uniqueness of solutions.

"Reality" thus changes with the nature of our physical theory, and with it the elements which can be measured. In the Wheeler–Feynman theory of electromagnetism, the electromagnetic field cannot be measured. This is because it is not there. It is not part of the theory. That is trivial. Less trivial may seem the understanding that the theory also says what elements are there and how those elements can be measured. In Maxwell–Lorentz theory, the electric field is measured, according to the theory, by its action on charges.

Here is another point one may think about from time to time. Although all variables needed to specify the physical theory are "real", there is nevertheless a difference. In a particle theory, the particle positions are primitive or primary variables, representing what may be called the *primitive ontology*.[13] They must be there: a particle theory without particle positions is inconceivable. Particle positions are what the theory is about. The role of all other variables is to say how the positions change. They are secondary variables, needed to spell out the law. We could also say that the particle positions are a priori and the other variables a posteriori. An example of the latter might be the electric field. In fact, secondary variables can be replaced by other variables or can even be dispensed with, as in Wheeler–Feynman electromagnetism. Another example which we did not touch upon at all is general relativity, which makes the Newtonian force obsolete. ■

2.6 On the Symplectic Structure of the Phase Space

With the understanding of tensors and forms, in particular differential forms, not only did mathematics move forward, but so did our our insight into physics. It was better appreciated what the objects "really" are. We have an example in our relativistic description of electromagnetism (2.43). The electric and magnetic field strengths are not vector fields, as one learns in school, but rather components of an antisymmetric second rank tensor. Mathematical abstraction helps one to get down to the basics of things, and that is satisfying. One such mathematical abstraction is symplectic geometry. We shall say a few things here mainly to make sure that further mathematical abstractions do not lead to a deeper understanding of the physics which we have presented so far.

Mathematically deeper than conservation of energy and volume is the symplectic structure of phase space, which goes hand in hand with Hamilton's formulation of mechanics [18]. Symplectic geometry needs a space of even dimensions, and classical physics provides that. Consider the phase space $\mathbb{R}^{2n}$ with coordinates $(q_1,\dots,q_n,p_1,\dots,p_n)$. Given $\mathbf{x},\mathbf{y}\in\mathbb{R}^{2n}$, let $q_i(\mathbf{x})$ be the projection of $\mathbf{x}$ on the q_ith coordinate axis. Then

$$\omega_i^2(\mathbf{x},\mathbf{y}) = q_i(\mathbf{y})p_i(\mathbf{x}) - q_i(\mathbf{x})p_i(\mathbf{y}) \tag{2.52}$$

[13] The notion and role of primitive ontology has long been ignored, but it has recently been revitalized and emphasized in [17].

is the area of the parallelogram generated by $\mathbf{x}, \mathbf{y}$ when projected into the (q_i, p_i) plane. Set

$$\omega^2(\mathbf{x}, \mathbf{y}) = \sum_{i=1}^{n} w_i^2(\mathbf{x}, \mathbf{y}) = \mathbf{x} \cdot (-I)\mathbf{y} = \left((-I)\mathbf{y}\right)^{\mathrm{t}} \mathbf{x}, \tag{2.53}$$

with

$$I = \begin{pmatrix} 0_n & +\mathsf{E}_n \\ -\mathsf{E}_n & 0_n \end{pmatrix},$$

where E_n is the n-dimensional unit matrix, 0_n is the n-dimensional zero matrix, and $\mathbf{z}^{\mathrm{t}}$ is the transpose of $\mathbf{z}$, i.e., the row vector, an element in the dual space of $\mathbb{R}^{2n}$. The 2-form ω^2, or equivalently the symplectic matrix I, defines the symplectic structure of phase space $\mathbb{R}^{2n}$ ($n = 3N$ for N particles) and gives (like a scalar product) an isomorphism between the vector space and its dual. From courses in analysis, we know that the gradient $\partial f / \partial \mathbf{x}$ of a function $f(\mathbf{x})$, $\mathbf{x} \in \mathbb{R}^d$, is in fact a dual element, i.e., the row vector which acts as a linear map on $\mathbf{h}$ according to $(\partial f / \partial \mathbf{x})\mathbf{h}$ = row times column (matrix multiplication by a vector). Now use ω^2 to identify $\partial f / \partial \mathbf{x}$ with a vector $\nabla_\omega f$ (using the Euclidean scalar product, this is the normal ∇f). Given $\mathbf{z} \in \mathbb{R}^{2n}$, the object $\omega^2(\cdot, \mathbf{z})$ is a linear form, i.e., a linear map

$$\begin{aligned} \mathbb{R}^{2n} &\longrightarrow \mathbb{R} \\ \mathbf{x} &\longmapsto \omega^2(\mathbf{x}, \mathbf{z}) = (-I\mathbf{z})^{\mathrm{t}} \mathbf{x} . \end{aligned}$$

which we wish to be equal to $\partial f / \partial \mathbf{x}$. We thus search for a $\mathbf{z}_f$ such that

$$\begin{aligned} & \frac{\partial f}{\partial \mathbf{x}} \mathbf{x} = (-I\mathbf{z}_f)^{\mathrm{t}} \mathbf{x}, \qquad \forall \mathbf{x}, \\ \Longleftrightarrow \quad & \left(\frac{\partial f}{\partial \mathbf{x}}\right)^{\mathrm{t}} = (\nabla f) = (-I\mathbf{z}_f) \\ \Longleftrightarrow \quad & \mathbf{z}_f = \nabla_\omega f = I(\nabla f) . \end{aligned}$$

It follows that (2.5) can be written in the form

$$\begin{pmatrix} \dot{\mathbf{q}} \\ \dot{\mathbf{p}} \end{pmatrix} = I(\nabla H) = I \begin{pmatrix} \dfrac{\partial}{\partial \mathbf{q}} H \\ \dfrac{\partial}{\partial \mathbf{p}} H \end{pmatrix} .$$

The Hamiltonian flow respects the symplectic structure. In particular, it is area preserving, which means the following. Let C be a closed curve in $\mathbb{R}^{2n}$ and define the enclosed area as the sum of the n areas which arise from projections of C onto the coordinate planes (q_i, p_i) [compare with (2.52) and (2.53)]. In the (q_i, p_i) plane, we have a curve C_i with area $A(C_i)$. The area can be transformed into a line integral by

Stokes' theorem in two dimensions:

$$\begin{aligned}\int_{A(C_i)} \mathrm{d}q_i\,\mathrm{d}p_i &= \int_{A(C_i)} \operatorname{curl}\begin{pmatrix} p_i \\ 0 \end{pmatrix} \mathrm{d}q_i\,\mathrm{d}p_i \\ &= \oint_{C_i} \begin{pmatrix} p_i \\ 0 \end{pmatrix} \cdot \begin{pmatrix} \mathrm{d}q_i \\ \mathrm{d}p_i \end{pmatrix} \\ &= \oint_{C_i} p_i\,\mathrm{d}q_i\,. \end{aligned} \tag{2.54}$$

In general,

$$A = \text{area of } C = \oint_C \mathbf{p}\cdot\mathrm{d}\mathbf{q} = \sum_i \oint_{C_i} p_i\,\mathrm{d}q_i\,.$$

Using differential forms,

$$\omega_i^2 = \mathrm{d}p_i \wedge \mathrm{d}q_i\,, \qquad \omega^2 = \sum_i \mathrm{d}p_i \wedge \mathrm{d}q_i\,,$$

and

$$\omega^2 = \mathrm{d}\omega^1\,, \qquad \omega^1 = \sum_i p_i\,\mathrm{d}q_i\,.$$

Equation (2.54) is nothing other than

$$\int_{A(C_i)} \mathrm{d}\omega^1 = \int_{C_i} \omega^1\,.$$

Transporting C with the Hamiltonian flow yields the area $A(t)$, and preservation of area means that $A(t) = A$. By change of variables, the integration over $A(t)$ can be expressed by $\mathbf{q}(t)$ and $\mathbf{p}(t)$ in the integral, so that

$$\begin{aligned}\frac{\mathrm{d}}{\mathrm{d}t}A(t) &= \frac{\mathrm{d}}{\mathrm{d}t}\oint_C \mathbf{p}(t)\,\mathrm{d}\mathbf{q}(t) = \oint_C \dot{\mathbf{p}}\,\mathrm{d}\mathbf{q} + \oint_C \mathbf{p}\,\mathrm{d}\dot{\mathbf{q}} \\ &= \oint_C \dot{\mathbf{p}}\,\mathrm{d}\mathbf{q} - \oint_C \dot{\mathbf{q}}\,\mathrm{d}\mathbf{p} \qquad \left[+\oint_C \mathrm{d}(\mathbf{q}\dot{\mathbf{p}}) = 0\right] \\ &= -\oint_C \frac{\partial H}{\partial \mathbf{q}}\,\mathrm{d}\mathbf{q} - \oint_C \frac{\partial H}{\partial \mathbf{p}}\,\mathrm{d}\mathbf{p} \\ &= -\oint_C \mathrm{d}H = 0\,. \end{aligned}$$

Furthermore, the volume in even-dimensional vector spaces can be thought of as arising from a product of areas (generalizing the area in $\mathbb{R}^2$ = width times length), i.e., products of two-forms (2.52) yield the volume form:

$$\omega = \mathrm{d}p_1 \wedge \mathrm{d}q_1 \wedge \ldots \wedge \mathrm{d}p_n \wedge \mathrm{d}q_n\,,$$

the Lebesgue measure on phase space (in form language). Then Liouville's theorem arises from preservation of area.

Transformations of coordinates $(\mathbf{q},\mathbf{p}) \stackrel{\psi}{\longrightarrow} (\mathbf{Q},\mathbf{P})$ are called canonical or symplectic if the Jacobi matrix $\nabla\psi$ is symplectic, which is a variation on the theme of the orthogonal matrix:

$$(\nabla\psi)^{\mathrm{t}} I \nabla\psi = I . \tag{2.55}$$

They preserve areas and satisfy the canonical equations

$$\begin{pmatrix} \dot{\mathbf{Q}} \\ \dot{\mathbf{P}} \end{pmatrix} = I \begin{pmatrix} \dfrac{\partial}{\partial \mathbf{Q}} \tilde{H} \\ \dfrac{\partial}{\partial \mathbf{P}} \tilde{H} \end{pmatrix} , \qquad \tilde{H} \circ \psi = H .$$

The Poisson bracket (2.11) is invariant under canonical transformations, because (2.11) can be written as

$$\{f,g\} = \nabla f \cdot I \nabla g ,$$

and if $f(\mathbf{Q},\mathbf{P})$, $g(\mathbf{Q},\mathbf{P})$, and $(\mathbf{Q},\mathbf{P}) = \psi(\mathbf{q},\mathbf{p})$ are given, since we have

$$\nabla(f \circ \psi) = \nabla\psi(\nabla f \circ \psi) ,$$

it follows that

$$\begin{aligned} \{f\circ\psi, g\circ\psi\} &= \nabla(f\circ\psi)\cdot I\nabla(g\circ\psi) \\ &= (\nabla f\circ\psi)\cdot(\nabla\psi)^{\mathrm{t}} I\nabla\psi(\nabla g\circ\psi) \\ &= (\nabla f\circ\psi)\cdot I(\nabla g\circ\psi) \qquad \text{[by (2.55)]} \\ &= \{f,g\}\circ\psi . \end{aligned}$$

Clearly,

$$\{q_i, p_i\} = \delta_{ij} , \qquad \{q_i, q_j\} = 0 , \qquad \{p_i, p_j\} = 0 ,$$

and variables which satisfy this are said to be canonical. Of particular interest are variables $(Q_1,\dots,Q_n,P_1,\dots,P_n)$, where $P_1,\dots,P_n$ do not change with time, and where the Hamilton function takes the form

$$\tilde{H}(\mathbf{Q},\mathbf{P}) = \sum_{i=1}^{n} \omega_i P_i .$$

Then $\dot{Q}_i = \omega_i$, i.e., $Q_i = \omega_i t + Q_{i,0}$, and Q_i is like the phase of a harmonic oscillator. Such (P_i, Q_i) are called action–angle variables. Systems which allow for such variables are said to be integrable, since their behavior in time is in principle completely under control, with their motion (in the new coordinates) being that of "uncoupled" harmonic oscillators. The solution can then be found by algebraic manipulation and integration. The Hamiltonian motions in $\mathbb{R}^2$ (H not time dependent, one particle in one space dimension) are integrable, since H itself does not change with time, and hence one may choose $P = H$. However, integrability is *atypical* for Hamiltonian systems.

References

1. J. Novik: Ann. Phys. **28**, 225 (1964)
2. F. Rohrlich: *Classical Charged Particles*, 3rd edn. (World Scientific Publishing Co. Pte. Ltd., Hackensack, NJ, 2007)
3. H. Spohn: *Dynamics of Charged Particles and the Radiation Field* (Cambridge University Press, 2004)
4. M.K.H. Kiessling: In: *Nonlinear Dynamics and Renormalization Group* (Montreal, QC, 1999), CRM Proc. Lecture Notes, Vol. 27 (Amer. Math. Soc., Providence, RI, 2001) pp. 87–96
5. A.O. Barut: *Electrodynamics and Classical Theory of Fields and Particles* (Dover Publications Inc., New York, 1980). Corrected reprint of the 1964 original
6. P. Dirac: Proc. Roy. Soc. A **178**, 148 (1938)
7. M. Born, L. Infeld: Commun. Math. Phys. A **144**, 425 (1934)
8. W. Appel, M.K.H. Kiessling: Ann. Physics **289** (1), 24 (2001)
9. G. Bauer, D. Dürr: Ann. Henri Poincaré **2** (1), 179 (2001)
10. A.D. Fokker: Z. Physik **58**, 386 (1929)
11. C. Gauss: A letter to Weber on 19 March 1845
12. K. Schwarzschild: Nachr. Ges. Wis. Gottingen **128**, 132 (1903)
13. H. Tetrode: Z. Physik **10**, 317 (1922)
14. J. Wheeler, R. Feynman: Rev. Mod. Phys. **17**, 157 (1945)
15. J.A. Wheeler, R.P. Feynman: Rev. Mod. Phys. **21**, 425 (1949)
16. A. Schild: Phys. Rev. **131**, 2762 (1962)
17. V. Allori, S. Goldstein, R. Tumulka, N. Zanghí: British Journal for the Philosophy of Science (2007)
18. F. Scheck: *Mechanics*, 3rd edn. (Springer-Verlag, Berlin, 1999). From Newton's laws to deterministic chaos. Translated from the German

Chapter 3
Symmetry

This chapter is more a footnote than a chapter on symmetry in physics. It has intentionally been kept very short and old-fashioned. We wish to say just as much as is needed to understand that Bohmian mechanics is a Galilean theory of nature. And we wish to emphasize certain features which may lie buried in more bulky accounts of symmetries in physics. Our emphasis is on the the role of ontology for symmetry. Ontology is the thing we must hold on to, because that is what the law of physics is for. At the end of the day, that is what determines which symmetries there are and how they act.

The invariance of a physical law under a transformation of variables entering the law defines a symmetry of that law. Invariance means this. Consider the set of solutions of the dynamical law, i.e., a set of histories $\mathbf{q}(t)$, $t \in \mathbb{R}$ (not necessarily positions of particles, but think of the variables which the theory is primarily concerned with, i.e., the primitive variables which relate directly to physical reality). Let $\mathscr{G}$ be a group of transformations. We do not specify them, but we suppose we know how $g \in \mathscr{G}$ acts on the trajectory $\mathbf{q}(t)$ to give a transformed trajectory which we denote by $(g\mathbf{q})(t)$, where

$$(g\mathbf{q})(t) := \big(g(\mathbf{q}_i(t))\big)_{i=1,\dots,N} \,.$$

The law is invariant if for every $g \in \mathscr{G}$ and every $\mathbf{q}(t)$, the transformed $(g\mathbf{q})(t)$ is also in the solution set.[1] The relevant thing to note is that, with the action of g on $\mathbf{q}$, there may be "strange" actions on secondary (or derived) variables in order for the law to be invariant. In other words the group action is not only described by the action on the primitive variables, but also by the action on the variables which are needed to formulate the law. We shall give examples below.

A symmetry can be a priori, i.e., the physical law is built in such a way that it respects that particular symmetry by construction. This is exemplified by spacetime

[1] There are two equivalent ways of having the transformations act. Thinking of particle trajectories, the transformation may act actively, changing the trajectories, or passively, in which case the trajectories remain unchanged but the coordinate system changes. These transformations are mutual inverses.

D. Dürr, S. Teufel, *Bohmian Mechanics*, DOI 10.1007/978-3-540-89344-8_3,

symmetries, because spacetime is the theater in which the physical law acts (as long as spacetime is not subject to a law itself, as in general relativity, which we exclude from our considerations here), and must therefore respect the rules of the theater. An example is Euclidean space symmetries like rotations. If the theater in which the physics takes place is Euclidean space (as in Newtonian mechanics), rotations must leave the physical law invariant, i.e., the physical law must respect the rotational invariance of Euclidean space.

The mathematical formulation of the law can also introduce new (secondary) symmetries. Examples of this are the so-called gauge symmetries, which arise when a secondary variable can be regarded as belonging to an equivalence class, defined by gauge transformations which do not affect the histories of the primitive variables.[2] The law is thus only specified up to a choice of gauge. A simple example is the potential function $V(\mathbf{q})$ in Newtonian mechanics, which can be changed by an additive constant without changing the "physics". Another example is given by A^μ in (2.44), where all A^μ which differ by $\partial f/\partial x_\mu$ yield the same dynamics for the charges, the dynamics being given by $F^{\mu\nu}$ which is insensitive to such differences.

Finding all symmetries of a given physical theory is technically important, since symmetries go hand in hand with conserved quantities (well known in Lagrangian formulations of physical laws in terms of Noether's theorem), which restrict the manifold of solutions. Energy (time-shift invariance), angular momentum (rotational invariance), and momentum (translational invariance) are examples in Newtonian physics.

One should not be afraid to classify symmetries according to their importance. Some are of a purely technical character, while some are fundamental. For example canonical transformations $(\mathbf{q},\mathbf{p}) \mapsto (\mathbf{Q},\mathbf{P})$, which leave Hamilton's equations invariant, are invariants of symplectic geometry. These are mathematical symmetries since the positions of particles are clearly primary or, if one so wishes, fundamental. Likewise the unitary symmetry in quantum mechanics. Here part of the description is encoded in the wave function, which in abstract terms is an element of a vector space. Since a vector space does not single out any basis, the coordinate representation[3] of the wave function is arbitrary. The arbitrariness of the choice of basis has become known as unitary symmetry, which is sometimes conceived of as a fundamental symmetry of nature. That is nonsense, of course, since "position" plays a fundamental role in our world and breaks the unitary symmetry in favor of the position representation of the wave function. But naturally, these are merely words unless one knows what position means in the physical theory, and to know that, the theory must contain position as a primitive notion.

---

[2] In quantum field theory, gauge transformations are viewed as a basic ingredient for finding good quantum field theories. However, that is not our concern here.

[3] This has to be taken with a grain of salt. For example, the so-called position or momentum representations of the wave function are not coordinate representations with respect to basis choices for the vector space. They are simply the function and its Fourier transform, which can however be connected to self-adjoint operators via the spectral decompositions. This is part of what will be explained in this book.

On the other hand the fundamental symmetries are clearly those of spacetime, since, as we said before, that is the theater in which the physical law acts. When one spells out the ontology specifying the primary objects the physical law is about, the physical law should be more or less clear, given the spacetime symmetries and metaphysical categories such as simplicity and beauty which the law should obey.

The Galilean spacetime symmetries are the Euclidean ones of three-dimensional space (translation, rotation, and reflection invariance) as well as time translation invariance and time reflection invariance – all parts of Galilean invariance. The latter expresses Galilean relativity, which uses the notion of inertial system to assert that the law must not change under change of inertial system. By definition (based on metaphysical insight), inertial systems can move with uniform velocities relative to one another, and the Galilean symmetries thus include one more symmetry transformation, the Galilean boost, which represents the change from one inertial frame to another in relative motion. In relativistic four-dimensional spacetime, things are simpler, because all symmetries are congruences of Minkowski spacetime given by the Minkowski metric.

To see symmetry and simplicity at work, let us consider a one-particle world and a law for the motion of that particle. Let us consider Newtonian mechanics. For example, *translations in space*. Consider N particles with positions $\mathbf{q}_i$, $i = 1, \ldots, N$. Then

$$g(\mathbf{q}_i) = \mathbf{q}'_i = \mathbf{q}_i + \mathbf{a} ,$$

and

$$m\ddot{\mathbf{q}}'_i = m_i\ddot{\mathbf{q}}_i = \mathbf{F}_i(\mathbf{q}) = \mathbf{F}_i(\mathbf{q}'_1 - \mathbf{a}, \ldots, \mathbf{q}'_N - \mathbf{a}) = \mathbf{F}_i(\mathbf{q}'_1, \ldots, \mathbf{q}'_N) ,$$

where the last equality uses the fact that the force is translation invariant, e.g.,

$$\mathbf{F}_i(\mathbf{q}_1, \ldots, \mathbf{q}_N) = \sum_{i \neq j = 1}^{N} \mathbf{F}(\mathbf{q}_i - \mathbf{q}_j) ,$$

otherwise translation invariance would not hold. Hence it follows that the translated trajectories obey the same law. Let us now consider just one particle, for simplicity of notation.

Let us look quickly at *time-shift invariance*, or $t \mapsto t + a$, i.e., $\mathbf{q}'(t) = \mathbf{q}(t + a)$ holds if $\mathbf{F}$ does not depend on time. Then there are *orthogonal transformations* R of $\mathbb{R}^3$, i.e., transformations which preserve the Euclidean scalar product so that $\det\mathsf{R} = \pm 1$, $\mathsf{R}\mathsf{R}^{\mathsf{t}} = \mathscr{E}_3$. For these,

$$\mathbf{q}' = \mathsf{R}\mathbf{q} ,$$

and

$$m\ddot{\mathbf{q}}' = m\mathsf{R}\ddot{\mathbf{q}} = \mathsf{R}\mathbf{F}(\mathbf{q}) = \mathsf{R}\mathbf{F}(\mathsf{R}^{\mathsf{t}}\mathbf{q}') = \mathbf{F}(\mathbf{q}') ,$$

if $\mathbf{F}$ transforms like a vector, that is, if

$$\mathbf{F}(\mathsf{R}^{\mathrm{t}}\,\cdot\,) = \mathsf{R}^{\mathrm{t}}\mathbf{F}(\,\cdot\,)\,.$$

For $\mathbf{F} = \mathbf{b} =$ constant, this does not hold, i.e., $\mathsf{R}\mathbf{b} \neq \mathbf{b}$. But $\nabla = (\partial/\partial\mathbf{q})^{\mathrm{t}}$ transforms like a vector (∇ is to be thought of as a column vector, while $\partial/\partial\mathbf{q}$ is a row vector):

$$\frac{\partial}{\partial\mathbf{q}'} = \frac{\partial}{\partial\mathbf{q}}\frac{\partial\mathbf{q}}{\partial\mathbf{q}'} = \frac{\partial}{\partial\mathbf{q}}\mathsf{R}^{\mathrm{t}} \quad\Longrightarrow\quad \nabla' = \left(\frac{\partial}{\partial\mathbf{q}}\mathsf{R}^{\mathrm{t}}\right)^{\mathrm{t}} = \mathsf{R}\nabla\,.$$

Now let V be a scalar function, i.e.,

$$V'(\mathbf{q}') := V(\mathsf{R}^{\mathrm{t}}\mathbf{q}') = V(\mathbf{q})\,,$$

which is R invariant: $V'(\mathbf{q}') = V(\mathbf{q}')$ or $V(\mathsf{R}\mathbf{q}) = V(\mathbf{q})$. Then for $\mathbf{F}(\mathbf{q}) = \nabla V(\mathbf{q})$,

$$m\ddot{\mathbf{q}}' = m\mathsf{R}\ddot{\mathbf{q}} = \mathsf{R}\nabla V(\mathbf{q}) = \nabla' V'(\mathbf{q}') = \mathbf{F}(\mathbf{q}')\,.$$

The Newtonian gravitational potential is a natural example, since it is solution of the simplest Galilean invariant equation, the potential equation $\Delta V = \nabla\cdot\nabla V = 0$, outside the mass distribution, with the boundary condition that V vanishes at infinity.

The special feature of Newtonian dynamics, i.e., to be of second order (with laws of the form $\ddot{\mathbf{q}} = \ldots$) has not played any role up to now. This happens when one considers the change to a relatively moving inertial system (a *Galilean boost*):

$$\mathbf{q}' = \mathbf{q} + \mathbf{u}t\,. \tag{3.1}$$

Clearly $\ddot{\mathbf{q}}' = \ddot{\mathbf{q}}$ and one sees how simply the invariance of the law under boosts (3.1) arises (for translation invariant $\mathbf{F}$). This seems to suggest that a first order theory where the particle law has the form $\dot{\mathbf{q}} = \ldots$ should have problems with Galilean relativity. One would be inclined to think that one ought to have an equation for $\ddot{\mathbf{q}}$ in order to make the law invariant. Here is a little elaboration on that idea.

Let us devise a Galilean theory for one particle. We start with a theory of second order:

$$\ddot{\mathbf{q}} = \mathbf{F}(\mathbf{q}) \overset{\substack{\text{translation}\\ \text{invariance}}}{\Longrightarrow} \mathbf{F} = \text{const.} \overset{\substack{\text{rotation}\\ \text{invariance}}}{\Longrightarrow} \mathbf{F} = 0\,,$$

whence $\ddot{\mathbf{q}} = 0$ is the only possibility, and that is Galilean invariant. Now try a first order theory:

$$\dot{\mathbf{q}} = \mathbf{v}(\mathbf{q},t) \overset{\substack{\text{translation}\\ \text{invariance}}}{\Longrightarrow} \mathbf{v} = \text{const.} \overset{\substack{\text{rotation}\\ \text{invariance}}}{\Longrightarrow} \mathbf{v} = 0\,,$$

whence $\dot{\mathbf{q}} = 0$ would be the law, i.e., no motion. But that is not Galilean invariant!

Thus one may wish to conclude that first order theories (which could be called Aristotelian, according to the Aristotelian idea that motion is only guidance toward a final destination) cannot be Galilean invariant. We already know that this cannot be right. The Hamilton–Jacobi formulation of Newtonian mechanics, which is a first order theory, must be Galilean invariant. We do not want to pursue this further here, but the following aspect of symmetry transformations is helpful if one wants to understand why Hamilton–Jacobi is Galilean invariant.

We want to discuss *time-reversal symmetry*: $t \mapsto -t$, $\mathbf{q}(t) \mapsto \mathbf{q}'(t) = \mathbf{q}(-t)$. Clearly, $\mathbf{q}'(t)$ is a solution of the Newtonian equations, since $(\mathrm{d}t)^2 = (-\mathrm{d}t)^2$. Again the form $\ddot{\mathbf{q}} = \ldots$ for the law is responsible for this invariance. But now move to a phase space description in terms of $(\mathbf{q}, \mathbf{p})$. We have

$$\begin{aligned} \dot{\mathbf{q}} &= \mathbf{p}/m \quad &\longmapsto \quad \dot{\mathbf{q}}' &= -\mathbf{p}/m\,, \\ \dot{\mathbf{p}} &= \mathbf{F} \quad &\longmapsto \quad \dot{\mathbf{p}}' &= -\mathbf{F}\,, \end{aligned}$$

and time-reversal invariance follows if $t \mapsto -t$ goes along with $(\mathbf{q}, \mathbf{p}) \mapsto (\mathbf{q}, -\mathbf{p})$ (clearly $\mathbf{p}' = -\mathbf{p}$ since velocities must be reversed).

It is worth saying this more generally. Let the phase point be given by X, and let $t \mapsto -t$ come along with $X \mapsto X^*$, where the asterisk denotes an involution $(X^*)^* = X$ (like multiplication by -1, or, as another example, complex conjugation). That is, the asterisk denotes a representation of time reversal. Invariance holds if $X'(t) = X^*(-t)$ is a solution of the dynamics whenever $X(t)$ is a solution. The important fact to note is that the way this operation acts depends on the role the variables play in the physical law. The primitive variables usually remain unchanged, while secondary variables, those whose role is "merely" to express the physical law for the primitive variables, may change in a "strange" way. A well-known example is Maxwell–Lorentz electrodynamics. The state is $X = (\mathbf{q}, \dot{\mathbf{q}}, \mathbf{E}, \mathbf{B})$ and

$$(\mathbf{q}, \dot{\mathbf{q}}, \mathbf{E}, \mathbf{B})^* = (\mathbf{q}, -\dot{\mathbf{q}}, \mathbf{E}, -\mathbf{B})\,,$$

which follows from the Lorentz force equation (2.37). It is clear that $\dot{\mathbf{q}} \mapsto -\dot{\mathbf{q}}$, but then $\mathbf{B}$ must follow suit to make the equation time-reversal invariant. The lesson is that some variables may need to be changed in a strange way for the law to be invariant. But as long as those variables are secondary, there is nothing to worry about.

Let us close with a final remark on time-reversal invariance. One should ask why we are so keen to have this feature in the fundamental laws when we experience the contrary. Indeed, we typically experience thermodynamic changes which are irreversible, i.e., which are not time reversible. The simple answer is that our platonic idea (or mathematical idea) of time and space is that they are without preferred direction, and that the "directed" experience we have is to be explained from the underlying time symmetric law. How can such an explanation be possible? This is at the same time both easy and confusing. Certainly the difference in scales is of importance. The symmetry of the macroscopic scale can be different from that of the microscopic scale, if the "initial conditions" are chosen appropriately. This will be further discussed in Sect. 4.2.

Chapter 4
Chance

Physical laws do not usually involve probabilities.[1] The better known physical theories are deterministic and defined by differential equations (like the Schrödinger equation), so that a system's evolution is determined by "initial data". In a deterministic theory (like Newtonian Mechanics or Maxwell–Lorentz electromagnetism), the physical world simply evolves like clockwork. Yet we experience *randomness* – *chance* seems to be the motor of many physical events, like the zigzag trajectories of a Brownian particle or the random dots left by particles which arrive at the screen in a two slit experiment. Chance determines whether a coin falls heads or tails – we cannot predict the actual outcome. How should we explain that in a deterministic world?

Chance, randomness, and probability are words one uses lightly, but their meaning is cloudy, and to many they seem unfit for the mathematical precision one expects of nature's laws. The way randomness manifests itself is something of an oxymoron, for there is *lawlike behavior* in random events, as attested by the law of large numbers, asserting for example that heads and tails *typically* come up equally often in the long run. On the other hand we describe coin tossing by *complete lack of knowledge* of what any single outcome will be. Heads or tails each come with probability 1/2. The main theme of this chapter will be to answer the following two summarizing questions, once asked by Marian von Smoluchowski [3]:

1. How can it be that the effect of chance is computable, i.e., that *random causes* can have lawlike effects?[2]

---

[1] An example of a physical law based on probability is provided by the so-called spontaneous collapse theory (or GRW theory). Here Schrödinger's equation is replaced by a jump process (or diffusion) equation, in which a macroscopic superposition of wave packets reduces quickly under the time evolution to a collapsed wave function, and only one of the packets remains – a possible remedy for the measurement problem of quantum mechanics [1], which may also relieve the tension between relativity and the nonlocality of quantum mechanics [2].

[2] Wie ist es möglich, daß sich der Effekt des Zufalls berechnen lasse, daß also *zufällige Ursachen gesetzmäßige Wirkungen* haben?

D. Dürr, S. Teufel, *Bohmian Mechanics*, DOI 10.1007/978-3-540-89344-8_4,

2. How can randomness arise if all events are reducible to deterministic laws of nature? Or in other words, how can *lawlike causes* have random effects?[3]

When correctly viewed, the answers are not difficult to find. Indeed, the answer to both is *typicality*, and this answer has in principle been known since the work of Ludwig Boltzmann (1844–1906). We say "correctly viewed" because, for various reasons, Boltzmann's probabilistic reasoning did not meet with the same acceptance as the calculus of Newton and Leibniz, for example.

The various reasons are difficult to sort out and this is certainly not the place to speculate about historical shortcomings. Some mathematical objections to Boltzmann's work (physically not serious at all) were clearly much overrated, and cast their shadows well into the twentieth century. The general idea behind Boltzmann's work, namely how to reason probabilistically within deterministic physics, was overshadowed by Boltzmann's explanation of *irreversibility*, the part of his work which became most famous and was most strongly attacked. However, the main attack against Boltzmann – which presumably hit more painfully than any technical bickering – was that Boltzmann reopened the pre-Socratic atomistic view of the world advocated by Democritus, Leucippus, and others – now with the rigor of mathematical language, reducing all phenomena to the motion of the unseen atoms. Boltzmann's contemporaries could easily dismiss this world view, which was essential to all of Boltzmann's thinking, by declaring: We do not believe in what we do not see. But for Boltzmann, as for the pre-Socratic physicists, what counted was the explanatory value, the gain of insight provided by atomism, explaining logically from few basic principles what we do experience with our gross senses – so never mind whether we can "see" atoms.

The attacks against Boltzmann's reductionism ended immediately with Einstein's and Smoluchowski's work on Brownian motion, the erratic motion of microscopically small particles suspended in a fluid, which are nevertheless visible through a microscope. This motion had been observed since the invention of the microscope in the 17th century, but it was Einstein and Smoluchowski who, using Boltzmann's reductionistic view, explained the erratic motion as arising from molecular collisions of the molecules in the fluid with the Brownian particles (see Sect. 5). The atoms in heated motion were visible after all. Boltzmann committed suicide in 1906 in Duino near Trieste, Italy.

We shall spend much time on this subject, and Boltzmann's view of it, because we shall apply the ideas to quantum mechanics, where probability commonly enters as an axiom. To see whether that axiom is sensible or not, it is good to understand first what probability is.

[3] Wie kann der Zufall entstehen, wenn alles Geschehen nur auf regelmäßige Naturgesetze zurückzuführen ist? Oder mit anderen Worten: Wie können *gesetzmäßige Ursachen eine zufällige Wirkung* haben?

4.1 Typicality

The answer to the question: "What is geometry" is rather simple. It is the doctrine of spatial extension and its elements are objects like points or straight lines. The answer to the question: "What is analysis" is also simple, but less intuitive, since it involves the doctrine of limiting processes and the handling of infinity. The elements are the continuum and the real numbers. What is a real number? A real number between 0 and 1 is an infinite sequence consisting of digits $0, 1, 2, \ldots, 9$ like $0.0564319\ldots$. In the binary representation, one only has the digits 0 and 1, and every number between 0 and 1 is a sequence of 0s and 1s, e.g.,

$$0.1001\ldots = 1 \times \frac{1}{2} + 0 \times \frac{1}{4} + 0 \times \frac{1}{8} + 1 \times \frac{1}{16} + \cdots .$$

What is probability theory? The doctrine of chance or randomness? But what are chance and randomness? They are what probability theory is about. Simply playing with different words is not helpful. What characterizations of chance can be found? Unpredictability, not knowing what will happen? As Henri Poincaré (1854–1912) once said [4]: "If we knew all, probability would not exist." But knowledge and ignorance are very complex notions and certainly not comparable to the primitivity of a straight line or a real number, and one cannot imagine how these complex notions could be seen as fundamental objects of a theory of probability. Most importantly, however, we wish to apply probability theory to a physical system, and its behavior obviously has nothing to do with what we know or do not know about the system. Yet physical systems do often behave in a random way (the famous erratic Brownian motion, for example) and this is what we wish to understand.

So what is probability theory? It is the doctrine of typical behavior (of physical systems). The meaning of typicality is easy to understand. Typically one draws blanks in a lottery, because there are so many more blanks than prizes, as can be ascertained by simple counting. Typically, in a run of N coin tosses, one obtains an irregular sequence of heads and tails, which is nevertheless regular in the numbers of heads and tails. In fact, these numbers are more or less equal if N is large enough. Why is this? It is because, for large N, the number of head–tail sequences of length N (which equals the number of sequences of 0s and 1s of length N) with more or less equal numbers of heads and tails is so enormously bigger than the number of sequences that have more heads than tails, or vice versa ("allowed" fluctuations are of the order of $\sqrt{N}$). In fact the number of all sequences is 2^N and the number of sequences with exactly K heads is given by the binomial coefficient $\binom{N}{K}$. For large N, one can use Stirling's formula

$$N! = \left(\frac{N}{\mathrm{e}}\right)^N \sqrt{2\pi N}\left[1 + \mathcal{O}\left(\frac{1}{N}\right)\right] \tag{4.1}$$

to evaluate the binomial coefficient, and a quick estimate shows that the maximum number is

$$\binom{N}{N/2} \approx \frac{2 \times 2^N}{\sqrt{2\pi N}}\,.$$

But summing over all possible sequences, we have

$$\sum_{K=0}^{N} \binom{N}{K} = 2^N\,,$$

so that the numbers of sequences with K such that $|K - N/2| \leq \sqrt{N}$ already sum up roughly to the total number 2^N. This is essentially the $\sqrt{N}$ fluctuation law, and one can imagine that for huge N the difference in numbers is exponential if K exceeds the size of the fluctuation (an easy exercise).

What about typicality in real physics? Consider a system of gas molecules in a container. Let us assume that, at this moment, the gas molecules fill the container in a homogeneous way. In our gross way of looking at things, we do not see the individual gas molecules, but we experience a homogeneous density of the gas in the container. We experience a *macrostate* of the gas. The detailed state of the gas, which lists all positions and momenta of all gas molecules, i.e., the phase space point of the gas system, is called a *microstate*. The microstate obviously changes with time, since the phase point wanders through phase space (see Fig. 2.3). But in our gross way of looking at things, we do not see that in the macrostate. So why does the macrostate remain? For the macrostate to change in a detectable way, the microstate has to wander into very particular regions of phase space. Why does the phase point (the microstate) not move into that particular region of phase space where, for example, all molecules are in the right half of the container? That we would of course feel with our gross senses. Suppose the container were a lecture hall and the air molecules all went into the right half of the lecture hall. Then all the students in the left half of the hall would be without air (see Fig. 4.1). So why are we lucky enough never to experience that? The answer is that it is *atypical*.

It is atypical because the number of microstates making up a macrostate which looks like the homogeneous density gas is so overwhelmingly larger than the number of microstates which make up a macrostate with a gross imbalance in the density profile. To get a quick feel for the sizes, number all the gas molecules $1,\ldots,N$, where $N \sim 10^{24}$, and distribute Rs (the molecule is on the right) and Ls (the molecule is on the left) over the gas molecules. Then it is again a question of counting 0–1 sequences of length N with N huge. In fact, these numbers give good estimates of the phase space volume of phase points corresponding to equipartition and non-equipartition of molecules.

This way of looking at probability is due to the physicist Ludwig Boltzmann (1844–1906), who perfected the kinetic gas theory of Rudolf Clausius (1822–1888) and James Clerk Maxwell (1831–1879). According to Boltzmann, what is happening is the typical scenario under the given constraints. Boltzmann's statistical reasoning is superior to physical theories, in the sense that it does not depend on the particular physical law. In the following we consider Newtonian mechanics in order

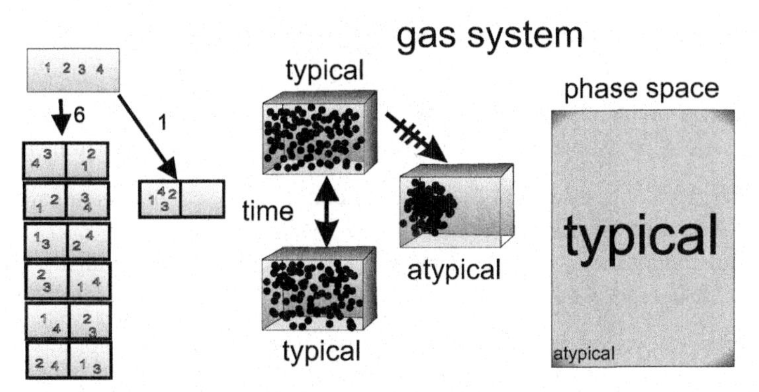

Fig. 4.1 There are 6 ways of distributing 4 objects between two containers so that there are 2 objects in each container. There is only 1 possibility for a 4–0 distribution. This difference in numbers (6 versus 1) grows extremely fast with the number of objects, like the gas molecules in the gas system. The typical phase points make up most of the phase space (*light grey*). Atypical phase points (corresponding to configurations where all molecules are in the left half of the container) make up only a very tiny fraction of the phase space (the *darker edges*). In fact, the region of atypicality is not concentrated as depicted, but is mixed up with the region of typicality as depicted in Fig. 4.9

to establish a connection with common notions, and later we shall apply the ideas to Bohmian mechanics to explain where quantum probability comes from.

Note in passing that the word "typical" is synonymous with "overwhelmingly many".[4] However, if one takes counting to heart, there is a slight difficulty here. The number of microstates of a gas system is uncountable, like the number of real numbers, since the positions and velocities vary over a continuum of values. In our example of air filling either half or the whole room, it is clear that the numbers of microstates are in both cases uncountable. How can we tell what is more and what is less in that case? We need a generalisation of counting, and that generalisation is the "size" or "volume" of the set of microstates. The mathematical doctrine of modern probability theory is based on that idea. Probability theory uses a natural notion of the content of sets, called a probability measure, where the emphasis should be put on measure rather than probability. Overwhelmingly many phase points will then mean that the relevant phase points make up a set of very large measure.

All this will be formalized later. For the moment it is more important to realize that something objective is being said about the physics of systems, when one says that a system behaves typically. It behaves in the way that the trajectories for the overwhelming majority of initial conditions of the system behave. Because of this we can make predictions for complex systems without the need to compute detailed trajectories. And it is exactly that role that chance plays in physics.

[4] Instead of "overwhelmingly many", Boltzmann also used the notion of "most probable". But the Ehrenfests [5], who described Boltzmann's ideas, used only the notion of "overwhelmingly many".

4.1.1 Typical Behavior. The Law of Large Numbers

Shortly before his death, the physicist Marian von Smoluchowski[5] (1872–1917) wrote the article [3], from which we quoted the two introductory questions above:

1. How can it be, that the effect of chance is computable, i.e., that *random causes* can have lawlike effects?
2. How can randomness arise if all events are reducible to deterministic laws of nature? Or in other words, how can *lawlike causes* have random effects?

Smoluchowski develops the answer to the first question. He does not answer the second question, and we shall answer it in the sense of Boltzmann.

Starting with the first question, let us dwell a while on what he meant by lawlike effects. In fact, he was referring to the law of the empirical mean, or in modern terms, the law of large numbers. The chance which can be calculated reveals itself in the predicted *relative frequencies*, i.e., in the predicted empirical means, which one obtains in the irregular outcomes of long runs of an experiment like coin tossing.

Smoluchowski discusses the physical conditions which allow for such a lawlike irregular sequence of outcomes in simple dynamical situations. Basic to these is instability of motion or as Smoluchowski puts it: *small cause, big effect*. Small fluctuations in the initial conditions yield completely different outcomes. But is it not surprising that amplification of small fluctuations should be responsible for the lawlike behavior of chance? Well, one must not forget that we look for lawlike behavior in a long irregular sequence of outcomes, and what instability can do here is to weaken influences of the past outcomes on the future outcomes. So in a certain sense, the more random the sequence is, the more simply we find the lawlike behavior we are looking for.

An example is provided by Galton's board (Francis Galton 1822–1911). A ball falls through an arrangement of n rows of nails fixed on a board, where the horizontal distance between two nails is only slightly bigger than the diameter of the ball (see Fig. 4.2). At the bottom of the board, one has n boxes, numbered from 1 to n, and every box marks a possible ending position of the ball. A ball ends its run in one of the boxes. After many runs, one obtains a distribution of the number of balls over the boxes. Suppose we let N balls drop, then if N is large, we can more or less say what the distribution of the number of balls will look like. The number of balls in box m will be about

$$\frac{1}{2^n}\binom{n}{m}N\,.$$

In other words, the empirical frequency for balls ending in box m is about $(1/2^n)\binom{n}{m}$. The usual argument is of course that there are 2^n possible trajectories: R(ight)–L(eft) sequences of length n with randomly distributed Rs and Ls, and the ball ends in box

[5] Smoluchowski discovered the molecular explanation of Brownian motion independently of Einstein, thereby opening the door for modern atomism.

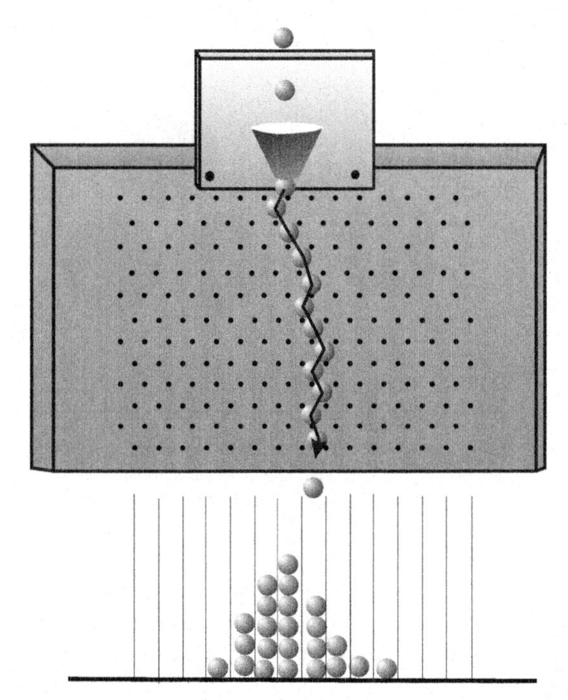

Fig. 4.2 Galton's Board. The figure shows the run of one ball

m if there are exactly m Rs in the sequence. There are $\binom{n}{m}$ such sequences, all sequences are equally probable, and hence $(1/2^n)\binom{n}{m}$ is the probability of ending up in box m. Thus in a series of many runs, by the law of large numbers, we get the relative frequencies as described. Equivalently, we can argue that the L–R decisions are independent, each occurring with probability $1/2$, and then ask what is the probability for m Rs in n independent identically distributed Bernoulli trials. These arguments are all quite accurate, but *they are not the end of the story*. The arguments use words like randomness, probability, and independence and it is the meaning of these words that we need to understand. What exactly is the status of the above argument? What exactly do the words mean?

Let us follow Smoluchowski. He assumes that the dynamics of the Galton board is Newtonian. Then he observes the following. The ball enters the Galton board with a "tiny uncertainty" concerning the position of its center (see Fig. 4.4). That means that the ball collides almost but not completely centrally with the first nail, so the ball goes either to the right or left nail slit. Suppose it goes to the right. Then the ball must pass the right nail slit. Now the whole secret of the physics of the Galton board lies in what happens during the passage through a nail slit. In fact, during this passage, an enormous number of (slightly inelastic) collisions take place between the ball and the two adjacent nails, as happens in a pinball machine! The effect of the many collisions is that very tiny changes in the incoming position and velocity of the ball, when it enters the slit, lead to drastically different outcomes for

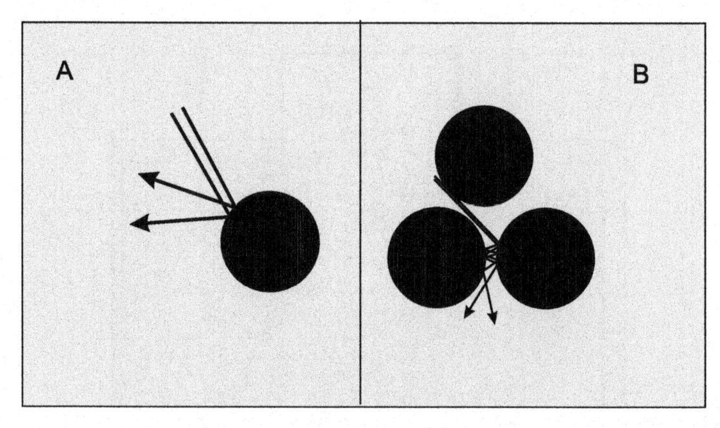

Fig. 4.3 (**A**) Two very close trajectories of a pointlike ball become separated by virtue of the collision with the round surface of the nail. (**B**) The effect is enhanced through the large number of collisions. The more collisions take place, the smaller the "initial uncertainty" need be so that in the end left or right is "unpredictable"

the outgoing directions upon leaving the nail slit. After hitting the next nail almost centrally, the ball can go to the left or to the right in the following row. Very tiny changes in the incoming position and velocity of the ball (entering the slit) result in drastically different outcomes: left or right.

Small cause, big effect! But why is this so? In fact we have a spreading of directions due to the convex boundaries of the ball and the nail. In Fig. 4.3, the ball is idealized as a point and the nail as a fat cylinder. The pointlike ball bounces off the cylinder surface according to the rules of elastic collisions (ignoring the fact that the collisions are in reality inelastic). We see that two very close initial trajectories are far apart after so many collisions. In the Galton board, "far apart" means that the following nail will be hit with a left-going or right-going bias. One can go even further. Suppose we let the two incoming trajectories get closer and closer, then left-outgoing may become right-outgoing and then again left-outgoing, then right-outgoing again and so on and so forth. Note that the particular shape of the initial distribution of displacements from the ideal central position becomes irrelevant, i.e., the particular details of the "initial randomness" with which every ball enters the board become irrelevant.

Let us formalize this a little more with the help of Fig. 4.4, which should be read from bottom to top. We look through a looking glass at the small initial uncertainty, i.e., the range δ over which the positions of the centers of the balls vary when the ball enters the hopper (this imprecision of the Galton board machine will be further scrutinized later). δ should be thought of as both the interval and its length. Let us assume for simplicity that the outlet of the hopper is also a nail slit. That way we have a self-similar picture at each row, which allows us to formalize the spreading in an idealized way. We do this in the following list:

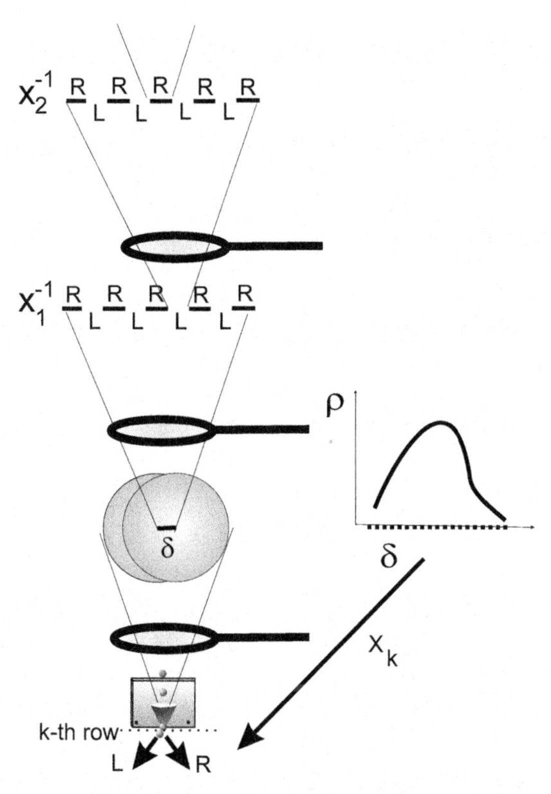

Fig. 4.4 Looking at the "initial randomness" of the Galton board through a magnifying glass. The initial uncertainty of the positions of the centers of the balls lies within δ, which denotes here an interval and also its length. δ is partitioned into cells by the functions $X_k : \delta \to \{R,L\}$, $k = 1,2,\ldots,n$, of the left or right moves of the ball when it hits a nail in row k. The partitions mix together in a perfect way. The shape of the distribution ρ of the points in δ has no influence on the outcomes X_k, because of the "spreading" character of the dynamics

1. Simplifying assumption. The trajectory of a ball through the Galton board is completely determined by the initial position of the center of the ball (with respect to the symmetry axis of the hopper) which may vary in δ. This means that every left–right direction X_k, $k = 1,2,\ldots,n$, upon hitting a nail of the kth row is a function of δ:

$$X_k : \delta \to \left\{0 \doteq L, 1 \doteq R\right\}.$$

In particular then, the end position $Y = \sum_{k=1}^{n} X_k$ of a ball is a function of δ.
2. X_k partitions δ into L–R cells. On points in an L cell $X_k = 0 \doteq L$, and likewise for R cells. In other words the cells are the pre-images $X_k^{-1}(R)$ and $X_k^{-1}(L)$ of X_k. The X_k are *coarse-graining functions* of the interval δ.
3. The partition of δ is very fine. Figure 4.4 shows $X_k^{-1}(R), X_k^{-1}(L)$, $k = 1,2$. The partition by X_1 is already very fine. The instability of the motion now "spreads"

the cells of X_1^{-1} further, i.e., it separates the points of every cell into L–R cells for X_2. Here there might be a source of confusion which we must be clear about. The X_k are coarse-graining functions, i.e., in this case they map cells (a continuum!) onto single points. The fine map which is given by the trajectories of the ball from the actual set of initial conditions, before entering the next nail slit, to new sets of initial conditions for entering the following slit is the one expanding in the sense of Fig. 4.3. This expansion is now encoded by the coarse-graining functions X_k in the finer and finer partitioning of the initial set δ into cells.

4. This mixing partitioning of the cells continues at every stage k.
5. The total size (added lengths of all intervals) of the L cells (R cells) is independent of the stage k and equals roughly $\delta/2$. For $\delta_k \in \{0,1\}$, let $|X_k^{-1}(\delta_k)|$ denote the length of the δ_k cells. Then from the foregoing, and looking at Fig. 4.4, we may expect the mixing character of the partitioning to yield the following (approximate) equality

$$|X_{k_1}^{-1}(\delta_{k_1})\cap\ldots\cap X_{k_j}^{-1}(\delta_{k_j})| \approx \prod_{n=1}^{j}|X_{k_n}^{-1}(\delta_{k_n})| \approx \left(\frac{\delta}{2}\right)^j . \tag{4.2}$$

This actually says that the coarse-graining functions form a family of *independent random variables*, with respect to the measure given by the interval length, or in view of point 6 below, with respect to any "decent" notion of measure for the intervals. Note in passing that random variables are nothing but, and only, coarse-graining functions, or if one prefers, coarse-graining variables!

6. Any "reasonable" weight ρ of the points of δ – the *initial randomness*, which we shall address below – will give the same picture. Intuitively this means that the details of this initial randomness are unimportant for the results of the Galton board. And this is as it should be, otherwise the results would not be stable under repairs to the board (which might be in order from time to time).
7. Theoretically (ignoring friction), the Galton board could be very large, i.e., the number of rows could be as large as we wish, and thus ideally the partitioning goes on forever.

Item 7 above asserts that the family of X_k, $k = 1,2,\ldots$, can be extended to arbitrary length. But can it in fact? How can one be sure? Of course, we can be sure. At least, nowadays we can, but at the beginning of the 20th century, this was a burning question: Do coarse-graining functions X_k, $k = 1,2,\ldots$, on an interval exist, partitioning the interval into finer and finer sets with an ideal mixing as we have described, going on *forever*? The answer lies at the heart of probability theory. The existence of such functions leads to the mathematical formulation of probability theory (as we teach it today). Their existence is intimately connected with the existence of real numbers, and we shall present probability theory in this genetic way.

Anticipating later sections, we shall already say what the functions are, namely the Rademacher functions (Hans Rademacher 1892–1962)

$$r_k : [0,1] \quad\longrightarrow\quad \{0,1\}\,, \tag{4.3}$$

which are simply the coordinate maps of the binary sequence representing the real number

$$x \in [0,1] , \qquad x = \sum_{k=1}^{\infty} r_k(x) 2^{-k} . \tag{4.4}$$

Let us quickly think about how these coarse-graining functions act: r_1 maps any number $x \in [0,1/2)$ to 0 and any $x \in [1/2,1]$ to 1. Hence r_1 partitions the interval $[0,1]$ into two equal parts. r_2 then partitions $[0,1/2)$ into $[0,1/4)$ and $[1/4,1/2)$ where all $x \in [0,1/4)$ are mapped to 0 and all $x \in [1/4,1/2)$ to 1. And so on and so forth. We see the mixing.

It is also useful to see the following analogy. Suppose we know that a number in $[0,1]$ starts with 0.0000. This means the number is between 0 and $1/2^4 = 1/16$. What can be concluded for the next digit? Nothing. It is either zero or one. Suppose it is also zero. Then the number is smaller than $1/32$, but we cannot conclude anything for the next digit. This goes on in the same way ad infinitum. And this corresponds perfectly to the way the ball runs through the nail slits. In particular, whatever direction it took before, nothing can be learned from that about the new direction. And nothing more detailed can be learned for the next ball's trajectory, no matter how close its initial condition is to the previous ball (as long as it is not identical), and so it will move "unpredictably".

Smoluchowski thus describes in this early paper what became known and celebrated much later as chaotic behavior. Following his reasoning, we have argued intuitively that it is very reasonable to assume that the left–right moves at each stage are statistically independent. In fact all we shall do later, when making this precise, is to define *independence of (identically distributed) random variables* by requiring the perfect mixing of partitions which are paradigmatically given by the Rademacher functions.

So we have argued that the probability computation of m hits in n trials seems appropriate, and on its basis we can *predict the numbers of balls in the boxes by the law of large numbers*. One should reflect for a moment on the law of large numbers. How does it read in fact, if we are to take the *microscopic* analysis and the coarse-graining functions seriously? We shall say more on that later!

That is as far as Smoluchowski went. We understand more or less how random causes (the initial uncertainty δ) can have lawlike effects. Because of the expansion due to instability, we have something like statistical independence (no matter exactly what the "initial uncertainty" is) and the law is this: the relative frequency of balls in box m is typically $\binom{n}{m}/2^n$.

But we must still face Smoluchowski's second question: Where does the random cause come from in the first place? What justifies the initial randomness? Let us be sure that we understand this question correctly. The first thought should be this. Suppose the first ball ends in box number 4. Why does the second ball not end up in the same box number 4, and in fact why do all the other balls not end up in the same box number 4? Impossible? Why? Think of a Galton board machine – a machine built with the precision of a Swiss clock, in which the mechanics is such

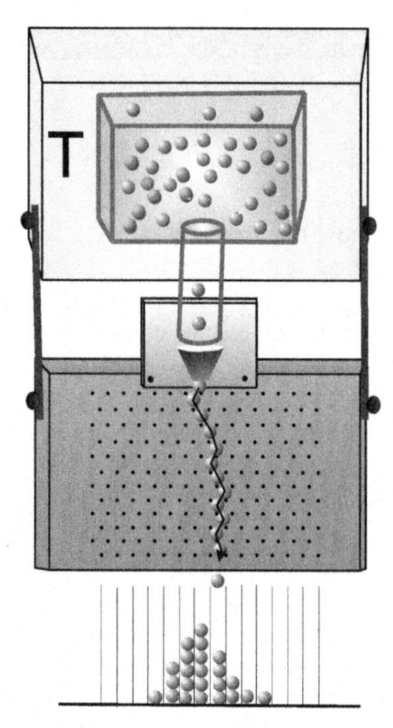

Fig. 4.5 A Galton board machine. We have a well insulated container at temperature T with a huge number of balls flying around in the container. These collide elastically with one another and with the inner walls of the container. Off and on, a ball drops through the opening and into the hopper of the Galton board. What are the predictions for the distributions of balls over the boxes at the bottom?

that balls are taken and sent with Swiss-made precision into the board. All the balls have exactly the same initial data as the first ball and all balls end up in the same box the first ball ended up in. That is clear. That is one way to always have the balls in the same box. There may of course be other ways to ensure that the balls are always in the same box. Now you say that in the real experiment there is some initial randomness. That is true, but why is the initial randomness *of such a kind* that not all balls end up in the same box?

To get a better grasp of this problem, we build another machine (see Fig. 4.5). A container of balls is kept at temperature T by a huge reservoir which is isolated from the rest of the world. The balls collide elastically with one another and with the walls of the container, and off and on a ball drops through the opening (slightly bigger than the diameter of a ball) at the bottom and drops through a somewhat bigger tube into the Galton board hopper. The container contains a huge number H of balls, a fraction of which makes up the large number N of balls which drop through the hole into the board. Now we have a machine which has all its randomness "inside" and we ask: What can we now assert about the relative frequency of balls ending up in the n boxes?

Let us be careful about this. What we have at some initial time (say the moment before the hole in the bottom of the container is opened) is one configuration, or better, one phase point $\omega = (\mathbf{q}_1, \ldots, \mathbf{q}_H, \mathbf{v}_1, \ldots, \mathbf{v}_H)$, where the $\mathbf{q}_k$ are the centers of the balls, taking values in the space occupied by the inner part of the container (ignoring for simplicity the fact that they have a nonzero extension and cannot overlap), and where the velocities can take arbitrary values. The collection of all possible ω is the phase space Ω. There is no longer any δ, so what are the X_k we had previously now functions of? We assume as before that the nails are infinitely heavy, i.e., that they do not move at all (in reality, the nails will oscillate and we will have material wear under collisions). Then the X_k must now depend on ω! But we need to be even more careful! Numbering the balls which drop through the board, we now have for the pth ball functions X_k^p, $k = 1, \ldots, n$. Its end position after leaving the board is described by the coarse-graining function $Y^p : \Omega \to \{0, 1, \ldots, n\}$ given by the sum

$$Y^p(\omega) = \sum_{k=1}^{n} X_k^p(\omega) .$$

If N balls drop through the board we have $Y^p(\omega) = \sum_{k=1}^{n} X_k^p(\omega)$, $p = 1, \ldots, N$. What now describes the distribution of balls over the boxes, i.e., what is the relative frequency we wish to predict? For that we need only write down the *empirical distribution* for end positions:

$$\rho_{\text{emp}}^N(\omega, x) = \frac{1}{N} \sum_{p=1}^{N} \chi_{\{x\}}\big(Y^p(\omega)\big) , \qquad x \in \{1, \ldots, n\} , \tag{4.5}$$

where $\chi_{\{x\}}(y)$ is the indicator function, equal to 1 if $y = x$ and zero otherwise. The empirical distribution is also a coarse-graining function taking on discrete fractional values. What is now a prediction for this function? If we can convince ourselves that the coarse-graining functions $Y^p(\omega)$ are stochastically independent (we must convince ourselves that they produce a Rademacher-like mixing partition, but instead of the interval δ, it is now the huge space Ω which is partitioned), we can prove a law of large numbers as follows. When N is large, the empirical distribution $\rho_{\text{emp}}^N(\omega, x)$ is typically the distribution we expect, namely

$$\rho_{\text{emp}}^N(\omega, m) \approx \frac{1}{2^n} \binom{n}{m} .$$

"Typically" means that the values are obtained for the overwhelming majority of ωs. But there is a catch. For stochastic independence we need a measure on Ω, measuring the sizes of the cells of the coarse-graining functions, and we need that measure to tell us what "typically" or equivalently "overwhelmingly many" means. Hence the assertion – the typicality assertion or law of large numbers – is made with respect to some measure, which will be the analogue of the length of an interval or the density ρ on δ, which we discussed in connection with Fig. 4.4. The appropriate measure will be discussed later – let us simply call it $\mathbb{P}^T$ for the moment. For concreteness we now write down the assertion of the law of large numbers in the familiar formal way, i.e., for all $\varepsilon > 0$ and all $\eta > 0$, and for N large enough,

$$\mathbb{P}_T\left(\left\{\omega \middle| \left|\rho_{\text{emp}}^N(\omega,m) - \frac{1}{2^n}\binom{n}{m}\right| < \varepsilon\right\}\right) > 1-\eta\,, \tag{4.6}$$

where

$$\frac{1}{2^n}\binom{n}{m} = \int_\Omega \rho_{\text{emp}}^N(\omega,m)\mathrm{d}\mathbb{P}_T(\omega)\,.$$

So "typically" means that the set of ωs, i.e., the set of initial phase points of the balls in the container which lead to the observed relative frequency of balls in the boxes has overwhelmingly large $\mathbb{P}_T$ measure. In statistical physics, $\mathbb{P}_T$ will commonly be assumed quite simply to be of such and such a form. This is called a statistical hypothesis, which must be argued for, a point we shall take up later.

Now let us be clear about what all this means for the experiment. When we start the experiment we have in the container one and only one configuration of balls, i.e., one phase space point ω in the phase space Ω of the container system and not a distribution of phase space points. This phase point will produce the $(1/2^n)\binom{n}{m}$ relative frequency of balls in the boxes. Why? Because overwhelmingly many such phase points, i.e., the typical phase point, would do that. The role of the statistical hypothesis is nothing other than to define typicality.

What about the proof? Now, if it were the case that the Y^i partitioned the phase space Ω in the mixing kind of way, and also that the X_k^i partitioned Ω in an equal-sized-cell kind of way (where size is now determined by the measure $\mathbb{P}_T$), then the law of large numbers would be immediate. In the formalized language of probability theory, the conditions simply mean independence of the random variables. We shall recall in the probability section the trivial proof of the law of large numbers when independence holds. This does not mean, however, that the law of large numbers is trivial. By no means! What is not at all trivial, better, what is outrageously difficult, is to prove that the conditions hold true, namely that the coarse-graining functions partition in just the right mixing kind of way. Establishing this is so exceedingly difficult that the proof of the law of large numbers (4.6) for the realistic Galton board is practically impossible.

One moral of the story is this. Independence is easily said and easily modeled, but extremely hard to establish for any realistic situation. This helps to explain the role and meaning of probability in physics. That is, one must first reduce probabilistic statements to their ultimate physical basis. Doing this, probability seems to go away, since one then deals with a purely analytical statement, just as in (4.6): the law of large numbers establishing what typical frequencies look like. *Much of the confusion about probability arises because the true depth of the law of large numbers as an extremely hard analytical assertion is not appreciated at all.*

We are almost done now. Only two questions remain: What is and what justifies $\mathbb{P}_T$? In a loose manner of speaking, reintroducing probabilities, one could ask: Where does this "initial randomness" come from? This loose manner of speaking is easily misleading, as it suggests that the law of large numbers is a probabilistic statement and $\mathbb{P}_T$ is a probability. Up until now, our presentation has nowhere made use of such an interpretation, and it will not do so. After this warning let us repeat the question: Where does the randomness come from? The most common answer goes like this. Systems like the one we constructed in Fig. 4.5 must be built, so they were not always closed, and there is some randomness coming from the outside. This is true. Someone must have built the machine and put the balls in the container, and someone must have heated up the container, or even mixed the balls up by shaking it, so randomness does seem to come from outside. But then take the larger system including the builder of the machine and whatever it requires. The typicality assertion is now shifted to a much larger phase space with a new measure $\mathbb{P}_{\mathrm{E+S}}$ (where $\mathrm{E+S}$ stands for environment plus system), and new coarse-graining functions on that larger phase space.

Thus we see that the question remains, only it is now for a larger system. What accounts for that randomness? Is it once again the outside of that larger system? Well then we go on and include the new outside, and we see that there seems to be no end to that. But there is an end! The end comes when we consider the largest closed system possible, the universe itself. For the universe the question now becomes rather embarrassing. There is no outside to which the source of randomness could be shifted. In fact, we have now arrived at the heart of the second of Smoluchowski's questions: How can randomness arise if all events are reducible to deterministic laws of nature? Or in other words, how can *lawlike causes* have random effects?

4.1.2 Statistical Hypothesis and Its Justification

We come now to Boltzmann's insight. What is the source of randomness when we have shifted the possible source to ever larger systems, until we reach the universe itself? Let us make two remarks. Firstly, if one thinks of randomness as probability, where probability is thought of as a notion which is based on relative frequencies in ensembles, as many people do, then that thought must end here. There is only one universe, our own, and sampling over an ensemble of universes gives no explanation for the occurrence of probability in our universe. Secondly, we did put up a big show, dramatically extending the question of the randomness from the little system to the whole universe in which the little system is only a part, as if that escape to the environment had any meaning. The question is, however, simple and direct. The physical laws are deterministic. How can there be randomness?

The answer can be made clear if we revisit the Galton board machine. After all, that could be a universe. In a typical universe we can have regular empirical frequencies, just the way we experience them in experiments, although the universe evolves deterministically. In a typical universe, things may look random, but they are not. In Boltzmann's words, for the overwhelming majority of universes it is true that, in ensembles of subsystems of a universe, the regular statistical patterns occur as we experience them.

So there is something to prove, namely, a law of large numbers for the universe. For example, for the Galton board machine as part of our universe, one would have to prove that in a typical universe[6] the balls fall into the boxes with just the right frequency. Such an assertion is usually called a prediction of the theory. Which measure defines a typical universe? We shall come to that! It is of course impracticable in proofs to have to deal with the whole universe, and one would therefore first see whether one could infer from the measure, which defines typicality, a measure of typicality for a subsystem, like the Galton board machine. That measure would be $\mathbb{P}_T$.

The introduction of such measures of typicality for subsystems of the universe is due to Boltzmann and Willard Gibbs (1839–1903). Gibbs preferred to call the measure of typicality an ensemble. After all, for subsystems of the universe, we can imagine having lots of copies of them, i.e., we can imagine many identical Galton board machines, so we can imagine having a true ensemble and we can sample, in a typical universe, the empirical distribution of ωs. Then $\mathbb{P}_T$ would arise as relative frequency itself.[7] However, this understanding is hopeless when it comes to the measure of typicality for the universe. It is therefore best to take the name "Gibbs ensemble" as synonymous with "measure of typicality". In the end it all reduces to one typical universe, as it must, since we have only one universe at our disposal, namely the one we live in. The best we can do is to prove an analytical statement about statistical regularities for ensembles of subsystems within a typical universe. A typical universe shows statistical regularities as we perceive them in a long run of coin tosses. It looks as if objective chance is at work, while in truth it is not. There is no chance. That is the basis of mathematical probability theory.

Is that all there is to this? Yes, more or less. We did answer Smoluchowski's second question, but there is a price to pay. The price is that we must prove something that is exceedingly difficult to prove and moreover we must answer the following question: *What measure tells us which sets of the phase space of the universe are overwhelmingly large and which sets are small?* In other words: Which measure defines the typical universe?

Now at first glance the measure of typicality does not seem to be unique. Recall that in our discussion of the Galton board, the distribution of points in δ (which could be such a measure of typicality) can be rather arbitrary as long as it does not vary on the very small cell scale. That holds for all measures of typicality. Typicality is really an equivalence class notion. All measures in a certain equivalence class (absolute continuity of measures would be an ideal class) define the same notion of

---

[6] Typicality with a grain of salt. A universe in which a Galton board experiment takes place might itself not be typical. We must eventually understand typicality conditionally, conditioned by macroscopic constraints which govern our universe, for example. This issue will resurface in Sect. 4.2, and when we justify Born's statistical law for the wave function in Chap. 11.

[7] There is a danger when too much emphasis is put on ensembles and distributions over ensembles. One is easily led astray and forgets that, for the molecules in a gas, while moving erratically around, one has at every instant of time one and only one configuration of gas molecules. The question is: What is the typical configuration?

typicality. Nevertheless there is a particularly nice representative of the equivalence class which we would like to introduce next.

The measures we are talking about are a kind of volume for the subsets of the tremendously high-dimensional phase space of the universe. Let us think of volume as a generalization of the volume of a ball or a cube to very high-dimensional phase spaces. Let us call the measure on phase space $\mathbb{P}$. What is a good $\mathbb{P}$? Let us recall what role it has to play. It must define typicality, but it must do so in the form of the law of large numbers for empirical distributions. That is, we must be able to consider ensembles of subsystems of the universe now, tomorrow, and at any time in principle, and it is intuitively clear that one particularly nice property of such a measure would be technically very welcome: it should not change with time, so that what is typical now is also manifestly typical at any other time.

The simplest requirement for that is that the volume measure we are looking for should not change with time, i.e., it must be a *stationary measure*. Apart from the technical advantage when proving things, the notion of typicality being timeless is appealing in itself. Time-independent typicality is determined by the physical law, i.e., time-independent typicality is given to us by physics. To see this more clearly, recall our discussion about stationarity in Remark 2.2 on p. 19. For a Hamiltonian universe, we have Liouville's theorem (2.12), which asserts that the phase space volume does not change with time under the Hamiltonian flow. The law of physics gives us a physical notion of typicality based on time independence. It is as reasonable to take that notion as relevant for our understanding of the universe as the law itself.

It is natural to guess that the statistical hypothesis to define typicality for subsystems should also appeal to stationarity. Boltzmann felt this way and, independently of Gibbs, introduced the measures which are known nowadays as the canonical ensembles or Gibbs ensembles. Indeed, the mathematical physicist Gibbs saw stationarity as a good requirement for a statistical hypothesis on which he based an axiomatic framework of statistical mechanics for subsystems. Gibbs' axiomatic framework is immediately applicable to the thermostatics (equilibrium thermodynamics) of subsystems, and explains his success, while Boltzmann's work was less widely recognized.

As already mentioned, Gibbs talks about distributions over ensembles and is not concerned with the actual phase point the system occupies. On that basis, he gave a justification for the use of Gibbs ensembles that is neither necessary nor sufficient, connected to the so-called mixing and convergence of non-equilibrium measures to equilibrium measures (see the next section). This is not necessary because typicality does the job, and it is not sufficient, because it is not linked to the actual trajectory of the system under consideration. A system is always in a particular configuration and never "in a probability distribution". Gibbs' view seems to deviate from typicality and Boltzmann's view of the world.[8] Historically Gibbs' view seems persistent, while Boltzmann's understanding that the role of the hypothetical Gibbs ensembles is to define typicality was lost.

[8] The Ehrenfests criticize Gibbs' view correctly [5].

We now look at some stationary measures relevant in Hamiltonian subsystems. We do that to sharpen our intuition for the quantum equilibrium analysis we shall undertake once we have introduced Bohmian mechanics. The analysis there is much simpler than what we do next. The important message is, however, that the basic structures are the same, something which is not recognized in textbooks.

4.1.3 Typicality in Subsystems: Microcanonical and Canonical Ensembles

Recall (2.26), viz.,

$$\frac{\partial}{\partial t}\rho(\mathbf{x},t) = -\mathbf{v}^H(\mathbf{x},t)\cdot\nabla\rho(\mathbf{x},t) .$$

We look for stationary solutions of this equation, i.e., we search for a density ρ for which the time derivative on the left-hand side is zero. According to Remark 2.2, a stationary density yields a stationary measure. Now the right-hand side of the equation is

$$\begin{aligned}\mathbf{v}^H\cdot\nabla\rho &= \left(\frac{\partial H}{\partial \mathbf{p}}\cdot\frac{\partial}{\partial \mathbf{q}} - \frac{\partial H}{\partial \mathbf{q}}\cdot\frac{\partial}{\partial \mathbf{p}}\right)\rho \\ &= \left(\dot{\mathbf{q}}\cdot\frac{\partial}{\partial \mathbf{q}} + \dot{\mathbf{p}}\cdot\frac{\partial}{\partial \mathbf{p}}\right)\rho(\mathbf{q},\mathbf{p}) = \frac{\mathrm{d}}{\mathrm{d}t}\rho\big(\mathbf{q}(t),\mathbf{p}(t)\big) ,\end{aligned}$$

which is the change in the function ρ along the system trajectories. When we ask for this to be zero, we ask for functions which are constant along trajectories. One such function is $H(\mathbf{q},\mathbf{p})$, which is just energy conservation. Hence every function $f(\mathbf{q},\mathbf{p}) \equiv f\big(H(\mathbf{q},\mathbf{p})\big)$ is conserved. Two functions play a particular role, as we shall explain. The simplest is

$$\rho = f(H) = \frac{\mathrm{e}^{-\beta H}}{Z(\beta)} , \tag{4.7}$$

where β is interpreted thermodynamically as $\beta = 1/k_\mathrm{B}T$ with T the temperature, and k_B is the Boltzmann constant. The latter is a dimensional factor, which allows one to relate thermodynamic units and mechanical units. We shall say more on that later. $Z(\beta)$ is the normalization.[9] The function defined by (4.7) determines a measure which is commonly called the canonical ensemble. Its role will become clear in a moment.

[9] We wish to define *typicality* with this measure and we wish to call the typical (i.e., the predicted) relative frequencies simply probabilities, as physicists normally do. For this purpose it is convenient to normalize the typicality measure like a probability measure.

The somewhat more complicated density is the following. It arises from the preserved volume measure on phase space, but taking into account the fact that energy is conserved. Energy conservation (2.9) partitions the phase space[10] Ω into surfaces of constant energy Ω_E :

$$\Omega_E = \{(\mathbf{q},\mathbf{p})|H(\mathbf{q},\mathbf{p}) = E\}\,, \qquad \Omega = \bigcup_E \Omega_E\,.$$

When we think of an isolated system its phase point will remain throughout the time evolution on the energy surface defined by the value of the system energy E. Which function defines a stationary density on the energy surface Ω_E ? Formally, this is clear:

$$\rho_E = \frac{1}{Z(E)}\delta\big(H(\mathbf{q},\mathbf{p}) - E\big)\,, \qquad Z(E) = \text{ normalization}\,. \tag{4.8}$$

This then defines the stationary measure – the so called microcanonical ensemble:

$$\mathbb{P}_E(A) = \int_A \frac{1}{Z(E)}\delta\big(H(\mathbf{q},\mathbf{p}) - E\big)\mathrm{d}^{3N}q\,\mathrm{d}^{3N}p\,, \quad A \subset \Omega_E\,. \tag{4.9}$$

The δ-function may seem scary for some readers. So let us rewrite the measure, introducing the area measure $\mathrm{d}\sigma_E$ on the surface. The point is that (4.8) is in general not simply the surface measure (when the energy surface is not a ball), but it has a nontrivial density. There are various ways to see this. Consider a surface given implicitly by $f(x_1,\ldots,x_n) = c$ and suppose that we can solve for $x_n = g_c(x_1,\ldots,x_{n-1})$, i.e., we can parameterize the surface by the vector

$$\mathbf{y} = \big(x_1,\ldots,x_{n-1},g_c(x_1,\ldots,x_{n-1})\big)\,.$$

Then we know from analysis that

$$\mathrm{d}\sigma_c = \frac{||\nabla f||}{|\partial_n f|}\mathrm{d}x_1\ldots\mathrm{d}x_{n-1}\,. \tag{4.10}$$

On the other hand, for any nice function h, by change of variables

$$x_n \longrightarrow y = f(x_1,\ldots,x_n)\,, \qquad x_n = g_y(x_1,\ldots,x_{n-1})\,,$$

we have

[10] In probability theory, it is common to call the phase space *probability space* denoted by Ω, but be aware that this space has nothing to do with probability, randomness, or chance, despite its name. Coarse-graining functions of phase space are called *random variables*, and they too have nothing to do with randomness per se.

$$\int h(x_1,\dots,x_n)\delta\big(f(x_1,\dots,x_n)-c\big)\mathrm{d}x_1\dots\mathrm{d}x_n$$

$$=\int h(x_1,\dots,g_y(x_1,\dots,x_{n-1}))\left|\frac{1}{\partial_n f(x_1,\dots,g_y(x_1,\dots,x_{n-1}))}\right|\delta(y-c)\mathrm{d}x_1\dots\mathrm{d}x_{n-1}\mathrm{d}y$$

$$=\int h\big(x_1,\dots,x_{n-1},g_c(x_1,\dots,x_{n-1})\big)\frac{1}{|\partial_n f|}\mathrm{d}x_1\dots\mathrm{d}x_{n-1}\,.$$

Hence we see that the δ density acts like a surface integral, and by comparison we find that

$$\delta(f-c)\mathrm{d}x_1\dots\mathrm{d}x_n=\frac{1}{\|\nabla f\|}\mathrm{d}\sigma_c\,. \tag{4.11}$$

Hence, applying this to our case, for the microcanonical ensemble $\mathbb{P}_E$ given by (4.9), we have the result

$$\mathbb{P}_E(A)=\frac{1}{\displaystyle\int_{\Omega_E}\frac{\mathrm{d}\sigma_E}{\|\nabla H\|}}\int_A\frac{\mathrm{d}\sigma_E}{\|\nabla H\|}\,,\qquad A\subset\Omega_E\,. \tag{4.12}$$

Here is a more pictorial way to arrive at that. Divide the volume element $\mathrm{d}^{3N}q\mathrm{d}^{3N}p$ of Ω into the surface element $\mathrm{d}\sigma_E$ on Ω_E and the orthogonal coordinate denoted by l, viz.,

$$\mathrm{d}^{3N}q\mathrm{d}^{3N}p=\mathrm{d}\sigma_E\,\mathrm{d}l\,.$$

The element on the left is invariant under the Hamiltonian flow, by Liouville's theorem. $\mathrm{d}H$ is also invariant. This is the l-coordinate difference between the two energy surfaces Ω_E and $\Omega_{E+\mathrm{d}E}$ (see Fig. 4.6), so that on the right-hand side, we should write

$$\mathrm{d}\sigma_E\frac{\mathrm{d}l}{\mathrm{d}H}\mathrm{d}H\,.$$

Then

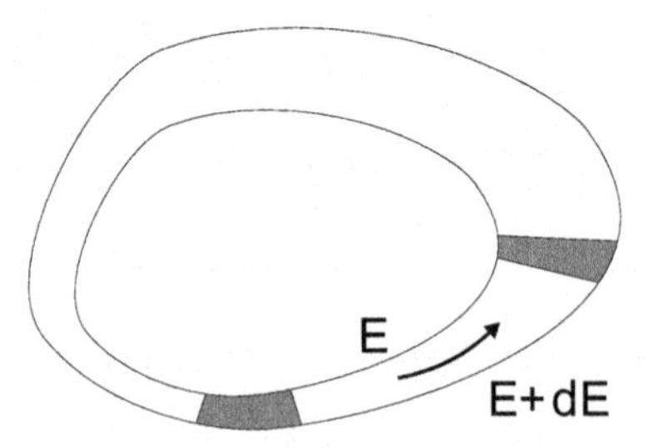

Fig. 4.6 Transport of phase space volume between two energy surfaces

$$\mathrm{d}\sigma_E \frac{\mathrm{d}l}{\mathrm{d}H}$$

must be invariant. It remains to be seen what $\mathrm{d}l/\mathrm{d}H$ is. Now ∇H is orthogonal to Ω_E, i.e., parallel to the coordinate line l,

$$\mathrm{d}H = \|\nabla H\| \mathrm{d}l \ ,$$

and hence

$$\frac{\mathrm{d}l}{\mathrm{d}H} = \frac{1}{\|\nabla H\|} \ .$$

Look again at the Fig. 4.6. The surface measure density $1/\|\nabla H\|$ takes into account the fact that the system trajectories moving between two energy shells must move faster when the shells get closer, since the phase space volume must be preserved by the flow lines. Accordingly, in the case of spherical energy surfaces ($\|\nabla H\| =$ constant), the surface Liouville measure is simply the surface measure.

Boltzmann's and Einstein's views (Einstein reworked Boltzmann's ideas for himself) is that the microcanonical measure, which is the projection of the natural Liouville measure onto the energy surfaces, is the typicality measure for the (Newtonian) universe. Why should this be natural? The universe is a closed thermodynamical system, since there is no energy exchange with an outside, i.e., the energy of the universe does not fluctuate. The distribution (4.7) has energy fluctuations, while (4.12) does not. So it is a natural choice. Why does (4.7) appear at all? We shall explain that.

Let us start with the simplest thermodynamical system, namely an ideal gas in a container. Ideal gas particles do not interact with one another and, assuming complete isolation, the gas system has only kinetic energy. For N equal mass particles with velocities $(\mathbf{v}_1, \ldots, \mathbf{v}_n) \in \mathbb{R}^{3N}$, the Hamiltonian is

$$H(\mathbf{q}, \mathbf{p}) = \frac{1}{2} \sum_{i=1}^{N} m\mathbf{v}_i^2 = \sum_{i=1}^{N} \frac{\mathbf{p}_i^2}{2m} \ ,$$

and $H = E$ are preserved surfaces under the ideal gas dynamics. The positions of the gas molecules are contained in the volume V. The microcanonical measure (4.8) for the gas system is then

$$\mathbb{P}_E(\mathrm{d}^{3N} q \mathrm{d}^{3N} p) = \frac{1}{V^N} \mathrm{d}^{3N} q \frac{1}{Z(E)} \delta \left(\sum_{i=1}^{N} \frac{\mathbf{p}_i^2}{2m} - E \right) \mathrm{d}^{3N} p \ , \tag{4.13}$$

where we now set

$$Z(E) = \int_{\mathbb{R}^{3N}} \delta \left(\sum_{i=1}^{N} \frac{\mathbf{p}_i^2}{2m} - E \right) \mathrm{d}^{3N} p \ ,$$

since for the ideal gas the volume normalization can be simply factorized.

Since the positions are uniformly distributed, we focus attention on the momentum, or what comes to the same up to a constant, the velocity distribution of that measure, which we obtain when we integrate all positions over the volume V. Then, with a slight abuse of notation, still denoting the normalization factor by $Z(E)$, we have

$$\mathbb{P}_E(\mathrm{d}^{3N}v) = \frac{1}{Z(E)}\delta\left(\sum_{i=1}^{N}\frac{m\mathbf{v}_i^2}{2} - E\right)\mathrm{d}^{3N}v\,. \tag{4.14}$$

The delta distribution forces a complicated dependence among the velocities. The velocity vector $(\mathbf{v}_1,\ldots,\mathbf{v}_N)$ lies on the spherical surface $S^{(3N)}_{\sqrt{2E/m}}$. Changing variables, referring to (4.11), and computing

$$\left\|\nabla\sum_{i=1}^{N}\mathbf{v}_i^2\right\| = 2\sqrt{\frac{2E}{m}}$$

on the spherical surface $S^{(3N)}_{\sqrt{2E/m}}$, we can express (4.14) in the form

$$\mathbb{P}_E(A) = \frac{\displaystyle\int_A \mathrm{d}\sigma^{(3N)}_{\sqrt{2E/m}}}{\left|S^{(3N)}_{\sqrt{2E/m}}\right|}\,, \qquad A\subset S^{(3N)}_{\sqrt{2E/m}}\,, \tag{4.15}$$

where $|S^{(3N)}_{\sqrt{2E/m}}|$ is the size of the surface $S^{(3N)}_{\sqrt{2E/m}}$ and $\mathrm{d}\sigma^{(3N)}_{\sqrt{2E/m}}$ its surface element. This is not very informative. Suppose we would like to know the so-called marginal distribution of only one velocity component, say $(\mathbf{v}_1)_x$, which we obtain by choosing

$$A = \left\{(\mathbf{v}_1,\ldots,\mathbf{v}_n)\in S^{(3N)}_{\sqrt{2E/m}}\,\middle|\,(\mathbf{v}_1)_x\in[a,b]\right\}\,,$$

or equivalently by integrating over all velocities in (4.14) except $(\mathbf{v}_1)_x\in[a,b]$. What would we get? Or suppose we asked for the marginal joint distribution of $(\mathbf{v}_1)_x,(\mathbf{v}_5)_y$, then what would we get?

Clearly, the answer is complicated, since all the velocities depend on each other. But intuitively, if the number N is huge (as it is in the gas system, of the order of Avogadro's number), the dependencies must become weak. Better still, if N gets large and with it the energy E, so that

$$\frac{E}{N} = \frac{1}{N}\sum_{i=1}^{N}\frac{m}{2}\mathbf{v}_i^2 = c\,,$$

we should expect the marginal distributions to attain characteristic forms depending on the value c, but independent of N. Before we show this, let us think about c. It does look like the empirical average of the single-particle kinetic energy. If the law of large numbers applies, we have

$$c = \frac{E}{N} \approx \frac{m}{2}\langle \mathbf{v}_1^2 \rangle_{E,N} := \int \frac{m}{2}\mathbf{v}_1^2 \frac{1}{Z(E)} \delta\left(\sum_{i=1}^{N} \frac{m\mathbf{v}_i^2}{2} - E\right) \mathrm{d}^{3N}v$$
$$=: \frac{m}{2}\int \mathbf{v}^2 f_c(\mathbf{v})\mathrm{d}^3 v \, .$$

Here f_c denotes the marginal distribution of the velocity of one particle in the situation where $E/N = c$ and N is large. This average is well known, and can be determined from a completely different view which is basic to all of Boltzmann's work, in fact to the whole of kinetic theory: the connection with thermodynamics! The gas in the container obeys the ideal gas law $pV = nRT$. The way we have formulated the law, it is already atomistic since it involves the number n of moles in the volume V. R is the gas constant p the pressure, V the volume, and T the temperature. Introducing Avogadro's number N_A, the number of molecules in the mole, we obtain

$$pV = Nk_\mathrm{B}T \, , \tag{4.16}$$

with Boltzmann's constant $k_\mathrm{B} = R/N_\mathrm{A}$ and $N = nN_\mathrm{A}$ the total number of gas molecules. Boltzmann's constant should be thought of as the heat capacity of one gas particle. This becomes clear when we put the above average in relation with the pressure of the gas. In kinetic theory the pressure arises from the gas molecules hitting the walls of the gas container:

$$p = \frac{\text{force}}{\text{area}} = \frac{\text{momentum transfer during } \Delta t}{\Delta t \times \text{area}} \, , \tag{4.17}$$

where Δt is a short time interval in which very many molecules collide with the wall area. Suppose the area A has normal vector in the x direction. Then the momentum transfer is $2mv_x$, since the particle is elastically reflected. The number of particles with x-components of the velocity in the range $[v_x, v_x + \mathrm{d}v_x]$ colliding with the area A and within Δt is roughly

$$N(v_x, \mathrm{d}v_x, \Delta t, A) = v_x \Delta t A f_c(v_x)\mathrm{d}v_x \frac{N}{V} \, .$$

Consider Fig. 4.7. $v_x \Delta t A$ is the volume of the cylinder (Boltzmann's collision cylinder), in which a particle with this x-component of velocity must be in order to collide with the area A in the given time interval. The "probability" for a particle to be in that cylinder is

$$\frac{N}{V} f_c(v_x)\mathrm{d}v_x \, ,$$

where N/V is the spatial density of particles and $f_c(v_x)$ is the density of the x-component of the velocity distribution. We thus obtain the above number denoted by

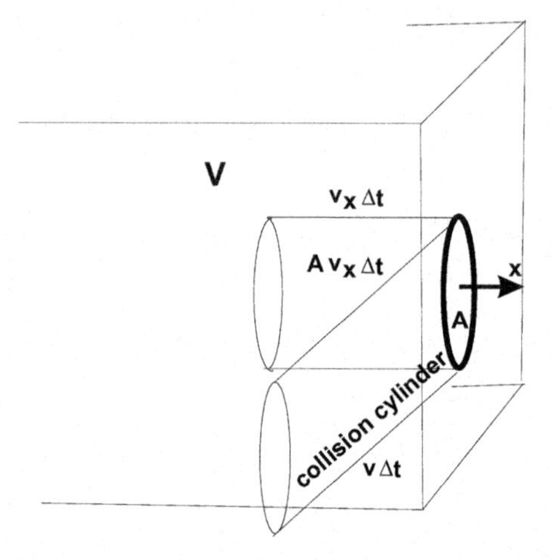

Fig. 4.7 A particle with velocity $\mathbf{v}$ must be in the collision cylinder to hit the wall within A and within time Δt. The volume of that cylinder is $v_x A \Delta t$

$N(v_x, \mathrm{d}v_x, \Delta t, A)$. Actually we do not need to know what $f_c(v_x)$ looks like, except that we require the symmetry $f_c(v_x) = f_c(-v_x)$. Each particle transfers the momentum $2mv_x$, so considering the symmetry and integrating over all relevant v_x, we get

$$\begin{aligned}\text{momentum transfer during } \Delta t &= 2m\frac{N}{V}\Delta t A \int_0^{\infty} v_x^2 f_c(v_x)\mathrm{d}v_x \\ &= m\frac{N}{V}\Delta t A \int_{-\infty}^{\infty} v_x^2 f_c(v_x)\mathrm{d}v_x \,.\end{aligned}$$

Hence, in view of (4.17), we obtain for the pressure

$$p = \frac{N}{V} m \int_{-\infty}^{\infty} v_x^2 f_c(v_x)\mathrm{d}v_x \,,$$

that is,

$$pV = N\langle m v_x^2 \rangle_{E,N} \,.$$

Comparing with (4.16), we see that $\langle m v_x^2 \rangle_{E,N} = k_{\mathrm{B}} T$. This important result is called the equipartition theorem. The average kinetic energy of a particle is

$$c = \left\langle \frac{m}{2}\mathbf{v}^2 \right\rangle_{E,N} = \frac{3}{2} k_{\mathrm{B}} T \,. \tag{4.18}$$

Why is this result important? The answer is that it connects two physical theories, thermodynamics and Newtonian mechanics! Thermodynamics does not know about molecules and Newtonian mechanics does not know about temperature. Yet they can be connected! And the constant which gets the dimensions right is k_{B}. In fact, k_{B} is as we already said essentially the heat capacity of an atom, since heat capacity is in general the proportionality between energy increase and temperature increase.

We now return to the question of what the marginal distribution f_c looks like. The answer is surprising! We wish to compute

$$\mathbb{P}_E(A) = \frac{\int_A \mathrm{d}\sigma^{(3N)}_{\sqrt{2E/m}}}{\left|S^{(3N)}_{\sqrt{2E/m}}\right|}, \quad \text{for } A = \left\{ (\mathbf{v}_1, \ldots, \mathbf{v}_n) \in S^{(3N)}_{\sqrt{2E/m}} \,\middle|\, (\mathbf{v}_1)_x \in [a,b] \right\}. \tag{4.19}$$

Let us simplify the notation. Let

$$S^{(n)}_r = \left\{ (x_1, \ldots, x_n) \in \mathbb{R}^n, \sum_{i=1}^n x_i^2 = r^2 \right\}$$

be the n-dimensional spherical surface, and let $|\cdot|$ denote the normalized surface measure on $S^{(n)}_r$. Furthermore, let

$$S^{(n)}_r(a,b) = \left\{ (x_1, \ldots, x_n) \in \mathbb{R}^n, \sum_{i=1}^n x_i^2 = r^2, a < x_1 < b \right\}$$

be the spherical zone defined by the interval (a,b) (see Fig. 4.8). We shall prove the following geometrical fact:

Lemma 4.1. *Let $\sigma^2 > 0$ and let $a < b \in \mathbb{R}$. Then the following holds:*

$$\lim_{n\to\infty} \frac{\left|S^{(n)}_{\sqrt{\sigma^2 n}}(a,b)\right|}{\left|S^{(n)}_{\sqrt{\sigma^2 n}}\right|} = \int_a^b \frac{\mathrm{e}^{-x^2/2\sigma^2}}{\sqrt{2\pi\sigma^2}} \,\mathrm{d}x\,. \tag{4.20}$$

Before proving that let us translate the result back in terms of (4.19). We have $n = 3N$, $E/N = 3k_\mathrm{B}T/2$, and thus

$$r^2 = \sigma^2 n = \frac{2E}{m} = \frac{nk_\mathrm{B}T}{m}\,,$$

that is, $\sigma^2 = k_\mathrm{B}T/m$. Therefore the marginal distribution we are after turns out to be the Maxwellian

$$\lim_{N\to\infty,\frac{E}{N}=\frac{3Tk_\mathrm{B}}{2}} \mathbb{P}_E\left(A = \left\{ (\mathbf{v}_1, \ldots, \mathbf{v}_n) \in S^{(3N)}_{\sqrt{2E/m}} \,\middle|\, (\mathbf{v}_1)_x \in [a,b] \right\} \right)$$

$$= \lim_{N\to\infty,\frac{E}{N}=\frac{3k_\mathrm{B}T}{2}} \frac{\int_A \mathrm{d}\sigma^{(3N)}_{\sqrt{2E/m}}}{\left|S^{(3N)}_{\sqrt{2E/m}}\right|} = \int_a^b \mathrm{d}v \frac{\exp\left(-\frac{1}{k_\mathrm{B}T}\frac{mv^2}{2}\right)}{\sqrt{\frac{2\pi k_\mathrm{B}T}{m}}}\,.$$

But this is the canonical distribution! It is a straightforward matter, once one understands the proof of the lemma, to conclude that the marginal distribution for more velocity components is the product of the single-component distributions. Furthermore, as an exercise, compute the mean kinetic energy of the particle! This must be $3k_BT/2$.

Proof of the Lemma. The most demanding step of the proof is to convince oneself that

$$|S_r^{(n)}| = r^{n-1}|S_1^{(n)}| \,. \tag{4.21}$$

Intuitively, this is clear if one thinks of the spherical surfaces in 2 and 3 dimensions. The straightforward way to get this rigorously is to introduce n-dimensional spherical coordinates and write down the parameterized surface element, or to simply recall the general formula (4.10), which in this case reads

$$\mathrm{d}\sigma_r^{(n)} = \frac{r}{\sqrt{r^2 - x_1^2 - \cdots - x_{n-1}^2}} \mathrm{d}x_1 \ldots \mathrm{d}x_{n-1} \,.$$

Now change variables $y_k = x_k/r$ to get

$$\mathrm{d}\sigma_r^{(n)} = \frac{r r^{n-1}}{r\sqrt{1 - y_1^2 - \ldots - y_{n-1}^2}} \mathrm{d}y_1 \ldots \mathrm{d}y_{n-1} = r^{n-1}\mathrm{d}\sigma_1^{(n)} \,,$$

from which (4.21) follows. Assuming $|a| < r$ and $|b| < r$ and recalling (4.11), we have

$$\begin{aligned}
|S_r^{(n)}(a,b)| &= \int_a^b \mathrm{d}x_1 \int_{-\infty}^{\infty} \mathrm{d}x_2 \ldots \int_{-\infty}^{\infty} \mathrm{d}x_n 2r\delta(x_1^2 + \ldots + x_n^2 - r^2) \\
&= \int_a^b \mathrm{d}x_1 \int_{-\infty}^{\infty} \mathrm{d}x_2 \ldots \int_{-\infty}^{\infty} \mathrm{d}x_n 2r\delta\left(x_2^2 + \ldots + x_n^2 - (r^2 - x_1^2)\right) \\
&= 2r \int_a^b \mathrm{d}x_1 \left| S^{(n-1)}_{\sqrt{r^2 - x_1^2}} \right| \frac{1}{2\sqrt{r^2 - x_1^2}} \qquad \text{[by (4.11)]} \\
&= 2r|S_1^{(n-1)}| \int_a^b \mathrm{d}x \sqrt{r^2 - x^2}^{(n-2)} \frac{1}{2\sqrt{r^2 - x^2}} \qquad \text{[by (4.21)]} \\
&= r|S_1^{(n-1)}| \int_a^b \mathrm{d}x \sqrt{r^2 - x^2}^{(n-3)} \,.
\end{aligned}$$

Once again using (4.21), we then obtain

$$\frac{|S_r^{(n)}(a,b)|}{|S_r^{(n)}|} = \frac{|S_1^{(n-1)}|}{|S_1^{(n)}|}\frac{1}{r^{n-2}}\int_a^b \left(r^2-x^2\right)^{(n-3)/2}\mathrm{d}x$$

$$= \frac{|S_1^{(n-1)}|}{|S_1^{(n)}|}\frac{1}{r}\int_a^b \left[1-\left(\frac{x}{r}\right)^2\right]^{(n-3)/2}\mathrm{d}x\,.$$

We now use the fact that the fraction on the left becomes unity for $a=-r$, $b=r$, which implies that the factor in front of the integral on the right is nothing but a normalization factor, whence

$$\frac{|S_r^{(n)}(a,b)|}{|S_r^{(n)}|} = \frac{\displaystyle\int_a^b \left[1-\left(\frac{x}{r}\right)^2\right]^{(n-3)/2}\mathrm{d}x}{\displaystyle\int_{-r}^r \left[1-\left(\frac{x}{r}\right)^2\right]^{(n-3)/2}\mathrm{d}x}\,.$$

Now comes a tiny piece of analysis. We wish to evaluate the expression for $r^2=\sigma^2 n$ as $n\longrightarrow\infty$, and for that we wish to pass the limit inside the integral. This would be done rigorously by appealing to Lebesgue dominated convergence, observing that $1-x\le \mathrm{e}^{-x}$ and hence

$$\left(1-\frac{x^2}{n\sigma^2}\right)^{(n-3)/2} \le \exp\left[-\frac{x^2}{n\sigma^2}\left(\frac{n-3}{2}\right)\right],$$

which is integrable for $n>3$. So we need only take the limit on the integrand

$$\lim_{n\longrightarrow\infty}\left(1-\frac{x^2}{n\sigma^2}\right)^{(n-3)/2} = \exp\left(-\frac{x^2}{2\sigma^2}\right),$$

and normalize the Gaussian. Thus the lemma is proven. If one feels uncomfortable with δ-functions, one can of course do without and compute the zone as in Fig. 4.8.

What moral should we draw from this? The measure of typicality for a subsystem of a large system, which is typical with respect to the microcanonical ensemble, should be the canonical ensemble, if interaction between the subsystem and the environment is small. So let us push this a bit further. Let us return to the general microcanonical measure (4.8) given by

$$\frac{1}{Z(E)}\delta\big(H(\mathbf{q},\mathbf{p})-E\big)\mathrm{d}^{3N}q\,\mathrm{d}^{3N}p = \frac{1}{\displaystyle\int_{\Omega_E}\frac{\mathrm{d}\sigma_E}{\|\nabla H\|}}\frac{\mathrm{d}\sigma_E}{\|\nabla H\|}\,,$$

and recall that for the ideal gas (4.13), where the volume factor V^{3N} is irrelevant, and setting $n=3N$,

$$Z(E)=\int_{\Omega_E}\frac{\mathrm{d}\sigma_E}{\|\nabla H\|}=\frac{1}{2}\sqrt{\frac{m}{2E}}\left|S^{(n)}_{\sqrt{2E/m}}\right| \sim \sqrt{\frac{2E}{m}}^{\,n-2},$$

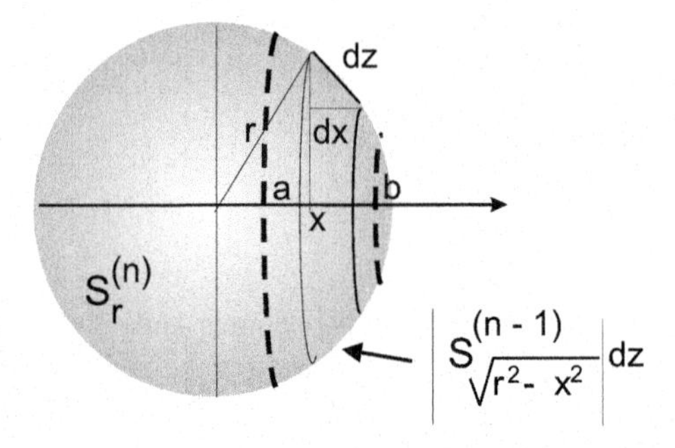

Fig. 4.8 Computing the spherical surface zone of a three-dimensional sphere of radius r, thereby recalling the famous identity that the (a,b) zone is the cylindrical surface of the cylinder with base size $2\pi r$ and height $b-a$. The base size of the cylinder is now $|S^{(n-1)}_{\sqrt{r^2-x^2}}|$ and its height is dz, yielding $|S_r^{(n)}(\mathrm{d}x)| = |S^{(n-1)}_{\sqrt{r^2-x^2}}|\mathrm{d}z$, and by similarity of the triangles with sides x, r and dx, dz, we have the proportionality $\mathrm{d}z : \mathrm{d}x = r : \sqrt{r^2-x^2}$

with $E/n = k_\mathrm{B}T/2$, whence $Z(E)$ is a huge exponential! We therefore make it smaller by taking the logarithm,

$$\ln Z(E) = \frac{n-2}{2}\ln E + \mathscr{O}(n) ,$$

then take the derivative with respect to E and use

$$\frac{E}{n} = \frac{k_\mathrm{B}T}{2}$$

to obtain

$$\frac{\mathrm{d}\ln Z(E)}{\mathrm{d}E} = \frac{n}{2E} - \frac{1}{E} \approx \frac{1}{k_\mathrm{B}T} . \tag{4.22}$$

This result suggests a microscopic definition of Clausius' entropy S for which the thermodynamic relation

$$\frac{\partial S}{\partial E} = \frac{1}{T}$$

will hold. So setting

$$S = k_\mathrm{B}\ln Z(E) \tag{4.23}$$

hits close to home. This is Boltzmann's setting, and we shall henceforth take this as valid in general (not only for the ideal gas).[11]

---

[11] A bit of unimportant history. Boltzmann never wrote this formula down, although it is inscribed on his tombstone. Planck wrote the formula (1900), introducing the Boltzmann constant k_B in his

But we want to focus first on the moral above. We imagine the subsystem to consist of N_1 particles with phase space coordinates

$$(q_1, p_1) = \left(\mathbf{q}_1^1, \ldots, \mathbf{q}_1^{N_1}, \mathbf{p}_1^1, \ldots, \mathbf{p}_1^{N_1}\right),$$

and the rest of the large system to consist of N_2 particles with phase space coordinates (q_2, p_2). So the phase points of the large system are naturally split into two coordinates: $(q, p) = \big((q_1, p_1), (q_2, p_2)\big)$. Suppose also that the Hamiltonian agrees with the splitting, i.e., $H(q, p) \approx H_1(q_1, p_1) + H_2(q_2, p_2)$, which means that the interaction energy between the subsystem and the environment is small compared to the energies E_1 and E_2. We shall assume that $N_2 \gg N_1$. Then as before for the ideal gas case, we write

$$\begin{aligned}
&\mathbb{P}_E\Big(\{((q_1, p_1), (q_2, p_2)) | (q_1, p_1) \in A\}\Big) \\
&\qquad = \frac{1}{Z(E)} \int_A \mathrm{d}^{3N_1} q_1 \, \mathrm{d}^{3N_1} p_1 \int \mathrm{d}^{3N_2} q_2 \, \mathrm{d}^{3N_2} p_2 \, \delta\left(H_1 + H_2 - E\right) \\
&\qquad = \frac{1}{Z(E)} \int_A \mathrm{d}^{3N_1} q_1 \, \mathrm{d}^{3N_1} p_1 \int \mathrm{d}^{3N_2} q_2 \mathrm{d}^{3N_2} p_2 \, \delta\big(H_2 - (E - H_1)\big) \\
&\qquad =: \frac{1}{Z(E)} \int_A \mathrm{d}^{3N_1} q_1 \, \mathrm{d}^{3N_1} p_1 Z_2\big(E - H_1(q_1, p_1)\big) \, .
\end{aligned}$$

As in the ideal gas case we wish to control $Z_2(E - H_1)/Z(E)$ for $E/N = \text{constant}$, when N gets large. Let us quickly check on the ideal gas to see how to proceed. Roughly $Z(E) \sim E^N$ (actually the exponent is $N/2$, but we simply call it N to ease notation) and $Z_2(E - H_1) \sim (E - H_1)^{N_2}$, where $N_2 \approx N$. Therefore, expanding Z_2 around E, which we think of as being large compared to the range of typical values of H_1, yields

$$\begin{aligned}
\frac{Z_2(E - H_1)}{Z(E)} &\approx \frac{(E - H_1)^N}{E^N} \\
&= \frac{E^N}{E^N} + \frac{N E^{N-1} H_1}{E^N} + \frac{1}{2} \frac{N(N-1) E^{N-2} H_1^2}{E^N} + \cdots \\
&= 1 + \frac{N}{E} H_1 + \frac{1}{2} \frac{N^2}{E^2} H_1^2 + \cdots \, .
\end{aligned}$$

Since E/N is approximately the mean energy of one particle, which is of the order of the typical value H_1 achieves, the important observation is now that all terms in the above expansion are of order 1. So this expansion is no good. The terms are too big. Therefore let us take the logarithm:

---

work on black body radiation (see Chap. 6). Einstein used that formula in reverse. Thinking of the size of regions in phase space as the probability for a microstate to be in that region, he wrote (we shall say more on this below) that the *probability of fluctuation* $\sim \exp(\Delta\text{entropy}/k_{\mathrm{B}})$.

$$\ln\frac{Z_2(E-H_1)}{Z(E)} \approx \ln\frac{(E-H_1)^N}{E^N} = N\ln\left(1-\frac{H_1}{E}\right)$$
$$= -N\frac{H_1}{E} - \frac{1}{2}N\frac{H_1^2}{E^2} + \cdots ,$$

where the second order term (and all following terms) is now of smaller order in $1/N$ than the first term. So this is looking better. Expanding in H_1 yields

$$\ln\frac{Z_2(E-H_1)}{Z(E)} = \ln Z_2(E-H_1) - \ln Z(E)$$
$$\approx \ln\frac{Z_2(E)}{Z(E)} - \frac{\partial \ln Z_2(E)}{\partial E}H_1$$
$$= \ln\frac{Z_2(E)}{Z(E)} - \frac{1}{k_B T}H_1 \qquad \text{[by (4.22)]}.$$

The first term on the right becomes the normalization factor, i.e., we obtain the normalized canonical distribution

$$\frac{Z_2(E-H_1)}{Z(E)} \approx \frac{e^{-H_1/k_B T}}{Z_1(T)} ,$$

or in other words

$$\mathbb{P}_E\Big(\{((q_1,p_1),(q_2,p_2))|(q_1,p_1)\in A\}\Big) \overset{N_2\gg N_1}{\approx} \int_A d^{3N_1}q_1 d^{3N_1}p_1 \frac{e^{-H_1/k_B T}}{Z_1(T)}$$
$$=: \int_A d^{3N_1}q_1\, d^{3N_1}p_1\, p_T(q_1,p_1) ,$$

which is the canonical distribution (4.7).

Since we have come this far now, let us briefly consider the Gibbs formalism to highlight the difference between the Gibbs entropy and the Boltzmann entropy. First note that the normalization $Z(T)$ of the canonical measure with Hamiltonian H (commonly referred to as the partition function) can be written in the form

$$Z(T) = \int d^{3N}q d^{3N}p\, e^{-H/k_B T}$$
$$= \int e^{-E/k_B T} dE \int d^{3N}q d^{3N}p\, \delta(E - H(q,p))$$
$$= \int dE\, Z(E) e^{-E/k_B T} .$$

Therefore the expectation value of E is

$$\langle E\rangle_T = \frac{1}{Z(T)}\int dE\, E Z(E) e^{-E/k_B T} =: \int E p_T(E) dE , \tag{4.24}$$

where

$$p_T(E) = \frac{Z(E)}{Z(T)} \mathrm{e}^{-E/k_\mathrm{B}T} \; .$$

The T-derivative of this is the heat capacity, which is readily computed from the definition of $Z(T)$:

$$C = \frac{\mathrm{d}\langle E \rangle_T}{\mathrm{d}T} = \frac{1}{k_\mathrm{B}T^2}\left(\langle E^2 \rangle_T - \langle E \rangle_T^2\right) =: \frac{1}{k_\mathrm{B}T^2}\langle \Delta E^2 \rangle_T \; .$$

Assume for simplicity that the heat capacity is constant, i.e., $\langle E \rangle_T = CT$. Then the above variance is simply $\langle \Delta E^2 \rangle_T = Ck_\mathrm{B}T^2$, and for the relative variance, we obtain the ratio between the heat capacity of an atom and that of the system:

$$\frac{\langle \Delta E^2 \rangle_T}{\langle E \rangle_T^2} = \frac{Ck_\mathrm{B}T^2}{C^2T^2} = \frac{k_\mathrm{B}}{C} \; .$$

This noteworthy and famous result says that, since $C \approx Nk_\mathrm{B}$ when there are N atoms in the system, the relative energy fluctuation is negligible when N gets large. This means that the typical phase points all lie in a relatively thin shell around the mean value $\overline{E} := \langle E \rangle_T$. This in turn suggests that typicality is as well expressed by the microcanonical ensemble on the surface $\Omega_{\overline{E}}$. This is known as the equivalence of ensembles. It can be sharpened to a rigorous statement by taking the thermodynamic limit of infinitely large systems $N \to \infty$, $V/N =$ constant.

Furthermore this suggests that it is reasonable to approximate the canonical distribution by a Gaussian with mean $\overline{E} := \langle E \rangle_T$ and variance $Ck_\mathrm{B}T^2$, i.e.,

$$p_T(E) = \frac{Z(E)}{Z(T)} \mathrm{e}^{-E/k_\mathrm{B}T} \approx \frac{\mathrm{e}^{-(E-\overline{E})^2/Ck_\mathrm{B}T^2}}{\sqrt{2\pi Ck_\mathrm{B}T^2}} \; .$$

Therefore,

$$p_T(\overline{E}) = \frac{Z(\overline{E})}{Z(T)} \mathrm{e}^{-\overline{E}/k_\mathrm{B}T} \approx \frac{1}{\sqrt{2\pi Ck_\mathrm{B}T^2}} \; ,$$

whence

$$Z(T) \approx \sqrt{2\pi Ck_\mathrm{B}T^2}\, Z(\overline{E}) \mathrm{e}^{-\overline{E}/k_\mathrm{B}T} \; ,$$

and therefore

$$\ln Z(T) \approx \ln Z(\overline{E}) - \frac{\overline{E}}{k_\mathrm{B}T} + \ln \sqrt{2\pi Ck_\mathrm{B}T^2} \; .$$

Then, using $C \approx N$,

$$\frac{\ln Z(T)}{N} \overset{N\,\text{large}}{\approx} \frac{\ln Z(\overline{E})}{N} - \frac{1}{k_{\mathrm{B}}T}\frac{\overline{E}}{N} .$$

In this sense,

$$-k_{\mathrm{B}}T\ln Z(T) = -k_{\mathrm{B}}T\ln Z(\overline{E}) + \overline{E} , \tag{4.25}$$

and in view of (4.23), this should be identified as the thermodynamic free energy relation

$$F = -TS + U .$$

We now reverse the argument. Take a microcanonically distributed system with energy E. Then the Boltzmann entropy (the *entropy*) is $S = k_{\mathrm{B}}\ln Z(E)$. For large systems, this value is well approximated by $k_{\mathrm{B}}\ln Z(\overline{E})$, with the canonically distributed system with $\langle E\rangle_T = \overline{E} = E$. By (4.25), we then have

$$\begin{aligned} S = k_{\mathrm{B}}\ln Z(E) &= k_{\mathrm{B}}\ln Z(T) + \frac{1}{T}\overline{E} = k_{\mathrm{B}}\left[\ln Z(T) + \frac{1}{k_{\mathrm{B}}T}\langle E\rangle_T\right] \\ &= -k_{\mathrm{B}}\int p_T \ln p_T \mathrm{d}^{3N}q\,\mathrm{d}^{3N}p \qquad [\text{by } (4.24)] , \end{aligned}$$

where we have reverted to the canonical density notation

$$p_T(q,p) = \frac{1}{Z(T)}\mathrm{e}^{-H(q,p)/k_{\mathrm{B}}T} .$$

The right-hand side of that equation is the famous Gibbs entropy

$$S_{\mathrm{G}} = -k_{\mathrm{B}}\int p_T \ln p_T \mathrm{d}^{3N}q\,\mathrm{d}^{3N}p ,$$

equal in value to Boltzmann's entropy (4.23) in equilibrium. Although the two values are equal, there is a huge difference between Boltzmann's entropy and the Gibbs entropy, a difference which becomes critical when we turn to the issue of non-equilibrium, irreversibility, and the second law. Boltzmann's entropy is naturally a function on phase space. It is constant on the set of typical phase points, where the size of the set of typical phase points is essentially $Z(E)$. The Gibbs entropy is not defined on phase space, but is rather a functional of distributions on phase space – a technically very abstract, but computationally useful tool.

4.2 Irreversibility

We understand how randomness enters into physics. It is how a typical universe appears. The balls in a Galton board behave in the random way they do because that is how they behave in a typical universe. The gas in a container fills the container in a homogeneous way because that is typical. One thing is strange, however: Why should there be a Galton board in a typical universe, why a gas container, and when we pick up that stone and throw it, how come we can compute its trajectory precisely, without any randomness at all? Are our macroscopic experiences not in fact non-random in many ways? Where is the place for all the typicality talk of the previous section in a universe like ours?

Unfortunately, this is a hard question, because the truthful answer is that we do not know. Our universe is *not* typical. For our universe there is no justification for typicality of microcanonical or canonical distributions, because they are false in general. For us it is typical to generate atypical situations. We can build a container with a piston which forces all molecules into one half of the container, and we can remove the piston and then we have a gas system where all gas molecules at that moment are in one half of the container and the other half is empty (at least we can do that in principle), as in Fig. 4.1. That is an atypical state. True, the gas did not do that by itself. We did it. But that would not help. Shifting the reason to the outside ends in atypicality for the largest system conceivable, the universe. Something needs to be explained.

A typical universe is an *equilibrium universe*, the equilibrium, to which our universe evolves (to a thermal death, as Clausius referred to it). Right now we are still very far from thermal death. Our universe is atypical or in non-equilibrium, and that is why we can build a Galton board or pick up a stone, or look at the moon circling the earth. But at the same time we experience the determined evolution to equilibrium at every moment and all over the place *via the second law of thermodynamics*. The law determines that thermal processes are directed, and that they are *irreversible*, i.e., the time-reversed process does not take place. The breaking of a glass is just such a thermal process – no one ever experienced a broken glass becoming whole again on its own. Never will a cold body make a warm body warmer on its own. The law says that thermal processes will always run in such a way that entropy increases!

This thermodynamic law is different from all the other laws of physics, which are time-reversal invariant. In thermodynamics there is no argument to back this law, no further insight to make the law plausible. It is a law that describes what we see. Boltzmann explained the law by reducing thermodynamics to Newtonian mechanics. That seems paradoxical, since Newtonian mechanics is time reversible in the sense that all mechanical processes can run both ways. Boltzmann's explanation is this.

4.2.1 Typicality Within Atypicality

Think of the atypical state of the gas in Fig. 4.1, where only half the container is occupied by gas molecules (suppose this has been achieved by a piston which was then removed). That is now an atypical state – no way around that. What happens next with the gas? Typically the gas system will move in such a way that the container becomes filled in a homogeneous way and remains in that equilibrium-like state "forever".[12] Of course, "typicality" is no longer our clear-cut equilibrium "typicality". It is typicality with respect to the measure on phase space which arises from conditioning under *macroscopic constraints* (like the piston which pushed the gas into one half of the container). In short, it is typicality under given macroscopic constraints. That is what remains of Boltzmann's typicality idea in our real atypical world. Things are as typical as possible under given constraints.

The typical process of the homogeneous filling of the container when the gas is freed from its macroscopic constraints is basic to the famous phenomenological Boltzmann equation, an equation which is not time-reversal invariant, but which can be derived from the time-reversible microscopic dynamics of gas molecules. The Boltzmann equation describes how the typical empirical density (of a constrained low density gas) changes in time to a typical density of equilibrium. That is a typical lawlike behavior (strongly analogous to the law of large numbers) on which the second law is based. The second law is in that sense a macroscopic law, which holds for a particular set of atypical initial conditions of the universe. We do not know by which fundamental physical law this set is selected, nor do we know exactly what the set looks like. All we know is that it is a very special set given by macroscopic constraints which we can to some extent infer from our knowledge about the present status of our universe (we shall expand a bit on that at the end of the chapter). But that should not in any way diminish our respect for Boltzmann's insight that the second law can be reduced to and therefore explained by fundamental microscopic laws of physics.

We shall say a bit more about the second law and entropy increase, and the way irreversible behavior can arise from reversible behavior, or in other words, how one can turn apples into pears. As we said, the key here is *special initial conditions*. We return to the notions of micro- and macrostate, since entropy is a thermodynamic notion. Entropy arises from the first and second laws of thermodynamics (when the latter is phrased in terms of the impossibility of a perpetuum mobile) as a thermodynamic state function $\mathscr{S}$ which reads in differential form

$$\mathrm{d}\mathscr{S} = \frac{1}{T}\left(\mathrm{d}\mathscr{E} + \mathscr{P}\mathrm{d}V\right), \tag{4.26}$$

with energy $\mathscr{E}$, pressure $\mathscr{P}$, and volume V.

A thermodynamic state (or macrostate) is determined by a very small number of thermodynamic variables: volume, density, temperature, pressure, energy, to name a few. According to kinetic gas theory, or in more modern terms, according to statistical mechanics, the extensive variables (those which grow with the size of the system) are functions on the phase space Ω of the system under consideration, i.e., extreme coarse-graining functions (random variables in other words), which do not fluctuate, because they come about as empirical means (the law of large numbers acting again). An example is the (empirical) density $\rho(\mathbf{x})$, the average number of

[12] The quotation marks on "forever" will become clear further down.

particles per volume:

$$\rho(\mathbf{x})\mathrm{d}^3x \approx \frac{1}{N}\sum_{i=1}^{N}\delta(\mathbf{x}-\mathbf{q}_i)\mathrm{d}^3x\,, \tag{4.27}$$

where N is the very large particle number of the system and the $\mathbf{q}_i$ are the actual particle positions. The values that this macroscopic ρ achieves, and which characterize the macrostate Ma, partition the phase space into large cells, so that very many phase points – microstates $\omega =:$ Mi(Ma) $\in \Omega$ – realize the macrostate. We call the size of the cell the phase space volume of the macrostate: $W(\mathrm{Ma})$. A simple mathematical example is provided by the Rademacher functions (4.3). Consider the empirical mean

$$\rho_n(x) = \frac{1}{n}\sum_{i=1}^{n} r_i(x) \ \in [0,1]\,. \tag{4.28}$$

This is certainly a macroscopic variable which coarse-grains the interval $[0,1]$. We shall prove in (4.2) the rather intuitive assertion that typically

$$\rho_n(x) \overset{n\,\text{large}}{\approx} \frac{1}{2}\,,$$

so that a very large subset (with respect to the natural notion of size as the "length" of the set) of points in $[0,1]$ (microstates) yields the macrostate corresponding to value $\approx 1/2$, in other words $W(\approx 1/2) \approx 1$. The macrostate corresponding to the value $1/4$, for example, would be realized by only a very tiny subset in $[0,1]$.

Entropy is extensive, but it was not originally realized as function on phase space. It was Boltzmann who achieved that. The entropy $\mathscr{S}$ of a system in the macrostate Ma with phase space volume $W(\mathrm{Ma})$ [= "number" of microstates Mi(Ma), which realize Ma] is determined by that "number" $W(\mathrm{Ma})$. This is in fact what (4.23) says, and the entropy becomes in this way a function on the microstates – a function which is constant on the cell corresponding to the macrostate.

The Boltzmann entropy is[13]

$$S_\mathrm{B}\big(\mathrm{Mi}(\mathrm{Ma})\big) = k_\mathrm{B}\ln W(\mathrm{Ma}) \qquad \big[=\mathscr{S}(\mathscr{E},V)\big]\,. \tag{4.30}$$

[13] $W(\mathrm{Ma})$ must actually be normalized by $N!$ for the entropy to become extensive. Consider the phase space volume of the following macrostate of an ideal gas in a volume V. We have N particles and they are distributed over *one-particle phase space cells* α_i, $i=1,\dots,m$, so that there are N_i particles in cell α_i with $|\alpha_i| = \alpha$, $i=1,\dots,m$. How large is that phase space volume? There are $N!/N_1!N_2!\cdots N_m!$ possibilities for such distributions, and for each possibility the particles have volume α to move in, whence

$$W(\mathrm{Ma}) = \frac{N!}{N_1!N_2!\cdots N_m!}\alpha^N\,. \tag{4.29}$$

With Stirling's formula (4.1) $n! \approx n^n$ and

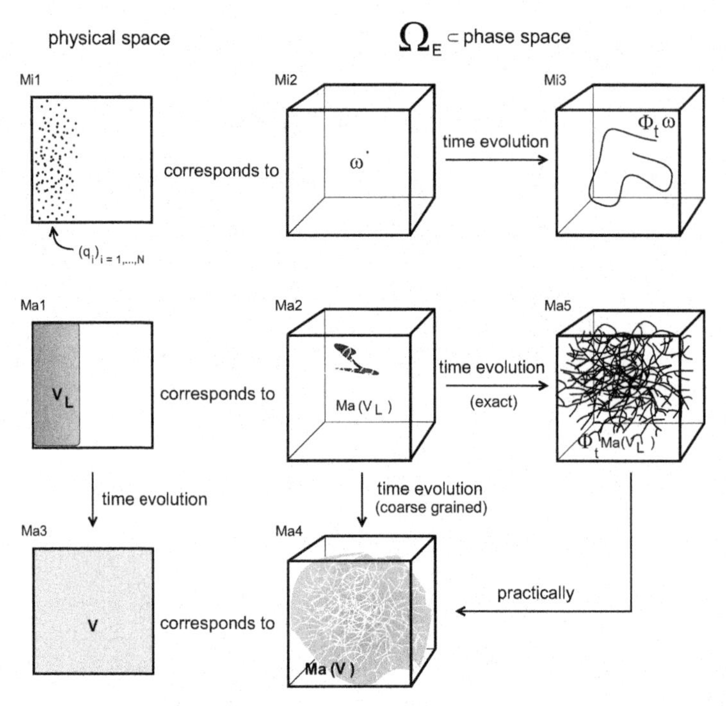

Fig. 4.9 The evolution to equilibrium depicted in physical space *on the left* and in phase space *on the right*: microscopically and macroscopically. Mi1: The system starts with all gas molecules on the left side of the container. Mi2: The corresponding phase point, the microstate, is ω. Ma1 shows the macrostate, of which Mi1 is one realization. Ma2 shows the set $\mathrm{Ma}(V_\mathrm{L})$ of all microstates which realize Ma1. (Not to scale! It would not be visible on the proper scale.) The time evolution on phase space is denoted by Φ_t. In Mi3(t), the trajectory $(\Phi_s\omega)_{s\in[0,t]}$ is depicted. In the course of time, the gas fills the whole volume. Ma3 shows the uniform density, as in equilibrium. Ma5 shows the time-evolved set $\Phi_t\mathrm{Ma}(V_\mathrm{L})$ in which $\Phi_t\omega$ lies. That set of microstates is perfectly mixed up with the equilibrium set and realizes for all practical purposes a macrostate, which is also realized by the huge set of equilibrium microstates $\mathrm{Ma}(V)$, i.e., for all practical purposes $\Phi_t\mathrm{Ma}(V_\mathrm{L})$ can be replaced by $\mathrm{Ma}(V)$. Nevertheless, the volume $|\mathrm{Ma}(V_\mathrm{L})|$ of $\mathrm{Ma}(V_\mathrm{L})$ is very small and, by Liouville's theorem, $|\mathrm{Ma}(V_\mathrm{L})| = |\Phi_t\mathrm{Ma}(V_\mathrm{L})| \ll |\mathrm{Ma}(V)|$

$$\frac{S}{k_\mathrm{B}} = \ln\binom{N}{N_1,\dots,N_m} = \ln\frac{N!}{N_1!\dots N_m!} \approx \left(N\ln N - \sum_{i=1}^m N_i \ln N_i\right)$$
$$= -\sum_{i=1}^m N_i(\ln N_i - \ln N) = -\sum_{i=1}^m N_i \ln\frac{N_i}{N}\,.$$

The thermodynamic entropy (as invented by Clausius) is extensive, which means it is additive. If we remove from a gas container a separating wall which separated two ideal gases, then the entropy of the "after system" will become the sum of the entropies of the "before systems". Our definition does not obey that, but it does when we cancel the N-dependence, i.e., when we consider $\hat{W} := W/N!$ and set $S = k_\mathrm{B}\ln\hat{W}$.

Do we now understand why entropy increases? Yes, by simple combinatorics. Lifting macroscopic constraints like taking away the piston which held the gas molecules on the left side of the container in Fig. 4.1, the new macrostates will be those which correspond to larger and larger phase space volumes, i.e., for which the number of microstates which realize the macrostate becomes larger and larger. This is all depicted in Fig. 4.9. The number of microstates does not change under the Hamiltonian time evolution, by Liouville's theorem. The gas extends over the whole container, yet the corresponding phase space region remains as small as the one it started in.

So does that mean that the entropy does not increase then? Where did our reasoning go wrong? Nowhere! We only have to accept that Boltzmann's definition (4.30) of the entropy is more subtle than we imagined at first sight. We must always consider the macrostate first! The *new macrostate* $\mathrm{Ma}(V)$ has a very much larger phase space volume $W\big(\mathrm{Ma}(V)\big)$, and the evolved microstate $\Phi_t\omega$ realizes this macrostate (for all practical purposes). There is a reason for the parenthetical addition of the phrase "for all practical purposes", which will be addressed later. But it is nothing *we* should be concerned with.

Here is a very simple mathematical example which may be helpful to understand the definition of entropy with regard to micro- and macrostates. Let $\Omega = [0,1]$ and consider the time-one map $Tx = 2x|_{\mathrm{mod}\,1}$ (see Fig. 4.10). The T action is most easily understood when we write $x \in [0,1]$ in the binary representation. Then

$$x = \sum_{k=1}^{\infty} r_k(x) 2^{-k} \quad \longrightarrow \quad Tx = \sum_{k=1}^{\infty} r_{k+1}(x) 2^{-k} .$$

In other words,

$$r_k(Tx) = r_{k+1}(x) . \tag{4.31}$$

The map is obviously not invertible (so the n-fold composition of the map defines irreversible dynamics, different from Hamiltonian dynamics). We look again at the macro-variable (4.28)

$$\rho = \frac{1}{n} \sum_{k=1}^{n} r_k : [0,1] \to [0,1] ,$$

and we shall show in Theorem 4.2 that for n large this is typically close to 1/2. We observe that a microstate $x \in [0, 1/2^n)$ corresponds to the macrostate Ma given by $\rho = 0$. Hence all microstates realizing Ma with $\rho = 0$ make up the set with $r_k = 0$ for $k = 1, 2, \ldots, n$. Therefore

$$S = k_{\mathrm{B}} \ln W\big(\mathrm{Ma}(\rho = 0)\big) = k_{\mathrm{B}} \ln \big|[0, 1/2^n)\big| = -k_{\mathrm{B}} n \ln 2 .$$

Now for n very large $x \in [0, 1/2^n)$ is atypical for a number in $[0,1]$, since the x starts with a lot of zeros. But for a conditionally typical $x \in [0, 1/2^n)$, i.e., typicality defined with respect to the *conditional measure* $2^n|A|$, $A \subset [0, 1/2^n)$, we should see,

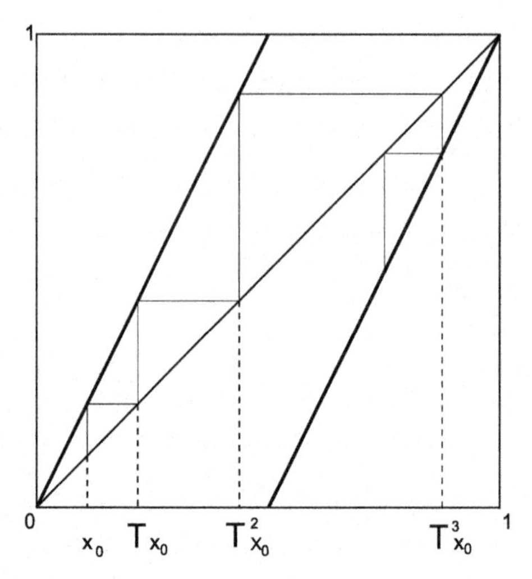

Fig. 4.10 The graph of the Bernoulli map $Tx = 2x|_{\text{mod}\,1}$ and iterations of an initial point x_0. Note that T is not invertible

from some large place in the binary expansion onwards, that zeros and ones occur in an irregular but lawlike manner, in the sense of the law of large numbers. In fact (we only look at discrete times, but that does no harm) with

$$T^j\left[0,\frac{1}{2^n}\right) = T\circ T\circ T\cdots\circ T\left[0,\frac{1}{2^n}\right) = [0,1)\,, \quad j\geq n\,,$$

we have that, typically with respect to the conditional measure,

$$\rho_j(x) := \rho(T^j x) = \frac{1}{n}\sum_{k=1}^{n} r_k(T^j x) = \frac{1}{n}\sum_{k=1}^{n} r_{k+j}(x) \overset{n\,\text{large}}{\approx} \frac{1}{2}\,, \quad j\geq n\,,$$

as we shall show in Theorem 4.2. But Ma corresponding to $\rho\approx 1/2$ has almost all of $[0,1]$ as its phase cell, and thus $S = k_{\text{B}}\ln(|[0,1)|) = 0$, which is of course much larger than the non-equilibrium value $-k_{\text{B}}n\ln 2$. That is the essence of the story.

For Boltzmann, the story did not end so happily. He could not cheat on the reversibility of the fundamental motion as we did. His map T_t is the Hamiltonian flow, which is time reversible, and this leads to extra complications which we have swept away by appeal to our catch phrase "for all practical purposes". Does the fact that the fundamental motion is time reversible change the qualitative picture we have shown in Fig. 4.9? Some physicists have answered that it does! On the other hand one can hardly fight the intuitive feeling that the picture is right. After all, there are typically more possibilities for the gas molecules to distribute themselves equally over the whole volume (i.e., phase space, see Fig. 4.1).

But the debate was nevertheless about this question: How can time reversible dynamics give rise to phenomenologically irreversible dynamics? This is a bit like a debate about whether something tastes bitter, when there is nothing fundamentally like "bitterness" in nature. But the criticism was in fact more technically oriented, with the question: How can one mathematically rigorously *derive* a time-irreversible equation of motion from time-reversible equations? And that is indeed what Boltzmann claimed to have achieved with his famous equation,[14] and the resulting H-theorem (see below).

Boltzmann met the criticism head on and rectified his assertions, where necessary, without sacrificing any of his ideas. The criticism was this. In the above mathematical example, under the T map, the conditionally typical point x eventually becomes unconditionally typical, since the zeros in front are all cut away after enough applications of T. Nothing in the point $T^j x$ recalls the special set x once lay in. We cannot turn the wheel back, nor the time for that matter. The map is not time reversible. In the gas example of Fig. 4.9, for all practical purposes, the macrostate does not distinguish the special phase point $\Phi_t \omega$ from typical phase points, but it is nevertheless special! To see that, all we need to do is to reverse all velocities in $\Phi_t \omega : \Phi_t(x,p) \to \Phi_t(x,-p)$. What does the gas then do? It returns peacefully, and without need for a piston to push it, into the left half of the container, which is atypical behavior at its very best. Reversing all velocities is like reversing time! Therefore, in a closed system, the gas carries its past history around with it all the time. For all practical purposes, we can forget that information and replace the phase point by a typical one, because practically speaking it is impossible to reverse the velocities of all the gas molecules.

But so what? This proves that there are in fact "bad initial conditions", for example, $\Phi_t(x,-p)$, for which Fig. 4.9 is not right. This was spotted by Josef Loschmidt, a friend and colleague of Boltzmann, and this "Umkehreinwand" (reversibility objection) led Boltzmann to recognize that his famous H-theorem (see below), which in its first publication claimed irreversible behavior for all initial conditions, was only true for typical initial conditions. Because, as Boltzmann immediately responded, these bad initial conditions are really very special, more atypical than necessary. They are not conditionally typical, conditioned on macroscopic constraints, which for example can be achieved by a piston which holds all gas molecules in the left half of a container. The time-irreversible Boltzmann equation (on which the H-theorem is based) holds for conditionally typical initial conditions and governs the time evolution of the conditional typical value of the empirical density [analogous to (4.27)]:

$$f(\mathbf{q},\mathbf{p}) \approx \begin{array}{c}\text{relative number of particles with}\\ \text{phase space coordinates around } (\mathbf{q},\mathbf{p})\end{array},$$

where the approximation holds in the sense of the law of large numbers. The conditional typical value is time dependent since, after removal of the macroscopic constraint, the phase points typically wander into the large phase space volume defining equilibrium. Boltzmann showed in his H-theorem that $H(t) =$

[14] For mathematically rigorous derivations of Boltzmann's equation in the sense of mathematical physics, the proof by Lanford is recommended [6]. See also [7].

$\int f(t,\mathbf{q},\mathbf{p})\ln f(t,\mathbf{q},\mathbf{p})\mathrm{d}^3q\mathrm{d}^3p$ increases (on average) as time goes by, toward an equilibrium value.

But the mathematician Ernst Zermelo went on to criticize Boltzmann's view of atomism, which is basic to Boltzmann's understanding of irreversibility, on the basis of a little theorem proved by Poincaré, known as the "Wiederkehreinwand" (recurrence objection). Poincaré's recurrence theorem is a very simple result for dynamical systems with a phase space of *finite* measure:

Theorem 4.1. Poincaré's Recurrence Theorem

Let $\big(\Omega,\mathscr{B}(\Omega),\Phi,\mathbb{P}\big)$ *be a dynamical system, that is,* Ω *is a phase space,* $\mathscr{B}(\Omega)$ *the* σ*-algebra,*[15] $\Phi:\Omega\to\Omega$ *a measurable*[16] *(time-one) map, and* $\mathbb{P}$ *a stationary measure, i.e.,* $\mathbb{P}(\Phi^{-1}A)=\mathbb{P}(A)$*,* $A\in\mathscr{B}(\Omega)$*. Assume the finite measure condition* $\mathbb{P}(\Omega)<\infty$ *and let* $M\in\mathscr{B}(\Omega)$*. Then for almost all* $\omega\in M$ *(i.e., except a set of* $\mathbb{P}$*-measure zero), the orbit* $(\Phi^n\omega)_{n\in\mathbb{N}}$ *visits* M *infinitely often.*

For simplicity, we prove the theorem for invertible Φ (as it was originally done for Hamiltonian flows) and we omit the proof of the assertion that the recurrence occurs infinitely often. Let N be the bad set,[17] i.e., the set of points in M which never return to M (and thus never to N). Then

$$\Phi^n(N)\cap M=\emptyset \quad \text{for all } n\geq 1\,,$$

and for $n>k$,

$$\Phi^n(N)\cap\Phi^k(N)=\Phi^k\left(\Phi^{n-k}(N)\cap N\right)=\emptyset\,.$$

Therefore the measures of the sets add and, with the finite measure condition and stationarity, which for invertible maps is equivalent to $\mathbb{P}(\Phi A)=\mathbb{P}(A)$, we obtain

$$\infty>\mathbb{P}\left(\bigcup_{n=0}^{\infty}\Phi^n(N)\right)=\sum_{n=0}^{\infty}\mathbb{P}(\Phi^n(N))=\sum_{n=0}^{\infty}\mathbb{P}(N)\,,$$

hence $\mathbb{P}(N)=0$.

There is no way to escape this fact. If the gas obeys the Hamiltonian law of motion then, when the gas started in the left half of the container, it will eventually return to the left half of the container – if the system remains isolated. So when the gas has expanded to fill the whole volume, it will not stay like that forever, but will return to the left half once in a while – for sure. Well, if one waits long enough, an equilibrium fluctuation will produce that too, and this was also Boltzmann's answer: If only one could live long enough to see that!

So what is the point here? The point is a mathematical technicality that must be taken into account when one tries to *derive* long-time irreversible behavior from

[15] See footnote 21 in Sect. 4.3.

[16] See (4.39).

[17] That N is measurable, i.e., $N\in\mathscr{B}(\Omega)$, is left as an exercise.

time-reversible dynamics – one must ensure that the Poincaré recurrence time is infinite. But how? By taking a limit of an infinitely large system, thereby violating the finite measure assumption. Think of our simple example of a typical $x \in [0,1]$ and the macro-variable (4.28) ρ which depends on n. For finite n, even for typical $x \in [0,1]$, $\rho(T^j x)$ will fluctuate to value zero for j large enough. There will always be a sequence of n consecutive zeros somewhere. But that can no longer happen if we first take $n \longrightarrow \infty$.

Boltzmann even gave a time scale on which a recurrence in the ideal gas situation is to be expected. That time is, not surprisingly, ridiculously long, longer than the cosmic time we imagine as the lifetime of our universe. Boltzmann's estimate was based on a new notion which he introduced and, which became a fashionable field in mathematics: ergodic theory.[18] According to Boltzmann, one should envision that a system runs over the whole of phase space, and that the time it spends in cells of macrostates is proportional to the cell size. The famous formula is, for the Hamiltonian system,

$$\lim_{t \to \infty} \frac{1}{t} \int_0^t \chi_A(\Phi_s \omega)\, \mathrm{d}s = \int_A \rho_E(\omega) \mathrm{d}\omega = \mathbb{P}_E(A) \,, \tag{4.32}$$

where the left-hand side is the fraction of time the system spends in A and the right-hand side is the microcanonical measure of A. This looks reasonable. Since the ratio of phase space volumes of equilibrium values to non-equilibrium values is huge, so are the time ratios. To recall the sizes, compare the numbers for equidistribution with a fluctuation of remarkable size in a gas of n particles, viz.,

$$\binom{n}{n/2} \quad \text{with} \quad \binom{n}{n(1-\varepsilon)/2} \,,$$

where n is something like 10^{24}. From (4.1), it is an easy exercise to show that the ratio of the left- and right-hand numbers is $\approx \exp(2n\varepsilon^2)$, i.e., the ratio of sojourn times in equilibrium to non-equilibrium $\approx \exp\left(10^{24}\varepsilon^2\right)$. Whatever microscopic time scale one uses, the resulting time spent in equilibrium is ridiculously large.

4.2.2 Our Atypical Universe

Boltzmann held the following picture of our universe. The part of the universe which lies behind the horizon of what is accessible to us right now is in equilibrium – an equilibrium sea. We – that is we, our solar system, the galaxies we see, the past of our universe as far as we can witness it – merely mirror a possibly giant equilibrium fluctuation – a non-equilibrium island in an equilibrium sea. That sounds reasonable. It does suggest that the "initial condition" of our universe is after all typical. No need to worry any further about it.

---

[18] "Ergoden" was Boltzmann's name for the microcanonical measure.

Nowadays we look much further into the universe than scientists did in Boltzmann's day, and we have witnesses of a much older past than in Boltzmann's day, but we do not see any glimpse of global equilibrium surrounding us at the new "visible" astronomical distances, and we find no glimpse of equilibrium in the "new" far distant past. What does that tell us? Does it tell us that the fluctuation turned out to be more violent than one had judged in Boltzmann's day? No, that would violate the typicality idea: no more atypical than necessary is Boltzmann's adage. Thus a question arises: How giant is the fluctuation? Is there a highly ordered (lower entropy) past of my present? But my present is already atypical and the moral of typicality is: no more atypical than necessary! Then the past of my present should look less atypical, i.e., the entropy of the universe should increase from NOW towards the past and the future. The NOW is the deepest point of the fluctuation!

Actually, we do not behave as if we believed in a fluctuation. Feynman calls Boltzmann's fluctuation hypothesis ridiculous [8, p. 129]. We believe (and Boltzmann did too) that there is a very special past, and in believing that, we are in fact reconstructing a logical past which makes sense and which proves that we are right in our belief. Paleontologists set out to find the missing link between ape and man and they eventually did – the Java man, or *Homo erectus*. Schliemann believed that Homer's Iliad had roots in a truly existing past, which was not just the words of the Iliad, and he found the remains of Troy. We probe the celestial depth by sending out the Voyager spacecraft with a golden disc containing Chuck Berry's Johnny B. Goode into space, hoping that, way out there, invisible and unknown to us, something will appreciate good old earthly sounds.

It is ridiculous to readjust a fluctuation after learning that non-equilibrium extended further than previously assumed. It does make sense to readjust the atypicality set of initial conditions if we become aware of further macroscopic constraints. That is actually how we think and behave. We are convinced that our initial condition is a very special one. We believe that the initial condition of the universe was carefully selected. So carefully, that the measure of that in numbers is way beyond our human scales (see, e.g., [9] for an estimate of the size of the special set, and for a nice drawing of a goddess carefully aiming with a sharp needle at the chosen initial configuration of our universe). What is the physics behind the selection? We do not know (see, e.g., [10] for speculations on that). That ignorance of ours deserves to be called an open problem: the problem of irreversibility.

4.2.3 Ergodicity and Mixing

Why should we need to know anything about ergodicity and mixing, if we are not interested in doing specific technical work in the theory of dynamical systems? In fact, it is not at all necessary. But since the time of Gibbs, these notions have been floating around as if they were fundamentally important for understanding the role of chance in physics, and in particular for justifying the use of equilibrium ensembles to compute the average values of thermodynamic quantities. To avoid those misconceptions, we shall address these notions briefly.

As we already said, ergodicity was introduced by Boltzmann for the purpose of computing times of sojourn from the formula (4.32). Boltzmann's idea here was that a phase point moves around and that its trajectory will eventually cover the energy surface in a dense manner. This allowed him to estimate the enormously long times (more or less the Poincaré recurrence time) a system stays in the overwhelmingly large equilibrium regions, i.e., the regions of typicality, in order to address people's worries about early returns. And that is all ergodicity is good for, apart from some other technical niceties.

A misconception one often encounters is that ergodicity justifies the use of equilibrium distributions for averaging thermodynamic quantities, because the measurement of thermodynamic quantities takes time, time enough for the phase point to wander about the energy surface, so that the "time average equals the ensemble average". But Boltzmann's understanding is obviously quite the opposite. Given the energy of a system, the overwhelming majority of its phase points on the energy surface are typical, i.e., equilibrium phase points, all of which look macroscopically more or less the same. Therefore, the value of any thermodynamic quantity is for all practical purposes constant on that energy surface. Averaging over the energy surface will merely reproduce that constant value, regardless of whether or not the system is ergodic.

The mathematical definition of ergodicity focusses on the dense covering of trajectories on the energy surface or more generally on phase space. Equation (4.32) is then a theorem when ergodicity holds. It is noteworthy that ergodicity is equivalent to the law of large numbers, and that independence is stronger than mixing, which is stronger than ergodicity. Indeed (4.32) is technically nothing but the law of large numbers, which we shall show later under independence conditions.

Definition 4.1. A dynamical system $\big(\Omega,\mathscr{B}(\Omega),\Phi,\mathbb{P}\big)$ (see Theorem 4.1) for which $\mathbb{P}(\Omega)=1$ is said to be ergodic if the invariant sets have measure zero or one:

$$A\in\mathscr{B}(\Omega) \quad \text{with} \quad \Phi^{-1}A=A \quad \Longrightarrow \quad \mathbb{P}(A)=0 \quad \text{or} \quad \mathbb{P}(A)=1\,. \tag{4.33}$$

Note that invariance of sets has been defined in accordance with the stationarity of the measure, namely in terms of the pre-image. If Φ is bijective we can equivalently define $\Phi A=A$ as an invariant set.

Birkhoff's ergodic theorem asserts that, for an ergodic system and for any (measurable[19]) f

$$\lim_{N\to\infty}\frac{1}{N}\sum_{n=0}^{N} f(\Phi_n\omega)=\int f(\omega)\,\mathrm{d}\mathbb{P}(\omega) \quad \text{for } \mathbb{P}\text{-almost all } \omega\,.$$

It is interesting to note that the proof has two parts. In the difficult part, only stationarity of the measure is required. From that alone, it follows that the limit on the

[19] See Sect. 4.3.

left-hand side exists for $\mathbb{P}$-almost all ω. The ergodic hypothesis enters only to determine the limit. The set of ω for which the limit exists is clearly invariant. If the limit takes different values on that set, then the pre-images of the values likewise define invariant sets, but by ergodicity, invariant sets have measure zero or one, i.e., there can be only one pre-image up to sets of measure zero, whence

$$\lim_{N\to\infty}\frac{1}{N}\sum_{n=0}^{N} f(\Phi_n\omega) = c \quad \text{for } \mathbb{P}\text{-almost all } \omega \,.$$

Integrating this equation with respect to $\mathbb{P}$, exchanging the limit on the left with the integration, we see by stationarity that $c = \int f \,\mathrm{d}\mathbb{P}(\omega)$.

Ergodicity gives uniqueness of stationary measures in the following way. Suppose $\big(\Omega, \mathscr{B}(\Omega), \Phi, \mathbb{P}\big)$ and $\big(\Omega, \mathscr{B}(\Omega), \Phi, \mathbb{Q}\big)$ are ergodic, i.e., in particular $\mathbb{P}$ and $\mathbb{Q}$ are stationary (like microcanonical and canonical ensembles). Then $\mathbb{Q} = \mathbb{P}$ or $\mathbb{Q}$ is concentrated on the zero measure sets of $\mathbb{P}$, and vice versa. To see this, suppose that $\mathbb{Q} \neq \mathbb{P}$, i.e., there exists a set B for which $\mathbb{Q}(B) \neq \mathbb{P}(B)$. Setting $f(\omega) = \chi_B(\omega)$, the ergodic theorem implies that

$$\lim_{N\to\infty}\frac{1}{N}\sum_{n=0}^{N} f(\Phi_n\omega) = \int f(\omega)\,\mathrm{d}\mathbb{P}(\omega) = \mathbb{P}(B) \quad \text{for } \mathbb{P}\text{-almost all } \omega \,,$$

and

$$\lim_{N\to\infty}\frac{1}{N}\sum_{n=0}^{N} f(\Phi_n\omega) = \int f(\omega)\,\mathrm{d}\mathbb{Q}(\omega) = \mathbb{Q}(B) \quad \text{for } \mathbb{Q}\text{-almost all } \omega \,,$$

but the right-hand sides are not equal. We must conclude that $\mathbb{Q}$-almost all ω is a null set of $\mathbb{P}$. The moral is that ergodic stationary measures are special among all stationary measures.

Let us go a little further. Up until now we have talked about a process in time, i.e., we sample at successive discrete times. But we can also think of sampling in space, e.g., measuring the particle number in an ideal gas in various spatially separated cells in a gas container. Let $f_i(\omega)$ be the particle configuration in cell i. Instead of moving from cell to cell, we can also stay in one cell and shift the entire configuration of all gas molecules (where it is best to think of an infinitely extended gas, see Fig. 4.11). We obtain a shift $\Phi : \Omega \longrightarrow \Omega$ (for simplicity, we consider cells which arise from dividing the x-axis, so that we need only shift in the x direction). If f denotes the configuration in cell 0, then $f(\Phi_i\omega)$ is the configuration in cell i.

It is intuitively clear that $\mathbb{P}_E$ is shift invariant in this sense, since that is how we imagine a homogeneous state to look. This gives rise to a dynamical system, but this time with a spatial translation flow Φ. However, our notions were already abstract, and everything transfers to spatial ergodicity and spatial ensembles. That abstraction is useful and pays off. Recall our discussion of the mixing partitions [compare (4.2)] on the Galton board. This characterizes stochastic independence and constitutes the condition under which the law of large numbers holds "trivially"

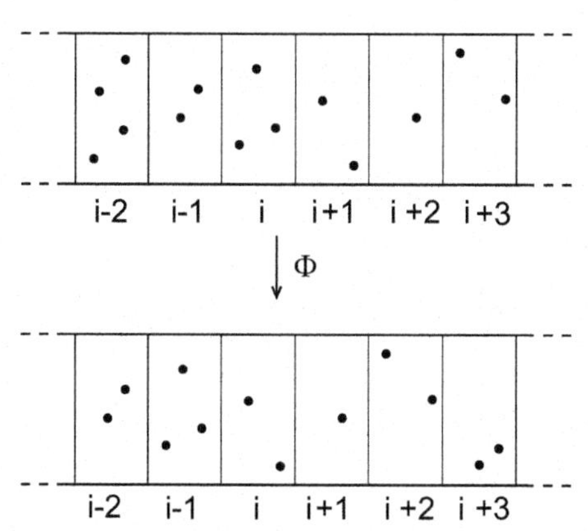

Fig. 4.11 Spatial shift

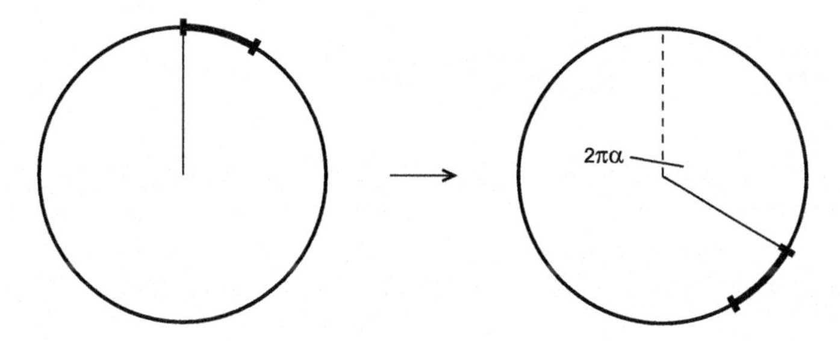

Fig. 4.12 Rotation on the circle

[compare (4.53)]. On the other hand, to see where ergodicity stands, we remark that the typical ergodic system is the following (see Fig. 4.12). Consider the circle $\Omega = \{\omega \in \mathbb{R}^2, |\omega| = 1\} \cong [0,1)$ with $\omega = (\cos 2\pi x, \sin 2\pi x)$, and

$$\begin{aligned} \Phi : [0,1) &\longrightarrow [0,1) \\ x &\longmapsto x + \alpha|_{\mathrm{mod}\,1} \,. \end{aligned}$$

Let $\mathbb{P} = \lambda = |\cdot|$ be the usual length. If α is rational, the dynamical system

$$\Big([0,1), \mathscr{B}\big([0,1)\big), \Phi, \lambda\Big)$$

is not ergodic (it is periodic), but if α is irrational, it is ergodic. This is most easily seen by using the equivalent definition of ergodicity: $\big(\Omega, \mathscr{B}(\Omega), \Phi, \mathbb{P}\big)$ is ergodic if every measurable bounded invariant function f on Ω is $\mathbb{P}$-almost everywhere constant, i.e.,

$$f \circ \Phi = f \quad \Longrightarrow \quad f = \text{const.} \quad \mathbb{P}\text{-almost everywhere}\,.$$

Let us decompose f into a Fourier series

$$f(x) = \sum_{n=-\infty}^{\infty} c_n \mathrm{e}^{inx2\pi} ,$$

with

$$\begin{aligned}(f \circ \Phi)(x) &= f\big((x+\alpha)|_{\mathrm{mod}\,1}\big) \\ &= \sum_{n=-\infty}^{\infty} c_n \mathrm{e}^{\mathrm{i}n2\pi(x+\alpha)} = \sum_{n=-\infty}^{\infty} c_n \mathrm{e}^{\mathrm{i}n2\pi\alpha} \mathrm{e}^{\mathrm{i}n2\pi x} \stackrel{!}{=} \sum_{n=-\infty}^{\infty} c_n \mathrm{e}^{\mathrm{i}n2\pi x} .\end{aligned}$$

Therefore

$$\sum_{n=-\infty}^{\infty} c_n \left(1 - \mathrm{e}^{\mathrm{i}n2\pi\alpha}\right) \mathrm{e}^{\mathrm{i}n2\pi x} = 0 ,$$

and also

$$c_n \left(1 - \mathrm{e}^{\mathrm{i}n2\pi\alpha}\right) = 0 ,$$

whence $c_n = 0$ or $1 - \mathrm{e}^{\mathrm{i}n2\pi\alpha} = 0$. But the second option cannot hold for irrational α, unless $n = 0$. If α is rational, for example, $\alpha = p/q$, choose $c_q = 1$ and all else null. f is then invariant and not constant.

Now let us look once more at Fig. 4.9, where trajectories in phase space expand so that a tiny concentrated set gets distorted and convoluted under the time evolution. This looks so different from the rather dull rotation on circles. Gibbs introduced the idea that equilibrium is "natural" because of a mixing process which takes place in phase space. He used the analogy of a drop of ink in water, which is mixed up very quickly by stirring the water and thus turning the water quickly light blue throughout. Likewise – or so the analogy went in the minds of many people – any region in phase space, no matter how small in volume, will spread out under the dynamics after a suitably long time to points that fill out a region that is distorted and convoluted in such a way as to be spread more or less uniformly throughout the corresponding energy surface. Then, for any reasonable function f, the uniform average of f over the energy surface and over the time-evolved set are more or less the same.

But the analogy is off target, since no one stirs in phase space. The system trajectories wander about and explore the relevant regimes of phase space for purely "entropic" reasons. The rough timescale over which trajectories mix all over phase space is again given by the Poincaré recurrence time. But since the phase space region of equilibrium is overwhelmingly large, a (typical) non-equilibrium phase point can enter rather quickly into the equilibrium region, so that the mixing time scale is typically reasonably short. That time scale governs the behavior of phase point functions like the empirical density. If one is interested in the time scale on which Ma1 goes to Ma3, then that is determined by (conditionally) typical trajectories. These can, for example, be read off from the Boltzmann equation for dilute gases, which describes the physically relevant transition to equilibrium. The concept of mixing in the sense of Gibbs, which concerns the evolution of a non-equilibrium phase space distribution to let us say the microcanonical ensemble does not play any role in that. In chap. 5 we shall discuss a much simpler equation, the heat equation of Brownian motion, which is, however, exactly in the spirit of Boltzmann. For further elaboration, see [11, 12], and also the classic [5], as well as the delightful essay entitled *The Pernicious Influence of Mathematics on Science*, by Jacob Schwartz in [13].

For the sake of completeness, we now give the definition of mixing:

Definition 4.2. $\big(\Omega, \mathscr{B}(\Omega), \Phi, \mathbb{P}\big)$ is called mixing if, for integrable functions f and g,

$$\int f(\Phi_n\omega)g(\omega)\,\mathrm{d}\mathbb{P}(\omega) \quad \stackrel{n\longrightarrow\infty}{\longrightarrow} \quad \int f(\omega)\,\mathrm{d}\mathbb{P}(\omega)\int g(\omega)\,\mathrm{d}\mathbb{P}(\omega)\,. \tag{4.34}$$

For sets A (water in a glass) and B (B for blue ink),

$$\mathbb{P}(\underbrace{A\cap\Phi_n B}_{\substack{\text{part of blue}\\ \text{in set } A}}) \quad \stackrel{n\longrightarrow\infty}{\longrightarrow} \quad \underbrace{\mathbb{P}(A)\mathbb{P}(B)}_{\substack{\text{proportional to}\\ \text{size of } A\\ \text{All lightblue}}}\,. \tag{4.35}$$

Let us see why mixing implies ergodicity. Let A be invariant, i.e., $\Phi(A) = A$. Then

$$\mathbb{P}(\Phi_n A\cap A) = \mathbb{P}(A\cap A) = \mathbb{P}(A)\,,$$

and by mixing,

$$\mathbb{P}(\Phi_n A\cap A) \quad \stackrel{n\longrightarrow\infty}{\longrightarrow} \quad \mathbb{P}(A)^2\,,$$

that is, $\mathbb{P}(A) = \mathbb{P}(A)^2$, so $\mathbb{P}(A) = 0$ or $\mathbb{P}(A) = 1$.

Mixing also implies the "transition to equilibrium" of distributions. To see this, let $\big(\Omega, \mathscr{B}(\Omega), \Phi, \mathbb{P}\big)$ be a mixing system and $\mathbb{Q}$ another measure, but which has a density with respect to $\mathbb{P}$, say $\rho(\omega) \geq 0$, with

$$\int \rho\,\mathrm{d}\mathbb{P}(\omega) = 1$$

and

$$\int f(\omega)\,\mathrm{d}\mathbb{Q}(\omega) = \int f(\omega)\rho(\omega)\,\mathrm{d}\mathbb{P}(\omega)\,. \tag{4.36}$$

Imagine $\mathbb{Q}$ being a "non-equilibrium" distribution. Then

$$\int f(\Phi_n\omega)\,\mathrm{d}\mathbb{Q}(\omega) \quad \stackrel{n\longrightarrow\infty}{\longrightarrow} \quad \int f(\omega)\,\mathrm{d}\mathbb{P}(\omega)\,,$$

which comes from (4.34) and (4.36), and

$$\int f(\Phi_n\omega)\,\mathrm{d}\mathbb{Q}(\omega) = \int f(\Phi_n\omega)\rho(\omega)\,\mathrm{d}\mathbb{P}(\omega)$$
$$\overset{n\longrightarrow\infty}{\longrightarrow} \int f(\omega)\,\mathrm{d}\mathbb{P}(\omega) \int \rho(\omega)\,\mathrm{d}\mathbb{P}(\omega)\,.$$

4.3 Probability Theory

Once (in Euclid's day) axioms in mathematics were self-evident assertions. Now axioms are either merely useful definitions or they are the completion of a long history of technical advances and thoughts. The axioms of probability theory are like that. The axioms, as they were formulated by Kolmogoroff in 1933, are the result of a long and highly technical development. Behind the axioms lies typicality and coarse-graining of points of a continuum. Underlying the axioms is Boltzmann's kinetic gas theory, Lebesgue's construction of a measure on subsets of the real numbers, the Rademacher functions, and finally Hilbert's quest for an axiomatization of the technical use of probability, which Hilbert – intrigued by Boltzmann's advances in kinetic gas theory – formulated in the year 1900 in his famous list of 23 problems as the sixth problem.

But of course none of this history is spelt out in the axioms. The axioms formulate the structure within which typicality arguments and coarse-graining can be phrased. Seen this way the prototype of a measure of typicality is the Lebesgue measure. This chapter is very much influenced by the writings of Mark Kac, in particular by his marvellous booklet [14].

4.3.1 Lebesgue Measure and Coarse-Graining

The Lebesgue integral is a center of mass integration. Let λ be the volume measure of subsets of the real numbers, i.e., the natural generalization of length of an interval. Then the Lebesgue integral is defined[20] by (see Fig. 4.13)

$$\int f\mathrm{d}\lambda \approx \sum_i f_i\lambda(U_i) = \sum_{i=0}^{\infty} \frac{i}{n}\lambda\left(f^{-1}\left[\frac{i}{n},\frac{i+1}{n}\right]\right)\,. \tag{4.37}$$

The main point is that the Lebesgue integral starts with a partition of the y-axis, and weights the values of f with the measure of the pre-image. Now, here is the catch. Suppose f is fluctuating wildly, beyond any stretch of the imagination. Then the pre-images of values of f can be awful sets – not at all obtainable in any constructive way from intervals. The question is: Do all subsets have a measure (a "length"), as

[20] First for non-negative functions, and then writing $f = f^+ - f^-$ with $f^\pm$ positive, one extends the definition to arbitrary (measurable) functions.

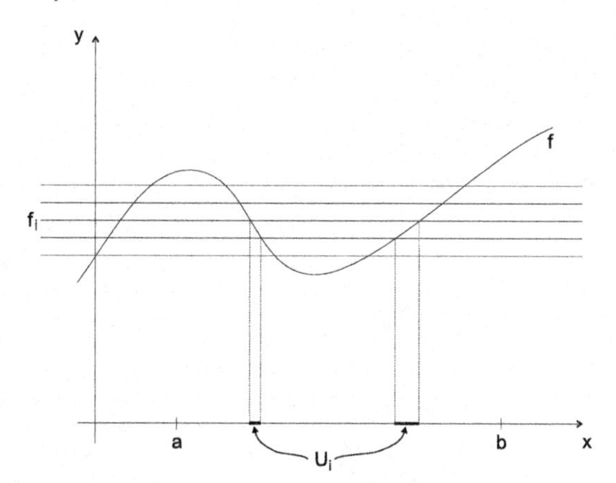

Fig. 4.13 In contrast to Riemann integration, where the x-axis is partitioned, in Lebesgue integration the y-axis is partitioned into cells of length $1/n$, and the values $f_i = i/n$ are weighted with the measure of the pre-images $U_i = f^{-1}[i/n,(i+1)/n]$

we intuitively think they would have? The answer is negative. There are sets which are not measurable.

This is an easy exercise in analysis. It forces us (it forced the mathematicians of the early 20th century) to introduce a family of measurable sets. One should think of these sets as "constructible" from intervals, and that family is called a Borel algebra $\mathscr{B}(\mathbb{R})$. This is a σ-algebra.[21] So f must be such that the pre-images of the intervals on the y-axis are measurable. Such f are called measurable. It is easy to make the above idea of integration precise. It is a rather dull exercise to establish that all sets in $\mathscr{B}(\mathbb{R})$ are measurable. The important fact is in any case that the intuitive notion of size of a set can be realized on the Borel algebra and is called Lebesgue measure.[22] The Lebesgue integral is the "measure" integral obtained with the Lebesgue measure.

The relevant structure for generalizing the notions of set size and integration is $\big(\mathbb{R},\mathscr{B}(\mathbb{R}),\lambda\big)$, or for a finite measure space with the measure normalized, $\big([0,1],\mathscr{B}([0,1]),\lambda\big)$. The generalization due to Kolmogoroff (the founder of axiomatic probability theory) is $\big(\Omega,\mathscr{B}(\Omega),\mathbb{P}\big)$, where in analogy with $\mathscr{B}([0,1])$ being generated by intervals on which the Lebesgue measure is "clear", the σ-algebra $\mathscr{B}(\Omega)$ is now generated by some family of subsets of some "arbitrary" phase space Ω on which the measure $\mathbb{P}$ is somehow clear and can be extended (in analogy with

---

[21] A σ-algebra $\mathscr{A}(\Omega)$ obeys:

(i) $\Omega \in \mathscr{A}$,
(ii) $A \in \mathscr{A} \Longrightarrow \bar{A} \in \mathscr{A}$,
(iii) $(A_i)_{i\in\mathbb{N}} \subset \mathscr{A} \Longrightarrow \bigcup A_i \in \mathscr{A}$.

[22] There is a mathematical distinction between measurable and Borel-measurable. That distinction comes from the fact that the Lebesgue measure can live on a larger algebra than the Borel algebra. However, the enlargement consists of sets of Lebesgue measure zero. So forget about that.

the Lebesgue measure) to the whole σ-algebra. But why is this triple the abstract skeleton of probability theory?

Recall the coarse-graining description of our Galton board machine in Fig. 4.5. The "fundamental" typicality was defined by the canonical measure on the phase space Ω of all the balls in the container. This measure is mapped by the coarse-graining function $Y^p : \Omega \rightarrow \{0,1,\ldots,n\}$ given by the sum $Y^p(\omega) = \sum_{k=1}^n X_k^p(\omega)$ to a discrete measure $\mathbb{P}'$ on the *new* phase space $\Omega' = \{0,1,\ldots,n\}$, namely the image space of Y^p. So let us forget about the "true" space and let us do our analysis on the image space $\big(\Omega',\mathscr{B}(\Omega'),\mathbb{P}'\big)$. Then the general idea is this. Let $\big(\Omega,\mathscr{B}(\Omega),\mathbb{P}\big)$ be a given *probability space*[23] and suppose we are interested only in a coarse-grained description of Ω by a coarse-graining function which is (very) many to one:

$$X : \Omega \longrightarrow \Omega' .$$

Through X, the image space becomes a probability space

$$\big(\Omega',\mathscr{B}(\Omega'),\mathbb{P}'\big) ,$$

with *image measure* $\mathbb{P}' = \mathbb{P}_X$, given by

$$\mathbb{P}_X(A) = \mathbb{P}\big(X^{-1}(A)\big) . \tag{4.38}$$

But this requires, and it is the only requirement a coarse-graining function must fulfill,[24] *measurability*:

$$X^{-1}(M) \in \mathscr{B}(\Omega) , \tag{4.39}$$

for all $M \in \mathscr{B}(\Omega')$. This allows two things. The coarse-graining function can be integrated with $\mathbb{P}$ (by analogy with the Lebesgue integral), and it can transport the "fundamental" probability space to the image space $\big(\Omega',\mathscr{B}(\Omega'),\mathbb{P}'\big)$ on which X, the coarse-graining function, is now the identity.

Such coarse-graining functions are called *random variables*, a horrible and misleading terminology, because there is nothing random about them. In Euclidean geometry the axioms, if one wishes to be axiomatic about it, are no big deal. They are obvious and readily implemented in proofs. But this is different in probability theory. The axioms are abstractions, ready to be used, but there is a price to pay. The price is that ontology sinks into oblivion. One easily forgets that there is a deeper theory of which we only describe the image. Coin-tossing is trivial when described by independent $1/2$–$1/2$ probabilities. But how about the true physical process?

The *expectation* value $\mathbb{E}(X)$ is

[23] That is just a name.

[24] Because non-measurable sets exist!

$$\begin{aligned}\mathbb{E}(X) &= \int X(\omega)\,\mathbb{P}(\mathrm{d}\omega) \\ &\approx \sum_{k=0}^{\infty} \frac{k}{n}\mathbb{P}\left(X^{-1}\left[\frac{k}{n},\frac{k+1}{n}\right]\right) \\ &= \sum_{k=0}^{\infty} \frac{k}{n}\mathbb{P}_X\left(\left[\frac{k}{n},\frac{k+1}{n}\right]\right) \\ &\approx \int \omega' \mathbb{P}_X(\mathrm{d}\omega')\ ,\end{aligned}$$

with (4.38). We shall see later that the expectation value predicts the empirical mean (hence the name) as the image probability predicts the relative frequencies, because the latter is $\mathbb{E}(\chi_A(x)) = \mathbb{P}_X(A)$, with $\chi_A(x)$ the indicator function of the set A.

X can be vector-valued: $\mathbf{X} = (X_1,\ldots,X_k)$, with $X_i : \Omega \longrightarrow \tilde{\Omega}$, $i = 1,\ldots,k$. Then $\mathbb{P}_{\mathbf{X}} = \mathbb{P}_{(X_1,\ldots,X_k)}$ is a measure on

$$\Omega' = \underbrace{\tilde{\Omega}\times\tilde{\Omega}\times\ldots\times\tilde{\Omega}}_{k\text{ times}}\ ,$$

and

$$\mathbb{P}_{(X_1,\ldots,X_k)}(M_1\times\ldots\times M_k) = \mathbb{P}\Big(\big\{X_1^{-1}(M_1)\cap X_2^{-1}(M_2)\cap\ldots\cap X_k^{-1}(M_k)\big\}\Big)\ . \tag{4.40}$$

$\mathbb{P}_{(X_1,\ldots,X_k)}$ is called the joint distribution of the X_i. Setting some of the $M_i = \tilde{\Omega}$, so that only $X_{i_1},\ldots,X_{i_l}$ remain specified, one calls the resulting distribution $\mathbb{P}_{X_{i_1},\ldots,X_{i_l}}$ a l-point distribution and $\mathbb{P}_{X_i}$ the marginal distribution.[25]

Let us stick with $\big([0,1),\mathscr{B}([0,1)),\lambda\big)$ for the moment and consider the coarse-graining Rademacher functions $r_k(x)$, i.e., the numbers in the binary representation of $x\in[0,1)$ (see Fig. 4.14). Take $r_1 : [0,1) \to \{0,1\}$, which is clearly coarse-graining, the image space being $\big(\{0,1\},\mathscr{P}(\{0,1\}),\mathbb{P}_{r_1}\big)$, where $\mathscr{P}(\{0,1\})$ is the power set of $\{0,1\}$ and

$$\mathbb{P}_{r_1}(0) = \lambda(\{r_1^{-1}(0)\}) = \lambda\left(\left[0,\frac{1}{2}\right)\right) = \frac{1}{2}\ ,$$

$$\mathbb{P}_{r_1}(1) = \lambda(\{r_1^{-1}(1)\}) = \lambda\left(\left[\frac{1}{2},1\right)\right) = \frac{1}{2}\ .$$

This is the probability space of the single coin toss: head = 0, tail = 1 and "a priori probabilities" are each 1/2. Now go on with $r_1(x),\ldots,r_n(x),\ldots$. It is clear from the graphs that, for any choice $(\delta_1,\delta_2,\delta_3)\in\{0,1\}^3$:

$$\mathbb{P}_{(r_1,r_2,r_3)}(\delta_1,\delta_2,\delta_3) = \lambda\left(r_1^{-1}(\delta_1)\cap r_2^{-1}(\delta_2)\cap r_3^{-1}(\delta_3)\right)\ ,$$

[25] Note that $X_i^{-1}(\tilde{\Omega}) = \Omega$.

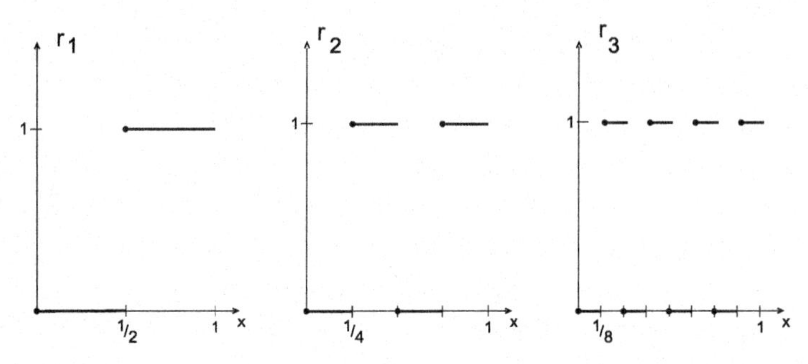

Fig. 4.14 The prototype of independent random variables as coarse-graining functions. Rademacher functions r_k given by the binary expansion of $x \in [0,1)$: $x = \sum_{k=1}^{\infty} r_k 2^{-k}$

and

$$\lambda\Big(\{x|r_1(x)=\delta_1\}\cap\{x|r_2(x)=\delta_2\}\cap\{x|r_3(x)=\delta_3\}\Big) = \frac{1}{8} = \left(\frac{1}{2}\right)^3$$
$$= \mathbb{P}_{r_1}(\delta_1)\mathbb{P}_{r_2}(\delta_2)\mathbb{P}_{r_3}(\delta_3)\,.$$

Likewise

$$\mathbb{P}_{(r_{i_1},\dots,r_{i_n})}(\delta_{i_1},\dots,\delta_{i_n}) = \left(\frac{1}{2}\right)^n = \prod_{k=1}^{n} \mathbb{P}_{r_{i_k}}(\delta_{i_k})\,, \tag{4.41}$$

and this property defines *independence of the random variables* $r_{i_1},\dots,r_{i_n}$. In general, independent random variables $X_1,\dots,X_n$ are defined by their joint distribution being a product:

$$\mathbb{P}_{(X_1,\dots,X_k)} = \prod_{i=1}^{k} \mathbb{P}_{X_i}\,.$$

It follows that the expectation value of products of functions $f_i = g_i(X_i)$ is the product of the expectation values:

$$\mathbb{E}(f_{i_1}\cdots f_{i_n}) = \mathbb{E}(f_{i_1})\cdots\mathbb{E}(f_{i_n})\,. \tag{4.42}$$

In particular,

$$\int \prod_{k=1}^{n} f_{i_k}\mathrm{d}\mathbb{P}(\omega) = \prod_{k=1}^{n}\int f_{i_k}\mathrm{d}\mathbb{P}(\omega)\,, \tag{4.43}$$

a property which practically never holds. (Choose two functions and see whether the integral, e.g., the usual Riemann integral, over their product equals the product of the integrals.) One may recall also (4.34).

The probability space generated by $(r_1,\dots,r_n)$ from $\big([0,1),\mathscr{B}([0,1)),\lambda\big)$ is the image space

$$\left(\{0,1\}^n,\mathscr{P}(\{0,1\}^n),\prod_{i=1}^{n}\mathbb{P}_{r_i}\right),$$

which is a perfect model for n-fold coin tossing. What is the "probability" that in n tosses one has tails exactly l times? On the *image space*, this is a simple combinatorial exercise. There are $\binom{n}{l}$ n-tuples with 1 at exactly l places and everywhere else zeros and each n-tuple has a priori probability $1/2^n$, and we add these to get the answer (adding the measure of disjoint sets to get the measure of the union set is intuitive and features as an axiom). But we can do this differently, taking the "fundamental" space seriously. And in the hope of gaining a better grasp of the difference, we go through the tedious complication of computing from "first principles". (This has been copied from [14].)

The problem is to compute the measure of the set (heads=0, tails=1):

$$A_l=\left\{x\,\middle|\,\sum_{k=1}^{n}r_k(x)=l\right\},\qquad \text{i.e.,}\quad \lambda(A_l)=\int\chi_{A_l}(x)\mathrm{d}x\,.$$

There is no need to be fancy here with the Lebesgue measure notation. Lebesgue integration is simply Riemann integration here. The trick is now to recognize that

$$\chi_{A_l}(x)=\frac{1}{2\pi}\int_0^{2\pi}\mathrm{d}y\,\mathrm{e}^{\mathrm{i}\left[\sum_{k=1}^{n}r_k(x)-l\right]y}\,.$$

We change the order of integration,

$$\lambda(A_l)=\frac{1}{2\pi}\int_0^{2\pi}\mathrm{d}y\int_0^1\mathrm{d}x\,\mathrm{e}^{\mathrm{i}\left[\sum_{k=1}^{n}r_k(x)-l\right]y}=\frac{1}{2\pi}\int_0^{2\pi}\mathrm{d}y\,\mathrm{e}^{-\mathrm{i}ly}\int_0^1\mathrm{d}x\prod_{k=1}^{n}\mathrm{e}^{\mathrm{i}r_k(x)y}\,,$$

and by virtue of the *very special property* (4.41) in the form (4.43), this becomes

$$\frac{1}{2\pi}\int_0^{2\pi}\mathrm{d}y\,\mathrm{e}^{-\mathrm{i}ly}\prod_{k=1}^{n}\int_0^1\mathrm{d}x\,\mathrm{e}^{\mathrm{i}r_k(x)y}\,.$$

But

$$\int_0^1\mathrm{d}x\,\mathrm{e}^{\mathrm{i}r_k(x)y}=\frac{1}{2}+\frac{1}{2}\mathrm{e}^{\mathrm{i}y}\,,$$

and we get the answer

$$\lambda(A_l)=\left(\frac{1}{2}\right)^n\frac{1}{2\pi}\int_0^{2\pi}\mathrm{d}y\,\mathrm{e}^{-\mathrm{i}ly}\left(1+\mathrm{e}^{\mathrm{i}y}\right)^n=\left(\frac{1}{2}\right)^n\binom{n}{l}\,,$$

with or without combinatorics, i.e., combinatorics has to come in anyway.

4.3.2 The Law of Large Numbers

What does it mean to say that the number

$$\left(\frac{1}{2}\right)^n \binom{n}{l}$$

is the probability for exactly l heads in n tosses? According to Boltzmann it means that typically, in a large number N of trials of n coin tosses, one finds that the relative fraction of outcomes with exactly l heads in n tosses is close to the above number. In other words, it is the theoretical prediction of the empirical distribution.

Let us go through a toy model for such a prediction concerning the probability of a single coin toss. The idea is to create an ensemble, tossing n coins simultaneously or tossing one coin n times. We use the Rademacher coarse-graining function r_1 as the map from the fundamental phase space to the outcome. We can even produce a dynamical picture by invoking the Bernoulli map T (see Fig. 4.10) with the property (4.31), which is in fact isomorphic to the shift map analogous to Fig. 4.11. To see this, we generalize to the infinite family $(r_1, \ldots, r_n, \ldots)$ of Rademacher functions. This is nice because it gives us the chance to explain the construction of more abstract probability spaces.

We need to construct the image Borel algebra $\mathscr{B}(\{0,1\}^{\mathbb{N}})$ over the image space of infinite 0–1 sequences. We know the "infinite product" measure $\mathbb{P}_{\mathscr{B}}$ on cylinder sets (playing the role of the intervals). These are sets specified by finitely many coordinates, while all other coordinates are free, like a cylinder extending freely over a base:

$$Z_{i_1,\ldots,i_n}(\delta_1,\ldots,\delta_n) = \left\{\omega \in \{0,1\}^{\mathbb{N}},\ \omega = (\ldots,\underbrace{\delta_1}_{i_1\text{th place}},\ldots,\underbrace{\delta_n}_{i_n\text{th place}},\ldots)\right\}.$$

This means that, in the infinite sequences, the i_1th to i_nth places are determined and all other places are free. The measure of the cylinder set is defined by

$$\mathbb{P}_{\mathscr{B}}\big(Z_{i_1,\ldots,i_n}(\delta_1,\ldots,\delta_n)\big) := \mathbb{P}_{(r_{i_1},\ldots,r_{i_n})}(\delta_1,\ldots,\delta_n) = \left(\frac{1}{2}\right)^n, \tag{4.44}$$

the product measure. The Borel algebra $\mathscr{B}(\{0,1\}^{\mathbb{N}})$ is now generated by cylinder sets $Z_{\ldots}(\ldots)$. The measure $\mathbb{P}_{\mathscr{B}}$, fixed on cylinder sets, extends to the algebra $\mathscr{B}(\{0,1\}^{\mathbb{N}})$ and is a product measure called *Bernoulli measure*. The shift Φ_s acts on the sequence space $\Omega = \{0,1\}^{\mathbb{N}}$, shifting to the left and cutting away entries left of the zero. We see immediately on cylinder sets that the measure is stationary with respect to the shift

$$\begin{aligned}\mathbb{P}_{\mathscr{B}}\big((\Phi_s)^{-1}Z_{i_1,\ldots,i_n}(\delta_1,\ldots,\delta_n)\big) &= \mathbb{P}_{\mathscr{B}}\big(Z_{i_1+1,\ldots,i_n+1}(\delta_1,\ldots,\delta_n)\big) = \left(\frac{1}{2}\right)^n \\ &= \mathbb{P}_{\mathscr{B}}\big(Z_{i_1,\ldots,i_n}(\delta_1,\ldots,\delta_n)\big)\,, \end{aligned} \tag{4.45}$$

and we conclude that the measure is invariant. It comes as no surprise that we now have two isomorphic (via binary representation) dynamical systems. The one on the left we call fundamental, while the one on the right we think of as being the description on the coarse-grained level:[26]

$$\Big([0,1),\mathscr{B}([0,1)),T,\lambda\Big) \cong \Big(\{0,1\}^{\mathbb{N}},\mathscr{B}(\{0,1\}^{\mathbb{N}}),\Phi_s,\mathbb{P}_{\mathscr{B}}\Big) \tag{4.46}$$

and

$$\begin{aligned} r_{i+1} &\cong k_{i+1}\text{th coordinate of}\,\omega \in \{0,1\}^{\mathbb{N}} \\ &= r_1(T^i x) \\ &= k_1(\Phi_s^i \omega)\,. \end{aligned} \tag{4.47}$$

We can interpret the random variables r_i or k_i as arising dynamically from T or from the shift Φ_s.

Now take (4.46) as the fundamental model of a coin-tossing experiment, with $r_1\big(T^k(x)\big)$ resulting from the kth toss. Suppose that in the first n tosses we obtained results $\delta_1,\ldots,\delta_n$. What does the theory predict for the $(n+1)$th toss? Put differently, suppose we know $r_1\big(T^i(x)\big)$, $i=1,\ldots,n$. What do we learn from that about x, so that we can be smart about our guess for the next result $r_1\big(T^{n+1}(x)\big)$? Nothing! No matter how often we toss, in this theory, we remain absolutely ignorant about future results.[27] In this sense a *typical* $x \in [0,1]$ represents *absolute uncertainty*.

Now here is a straightforward generalization of the foregoing:

Definition 4.3. A sequence $(X_i)_{i\in\mathbb{Z}}$ ($\mathbb{Z}$ is more comfortable than $\mathbb{N}$) of identically distributed independent random variables on $\big(\Omega,\mathscr{B}(\Omega),\mathbb{P}\big)$ is called a Bernoulli sequence. The X_i are identically distributed if $\mathbb{P}_{X_i}=\mathbb{P}_{X_0}$.

Every Bernoulli sequence represents a dynamical system. Let $X_0 \in E$. Construct

$$\big(E^{\mathbb{Z}},\mathscr{B}(E^{\mathbb{Z}}),\Phi_s,\mathbb{P}_{\mathscr{B}}\big)\,,$$

with the Bernoulli shift

$$\Phi_s : E^{\mathbb{Z}} \longrightarrow E^{\mathbb{Z}}\,,$$

a shift of one place to the left, and

$$\mathbb{P}_{\mathscr{B}} = \prod_{i\in\mathbb{Z}} \mathbb{P}_{X_i}\,.$$

On the new space X_i becomes $k_i\big((e_n)_{n\in\mathbb{Z}}\big)$, where $e_n \in E$ for all $n \in \mathbb{Z}$, or equivalently $k_0\Big(\Phi_s^i\big((e_n)_{n\in\mathbb{Z}}\big)\Big)$. With $(X_i)_{i\in\mathbb{N}}$, then $\big(f(X_i)\big)_{i\in\mathbb{N}}$ is also a Bernoulli sequence.

[26] There is of course the one-to-one map of the binary expansion between the two phase spaces, since the phase space on the right contains infinitely extended sequences.

[27] This sounds almost like the irreducible probability in quantum mechanics, does it not?

Finally we can arrive at the empirical distribution $\rho_{\text{emp}}^{(N)}$, the object of interest our mathematical theory of probability should predict. The prediction will be of Boltzmann-type certainty, i.e., not certain but typical. The overwhelming majority of ωs will agree upon what the empirical distribution will be. The *empirical distribution* or relative frequency is a random variable, a density for that matter. For the Rademacher functions,

$$\begin{aligned}\rho_{\text{emp}}^{(N)}(y,x) &= \frac{1}{N}\sum_{k=1}^{N}\delta\big(r_k(x)-y\big)\\ &= \frac{1}{N}\sum_{k=0}^{N-1}\delta\Big(r_1\big(T^k(x)\big)-y\Big) \qquad \text{[by (4.47)]}\,,\end{aligned}$$

or equivalently on the image space,

$$\rho_{\text{emp}}^{(N)}(y,\omega) = \frac{1}{N}\sum_{k=0}^{N-1}\delta\Big(k_1\big(\Phi_s^k(\omega)\big)-y\Big) \qquad \text{[by (4.47)]}\,.$$

If one feels uncomfortable with the δ-density, one can simply integrate a function of interest, e.g., the indicator function of a set (which yields the relative frequency of ending up in that set) or simply the identity, yielding the empirical mean of the random variable:

$$\rho_{\text{emp}}^{(N)}(f,x) := \int f(y)\rho_{\text{emp}}^{(N)}(y,x)\mathrm{d}y \tag{4.48}$$

$$= \frac{1}{N}\sum_{k=1}^{N} f\big(r_k(x)\big) \tag{4.49}$$

$$= \frac{1}{N}\sum_{k=0}^{N-1} f\Big(r_1\big(T^k(x)\big)\Big) \tag{4.50}$$

$$= \frac{1}{N}\sum_{k=0}^{N-1} f\Big(k_1\big(\Phi_s^k(\omega)\big)\Big)\,. \tag{4.51}$$

This not only looks like, but actually is an ergodic mean [compare (4.32)], and since independence is stronger than ergodicity, we could conclude from Birkhoff's theorem that

$$\rho_{\text{emp}}^{(N)}(f) \overset{N\longrightarrow\infty}{\longrightarrow} \mathbb{E}\big(\rho_{\text{emp}}^{(N)}(f)\big) = \mathbb{E}\big(f(r_1)\big) = \frac{1}{2}f(1)+\frac{1}{2}f(0)\,, \tag{4.52}$$

for almost all x, i.e., the exceptional set will have Lebesgue measure zero, equivalently for $\mathbb{P}_{\mathscr{B}}$-almost all ω. We rephrase this for $f(y)=\chi_{\{1\}}(y)$, observing that r_k takes only values zero or one, in the form:

Theorem 4.2. *For all* $\varepsilon > 0$,

$$\lambda\left(\left\{x\in[0,1)\,\middle|\,\lim_{N\to\infty}\left|\frac{1}{N}\sum_{k=0}^{N} r_k(x)-\frac{1}{2}\right|>\varepsilon\right\}\right)=0\,.$$

This theorem, which has a rather simple proof when one knows a bit of Lebesgue integration theory, says that the typical number in the interval $[0,1)$ is a *normal number*, meaning that the binary expansion[28] yields a 0–1 sequence with relative frequency 1/2 of 1s. It is *the* fundamental theorem about numbers in the continuum and the Lebesgue measure. It marks the beginning of probability theory as a purely mathematical discipline, and it is moreover the prototype of the prediction for empirical densities we can only hope for in physically relevant situations.

We shall be content with a weaker assertion, namely where the limit is taken outside the measure. In mathematics this is then called the *weak law of large numbers*, while Theorem 4.2 is referred to as the strong law of large numbers. We prove the weak law in the abstract setting:

Theorem 4.3. *Let $(X_i)_{i\in\mathbb{N}}$ be a Bernoulli sequence of identically distributed random variables X_i on $\big(\Omega,\mathscr{B}(\Omega),\mathbb{P}\big)$. Let $\mathbb{E}(X_i^2)=\mathbb{E}(X_0^2)<\infty$. Then, for any $\varepsilon>0$,*

$$\mathbb{P}\left(\left\{\omega\left|\left|\frac{1}{N}\sum_{k=1}^{N}X_k(\omega)-\mathbb{E}(X_0)\right|>\varepsilon\right.\right\}\right)\leq\frac{1}{N\varepsilon^2}\left[\mathbb{E}(X_0^2)-\mathbb{E}(X_0)^2\right]. \tag{4.53}$$

Let us make a few remarks:

(i) For simplicity of notation and without loss of generality, we have used X_i instead of $f(X_i)$, where f is some function of interest.

(ii) Under the conditions assumed here (ergodicity is sufficient, we assume independence), the theoretical prediction for the empirical distribution is given by the theoretical expectation of the empirical distribution. Here, because of "stationarity" or identical distribution,

$$\mathbb{E}\left(\frac{1}{N}\sum_{k=1}^{N}X_k(\omega)\right)=\frac{1}{N}\sum_{k=1}^{N}\mathbb{E}(X_k)=\frac{1}{N}N\mathbb{E}(X_0)=\mathbb{E}(X_0)\,.$$

(iii) Taking the limit $N\longrightarrow\infty$, we obtain zero on the right-hand side.

(iv) What does this theorem say? The typical value for the empirical distribution is its theoretical expectation. The latter is commonly referred to as "probability".

Rephrasing this in a formula, the theorem asserts that, for a Bernoulli sequence $X_0,\ldots,X_N$,

$$\mathbb{P}\left(\left\{\omega\left|\left|\rho_{\mathrm{emp}}^{(N)}(f,\omega)-\int f(\omega)\mathrm{d}\mathbb{P}_{X_0}\right|>\varepsilon\right.\right\}\right)\overset{N\longrightarrow\infty}{\longrightarrow}0\,. \tag{4.54}$$

Proof. Start with the set whose measure is taken and rewrite it

[28] It is easy to see that this is true for every p-adic expansion.

$$A_\varepsilon^N = \left\{ \omega \middle| \left| \frac{1}{N} \sum_{k=1}^{N} X_k(\omega) - \mathbb{E}(X_0) \right| > \varepsilon \right\} \tag{4.55}$$

$$= \left\{ \omega \middle| \left[\frac{1}{N} \sum_{k=1}^{N} \left[X_k - \mathbb{E}(X_k) \right] \right]^2 > \varepsilon^2 \right\} . \tag{4.56}$$

Then observe that

$$\mathbb{P}(A_\varepsilon^N) = \mathbb{E}\big(\chi_{A_\varepsilon^N}\big) = \int \chi_{A_\varepsilon^N}(\omega)\,\mathrm{d}\mathbb{P}(\omega) .$$

It is always good to have an integral over a function, because that can be estimated by a bound on the function. In this case, we use a trivial upper bound, which obviously holds true in view of the above rewriting of the set A_ε^N:

$$\chi_{A_\varepsilon^N}(\omega) \leq \frac{\left[\frac{1}{N} \sum_{k=1}^{N} \left[X_k - \mathbb{E}(X_k) \right] \right]^2}{\varepsilon^2} .$$

Hence,

$$\begin{aligned}
\mathbb{P}(A_\varepsilon^N) &\leq \int \frac{\left[\frac{1}{N} \sum_{k=1}^{N} \left[X_k - \mathbb{E}(X_k) \right] \right]^2}{\varepsilon^2} \mathrm{d}\mathbb{P}(\omega) \\
&= \frac{1}{N^2\varepsilon^2} \mathbb{E}\left(\left[\sum_{k=1}^{N} X_k - \mathbb{E}(X_k) \right]^2 \right) \\
&= \frac{1}{N^2\varepsilon^2} \mathbb{E}\left(\sum_{k=1}^{N} \left[X_k - \mathbb{E}(X_k) \right]^2 \right) + \frac{1}{N^2\varepsilon^2} \mathbb{E}\left(\sum_{\substack{k,j=1 \\ k \neq j}}^{N} \left[X_k - \mathbb{E}(X_k) \right] \left[X_j - \mathbb{E}(X_j) \right] \right) ,
\end{aligned}$$

and using the fact that the variables are identically distributed and *independent* (4.42), we continue

$$\begin{aligned}
&= \frac{1}{N\varepsilon^2} \mathbb{E}\left(\left[X_0 - \mathbb{E}(X_0) \right]^2 \right) + \frac{1}{N^2\varepsilon^2} \sum_{\substack{k,j=1 \\ k \neq j}}^{N} \mathbb{E}\big(X_k - \mathbb{E}(X_k) \big) \mathbb{E}\big(X_j - \mathbb{E}(X_j) \big) \\
&= \frac{1}{N\varepsilon^2} \left[\mathbb{E}(X_0^2) - \mathbb{E}(X_0)^2 \right] .
\end{aligned}$$

We have also used the trivial fact that $\mathbb{E}\big(X - \mathbb{E}(X)\big) = 0$.

This computation is instructive and should be thoroughly mastered.[29]

[29] The first inequality is Tchebychev's. As trivial as the inequality is, it is very useful.

References

1. A. Bassi, G. Ghirardi: Phys. Rep. **379** (5–6), 257 (2003)
2. R. Tumulka: J. Stat. Phys. **125** (4), 825 (2006)
3. M. Smoluchowski: Die Naturwissenschaften **17**, 253 (1918)
4. H. Poincaré: *Wissenschaft und Hypothese* (Teubner Verlag, Leipzig, 1914)
5. P. Ehrenfest, T. Ehrenfest: *The Conceptual Foundations of the Statistical Approach in Mechanics*, English edn. (Dover Publications Inc., New York, 1990). Translated from the German by Michael J. Moravcsik, with a foreword by M. Kac and G.E. Uhlenbeck
6. O.E. Lanford III: In: *Dynamical Systems, Theory and Applications* (Recontres, Battelle Res. Inst., Seattle, Wash., 1974) (Springer, Berlin, 1975), pp. 1–111. Lecture Notes in Physics Vol. 38
7. H. Spohn: *Dynamics of Interacting Particles* (Springer Verlag, Heidelberg, 1991)
8. R. Feynman: *The Character of Physical Law* (MIT Press, Cambridge, 1992)
9. R. Penrose: *The Emperor's New Mind* (The Clarendon Press, Oxford University Press, New York, 1989). Concerning computers, minds, and the laws of physics, with a foreword by Martin Gardner
10. R. Penrose: *The Road to Reality* (Alfred A. Knopf Inc., New York, 2005). A complete guide to the laws of the universe
11. J. Bricmont: In: *The Flight from Science and Reason* (New York, 1995), Ann. New York Acad. Sci., Vol. 775 (New York Acad. Sci., New York, 1996) pp. 131–175
12. S. Goldstein: In: *Chance in Physics: Foundations and Perspectives*, Lecture Notes in Physics, Vol. 574, ed. by J. Bricmont, D. Dürr, M.C. Galavotti, G. Ghirardi, F. Petruccione, N. Zanghí (Springer, 2001)
13. M. Kac, G.C. Rota, J.T. Schwartz: *Discrete Thoughts*. Scientists of Our Time (Birkhäuser Boston Inc., Boston, MA, 1986). Essays on mathematics, science, and philosophy
14. M. Kac: *Statistical Independence in Probability, Analysis and Number Theory*. The Carus Mathematical Monographs, No. 12 (Published by the Mathematical Association of America. Distributed by John Wiley and Sons, Inc., New York, 1959)

Chapter 5
Brownian motion

Brownian motion has been observed since the invention of the microscope (circa 1650). The biologist Robert Brown (1773–1858) published systematic experimental studies of the erratic motion of pollen and of other microscopically visible grains swimming in drops of water. He called them primitive molecules, and it was unclear what made them move. In 1905, Albert Einstein predicted a *diffusive* motion of mesoscopic particles (i.e., macroscopically very small, but still visible through a microscope) immersed in a fluid, by adopting Boltzmann's view of atomism. When the molecules in the fluid undergo heat motion and hit the mesoscopic particle in a random manner, they force the particle to move around erratically. He suggested that this might be the already known Brownian motion.

We shall give Einstein's argument, which has surprisingly little to do with the microscopic motion of atoms, and illustrates his genius once again. Independently of Einstein, Smoluchowski explained Brownian motion as arising from molecular collisions. His argument is stochastic and is based on the $\sqrt{N}$ "density fluctuations", thereby disproving claims by his contemporaries that the molecular collisions, if happening at all, would average out to zero net effect. We shall also discuss this for a toy example.

Brownian motion proved to the unbelievers of Boltzmann's time that atoms do in fact exist. The quantitative prediction – diffusive motion – was experimentally verified by Perrin, who received the Nobel prize for that. Although based on the atomistic nature of matter, Brownian motion is not a quantum phenomenon. The reader should therefore ask why we need to know more about Brownian motion, when we have long since accepted the atomistic view of nature? The answer is that it is a physical phenomenon of the uttermost importance, because it bridges a gap between the microscopic and the macroscopic worlds, very much as Boltzmann's equation does, except that Brownian motion is simpler. It is good to see the transition from microscopic to macroscopic dynamics at work. It shows that macroscopic motion can look totally different (diffusive and irreversible) from microscopic motion (ballistic and reversible). Finally, the heat equation which governs the probability of Brownian motion is Schrödinger's equation with imaginary time, a technical feature that is sometimes put to use.

D. Dürr, S. Teufel, *Bohmian Mechanics*, DOI 10.1007/978-3-540-89344-8_5,

5.1 Einstein's Argument

This is taken from [1]. Imagine very many Brownian particles in a fluid. Very many, but not enough to form a crowd, i.e., a low density dilution. The Brownian particles themselves do not meet. The dilution can be treated as an ideal gas. For simplicity imagine that the density depends only on the x-coordinate. We thus have an ensemble of independent Brownian particles in the fluid, and according to the law of large numbers, the empirical density (Boltzmann's view!) is given by the probability density, i.e.,

$$\rho(x,t) = \frac{N(A\mathrm{d}x,t)}{A\mathrm{d}x} ,$$

where A is the area perpendicular to the x-axis and N is the number of Brownian particles in the volume Adx (see Fig. 5.1).

The aim is to derive an equation for the probability density ρ, the determination of which has been transferred by the law of large numbers to that of a thermodynamic quantity, which is all Einstein uses. The microscopic picture ends here. So far, this is all there is in Boltzmann's way of thinking. Now comes Einstein's innovation. Let m_{B} be the mass of the Brownian particle. Then $\rho(x,t)m_{\mathrm{B}} = \nu(x,t)$ gives the gram-moles per unit volume. The mass flux in the x-direction is governed by the continuity equation

$$\frac{\partial \nu(x,t)}{\partial t} = -\frac{\partial j}{\partial x} ,$$

where j is the mass flux through A. We now make a phenomenological ansatz, to be justified shortly. We shall assume that the flux is proportional to the gradient of ν and that the flux goes from high concentration to low concentration, bringing in the minus sign (the factor 1/2 is chosen for later convenience):

$$j = -\frac{1}{2}D\frac{\partial \nu}{\partial x} , \tag{5.1}$$

whence we obtain the desired equation

$$\frac{\partial \nu(x,t)}{\partial t} = \frac{1}{2}D\frac{\partial^2 \nu}{\partial x^2} . \tag{5.2}$$

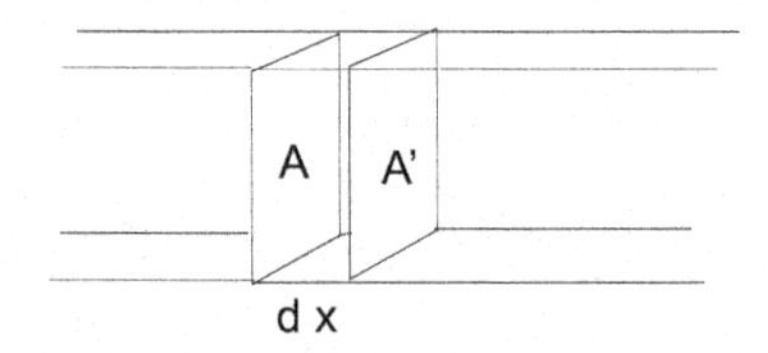

Fig. 5.1 Formulating the problem of Brownian motion

The constant D needs to be determined. The determination of D, and in fact the "derivation" of (5.1), is where Einstein's genius shows through.

We first derive the osmotic force acting on a Brownian particle. Let p and p' be the pressures exerted by the Brownian gas on the areas A and A', respectively. The pressure difference is known as the osmotic pressure, and the osmotic force per unit volume ($A\mathrm{d}x$) is (ignoring the time dependence to simplify the notation):

$$\frac{(p-p')A}{A\mathrm{d}x} = \frac{\mathrm{d}p}{\mathrm{d}x} .$$

Now p obeys the gas law $pA\mathrm{d}x = n(x)RT$, so $p = \overline{n}(x)RT$, with $\overline{n}(x)$ the number of moles per unit volume, which depends here on x. Hence

$$\frac{(p-p')A}{A\mathrm{d}x} = RT\frac{\partial \overline{n}(x)}{\partial x} .$$

From this we deduce the osmotic force F per particle, observing that, in terms of Avogadro's number N_A, the total osmotic force $(p-p')A = Fn(x)N_A$. Hence,

$$F = \frac{1}{\overline{n}(x)} k_B T \frac{\partial \overline{n}(x)}{\partial x} ,$$

where we have used the definition of Boltzmann's constant $k_B = R/N_A$.

The osmotic force accelerates the Brownian particle but the particle also suffers friction $F_R = -\gamma v$, where γ is the friction coefficient and v the velocity. In equilibrium the osmotic force equals the friction force, so that

$$-\gamma v = \frac{1}{\overline{n}(x)} k_B T \frac{\partial \overline{n}(x)}{\partial x}$$

and

$$-\overline{n}(x)v = \frac{k_B T}{\gamma}\frac{\partial \overline{n}(x)}{\partial x} ,$$

and since

$$\overline{n} \sim \frac{\nu}{m_B} ,$$

we obtain

$$-\nu(x)v = \frac{k_B T}{\gamma}\frac{\partial \nu(x)}{\partial x} .$$

So what has been achieved? We have derived (5.1), i.e., we have determined D! The reason is simply that $\nu(x)v = j$, the mass flux through A. Hence,

$$\frac{1}{2}D = \frac{k_B T}{\gamma} ,$$

the *Einstein relation*, one of the most famous formulas of kinetic theory. Here is one reading of it. The fluid molecules are responsible for pushing the Brownian particle, and at the same time, by the very same effect, namely collisions, they slow the particle down. Fluctuation (D, which is measurable!) and dissipation (γ, which is measurable!) have one common source: the molecular structure of the fluid. It is no wonder then that they are related. The greatness of Einstein's contribution is, however, that it is directly aimed at the determination of the diffusion constant in terms of *thermodynamic (i.e., measurable) quantities*. We give a realistic example for D below.

Let us immediately jump to three dimensions and rephrase the result (5.2) as the diffusion equation for $p(\mathbf{x},t;\mathbf{x}_0)$, where $p(\mathbf{x},t;\mathbf{x}_0)\mathrm{d}^3x$ is the probability of finding the particle within d^3x around $\mathbf{x}$ at time t, when the particle is put into the fluid at the position $\mathbf{x}_0$ at time 0:

$$\frac{\partial}{\partial t}p(\mathbf{x},t;\mathbf{x}_0) = \frac{1}{2}D\frac{\partial^2}{\partial \mathbf{x}^2}p(\mathbf{x},t;\mathbf{x}_0)\ , \tag{5.3}$$

with

$$p(\mathbf{x},0;\mathbf{x}_0) = \delta(\mathbf{x}-\mathbf{x}_0)\ . \tag{5.4}$$

Let us make some remarks on the equation and its solution. The initial condition (5.4) defines the *fundamental solution* of (5.3), which generates solutions for any initial distribution by integration:

$$\rho(\mathbf{x},t) = \int p(\mathbf{x},t;\mathbf{x}_0)\rho(\mathbf{x}_0)\mathrm{d}^3x_0\ .$$

It also solves (5.3) and is the probability density for finding the particle at $\mathbf{x}$ at time t when at time 0 the initial distribution of the position is $\rho(\mathbf{x}_0)$, with $\int\rho(\mathbf{x}_0)\mathrm{d}^3x_0 = 1$. Note that (5.3) is not time-reversal invariant. The equation describes the time-irreversible spreading of diffusive motion. Temperature does spread in that way through a medium. Hence the name *heat equation*. One solves (5.3) by Fourier decomposition, viz.,

$$\hat{p}(\mathbf{k},t;\mathbf{x}_0) = (2\pi)^{-3/2}\int \mathrm{e}^{-\mathrm{i}\mathbf{k}\cdot\mathbf{x}}p(\mathbf{x},t;\mathbf{x}_0)\mathrm{d}^3x\ ,$$

and obtains

$$\begin{aligned}\hat{p}(\mathbf{k},t;\mathbf{x}_0) &= \mathrm{e}^{-k^2Dt/2}\hat{p}(\mathbf{k},0;\mathbf{x}_0)\\ &= (2\pi)^{-3/2}\mathrm{e}^{-k^2Dt/2}\mathrm{e}^{-\mathrm{i}\mathbf{k}\cdot\mathbf{x}_0}\ .\end{aligned} \tag{5.5}$$

Hence,

$$p(\mathbf{x},t;\mathbf{x}_0) = \frac{1}{(2\pi)^3}\int \mathrm{e}^{\mathrm{i}\mathbf{k}\cdot\mathbf{x}}\mathrm{e}^{-k^2Dt/2}\mathrm{e}^{-\mathrm{i}\mathbf{k}\cdot\mathbf{x}_0}\mathrm{d}^3k\ , \tag{5.6}$$

a straightforward Gaussian integration:

$$\begin{aligned} p(\mathbf{x},t;\mathbf{x}_0) &= \frac{1}{(2\pi)^3}\int e^{i\mathbf{k}\cdot(\mathbf{x}-\mathbf{x}_0)}e^{-k^2Dt/2}d^3k \\ &= \frac{1}{(2\pi)^3}\left(\frac{2}{Dt}\right)^{3/2}\exp\left[-\frac{(\mathbf{x}-\mathbf{x}_0)^2}{2Dt}\right]\int e^{-y^2}d^3y \\ &= (2\pi Dt)^{-3/2}\exp\left[-\frac{(\mathbf{x}-\mathbf{x}_0)^2}{2Dt}\right], \end{aligned} \tag{5.7}$$

where one should know that

$$\int e^{-y^2}d^3y = \int e^{-x^2}dx\int e^{-y^2}dy\int e^{-z^2}dz = \pi^{3/2}. \tag{5.8}$$

In one dimension,

$$p(x,t;x_0) = \frac{1}{\sqrt{2\pi Dt}}e^{-(x-x_0)^2/2Dt}. \tag{5.9}$$

Imagine now the zigzag paths $\mathbf{X}(t)$, $t \in [0,\infty)$ of the Brownian particle. The distribution $p(\mathbf{x},t;\mathbf{x}_0)$ is the image distribution of the coarse-graining variable $\mathbf{X}(t)$, the position of the Brownian particle at time t. The position is *diffusive*, which means that the expectation value of $\mathbf{X}^2(t)$ is proportional to t with proportionality given by D:

$$\begin{aligned} \mathbb{E}\Big(\big(\mathbf{X}(t)-\mathbf{x}_0\big)^2\Big) &= \frac{1}{\sqrt{2\pi Dt}^3}\int(\mathbf{x}-\mathbf{x}_0)^2e^{-(\mathbf{x}-\mathbf{x}_0)^2/2Dt}d^3x \\ &= \frac{3}{\sqrt{2\pi Dt}^3}\left(\int e^{-z^2/2Dt}dz\right)^2\int y^2e^{-y^2/2Dt}dy \\ &= \frac{3(2\pi)Dt}{\sqrt{2\pi Dt}^3}(-2Dt)\frac{d}{d\alpha}\int e^{-\alpha y^2/2Dt}dy\bigg|_{\alpha=1} \\ &= 3Dt\,. \end{aligned}$$

It was this diffusive behavior $\langle X^2(t)\rangle \sim t$ that Perrin verified. D can be determined (Stokes' friction law for the friction of a ball in a fluid brings in the viscosity) and that can also be checked quantitatively. For a Brownian ball with radius a in a fluid with viscosity η,

$$D = \frac{k_B T}{6\pi\eta a}.$$

The friction can be measured (as can η and a), and T can be measured. Then since D can be determined from the measured variance of $X(t)$, one can determine k_B or Avogadro's number [2].

This is a masterpiece of theoretical thought, comparable to Boltzmann's derivation of the Boltzmann equation. While the latter is far more complicated, because it does not focus on the special Brownian particles but on the fluid itself, resulting in a nonlinear equation for the phase space density, the former allows a *direct view* of the molecular motion via the mesoscopic Brownian particle as mediator. This might explain why physicists were converted to atomism only after the work of Einstein and Smoluchowski had become known in 1905/1906.

5.2 On Smoluchowski's Microscopic Derivation

The microscopic dynamics of the fluid molecules is time-reversal invariant. The motion of the Brownian particle is governed by a time-irreversible equation. The idea that the Brownian particle is kicked from all sides by the fluid molecules suggests that the effects should balance out, whence no net motion should be visible. So why do the collisions have a nonzero net effect? Why is the motion diffusive? Why is the probability Gaussian? Smoluchowski gives answers based on *microscopic* principles. We do not present his work (see, for example, the collection [3]), but rather jump to a toy model which is in the same spirit, and quite adequate to understand the basic idea. It is a purely mathematical model of Brownian motion, which we phrase in physical terms, but it lacks essentially all of the physical ingredients which make Brownian motion (diffusion) a very difficult problem of mathematical physics when one starts from first principles and aims for rigor.

The model is one-dimensional. All molecules move on the x-axis. At time 0, the system has the following state. It consists of infinitely many identical particles $\{q_i, v_i\}_{i\in\mathbb{Z}}$, $q_i = il$ (i.e., $q_i \in l\mathbb{Z}$), $v_i \in \{-v, v\}$, where one should think of l as representing the mean distance between the molecules in a real fluid. The v_i are independently drawn from $\{-v, v\}$ with probability 1/2. The phase space is $\Omega = \{\omega | \omega = (il, \pm v)_{i\in\mathbb{Z}}\}$ with product probability $\mathbb{P}_{\mathscr{B}}$. The particles interact via elastic collisions. Since they all have equal masses, the worldline picture of the system is simply given by spacetime trajectories which cross each other as shown in Fig. 5.2.

Focus on the particle starting at $X = 0$, i.e., color it brown and follow the brown trajectory X_t, which is a zigzag path. We call it the Brownian particle. All other particles represent the molecules of the fluid. $(X_t)_{t\geq 0}$ is a coarse-graining function depending on ω, and $\big(X_t(\omega)\big)_{t\geq 0}$ is a stochastic process.

What is non-equilibrium about that? The answer is: putting the Brownian particle at the origin at time zero. What is artificial in the model?

- The dynamics is not given by differential equations.
- The system is infinite, which must eventually be the case for sharp mathematical statements. Poincaré cycles must be "broken".
- The brown particle is only distinguished by color, not by its dynamics. It is in reality very much bigger and heavier.
- The fluid particles should be distributed according to the canonical distribution.

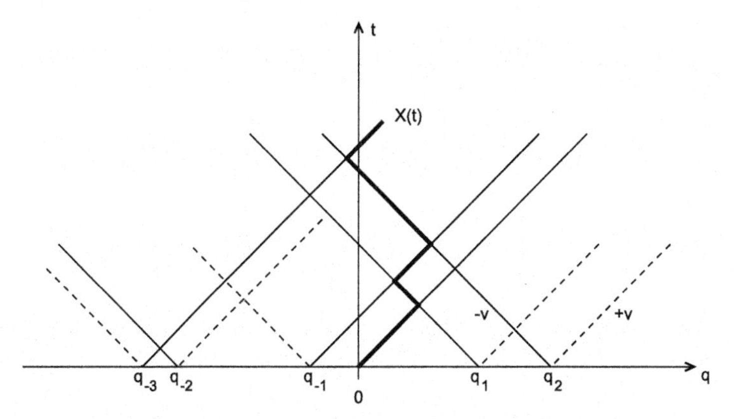

Fig. 5.2 A microscopic Brownian path. Note that the initial positions of the "bath particles" are drawn randomly. They may be distributed according to a Poisson distribution, but in the text we handle the simpler case where the particles start on a lattice with spacing l

Having said all this, the model nevertheless shows the characteristic macroscopic behavior we are after, without all the extreme complications a realistic model would introduce.

So let us move on to $X_t(\omega)$, the brown path. It can be described as follows. Every $\Delta t = l/2v$, there is a change of direction with probability 1/2, and

$$X_t \approx \sum_{1 \le k \le [t/\Delta t]} \Delta x_k \,, \tag{5.10}$$

where the Gauss bracket $[a]$ means the greatest whole number $\le a$, and $\Delta x_k \in \{-l/2, l/2\}$ are independently identically distributed random variables assuming values $\Delta x_k = \pm l/2$ with probability 1/2. The approximation in (5.10) concerns the last time interval $\big[[t/\Delta t]\Delta t, t\big]$ and is unimportant. This is simply a nuisance to carry around and we shall ignore it here. X_t is thus a sum of independent random variables. The expectation value is

$$\begin{aligned} \mathbb{E}(X_t) &= \mathbb{E}\left(\sum_{1 \le k \le [t/\Delta t]} \Delta x_k\right) \\ &= \sum_{1 \le k \le [t/\Delta t]} \mathbb{E}(\Delta x_k) \\ &= \left[\frac{t}{\Delta t}\right]\left[\frac{1}{2}\frac{(-l)}{2} + \frac{1}{2}\frac{l}{2}\right] \\ &= 0 \,, \end{aligned} \tag{5.11}$$

and its variance (which we should know by now, having proven the law of large numbers) is

$$\begin{aligned}\mathbb{E}(X_t^2) &= \sum_{1\leq k\leq[t/\Delta t]} \mathbb{E}(\Delta x_k^2)\\ &= \left[\frac{t}{\Delta t}\right]\left(\frac{1}{2}\frac{l^2}{4}+\frac{1}{2}\frac{l^2}{4}\right)\\ &= \frac{l^2}{4}\left[\frac{t}{\Delta t}\right]\\ &\sim t\,. \end{aligned} \tag{5.12}$$

Hence heuristically $X(t) \sim \sqrt{t}$. This is what we are interested in. We do not care about single collisions on the molecular scale, but only about the macroscopic growth of brown's position. We investigate that by *scaling*. We go to a larger scale in space and time. The interesting thing is to see how this is done. We go on the *diffusive scale*, i.e., when time gets enlarged by a factor $1/\varepsilon$. Then when ε gets small, the Brownian motion will vanish from sight. To maintain eye contact, we must rescale its position by a factor $\sqrt{\varepsilon}$. For $\varepsilon \to 0$,

$$X_t^\varepsilon = \varepsilon X_{t/\varepsilon^2}\,. \tag{5.13}$$

One second for X_t^ε corresponds to $1/\varepsilon^2$ seconds for the brown path, which means $\sim 1/\varepsilon^2$ collisions have occurred. Moreover,

$$\mathbb{E}\big((X_t^\varepsilon)^2\big) = \varepsilon^2\mathbb{E}(X_{t/\varepsilon^2}^2) \sim \varepsilon^2\frac{t}{\varepsilon^2} = t\,.$$

What is the distribution $p^\varepsilon(x,t)$ of X_t^ε? We get this via the Fourier transform of the distribution $\hat{p}^\varepsilon(k,t)$. This is so important that it has a name all of its own, viz., the characteristic function:

$$(2\pi)^{3/2}\hat{p}^\varepsilon(k,t) = \int \mathrm{e}^{-\mathrm{i}kx}p^\varepsilon(x,t)\mathrm{d}x = \mathbb{E}(\mathrm{e}^{-\mathrm{i}kX_t^\varepsilon})\,.$$

We see immediately why this is useful, because introducing (5.10) yields

$$
\begin{aligned}
(2\pi)^{3/2}\hat{p}^{\varepsilon}(k,t) &= \mathbb{E}\left[\exp\left(-ik\sum_{n\leq[t/\varepsilon^2\Delta t]}\varepsilon\Delta x_n\right)\right] \\
&= \mathbb{E}\left(\prod_{n\leq[t/\varepsilon^2\Delta t]}\mathrm{e}^{-ik\varepsilon\Delta x_n}\right) \\
&= \prod_{n\leq[t/\varepsilon^2\Delta t]}\mathbb{E}(\mathrm{e}^{-ik\varepsilon\Delta x_n}) \\
&= \prod_{n\leq[t/\varepsilon^2\Delta t]}\mathbb{E}\left(1-ik\varepsilon\Delta x_n-\frac{1}{2}k^2\varepsilon^2(\Delta x_n)^2+o(\varepsilon^2)\right) \\
&= \prod_{n\leq[t/\varepsilon^2\Delta t]}\left[1-\frac{1}{2}k^2\varepsilon^2\frac{l^2}{4}+o(\varepsilon^3)\right].
\end{aligned}
$$

The third equality is obtained because, by independence, expectation and product can be interchanged, and the fourth equality is obtained by expanding the exponential. In addition, we have used (5.11) and (5.12). Choosing a sequence of ε so that $t/\varepsilon^2\Delta t = N \in \mathbb{N}$,

$$
(2\pi)^{3/2}\hat{p}^{\varepsilon}(k,t) = \left[1-\frac{1}{N}\frac{k^2l^2t}{8\Delta t}+o\left(\frac{1}{N}\right)\right]^N .
$$

For $N \to \infty$, i.e., $\varepsilon \to 0$, this converges to

$$
(2\pi)^{3/2}\hat{p}(k,t) = \exp\left(-\frac{k^2l^2}{8\Delta t}t\right),
$$

and observing that $\Delta t = l/2v$, we obtain

$$
\hat{p}(k,t) = (2\pi)^{-3/2}\exp\left(-\frac{k^2lv}{4}t\right).
$$

Comparing with (5.5), we see that

$$
D = \frac{1}{2}lv ,
$$

whence

$$
p(x,t) = \frac{1}{\sqrt{2\pi Dt}}\mathrm{e}^{-x^2/2Dt} = \frac{1}{\sqrt{\pi lvt}}\mathrm{e}^{-x^2/lvt} . \tag{5.14}
$$

Setting $\rho = 1/l$ as the fluid density and $\eta = v\rho$ as the viscosity, this allows for a somewhat artificial macroscopic interpretation of the diffusion coefficient, namely $D = v^2/2v\rho \sim k_{\mathrm{B}}T/\eta$. This is physically reasonable. The diffusion is greater when the fluid is less viscous and when the "temperature" increases. One should observe that the limit $\varepsilon \to 0$ means that we consider the microscopic motion in infinite time, whence the system must consist of infinitely many particles and extend infinitely in space to avoid cycles. It is only in the limit that irreversibility becomes "true".

What we have done mathematically is to "prove" what is known as the *central limit theorem* for independent variables.[1]

5.3 Path Integration

In (5.14), we only looked at the scaled position at time t. What can be said about the scaled process $(X_t^\varepsilon)_{t\in[0,\infty)}$? We look at cylinder sets in the set of continuous functions on the interval $[0,T]$, $T<\infty$, denoted by $C([0,T])$. For this purpose we consider, at arbitrarily chosen times $t_1,\dots,t_n$, what we call gates $\Delta_i\subset\mathbb{R}$, $i=1,\dots,n$, through which the functions are required to pass (see Fig. 5.3). This defines a set of functions, namely the cylinder set

$$Z_{t_1,\dots,t_n}(\Delta_1,\dots,\Delta_n)=\Big\{\omega(t)\in C([0,T])\Big|\omega(t_1)\in\Delta_1,\dots,\omega(t_n)\in\Delta_n\Big\},$$

the measure of which is induced by the process X_t^ε:

$$\mathbb{P}^\varepsilon\left(Z_{t_1,\dots,t_n}(\Delta_1,\dots,\Delta_n)\right)=\mathbb{P}_{\mathscr{B}}\Big(\{\omega|X_{t_1}^\varepsilon(\omega)\in\Delta_1,\dots,X_{t_n}^\varepsilon(\omega)\in\Delta_n\}\Big). \tag{5.15}$$

This can be read as follows. A trajectory which lies in $Z_{t_1,\dots,t_n}(\Delta_1,\dots,\Delta_n)$ starts at zero and goes to $X_1\in\Delta_1$, and from there to $X_2\in\Delta_2$, and so forth. We can view (5.14) as the probability that a trajectory goes from 0 to x_1 in time t_1:

$$p(x_1,t_1;0)=\frac{1}{\sqrt{2\pi Dt_1}}\mathrm{e}^{-x_1^2/2Dt_1}.$$

Consider now the process X_t^ε starting in x_1 at time t_1. It is intuitively clear that the distribution of $X_{t_2}^\varepsilon$ is

$$p(x_2,t_2;x_1,t_1)=\frac{1}{\sqrt{2\pi D(t_2-t_1)}}\mathrm{e}^{-(x_2-x_1)^2/2D(t_2-t_1)}.$$

Putting this together, we expect an n-dimensional Gaussian distribution for (5.15) in the limit $\varepsilon\to 0$:

---

[1] It is easily made mathematically rigorous and the assertion is as follows. Let $(X_i)_{i\in\mathbb{N}}$ be a sequence of identically independently distributed random variables with $\mathbb{E}(X_i)=0$, $\mathbb{E}(X_i^2)=\sigma^2\neq 0$ on $(\Omega,\mathscr{B}(\Omega),\mathbb{P})$ and assume (for an easy proof) that $\mathbb{E}(|X|^3)$ is finite. Then for $S_n=\sum_{i=1}^n X_i$,

$$\lim_{n\to\infty}\mathbb{P}\left(\left\{\omega\Big|\frac{1}{\sqrt{n}}S_n(\omega)\in[a,b]\right\}\right)=\int_a^b\frac{1}{\sqrt{2\pi\sigma}}\mathrm{e}^{-x^2/2\sigma}\mathrm{d}x.$$

This tells us how big fluctuations about the typical value are, which is the $\sqrt{N}$ law.

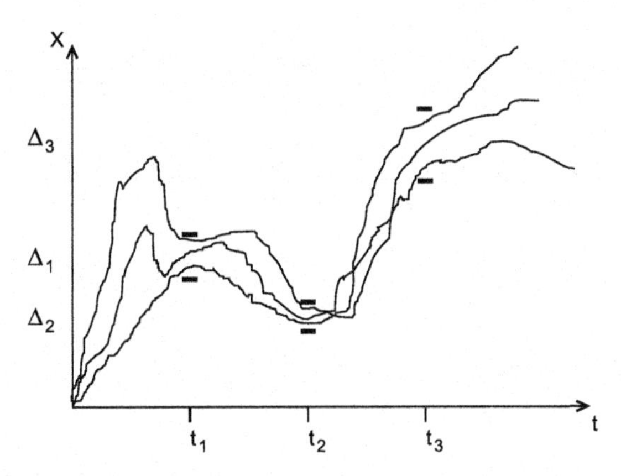

Fig. 5.3 Cylinder set

$$
\begin{aligned}
&\mathbb{P}\big(Z_{t_1,\dots,t_n}(\Delta_1,\dots,\Delta_n)\big) \qquad (5.16)\\
&=\int_{\Delta_1}\mathrm{d}x_1\int_{\Delta_2}\mathrm{d}x_2\dots\int_{\Delta_n}\mathrm{d}x_n\frac{\exp\left[-\frac{x_1^2}{2Dt_1}\right]}{\sqrt{2\pi Dt_1}}\frac{\exp\left[-\frac{(x_2-x_1)^2}{2D(t_2-t_1)}\right]}{\sqrt{2\pi D(t_2-t_1)}}\cdots\frac{\exp\left[-\frac{(x_n-x_{n-1})^2}{2D(t_n-t_{n-1})}\right]}{\sqrt{2\pi D(t_n-t_{n-1})}}\,.
\end{aligned}
$$

The cylinder sets generate the Borel algebra $\mathscr{B}(C[0,T])$.

One needs a bit of reflection to convince oneself that $\mathscr{B}(C[0,T])$ is the Borel algebra generated by open sets in the uniform topology $|\omega|_\infty = \sup_{t\in[0,T]}|\omega(t)|$. The corresponding measure on function space $C([0,T])$ is then the so-called Wiener measure μ_W, which is a Gaussian measure. The process $(W_t)_{t\in[0,\infty)}$, the distribution of which is given by (5.16), is called a Wiener process or Brownian motion process. Having a measure on function space, we can integrate functions $C([0,T]) \mapsto \mathbb{R}^n$ (analogously to Lebesgue integration). This integration is sometimes referred to as functional integration, Feynman–Kac path integration, or simply path integration.

References

1. A. Einstein: *Investigations on the Theory of the Brownian Movement* (Dover Publications Inc., New York, 1956). Edited with notes by R. Fürth, translated by A.D. Cowper
2. E. Kappler: Annalen der Physik **11**, 233 (1931)
3. A. Einstein, M.v. Smoluchowski: *Untersuchungen über die Theorie der Brownschen Bewegung/Abhandlung über die Brownsche Bewegung und verwandte Erscheinungen* (Verlag Harry Deutsch, New York, 1997). Reihe Ostwalds Klassiker 199

Chapter 6
The Beginning of Quantum Theory

We shall keep this short. Classical physics, by which we mean Newton's laws and the Maxwell–Lorentz theory (see Chap. 2), fails to explain many atomistic phenomena, apart from the coarse-grained phenomena we discussed in previous chapters. This means that attempts to explain atomistic detail look artificial, and are no longer believable. In other words, straightforward application of classical physics yields descriptions that contradict experience in certain situations. In that sense, for example, Newtonian mechanics is superior to the theory of epicycles, which was invented to save Ptolemy's geocentric astronomy, because the Newtonian explanation of the motion of heavenly bodies is straightforward and reduces to a single equation: Newton's equation. The failures of classical physics in the atomistic regime are mirrored in the names of certain effects, such as the anomalous Zeeman effect. This refers to the complicated splitting of spectral lines in magnetic fields that was found experimentally, while Lorentz's classical analysis led to the "normal" Zeeman effect.

A famous example leading to quantum mechanics is black body radiation. This is the radiation in a box, the walls of which are kept at a fixed temperature T, so that a thermodynamic equilibrium between radiation and walls is achieved. The distribution of the energy H of the radiation is then given by the canonical distribution (4.7):

$$\frac{\exp\left(-\dfrac{1}{k_{\mathrm{B}}T}H\right)}{Z(T)} . \tag{6.1}$$

In view of (2.44) with $j^{\mu} = 0$ (and $c = 1$), we have

$$\left(\frac{\partial^2}{\partial t^2} - \Delta\right) A^{\mu} = 0 , \tag{6.2}$$

and with Fourier transformation,

D. Dürr, S. Teufel, *Bohmian Mechanics*, DOI 10.1007/978-3-540-89344-8_6,

$$A^\mu(\mathbf{x},t) = \frac{1}{(2\pi)^{3/2}} \int \mathrm{d}^3 k \mathrm{e}^{-\mathrm{i}\mathbf{k}\cdot\mathbf{x}} \hat{A}^\mu(\mathbf{k},t) , \tag{6.3}$$

the Fourier modes are found from (6.2) to be

$$\ddot{\hat{A}}^\mu(\mathbf{k},t) = -k^2 \hat{A}^\mu(\mathbf{k},t) . \tag{6.4}$$

For $\mathbf{k}$ fixed, (6.4) is the harmonic oscillator equation with oscillator frequency $\omega(\mathbf{k}) = \|\mathbf{k}\|$, i.e.,

$$\ddot{q} = -k^2 q \quad (= -\omega^2 q) . \tag{6.5}$$

Therefore electromagnetic radiation consists of uncoupled harmonic oscillators.[1] Every $\mathbf{k}$-mode $\hat{A}^\mu(\mathbf{k},t)$ has frequency $\omega = \|\mathbf{k}\|$. The functional dependence $\omega(\mathbf{k})$ is called a *dispersion relation.*

Let us now consider one oscillator (6.5) and determine its mean energy at temperature T:

$$H = \frac{1}{2m} p^2 + \frac{1}{2} m\omega^2 q^2$$

and

$$\begin{aligned} E(\omega) &= \frac{\int \mathrm{e}^{-\beta H} H \mathrm{d}q \mathrm{d}p}{\int \mathrm{e}^{-\beta H} \mathrm{d}q \mathrm{d}p} \\ &= -\frac{\mathrm{d}}{\mathrm{d}\beta} \ln \left\{ \int \exp\left[-\beta \left(\frac{1}{2m} p^2 + \frac{1}{2} m\omega^2 q^2 \right) \right] \mathrm{d}q \mathrm{d}p \right\} \\ &= -\frac{\mathrm{d}}{\mathrm{d}\beta} \ln \frac{2\pi}{\omega} \frac{1}{\beta} = \frac{1}{\beta} = k_{\mathrm{B}} T . \end{aligned}$$

Hence every ω-mode has the average energy $k_{\mathrm{B}} T$ and the number of modes in $[\omega, \omega + \mathrm{d}\omega]$ is given by the number of $\mathbf{k}' \in [\mathbf{k}, \mathbf{k} + \mathrm{d}\mathbf{k}]$ which is $4\pi k^2 dk$, whence the number of modes is $4\pi\omega^2 \mathrm{d}\omega$. We thus obtain the average energy $U(\omega, T)$ of radiation at temperature T:

$$U(\omega, T)\mathrm{d}\omega \sim k_{\mathrm{B}} T \omega^2 \mathrm{d}\omega , \tag{6.6}$$

which is called the Rayleigh–Jeans distribution. Experimentally, one finds something else, something more reasonable for large ω. What one finds is what the physicist Wien had predicted in (1896) when he was working on the Stefan–Boltzmann law, according to which there is a function f such that

$$U(\omega, T)\mathrm{d}\omega \sim \omega f\left(\frac{\omega}{T}\right) \omega^2 \mathrm{d}\omega .$$

Wien argued that

[1] j^μ in (2.44) can be thought of as generating oscillations.

$$f\left(\frac{\omega}{T}\right) = \hbar e^{-\hbar\omega/k_B T} ,$$

so that

$$U(\omega,T)d\omega \sim \hbar\omega e^{-\hbar\omega/k_B T}\omega^2 d\omega \qquad \text{(Wien's law)} . \tag{6.7}$$

The new constant $\hbar = h/2\pi$ is to be determined by experiment. Planck then found an interpolation between Wien and Rayleigh–Jeans which in fact reproduced the observed $E(\omega)$ in its totality. To his own discontent he had to assume in an ad hoc manner that the oscillators he hypothesised in the walls of the black body could only absorb and emit discrete energies $n\hbar\omega$, $n = 0, 1, 2, \ldots$.

The whole paper was very involved. But then Einstein came into the picture. He took the field – the radiation – as physically real, and assumed that it satisfied a different law from what had previously been thought. But, with typical genius, he noted that we do not need to know what the law looks like. In fact it suffices to suppose that the energy of radiation of frequency ω can only be an integer multiple of $\hbar\omega$. Then everything is suddenly trivial. We take the usual thermodynamical statistics, but with a different dynamics of which we only know $E_n(\omega) = n\hbar\omega$. Then,

$$E(\omega) = -\frac{d}{d\beta}\ln\sum_{n=0}^{\infty} e^{-\beta E_n} = -\frac{d}{d\beta}\ln\sum_{n=0}^{\infty} e^{-\beta n\hbar\omega} \tag{6.8}$$

$$= -\frac{d}{d\beta}\ln\left(\frac{1}{1-e^{-\beta\hbar\omega}}\right) = \frac{\hbar\omega e^{-\beta\hbar\omega}}{1-e^{-\beta\hbar\omega}} , \tag{6.9}$$

which gives Planck's celebrated radiation formula

$$U(\omega,T)d\omega \sim \frac{\hbar\omega e^{-\hbar\omega/k_B T}}{1-e^{-\hbar\omega/k_B T}}\omega^2 d\omega . \tag{6.10}$$

For $k_B T \gg \hbar\omega$ (high temperature, small frequency), we obtain approximately

$$E(\omega) \approx \frac{\hbar\omega\left(1-\dfrac{\hbar\omega}{k_B T}\right)}{\hbar\omega} k_B T \approx k_B T .$$

When inserted in (6.10), this yields (6.6). And for $k_B T \ll \hbar\omega$, we have

$$E(\omega) \approx \hbar\omega e^{-\hbar\omega/k_B T} ,$$

from which we get Wien's law (6.7).

The new description of the electromagnetic field in terms of energy quanta – *photons* of energy $\hbar\omega$ – also provided a straightforward explanation of the photoelectric effect, discovered by Hertz in 1880. Einstein did that in the second of his fundamental 1905 works, and received the Nobel prize for it. A strange thing about this is that neither Einstein (and he did not pursue the question any further, but moved on to gravitation) nor ourselves know today what "photons" really are. Are they particles? Are they extended objects? Are they anything at all?

A similar ansatz was given by Bohr in 1913 for the angular momentum of an electron moving around the nucleus. He assumed that

$$L = n\hbar\,, \quad n \in \mathbb{N}\,. \tag{6.11}$$

The Bohr–Sommerfeld quantization condition for the action of periodic orbits was similar:

$$\oint p\mathrm{d}q = n\hbar\,. \tag{6.12}$$

This yields the spectral lines of hydrogen and hydrogen-like atoms with surprising precision and simplicity.

This is all well known and has been retold often enough, so let us now move on to Louis-Victor de Broglie's ideas back in 1923. The notion of plane wave $\mathrm{e}^{\mathrm{i}\mathbf{k}\cdot\mathbf{x}}$, $\mathbf{k} \in \mathbb{R}^3$ is clear. A wave packet is a superposition of plane waves centered around a value $\mathbf{k}_0$:

$$f(\mathbf{x}) = \int \mathrm{e}^{\mathrm{i}\mathbf{k}\cdot\mathbf{x}} \hat{f}_{\mathbf{k}_0}(\mathbf{k})\mathrm{d}^3k\,, \tag{6.13}$$

with $\hat{f}_{\mathbf{k}_0}(\mathbf{k})$ centered around $\mathbf{k}_0$, as in

$$\hat{f}_{\mathbf{k}_0}(\mathbf{k}) = \frac{1}{(2\pi)^{3/2}\sigma^3}\mathrm{e}^{-(\mathbf{k}_0-\mathbf{k})^2/2\sigma^2}\,. \tag{6.14}$$

Then

$$f(\mathbf{x}) = \mathrm{e}^{\mathrm{i}\mathbf{k}_0\cdot\mathbf{x}}\mathrm{e}^{-x^2\sigma^2/2}\,.$$

This means that f is concentrated on a region of size $1/\sigma$. Nothing special about that. The wavy character appears when we consider the time evolution of waves. What is special is that the frequency ω is a function of $\mathbf{k}$. The wave character lies in the dispersion relation $\omega(\mathbf{k})$, a relation basic to all wave phenomena. It regulates the spreading of wave packets which group around different wavelengths, a phenomenon of great importance for Bohmian mechanics.

The dispersion relation contains a way of relating wave phenomena to Newtonian notions like momentum and energy. This will be further elaborated in Sect. 9.4. The dispersion relation of electromagnetic waves in vacuum is $\omega^2 = c^2k^2$, and this means that packets of light waves around different wavelengths do not separate, i.e., they do not spread. Electromagnetic waves in matter behave differently, for one has phenomenologically different dispersion relations.

A characteristic property of waves is the ability to interfere, i.e., the superposition of waves creates new wavy pictures with new amplitudes and new nodal areas. This is most relevant when waves in a wave packet separate.

The time evolution of (6.13) is

$$f(\mathbf{x},t) = \int \mathrm{e}^{\mathrm{i}[\mathbf{k}\cdot\mathbf{x}-\omega(\mathbf{k})t]} \hat{f}_{\mathbf{k}_0}(\mathbf{k}) \mathrm{d}^3 k \,. \tag{6.15}$$

We focus on $(\mathbf{x},t)$ values for which $f(\mathbf{x},t)$ is large. To this end we expand the *phase* $S(\mathbf{k})$ in the integral of (6.15) around $\mathbf{k}_0$:

$$\begin{aligned} S(\mathbf{k}) &= \mathbf{k}\cdot\mathbf{x} - \omega(\mathbf{k})t \\ &= \mathbf{k}_0\cdot\mathbf{x} - \omega(\mathbf{k}_0)t + (\mathbf{k}-\mathbf{k}_0)\cdot\mathbf{x} - (\mathbf{k}-\mathbf{k}_0)\cdot\nabla\omega(\mathbf{k}_0)t \\ &\quad -\frac{1}{2}\left[(\mathbf{k}-\mathbf{k}_0)\cdot\omega''(\mathbf{k}_0)(\mathbf{k}-\mathbf{k}_0)\right]t + \cdots \,, \end{aligned}$$

with the Hessian $\omega''(\mathbf{k}_0)$ (a matrix of second derivatives). $S(\mathbf{k})$ varies most in the linear $\mathbf{k}$ term, so we choose $\mathbf{x}$ and t in such a way that this term vanishes. Then for such values $(\mathbf{x},t)$, there will be little "destructive interference" and $f(\mathbf{x}',t')$ will have maximal amplitude around $(\mathbf{x},t)$. This idea underlies the so-called *stationary phase argument*, and a great many quantum phenomena are explained by it. It will appear over and over again, for example in (9.18) and Remarks 15.8 and 15.9.

One defines the *group velocity* of the wave packet[2] $f(\mathbf{x},t)$ by

$$\mathbf{v}_\mathrm{g} := \frac{\mathbf{x}}{t} = \frac{\partial w}{\partial \mathbf{k}}(\mathbf{k}_0) \,, \tag{6.16}$$

so that $f(\mathbf{v}_\mathrm{g}t,t)$ always has maximal amplitude. For stationary points $(\mathbf{x},t)$, the phase $S(\mathbf{k})$ has no linear $\mathbf{k}$ term, and introducing this in (6.15) yields

$$f(\mathbf{x},t) \approx \mathrm{e}^{\mathrm{i}\mathbf{k}_0\cdot\mathbf{x}-\omega(\mathbf{k}_0)t} \int \exp\left\{-\frac{\mathrm{i}}{2}\left[(\mathbf{k}-\mathbf{k}_0)\cdot\omega''(\mathbf{k}_0)(\mathbf{k}-\mathbf{k}_0)\right]t\right\} \hat{f}_{\mathbf{k}_0}(\mathbf{k})\mathrm{d}^3 k \,. \tag{6.17}$$

In view of (6.14) we see that the $\mathbf{k}$ width has changed. For simplicity replace $\omega''(\mathbf{k}_0)$ by $\gamma\mathsf{E}_3$, where E_3 is the 3×3 identity matrix, and note that the $\mathbf{k}$ width becomes

$$\sqrt{\Re\frac{1}{\sigma^{-2}+\mathrm{i}t\gamma}} = \sqrt{\frac{1}{\sigma^{-2}+\sigma^2 t^2\gamma^2}} \,.$$

The position width [of (6.13) with (6.14)] is the reciprocal, i.e.,

$$\sqrt{\sigma^{-2}+\sigma^2 t^2\gamma^2} \approx \sigma\gamma t \,, \quad \text{for } t\to\infty \,. \tag{6.18}$$

This is referred to as the spreading of the wave packet. It spreads faster for smaller initial position widths (σ large). We merely mention this in passing.

---

[2] The group velocity is the "phenomenological" velocity of a wave. It is determined by the motion of the maximal amplitude which is influenced by interference. The *phase velocity* of an almost plane wave is determined by the $(\mathbf{x},t)$ pairs of constant phase, $\mathbf{k}_0\cdot\mathbf{x}-\omega(\mathbf{k}_0)t = 0$.

Let us go on with de Broglie's idea. In spacetime, electromagnetic radiation moves along light cones. That is to say, the trajectories of what we would like to call photons, if they exist, should lie on light cones, where $\mathrm{d}s^2 = 0$. Thus proper time is no good for parameterizing photon trajectories. To get a consistent picture of these photons as particles in some sense, the energy–momentum relation must read $E^2/c^2 - p^2 = 0$, i.e., the rest mass of photons must be zero. A photon with frequency ω has $E = \hbar\omega$ and hence $\hbar^2\omega^2/c^2 - p^2 = 0$.

Since the dispersion relation for electromagnetic waves is $\omega = c|\mathbf{k}|$, it is natural to set $\mathbf{p} = \hbar\mathbf{k}$. Hence the energy–momentum four-vector for photons would be

$$\left(\frac{\hbar\omega}{c}, \hbar\mathbf{k}\right) .$$

Let us connect with Hamilton's idea of reading Newtonian mechanics by analogy with the geometric optics of a wave theory, and Einstein's idea of introducing photons into a wave theory. Then we arrive at de Broglie's 1923 idea of attempting once again to unite wave and particle, which gave the de Broglie matter waves.

It should be noted that this early attempt differed from Hamilton's. De Broglie had in mind one wave per particle. The basic idea is this. Insert $E = \hbar\omega$ and $\mathbf{p} = \hbar\mathbf{k}$ into the energy–momentum relation for a particle of rest mass m, viz.,

$$\frac{E^2}{c^2} - p^2 = m^2c^2 ,$$

and out comes the dispersion relation

$$\omega(k) = \sqrt{\frac{m^2c^4}{\hbar^2} + k^2c^2} . \tag{6.19}$$

This is almost too good to be true. Bohr's quantization condition emerges at once, requiring the "electron wave" circling the nucleus to be a standing wave. Then with $\lambda = 2\pi/k$ as wavelength, put

$$2\pi r = n\lambda = n\frac{2\pi}{k} = n\frac{2\pi\hbar}{p} ,$$

which yields

$$L = rp = n\hbar .$$

But a particle should be associated with a wave packet around some $\mathbf{k}_0$. The particle's velocity should be the group velocity $\mathbf{v}_\mathrm{g}$ given by (6.16). Is that consistent? The momentum was already fixed to be $\mathbf{p} = \hbar\mathbf{k}_0$, while on the other hand

$$\hbar\mathbf{k}_0 = \mathbf{p} = \tilde{m}\mathbf{v}_\mathrm{g} = \tilde{m}\frac{\partial w}{\partial \mathbf{k}}(\mathbf{k}_0) \tag{6.20}$$

should hold. But (6.19) is already fixed. It is rather surprising that (6.19) solves the equation (6.20). According to (6.19), we get

$$\tilde{m}\frac{\partial w}{\partial \mathbf{k}}(\mathbf{k}_0) = \frac{\tilde{m}\mathbf{k}_0 c^2}{\omega(\mathbf{k}_0)} = \frac{\tilde{m}c^2}{\hbar\omega(\mathbf{k}_0)}\hbar\mathbf{k}_0 = \frac{\tilde{m}c^2}{E}\mathbf{p} = \mathbf{p}\,.$$

Equation (6.19) is the relativistic dispersion relation for a particle of mass m. Making the Newtonian approximation yields the Galilean dispersion relation (dropping the constant phase)

$$\omega(\mathbf{k}) = \frac{1}{2}\frac{\hbar k^2}{m}\,. \tag{6.21}$$

This is the key to Schrödinger's equation.

The careful reader will have noticed that we did not spell out exactly how wave and particle were brought together. Louis de Broglie had various ideas about that. He tried one idea – after the advent of Schrödinger's wave equation, which he subsequently used – and wrote the equations of Bohmian mechanics on the blackboard at the famous fifth Solvay conference in 1927. His colleagues did not like it and gave him a hard time, especially Wolfgang Pauli, who actually reformulated de Broglie's suggestion as in (7.16). Discouraged by the general animosity of his colleagues, he did not pursue his idea any further, and it was not until 1952 that David Bohm [1] published the equations again without being aware of the Solvay history (more details of this history can be found in [2] and [3]). However, Bohm did not only write down the equations, which incidentally are quite obvious. He also analyzed the new theory, explaining non-relativistic quantum theory. All mysteries evaporated under Bohm's analysis, except maybe one: Where does quantum randomness come from?

References

1. D. Bohm: Physical Review **85**, 166 (1952)
2. J.T. Cushing: *Quantum Mechanics*. Science and Its Conceptual Foundations (University of Chicago Press, Chicago, IL, 1994). Historical contingency and the Copenhagen hegemony
3. G. Bacciagaluppi, A. Valentini: To be published by Cambridge University Press 2008 (2006) p. 553

Chapter 7
Schrödinger's Equation

Over the period 1925–1926, Werner Heisenberg, Max Born, Pasqual Jordan, Paul Dirac and Erwin Schrödinger discovered "modern" quantum theory almost simultaneously. Schrödinger's first steps were rather different from Heisenberg's. Schrödinger turned de Broglie's 1923 idea of matter waves into a mathematical theory connecting them to the eigenvalue problem of partial differential operators – a prospering topic in mathematical physics at the time: eigenmodes and discrete eigenvalues fitted well with the discreteness of spectral lines. Schrödinger found the partial differential equation which governed all that.

7.1 The Equation

Consider the de Broglie wave packet

$$\psi(\mathbf{q},t) = \int f(\mathbf{k}) \mathrm{e}^{\mathrm{i}[\mathbf{k}\cdot\mathbf{q} - \omega(\mathbf{k})t]} \mathrm{d}^3 k \, .$$

Introducing the dispersion relation (6.21), viz.,

$$\omega(\mathbf{k}) = \hbar k^2 / 2m \, , \tag{7.1}$$

we realize that differentiation with respect to time can also be achieved by differentiating with respect to position:

$$\mathrm{i}\frac{\partial \psi}{\partial t} = -\frac{\hbar}{2m}\frac{\partial^2}{\partial \mathbf{q}^2}\psi \, .$$

In a manner of speaking this equation governs the freely moving wave packet of a particle of "mass" m.

Now according to de Broglie $\hbar\omega = E$, which suggests an analogy with the Hamiltonian of a freely moving particle:

D. Dürr, S. Teufel, *Bohmian Mechanics*, DOI 10.1007/978-3-540-89344-8_7,

$$\hbar\omega(\mathbf{k}) = \frac{(\hbar\mathbf{k})^2}{2m} = \frac{\mathbf{p}^2}{2m} = H(\mathbf{p}) .$$

We may therefore view

$$-\frac{\hbar^2}{2m}\frac{\partial^2}{\partial\mathbf{q}^2}$$

as the analogue of $H(\mathbf{p})$. Extending this to the general form of the Hamiltonian

$$H = \frac{p^2}{2m} + V(\mathbf{q}) ,$$

it seems plausible to add $V(\mathbf{q})$ to $-(\hbar^2/2m)(\partial^2/\partial\mathbf{q}^2)$, leading to

$$\mathrm{i}\hbar\frac{\partial}{\partial t}\psi(\mathbf{q},t) = \left[-\frac{\hbar^2}{2m}\frac{\partial^2}{\partial\mathbf{q}^2} + V(\mathbf{q})\right]\psi(\mathbf{q},t) .$$

The nice thing about the analogy is that the Hamiltonian is a function on phase space. Therefore, extending the analogy to an N-particle system, we arrive at

$$\mathrm{i}\hbar\frac{\partial}{\partial t}\psi(\mathbf{q},t) = \left[\sum_{i=1}^{N} -\frac{\hbar^2}{2m_i}\frac{\partial^2}{\partial\mathbf{q}_i^2} + V(\mathbf{q})\right]\psi(\mathbf{q},t) . \tag{7.2}$$

Here $\mathbf{q} = (\mathbf{q}_1, \dots, \mathbf{q}_N)$, and the wave function ψ for an N-particle system comes out "naturally" as a function of *configuration* and time, i.e., $\psi(\mathbf{q},t) = \psi(\mathbf{q}_1, \dots, \mathbf{q}_N, t)$. Equation (7.2) is the celebrated Schrödinger wave equation for an N-particle system.

Schrödinger's way of arriving at his equation was different from ours, but by no means more straightforward. The differential operator (including the potential as a multiplication operator) on the right-hand side of (7.2) is commonly called the Hamilton operator. The main reason for this name is that it is built from the old Hamilton function, but with $\mathbf{p}_i$ replaced by the momentum operator

$$\hat{\mathbf{p}}_i := -\mathrm{i}\hbar\frac{\partial}{\partial\mathbf{q}_i} ,$$

while $\mathbf{q}_i$ becomes the position multiplication operator $\hat{\mathbf{q}}_i := \mathbf{q}_i$: multiply by $\mathbf{q}_i$. For some, the essence of quantum physics is to "put hats" on classical observables to turn them into operators.

Why is it nice that the wave function is a function on configuration space, when this is so very different from de Broglie's original idea of having a particle and a wave – one wave per particle? The answer is that this turns out to be the correct description of nature. Spectral lines for many-electron atoms result correctly from ground states and excited states,[1] just because the corresponding wave functions are bona fide functions on configuration space, i.e., they do not factorize into "single-particle" wave functions. Schrödinger's wave function was, and will always be, a function on configuration space.

It was recognized, mainly by Schrödinger and Einstein, that this fact might be a revolution in physics. It is in fact *the* revolution brought upon us by quantum mechanics. Not wave and/or particle, not operator observable, not uncertainty principle, and certainly not philosophical talk, even when filled with good sense (if that is possible). These are no revolution with respect to good old classical physics. What *is* new is that the description of nature *needs* a function on the *configuration space* of all particles in the system. And why is that revolutionary? The point is that such a description involves *all* particles in the universe at once, whence all particles are "entangled" with each other, and there is no obvious reason why particles that are very far apart from each other should become disentangled. *Entanglement* was the word Schrödinger used, acknowledged and celebrated almost a century later. *Entanglement* is the source of *nonlocality*. We shall discuss that in the chapter on nonlocality.

Now here is a less profound assertion: the Schrödinger equation is the simplest Galilean invariant wave equation one can write down. That is in the end how we should view that equation. Never mind how we came to it! Mathematicians may want to argue about the terminology "simple", and others may argue about Galilean invariance, although physicists do not usually argue with that. What is there to argue about Galilean invariance? Clearly, V has to be a Galilean invariant potential, so there is nothing to argue about there. What may seem puzzling, however, is that the Schrödinger equation contains a first order derivative in time, so what about time-reversal invariance? Another disquieting observation is the imaginary unit i multiplying the time derivative. That means that ψ will in general be complex. As if complex numbers could suddenly achieve physical reality! They do in fact, and we shall come to that.

Before that we begin with the simple things. We may forget about the potential, which must be a Galilean invariant expression and nothing more needs to be said about that. Let us look for simplicity at the one-particle equation

$$\mathrm{i}\hbar\frac{\partial\psi}{\partial t}(\mathbf{q},t) = -\frac{\hbar^2}{2m}\Delta\psi(\mathbf{q},t)\,. \tag{7.3}$$

First check translation invariance in space ($\mathbf{q}' = \mathbf{q}+\mathbf{a}$) and time ($t' = t+s$), as well as rotation invariance ($\mathbf{q}' = R\mathbf{q}$, $t' = t$). The invariance is easy to see. Both

$$\psi'(\mathbf{q}',t') := \psi(\mathbf{q}'-\mathbf{a},t'-s)$$

and

[1] V can be taken as sum of Coulomb potentials between the electrons surrounding the nucleus and the Coulomb potential between the electrons and the nucleus, "binding" the electrons to the nucleus.

$$\psi'(\mathbf{q}',t') := \psi(R^{\mathrm{t}}\mathbf{q}',t')$$

satisfy (7.3) in the primed variables. (Note that $\Delta = \nabla^{\mathrm{t}}\nabla = \nabla\cdot\nabla$ is a scalar and hence automatically rotation invariant.)

In order to see time reversibility, one must open up a new box. Suppose $t' = -t$, $\mathbf{q}' = \mathbf{q}$, and suppose that $\psi'(\mathbf{q}',t') = \psi(\mathbf{q}',-t')$. Equation (7.3) contains only one derivative with respect to time, and therefore the left-hand side of (7.3) changes sign, while the right-hand side remains unchanged. This is not good. Chapter 3 helps us to move on. To $t \mapsto -t$ and $(\mathbf{q}' = \mathbf{q})$, we adjoin complex conjugation $\mathrm{i} \mapsto -\mathrm{i}$, i.e., $\psi \mapsto \psi^*$. Then setting

$$\psi'(\mathbf{q}',t') = \psi^*(\mathbf{q}',-t') ,$$

we see that ψ' satisfies (7.3) (in primed variables). How is this? We take the complex conjugate of (7.3) and see that there is an extra minus sign. But hold on! What if there is a potential function $V(\mathbf{q})$ in the equation, as there generally will be? Then taking the complex conjugate of the equation (7.2), the resulting equation contains V^*. Time reversal holds only if $V = V^*$, i.e., if V is real!

Now, the way we argued that the Schrödinger equation comes about, it seems natural that V, as the classical potential, should be real. But once the equation is there, the meaning of V must emerge from the analysis of the new theory, and V could in principle be any Galilean invariant function. Since ψ is complex, why not V? Here we now see that V must be real for the "theory" to be time-reversal invariant.

We come now to Galilean boosts when the coordinate system is changed to a frame which moves with relative velocity $-\mathbf{u}$. Then

$$\mathbf{q}' = \mathbf{q}+\mathbf{u}t \quad (\mathbf{v}' = \mathbf{v}+\mathbf{u}) , \qquad t' = t . \tag{7.4}$$

Suppose

$$\psi'(\mathbf{q}',t') = \psi(\mathbf{q}'-\mathbf{u}t',t') . \tag{7.5}$$

But then (in the following $\mathbf{q}$ stands for $\mathbf{q}'-\mathbf{u}t'$)

$$\begin{aligned}
\mathrm{i}\hbar\frac{\partial}{\partial t'}\psi'(\mathbf{q}',t') &= \mathrm{i}\hbar\frac{\partial}{\partial t'}\psi(\mathbf{q}'-\mathbf{u}t',t') \\
&= \mathrm{i}\hbar\frac{\partial}{\partial t'}\psi(\mathbf{q},t') - \mathrm{i}\hbar\mathbf{u}\cdot\nabla\psi(\mathbf{q},t') \\
&= -\frac{\hbar^2}{2m}\Delta\psi(\mathbf{q},t') - \mathrm{i}\hbar\mathbf{u}\cdot\nabla\psi(\mathbf{q},t') \\
&= -\frac{\hbar^2}{2m}\Delta'\psi'(\mathbf{q}',t') - \mathrm{i}\hbar\mathbf{u}\cdot\nabla'\psi'(\mathbf{q}',t'),
\end{aligned} \tag{7.6}$$

and we obtain an extra term in the Schrödinger equation. The transformation of ψ under boosts must be more complicated than (7.5). No wonder you might say. It is after all a wave which we wish to boost in this way.

So let us consider the de Broglie plane wave $e^{i[\mathbf{k}\cdot\mathbf{q}-\omega(\mathbf{k})t]}$, with dispersion

$$\omega(\mathbf{k}) = \frac{1}{2}\frac{\hbar k^2}{m} .$$

Applying (7.5), setting $\mathbf{k}_u = m\mathbf{u}/\hbar$, and doing simple binomial analysis using the dispersion, we obtain

$$\begin{aligned}
&e^{i[\mathbf{k}\cdot(\mathbf{q}'-\mathbf{u}t')-\omega(\mathbf{k})t']} \\
&\quad = \exp\left\{i\left[\mathbf{k}\cdot\mathbf{q}' - \left(\mathbf{k}\cdot\frac{\hbar}{m}\mathbf{k}_u + \omega(\mathbf{k})\right)t'\right]\right\} \\
&\quad = \exp\left\{i\left[\mathbf{k}\cdot\mathbf{q}' - \omega(\mathbf{k}+\mathbf{k}_u)t' + \omega(\mathbf{k}_u)t'\right]\right\} \\
&\quad = \exp\left\{i\left[(\mathbf{k}+\mathbf{k}_u)\cdot\mathbf{q}' - \omega(\mathbf{k}+\mathbf{k}_u)t'\right]\right\}\exp\left\{-i\left[\mathbf{k}_u\cdot\mathbf{q}' - \omega(\mathbf{k}_u)t'\right]\right\} .
\end{aligned}$$

This suggests boldly putting, rather than (7.5),

$$\psi'(\mathbf{q}',t') = \Phi_{\mathbf{k}_u}(\mathbf{q}',t')\psi(\mathbf{q}'-\mathbf{u}t',t') , \tag{7.7}$$

with the plane wave

$$\Phi_{\mathbf{k}_u}(\mathbf{q}',t') = e^{i[\mathbf{k}_u\cdot\mathbf{q}'-\omega(\mathbf{k}_u)t']} . \tag{7.8}$$

Instead of (7.6), we now have (recalling that $\mathbf{q}$ stands in the following for $\mathbf{q}'-\mathbf{u}t'$)

$$\begin{aligned}
i\hbar\frac{\partial}{\partial t'}\psi'(\mathbf{q}',t') &= i\hbar\frac{\partial}{\partial t'}\left\{e^{i[\mathbf{k}_u\cdot\mathbf{q}'-\omega(\mathbf{k}_u)t']}\psi(\mathbf{q}'-\mathbf{u}t',t')\right\} \\
&= e^{i[\mathbf{k}_u\cdot(\mathbf{q}+\mathbf{u}t')-\omega(\mathbf{k}_u)t']}\left[\hbar\omega\psi(\mathbf{q},t') - \frac{\hbar^2}{2m}\Delta\psi(\mathbf{q},t') - i\hbar\mathbf{u}\cdot\nabla\psi(\mathbf{q},t')\right] ,
\end{aligned}$$

which equals $-\hbar^2\Delta'\psi'(\mathbf{q}',t')/2m$, as can be seen from the following:

$$\begin{aligned}
&-\frac{\hbar^2}{2m}\Delta'\psi'(\mathbf{q}',t') \\
&\quad = -\frac{\hbar^2}{2m}\Delta'\left[\Phi_{\mathbf{k}_u}(\mathbf{q}',t')\psi(\mathbf{q}'-\mathbf{u}t',t')\right] \\
&\quad = -\frac{\hbar^2}{2m}\left[\Delta'\Phi_{\mathbf{k}_u}(\mathbf{q}',t')\right]\psi(\mathbf{q}'-\mathbf{u}t',t') - \frac{\hbar^2}{2m}\Phi_{\mathbf{k}_u}(\mathbf{q}',t')\Delta'\psi(\mathbf{q}'-\mathbf{u}t',t') \\
&\qquad - \frac{\hbar^2}{m}\left[\nabla'\Phi_{\mathbf{k}_u}(\mathbf{q}',t')\right]\cdot\left[\nabla'\psi(\mathbf{q}'-\mathbf{u}t',t')\right] \\
&\quad = e^{i[\mathbf{k}_u\cdot(\mathbf{q}+\mathbf{u}t')-\omega(\mathbf{k}_u)t']}\left[\hbar\omega\psi(\mathbf{q},t') - \frac{\hbar^2}{2m}\Delta\psi(\mathbf{q},t') - \frac{i\hbar^2\mathbf{k}_u}{m}\cdot\nabla\psi(\mathbf{q},t')\right] ,
\end{aligned}$$

where we have made use of $\mathbf{u} = \hbar\mathbf{k}_u/m$ and (7.1).

The transformation (7.7), viz.,

$$\psi'(\mathbf{q}',t') = \mathrm{e}^{\mathrm{i}\mathbf{k}_u\cdot\mathbf{q}'}\psi(\mathbf{q}'-\mathbf{u}t',t')\mathrm{e}^{-\mathrm{i}\omega(\mathbf{k}_u)t'} ,$$

contains the phase factor $\mathrm{e}^{-\mathrm{i}\omega(\mathbf{k}_u)t'}$ which, if ignored, would result in an additive constant in the wave equation, i.e., the potential V (when we put that back in) gets "gauged" by the constant $\hbar\omega(\mathbf{k}_u)$, which should not have any effect on the physics.[2] This would imply that the physics is described by an equivalence class of wave equations which all differ by constants added to the potential. Put another way, wave functions which differ by a position-independent phase factor describe the same physics. In technical terms, we can say that the Galilean group is represented in the projective space consisting of the rays

$$(\psi) := \{c\psi, c\in\mathbb{C}, c\neq 0\} .$$

A common argument to support this goes as follows. The translation $T_\mathbf{a}$ and boost $B_\mathbf{u}$ are transformations which commute in the Galilean group, whereas acting on waves, and ignoring the phase factors $\mathrm{e}^{-\mathrm{i}\omega(\mathbf{k}_u)t}$,

$$B_\mathbf{u}T_\mathbf{a} \quad \text{yields} \quad \mathrm{e}^{\mathrm{i}\mathbf{k}_u\cdot\mathbf{q}'}\psi(\mathbf{q}'-\mathbf{a}-\mathbf{u}t',t') = \psi'(\mathbf{q}',t') ,$$

$$T_\mathbf{a}B_\mathbf{u} \quad \text{yields} \quad \mathrm{e}^{-\mathrm{i}\mathbf{k}_u\cdot\mathbf{a}}\mathrm{e}^{\mathrm{i}\mathbf{k}_u\cdot\mathbf{q}'}\psi(\mathbf{q}'-\mathbf{a}-\mathbf{u}t',t') = \mathrm{e}^{-\mathrm{i}\mathbf{k}_u\cdot\mathbf{a}}\psi'(\mathbf{q}',t') ,$$

so that we should view $\psi'(\mathbf{q}',t')$ and $\mathrm{e}^{-\mathrm{i}\mathbf{k}_u\cdot\mathbf{a}}\psi'(\mathbf{q}',t')$ as equivalent.

The way the wave function transforms may be a bit upsetting. Why should a fundamental description of nature change its appearance so drastically when the coordinate frame is changed? Maybe it would all become more understandable if we knew what role the ψ function really played? Perhaps the ψ function is not the primary ontological variable and its role is understood only in connection with some primary variables still to be determined, like the **B** field in electromagnetism, which changes its sign under time reversal? In the latter case, we understand that this must be so by the role the **B** field plays in the dynamics of charged particles.

Schrödinger's equation is a wave equation, a wave evolving in time. However, the first use of the equation was to explain the stationary electronic states of an atom, i.e., wave functions which depend only trivially on time (through a simple phase factor $\mathrm{e}^{\mathrm{i}Et/\hbar}$). They are solutions of the so-called stationary Schrödinger equation. The contact made with the real world is very indirect, via spectral lines. But what, we must ask, is the physical meaning of the wave function? What role does it play? What does a traveling (i.e., non-stationary) wave mean?

[2] Whether it should or should not have any effect is of course not decidable on a priori grounds. We should really withhold any definitive claim until we have completely understood what the new theory is.

7.2 What Physics Must Not Be

Heisenberg also wanted a theory behind the spectral lines, but he did not think of waves at all. What he did was in a sense much more ingenious, by seeing the discrete spectral energies emerging from a new calculus of arrays of numbers – the matrix calculus of infinitely extended matrices. In that view, quantization meant that the Poisson bracket of Hamiltonian mechanics is to be replaced by the commutator $[\cdot,\cdot]/i\hbar$ and thereby forming an operator algebra. The important output from this was the famous Heisenberg uncertainty principle, which in a lax manner of speaking says that one cannot simultaneously determine the position and momentum of a particle. This became the principle of indeterminism in the new physics.

After Schrödinger had established the equivalence between his wave mechanics and Heisenberg's matrix mechanics, everything eventually found its place in Dirac's formalism. Dirac established the powerful formalism of quantum mechanics, by adopting a notation for infinite-dimensional vectors whereby the vectors can be effortlessly represented relative to a suitable "orthonormal basis". This is a technical tool of great power, although the terminology "orthonormal basis" is in most cases mathematically incorrect. But having said this, the final result of such formal computations is always correct. Dirac's formalism is what one usually learns in courses on quantum mechanics.

But let us return to 1926. Heisenberg did not like Schrödinger's waves, and hoped that this path would eventually turn into a mere meander. Born, however, immediately saw the descriptive power of the time-dependent Schrödinger equation, and applied it to scattering. In this application he discovered that the wave function ψ has an empirical meaning as a "probability amplitude". In fact, through its modulus squared $|\psi(\mathbf{x},t)|^2$, the wave function, which is generally a complex function, delivers the theoretical prediction for the empirical density of finding the particle at position $\mathbf{x}$ at time t.

There are two famous quotes from Born's two papers. The first [1] is an announcement, and the second [2], an elaboration on it. The quotes show that Born had an absolutely correct, and one must say frankly ingenious, intuition about the meaning of the wave function, namely that it guides the particles and that it determines the probabilities of particle positions. This probability was understood as irreducible[3] (backed, of course, by the uncertainty principle). The intrinsic randomness fascinated the scientific community for decades, and the second quote in particular shows that "dark" discussions on that were taking place. In truth, quantum randomness is good old Boltzmannian statistical equilibrium, albeit for a new mechanics: Bohmian mechanics. But more on that later. Here is the quote from [1]:

> I want to tentatively follow the idea: The guiding field, represented by a scalar function of the coordinates of all particles taking part and the time, evolves according to Schrödinger's differential equation. Momentum and energy will however be transferred as if corpuscles (electrons) were indeed flying around. The trajectories of these corpuscles are only restricted by energy and momentum conservation; apart from that the choice of a particular

[3] Born received the Nobel prize for this discovery, although very late and mainly because of repeated recommendations by Einstein.

> trajectory is only determined by probability, given by the values of the function ψ. One could, somewhat paradoxically, summarize this as follows: The motion of the particles follows probabilistic laws, while the probability itself evolves according to a causal law.[4]

In the main paper [2], Born surprisingly felt the need to relativize the strong indeterminism:

> In my preliminary note I have strongly emphasized this indeterminism, since that seems to me to best conform with the praxis of experimenter. But anybody who is not satisfied with that is free to assume that there are more parameters, which have so far not been introduced into the theory, and which determine the individual event. In classical physics these are the "phases" of the motion, for example the coordinates of the particles at a certain moment. It seemed to me at first improbable that one could casually introduce variables which correspond to such phases into the new theory; but Mr. Frenkel told me that this may be possible after all. In any case, this possibility would not change the practical indeterminism of the collision processes, since one cannot give the values of the phases; it must lead to the same formulas, like the "phaseless" theory proposed here.[5]

Born is close to Bohmian mechanics in his preliminary note, although his naive insistence on the Newtonian momentum and energy conservation for the particle trajectories[6] is completely gratuitous and is not in fact correct for Bohmian mechanics. Why should the new mechanics, which is based on a guidance principle, obey Newtonian principles?

In the second quote he says that he thought it improbable (meaning something like impossible) that particle trajectories could be introduced in a casual (meaning natural) way. But that is exactly what can be done, and it is trivial! The reference to Mr. Frenkel is amusing, and historians may find pleasure in finding out what the Frenkel story was about, and what he had in mind.

Why are the quotes so important? Because they show the advent of a new theory of physics, supplementing Schrödinger's wave function description by an idea

[4] Ich möchte also versuchsweise die Vorstellung verfolgen: Das Führungsfeld, dargestellt durch eine skalare Funktion der Koordinaten aller beteiligten Partikeln und der Zeit, breitet sich nach der Schrödingerschen Differentialgleichung aus. Impuls und Energie aber werden so übertragen, als wenn Korpuskeln (Elektronen) tatsächlich herumfliegen. Die Bahnen dieser Korpuskeln sind nur so weit bestimmt, als Energie- und Impulssatz sie einschränken; im übrigen wird für das Einschlagen einer bestimmten Bahn nur eine Wahrscheinlichkeit durch die Werteverteilung der Funktion ψ bestimmt. Man könnte das, etwas paradox, etwa so zusammenfassen: Die Bewegung der Partikeln folgt Wahrscheinlichkeitsgesetzen, die Wahrscheinlichkeit selbst aber breitet sich im Einklang mit dem Kausalgesetz aus.

[5] Ich habe in meiner vorläufigen Mitteilung diesen Indeterminismus ganz besonders betont, da er mir mit der Praxis des Experimentators in bester Übereinstimmung zu sein scheint. Aber es ist natürlich jedem, der sich damit nicht beruhigen will, unverwehrt, anzunehmen, da es weitere, noch nicht in die Theorie eingeführte Parameter gibt, die das Einzelereignis determinieren. In der klassischen Mechanik sind dies die "Phasen" der Bewegung, z.B. die Koordinaten der Teilchen in einem bestimmten Augenblick. Es schien mir zunächst unwahrscheinlich, daß man Größen, die diesen Phasen entsprechen, zwanglos in die neue Theorie einfügen könne; aber Herr Frenkel hat mir mitgeteilt, da dies vielleicht doch geht. Wie dem auch sei, diese Möglichkeit würde nichts an dem praktischen Indeterminismus der Stoßvorgänge ändern, da man ja die Werte der Phasen nicht angeben kann; sie muß übrigens zu denselben Formeln führen, wie die hier vorgeschlagene "phasenlose" Theorie.

[6] As if the notion of particle alone forced Newtonian behavior.

which can be trivially brought to completion. It could easily have been done by Born himself. What were Born's grounds for bringing in probability at all? Of course, there was Heisenberg's uncertainty principle, but that alone was not sufficient, as Schrödinger had pointed out to him. There was an extra equation (also derived by Schrödinger) – an identity which follows from Schrödinger's equation – which corroborated the interpretation of the wave function as determining the probabilities. From that equation the guided trajectories are transparently obvious, as will be shown in Sect. 7.3. But apart from Born, either Einstein or Schrödinger could have done that.

In fact the equation for the trajectories, albeit interpreted as fluid lines, was immediately seen by Erwin Madelung (a mathematical physicist and friend of Born) [3], and at the 1927 Solvay conference de Broglie introduced the same Madelung fluid lines as particle trajectories. But this was ridiculed by all the other participants. Einstein in particular found that a guiding wave on configuration space made no sense whatsoever: physics *must* not be like that. However, Einstein had no problem with probability being on configuration space, just as the classical canonical ensembles are measures on many-particle phase space, so that the statistical part of Born's thesis was fine for him. Einstein nevertheless felt, and Schrödinger likewise, that the wave function guiding particles on configuration space meant a revolution in physics – if that turned out to be a true feature of nature.

It is very difficult these days to appreciate what really blocked physicists' minds in this context, preventing them from focusing on the obvious – a new mechanics, even if it did turn out to be revolutionary. In our own time people come up with all kinds of crazy ideas in physics and are applauded for it. So why did nobody probe Born's guidance idea and see what it could achieve? Why only Mr. Frenkel and nobody else? What really prevented anyone from saying loud and clear that particles exist and move (what else can they do when they exist)? This is what historians should work to find out, because what happened is really beyond understanding. But it may take many good historians of science to sort out the mess.

Ignoring de Broglie's attempt in the 1927 Solvay conference, and ignoring Mr. Frenkel's and Born's feeble attempts, the question: *What is quantum mechanics really about?* never quite surfaced as a clear-cut and burning question in the minds of physicists, with the exceptions of Einstein and Schrödinger. Instead there was a muddle of philosophical talk about what Heisenberg's findings really meant for physics, and whether Schrödinger's wave function provided a *complete* description of a physical system, in the sense that nothing more is needed (except talk perhaps). Early on, Einstein and Bohr were the leading figures in that discussion, and they were both right and wrong in some ways.

Schrödinger originally thought that the wave function provided a complete description, in that it described matter, but in the end he could not adhere to that view, because the wave spreads while atomistic matter in all experiments came out pointlike. Let a wave impinge on a double slit. Then on the screen at the other side of the slit, a black point appears randomly (in time and space).

More dramatically Einstein and Schrödinger were concerned about the linearity of the wave equation and entanglement. If the wave function is the complete description of the physics, this leads to the measurement problem (known most prominently as Schrödinger's cat paradox), which we talked about in the introduction and shall do again in the next section. Schrödinger and Einstein (among a few others) had to hold on to the view that the quantum mechanical description of nature is *incomplete*.

In opposition to this, Bohr, Heisenberg, Dirac, Pauli, and most physicists held that the new findings necessitated a new philosophical view of nature. The new philosophical view was that the question "What is going on in nature?" is unphysical, unscientific, and uneverything. In other words, the incompleteness idea became rather heavily outlawed. The new philosophical view also endorsed the understanding that one does not necessarily mean what one says. For example, when one says that the particle hits the screen (where the black spot appears in the two slit experiment), one does not mean that there is a particle hitting the screen, but rather something else. However, this something else cannot be expressed other than by saying that a particle hits the screen.

It was as if our human capabilities were not far-reaching enough to actually grasp and describe what really happens when we say that a particle hits the screen. The mystery which historians have to work out is why there was a need to forbid the meaning that a particle hits the screen, when it does seem that a particle hits the screen, especially when this is theoretically very simple to describe. Bohmian mechanics does it with consummate ease. The great mystery is why the majority of physicists took to heart the idea that physics *must* not be about ontology. But what else could it be about?

Perhaps the time has come for something positive. In the end all the founding fathers of quantum mechanics were right to some extent. Bohr's insistence that "observables" only have meaning in connection with an experiment and represent no properties of a system is largely justified. Einstein and Schrödinger were right in their conviction that the wave function cannot represent the complete description of the state of a system. Born was right in seeing that, at the end of the day, the empirical feature emerging from the complete state description is the $|\psi|^2$ statistics. Bohmian mechanics combines all these views into one theory in a *surprisingly trivial manner*.

Remark 7.1. On Solutions of Schrödinger's Equation
The Schrödinger equation is a linear partial differential equation. As such it does not conceal any of the exciting features which make nonlinear partial differential equations so appealing to mathematicians: shock waves, explosions, and so on. And yet the theory of classical solutions of the Schrödinger equation, i.e., solutions which are nice differentiable functions and which solve the Schrödinger equation in just the way a normal person would expect, is not usually textbook material. Mathematical physics focuses more on the Hilbert space theory of solutions which is connected to the self-adjointness of the Schrödinger operator and the unitarity of the time evolution. We also do that in Chap. 14. In Bohmian mechanics, which we discuss in the next chapter, the wave function must be differentiable, and so one needs classical solutions of the Schrödinger equation. The relevant assertions about the classical solutions of Schrödinger's equation can be found in [4, 5]. ■

7.3 Interpretation, Incompleteness, and $\rho = |\psi|^2$

In books and seminars on quantum mechanics, there is so much talk about *interpretation*. One talks about *interpretations of quantum mechanics*: Copenhagen, many worlds, Bohmian , and so on. As if the laws of quantum mechanics were a Delphic oracle which required high priests to be deciphered. What is special about quantum mechanics as compared to Newtonian mechanics, where only a few scientists (influenced by quantum mechanics) would insist that Newtonian mechanics needed an interpretation? Newton certainly did not think this way, and nor did Leibniz (actually the equations in the form we are used to seeing them were written by Leibniz).

Interpretation has become a multipurpose concept in quantum mechanics. The interpretation of the wave functions is that $|\psi|^2$ is a probability density. That is interpretation in the good sense. We shall look at that in this section. Bohmian mechanics is often said to be an interpretation of quantum mechanics, which should indicate redundancy, i.e., that it is *merely* an interpretation. But Bohmian mechanics is not an interpretation of anything. We shall come to that in the next chapter. It should be clear from the above quotes that Born would also have called the theory with trajectories (with "phases") a new theory. On the other hand, in some very vague sense, Bohmian mechanics is an interpretation of quantum mechanics. It is a complete theory where nothing is left open, and above all, it does not need an interpretation. It is a theory of nature, and it has a precise link to quantum mechanics. Indeed, it explains its rules and formalisms, so when someone says that the momentum and the position of a particle are non-commutative operators, which does sound like a Delphic oracle, Bohmian mechanics fills in all the ideas needed to see what this could possibly mean.

All interpretations in quantum mechanics are linked to one essential question: What is the role of the wave function in physics? As already mentioned, Schrödinger originally thought that the wave function represented the stuff the world was made of – a matter wave, so to speak. Indeed, it is a matter wave, on a strangely high-dimensional space, the configuration space of all the particles as it were. But since there are no particles, only a wave, it is not the configuration space of particles, but simply a curious high-dimensional space.

One could just live with that. But other things speak against the "interpretation" of the wave function as a matter wave. Even thinking of the wave packet of a single electron, that wave packet spreads according to the dispersion relation, whereas spread out electrons are not observed. They are always pointlike. Send an electron wave through a slit, and the wave evolves after the slit according to Huygens' principle into a spherical wave. Yet on the photographic plate somewhere behind the slit, we see a black spot where the particle has arrived, and not the gray shade of the extended wave.

The idea of matter waves becomes grotesque when one considers measurement situations and the measurement apparatus is itself described by a matter wave. Since the Schrödinger evolution is linear, superpositions of wave functions evolve into superpositions. We arrive at Schrödinger's cat. We do not really need to repeat this here, but let us do so anyhow, just to make the point. Suppose a system is described by two wave functions φ_1 and φ_2, and an apparatus is built such that its interaction with the system (the "measurement") results in the following situation: the pointer points out 1 if the system is described by the wave function φ_1 or 2 if the system is described by φ_2. That is, the apparatus is described by the two wave functions Ψ_1 and Ψ_2 (when the pointer points to 1 or 2, respectively) and the pointer 0 wave function Ψ_0, so that

$$\varphi_i \Psi_0 \quad \overset{\text{Schrödinger evolution}}{\longrightarrow} \quad \varphi_i \Psi_i \,. \tag{7.9}$$

However, the Schrödinger evolution is linear and it follows from (7.9) that, if the wave function of the system is

$$\varphi = c_1 \varphi_1 + c_2 \varphi_2 \,, \qquad c_1, c_2 \in \mathbb{C} \,,$$

then one obtains

$$\varphi \Psi_0 = (c_1 \varphi_1 + c_2 \varphi_2) \Psi_0 \quad \overset{\text{Schrödinger evolution}}{\longrightarrow} \quad c_1 \varphi_1 \Psi_1 + c_2 \varphi_2 \Psi_2 \,. \tag{7.10}$$

This is a bizarre matter wave on the right where the apparatus points *simultaneously* to 1 and 2. This conflicts with the way the apparatus was built, or if one prefers, with what our experience tells us about the way pointers show facts: either 1 or 2.

Schrödinger was well aware of this, as we explained in the introductory chapter.[7] The conclusion is that either the Schrödinger evolution is not right, or the description is incomplete. The Schrödinger evolution not being right lends support to a serious competitor of Bohmian mechanics, namely the dynamical reduction theory or GRW theory.[8] The description not being complete leads straightforwardly to Bohmian mechanics.

Let us now return to Born. We quoted above from his scattering paper, and the question is: On what grounds could Born corroborate his intuition for the probabilistic interpretation of the wave function? With Schrödinger's help,[9] an identity was established for any solution of the Schrödinger equation. This identity involves the "density" $|\psi(\mathbf{q},t)|^2 = \psi^*(\mathbf{q},t)\psi(\mathbf{q},t)$ and has the form of a continuity equation, called the quantum flux equation:

$$\frac{\partial |\psi|^2}{\partial t} + \nabla \cdot \mathbf{j}^\psi = 0 \,, \tag{7.11}$$

[7] A helpful discussion of the possibility of "assigning matter to the wave function" can be found in [6].

[8] GRW stands for Ghirardi, Rimini, and Weber, who formulated a nonlinear random evolution law for wave functions in such a manner that, in measurement situations, the theory reproduces the correct Born statistical law. See [7] for an extensive overview. It is remarkable that this nonlocal random collapse theory can be formulated in a Lorentz invariant way [8].

[9] Born's first idea was that $|\psi|$ was the probability density, but Schrödinger pointed out that $|\psi|$ does not satisfy a continuity equation, while $|\psi|^2$ does. This meant that, with the latter choice, probability would be conserved.

where $\nabla = (\nabla_k)_{k=1,\dots,n} := (\partial_{\mathbf{q}_1}, \partial_{\mathbf{q}_2}, \dots, \partial_{\mathbf{q}_N})$ and $\mathbf{j}^\psi = (\mathbf{j}_1^\psi, \dots, \mathbf{j}_N^\psi)$ is the so-called quantum flux with

$$\mathbf{j}_k^\psi = \frac{\hbar}{2\mathrm{i}m_k}(\psi^*\nabla_k\psi - \psi\nabla_k\psi^*) = \frac{\hbar}{m_k}\Im\psi^*\nabla_k\psi\,, \tag{7.12}$$

or in configurational terms, introducing the mass matrix m,

$$\mathbf{j}^\psi = \frac{\hbar}{2\mathrm{i}}m^{-1}(\psi^*\nabla\psi - \psi\nabla\psi^*) = \hbar m^{-1}\Im\psi^*\nabla\psi\,. \tag{7.13}$$

What one usually observes is that, integrating (7.11) over the whole of (configuration) space and using Gauss' theorem, the integral over the divergence term yields zero when ψ falls off fast enough at spatial infinity, whence

$$\int \frac{\partial|\psi|^2}{\partial t}\mathrm{d}^{3N}q = -\int \nabla\cdot\mathbf{j}^\psi \mathrm{d}^{3N}q = -\lim_{R\to\infty}\int_{B_R}\mathbf{j}^\psi\cdot\mathrm{d}\sigma = 0\,, \tag{7.14}$$

where B_R is a ball of radius R. The vanishing of the flux through an infinitely distant surface will later be taken up in rigorous mathematical terms when we discuss the self-adjointness of the Schrödinger operator.

So $\int q|\psi(\mathbf{q},t)|^2\mathrm{d}^{3N}q$ does not change with time, and

$$\int q|\psi(\mathbf{q},t)|^2\mathrm{d}^{3N}q = \int q|\psi(\mathbf{q},0)|^2\mathrm{d}^{3N}q\,,$$

which is usually expressed by saying that probability does not get lost. The reader who has absorbed Chap. 2 will have little problem in rephrasing (7.11) as a bona fide continuity equation. Just write

$$\nabla\cdot\mathbf{j}^\psi = \nabla\cdot\frac{\mathbf{j}^\psi}{|\psi|^2}|\psi|^2 =: \nabla\cdot\mathbf{v}^\psi|\psi|^2\,, \tag{7.15}$$

so that we have a vector field

$$\mathbf{v}^\psi(\mathbf{q},t) = \frac{\mathbf{j}^\psi(\mathbf{q},t)}{|\psi(\mathbf{q},t)|^2} = \hbar m^{-1}\Im\frac{\nabla\psi}{\psi}(\mathbf{q},t)\,, \tag{7.16}$$

along which the density $\rho(\mathbf{q},t) = |\psi(\mathbf{q},t)|^2$ is transported, i.e., we have the continuity equation

$$\frac{\partial|\psi|^2}{\partial t} + \nabla\cdot\left(\mathbf{v}^\psi|\psi|^2\right) = 0\,. \tag{7.17}$$

The integral curves along this vector field are the trajectories Born might have had in mind. Particles moving along these trajectories are guided by the wave function, since the vector field is induced by the wave function. But unfortunately Born insisted on momentum and energy conservation, which do not hold here, because the integral curves are in fact the Bohmian trajectories. Why was this not accepted and the theory thereby completed? No one knows. In any case $\rho(\mathbf{q},t) = |\psi(\mathbf{q},t)|^2$ became the accepted interpretation of the wave function: it determines the probability of finding the system in the configuration $\mathbf{q}$ when its wave function is $\psi(\mathbf{q},t)$.

And here enters another confusing issue. There was a debate about whether the right word was "finding" or "is", because the latter would insist on particles actually being there, i.e., the system *is* in the configuration q with probability $|\psi(\mathbf{q},t)|^2$, and the former would not say anything of substance. However, the source of the debate can be located somewhere else, namely, in the measurement formalism of quantum mechanics. Bohmian mechanics will explain that it is correct to say "is" for the positions of the system particles, but that it is *not* correct to say "is" for other "observables". This sounds deeper than it really is, but the clarification of this point is absolutely essential for a rational understanding.

We now compute the identity (7.11) with $\mathbf{j}^\psi$ given by (7.12). Starting with Schrödinger's equation

$$\mathrm{i}\hbar\frac{\partial\psi}{\partial t} = -\sum\frac{\hbar^2}{2m_k}\Delta_k\psi + V\psi\,,$$

since V is a real function (by time reversibility!), complex conjugation yields

$$-\mathrm{i}\hbar\frac{\partial\psi^*}{\partial t} = -\sum\frac{\hbar^2}{2m_k}\Delta_k\psi^* + V\psi^*\,. \tag{7.18}$$

Multiply the first equation by ψ^* and the second by ψ. The resulting equations have the same $V\psi\psi^*$ term on the right, so subtract the equations and observe that the time derivative terms add to a time derivative of $|\psi|^2$. Hence,

$$\begin{aligned}
\mathrm{i}\hbar\frac{\partial|\psi|^2}{\partial t} &= -\sum\frac{\hbar^2}{2m_k}(\psi^*\Delta_k\psi - \psi\Delta_k\psi^*) \\
&= -\sum\frac{\hbar^2}{2m_k}\nabla_k(\psi^*\nabla_k\psi - \psi\nabla_k\psi^*) \qquad \text{(by the product rule)} \\
&= -\mathrm{i}\hbar\nabla\cdot\mathbf{j}^\psi\,,
\end{aligned} \tag{7.19}$$

with $\mathbf{j}^\psi$ given by (7.12).

So what happens next? The reader who has come this far and who appreciates the statistical mechanics we prepared in previous sections may find the situation thrilling. Many analogies may come to mind, and many questions: What is this particle theory whose trajectories are so clearly stated in the trivial rewriting (7.17)? What is the status of $\rho = |\psi|^2$ in this theory? And what about the uncertainty principle? Is that an extra metaphysical principle for the new theory with trajectories? Does $\rho = |\psi|^2$ bear any analogy with the Liouville measure of classical statistical mechanics? What new things can be learned from the new particle theory? What about entanglement? So many exciting questions to ask, and so many easy answers to give! The following chapters will address all these questions. The simplicity and ease with which all these issues are explained is stunningly beautiful.

References

1. M. Born: Z. Phys. **37**, 863 (1926)
2. M. Born: Z. Phys. **38**, 803 (1926)
3. E. Madelung: Zeitschrift für Physik **40** (3/4), 322 (1926)
4. K. Berndl, D. Dürr, S. Goldstein, G. Peruzzi, N. Zanghì: Commun. Math. Phys. **173** (3), 647 (1995)
5. K. Berndl: Zur Existenz der Dynamik in Bohmschen Systemen. Ph.D. thesis, Ludwig-Maximilians-Universität München (1994)
6. V. Allori, S. Goldstein, R. Tumulka, N. Zanghí: British Journal for the Philosophy of Science (2007)
7. A. Bassi, G. Ghirardi: Phys. Rep. **379** (5–6), 257 (2003)
8. R. Tumulka: J. Stat. Phys. **125** (4), 825 (2006)

Chapter 8
Bohmian Mechanics

Bohmian mechanics is the new mechanics for point particles. In the equations for Bohmian mechanics there are parameters $m_1,\dots,m_N$ which we shall call masses. We do so because in certain physical situations the particles will move along Newtonian trajectories and then these masses are Newtonian masses, and there is no point in inventing new names here. Although the theory is not at all Newtonian, it is nevertheless close in spirit to the Hamilton–Jacobi theory and an implementation of Born's guiding idea. The theory is in fact the minimal non-trivial Galilean theory of particles which move. We already gave the defining ingredients in the last chapter. Now we shall spell things out in detail.

An N-particle system with "masses" $m_1,\dots,m_N$ is described by the positions of its N particles $\mathbf{Q}_1,\dots,\mathbf{Q}_N$, $\mathbf{Q}_i \in \mathbb{R}^3$. The mathematical formulation of the law of motion is on configuration space $\mathbb{R}^{3N}$, which is the set of configurations $\mathbf{Q} = (\mathbf{Q}_1,\dots,\mathbf{Q}_N)$ of the positions. The particles are guided by a function

$$\begin{aligned} \psi : \mathbb{R}^{3N} \times \mathbb{R} &\longrightarrow \mathbb{C} \\ (\mathbf{q},t) &\longmapsto \psi(\mathbf{q},t) \end{aligned}$$

in the following way. ψ defines a vector field $\mathbf{v}^{\psi}(\mathbf{q},t)$ on configuration space

$$\mathbf{v}^{\psi}(\mathbf{q},t) = \hbar m^{-1} \Im \frac{\nabla \psi}{\psi}(\mathbf{q},t) , \tag{8.1}$$

where m is the mass matrix as in (2.3) and the possible particle trajectories are integral curves along the vector field:

$$\frac{\mathrm{d}\mathbf{Q}}{\mathrm{d}t} = \mathbf{v}^{\psi}\big(\mathbf{Q}(t),t\big) . \tag{8.2}$$

In other words the kth particle trajectory $\mathbf{Q}_k(t)$ obeys

$$\frac{\mathrm{d}\mathbf{Q}_k}{\mathrm{d}t} = \frac{\hbar}{m_k} \Im \frac{\nabla_k \psi}{\psi}(\mathbf{Q},t) , \qquad k = 1,\dots,N . \tag{8.3}$$

D. Dürr, S. Teufel, *Bohmian Mechanics*, DOI 10.1007/978-3-540-89344-8_8,

ψ obeys the Schrödinger equation

$$i\hbar\frac{\partial\psi}{\partial t}(\mathbf{q},t) = -\sum\frac{\hbar^2}{2m_k}\Delta_k\psi(\mathbf{q},t) + V(\mathbf{q})\psi(\mathbf{q},t)\,, \tag{8.4}$$

where V is a Galilean invariant function. The operator on the right of the Schrödinger equation is often called the Schrödinger Hamiltonian or simply the Hamiltonian. We shall use this name in the sequel.

Equations (8.1)–(8.4) define Bohmian mechanics. Looking at them, one grasps immediately that this mechanical theory is different from Newtonian mechanics. Given the two equations, it is now only a matter of analyzing the theory to see what it says about the world. We stress this point: what is required now is not philosophy, and not interpretation, but the very thing that physicists are supposed to do best. The task is to analyse the equations, and only the equations, and see what they say about our world. We shall do that in the following chapters, but first we allow ourselves an interlude.

Everybody (except perhaps mathematicians) must admit that equation (8.1) looks a bit ugly. The vector field was already introduced in the previous chapter [see (7.16)]. So in principle we know "where it comes from". It comes from interpreting the squared modulus of the wave function as a probability density, whose transport is given by the quantum flux equation (7.11). Bohmian trajectories are nothing but the flux lines along which the probability gets transported. The velocity field is thus simply the tangent vector field of the flux lines. That is the fastest way, given the Schrödinger equation (or for that matter any wave equation which allows for a conserved current) to define a Bohmian theory, a way which was often chosen by John Bell [1] to introduce Bohmian mechanics.[1]

But one must admit that this does not say much to support the integrity of the velocity field, since as we understand it probability is a secondary concept, arising only when one analyzes typical behavior. Nevertheless the reader who is eager to see the theory at work, and who is happy with the mechanics as given, may skip the next section, whose only purpose is to explain that Bohmian mechanics follows from arguments of minimality, simplicity, and symmetry.[2]

---

[1] Bell's way of defining Bohmian mechanics via the current can be generalized to more general Hamiltonians (i.e., more general than the usual Schrödinger Hamiltonian) which appear in quantum field theory [2, 3].

[2] The quantum flux as defined by the quantum flux equation (7.11) is not uniquely defined. It can be changed by adding the curl of a vector field, because the divergence of such an object is zero. This then leads to the idea that the Bohmian trajectories, when defined as flux lines, are not uniquely defined, i.e., there exist other versions of Bohmian mechanics [4]. However, these are neither simple nor minimal in any sense of the words. The flux is nevertheless a good guidance principle for generalizing Bohmian mechanics to more general Schrödinger equations, which may involve higher order terms.

8.1 Derivation of Bohmian Mechanics

We wish to find a theory of particles from “first principles”. Moving particles agree with our experience of the microscopic world. In a double slit experiment, a particle is sent onto a double slit and later caught on a screen. We see the spot on the screen (where the particle hits the screen), and since the particle came from somewhere else it had to move.

Particles have coordinates $\mathbf{Q}_k \in \mathbb{R}^3$. An N-particle system is then given by a collection $\mathbf{Q} = (\mathbf{Q}_1, \ldots, \mathbf{Q}_N) \in \mathbb{R}^{3N}$. Particle positions are the primary variables we need to be concerned with. Naturally, the particles move according to some law. The simplest possibility is to prescribe the velocities

$$\frac{\mathrm{d}\mathbf{Q}}{\mathrm{d}t} = \mathbf{v}(\mathbf{Q},t) , \tag{8.5}$$

where $\mathbf{v}$ is a vector field on configuration space $\mathbb{R}^{3N}$ (analogous to the Hamiltonian vector field on phase space). For simplicity we begin with one particle. We need to find a Galilean covariant expression for the velocity vector, and the most demanding transformation for a vector is rotation. From Chap. 3, we recall that the gradient of a scalar function transforms just right [see (2.35)]. This gives the idea that the velocity field should be generated by a function ψ:

$$\mathbf{v}^{\psi}(\mathbf{q},t) \sim \nabla\psi(\mathbf{q},t) . \tag{8.6}$$

Another symmetry which may be informative when we prescribe only the velocity is time-reversal invariance. The velocity changes sign when $t \mapsto -t$. How can one cope with that? We are already prepared to have complex conjugation do the job. Let ψ be complex, then with $t \mapsto -t$, let $\psi(\mathbf{q},t) \mapsto \psi^*(\mathbf{q},-t)$ and consider only the imaginary part in (8.6):

$$\mathbf{v}^{\psi}(\mathbf{q},t) \sim \Im\nabla\psi(\mathbf{q},t) . \tag{8.7}$$

That takes care of time-reversal invariance.

Now consider a Galilean boost $\mathbf{v}$: $\mathbf{v}^{\psi} \mapsto \mathbf{v}^{\psi} + \mathbf{u}$, i.e.,

$$\Im\nabla\psi + \mathbf{u} = \Im\nabla'\psi' .$$

This suggests putting $\psi' = \mathrm{e}^{\mathrm{i}\mathbf{q}'\cdot\mathbf{u}}\psi$ as the simplest possibility. However, it leads to an adjustment of the velocity, namely,

$$\mathbf{v}^{\psi}(\mathbf{q},t) = \alpha\Im\frac{\nabla\psi}{\psi}(\mathbf{q},t) \tag{8.8}$$

and

$$\psi' = \mathrm{e}^{\mathrm{i}\mathbf{q}'\cdot\mathbf{u}/\alpha}\psi , \tag{8.9}$$

where α is a real constant with dimensions of [length^2/time]. $\mathbf{v}^\psi$ is therefore homogeneous of degree 0 as a function of ψ, i.e., $\mathbf{v}^{c\psi} = \mathbf{v}^\psi$ for all $c \in \mathbb{C}$.

So far so good, but how should we choose ψ? Its role for the motion of particles is clear, but what determines the function? We need a Galilean invariant law for that. The simplest law that comes to mind is the Poisson equation for the gravitational potential. However, one may find many arguments as to why ψ should be a *dynamical* object itself. For example, the fact that ψ transforms "strangely" under time reversal suggests that ψ should obey an equation which contains time. We are therefore led to a minor extension of the Poisson equation to the simplest Galilean invariant dynamical equation for a complex function, observing that time reversal goes along with complex conjugation:

$$\mathrm{i}\frac{\partial \psi}{\partial t}(\mathbf{q},t) = \beta \Delta \psi(\mathbf{q},t) .$$

In the last chapter we discussed the equation with $\beta = -\hbar/2m$. The Galilean boost to a frame with relative velocity $\mathbf{u}$ was implemented using the extra factor $\mathrm{e}^{\mathrm{i}m\mathbf{u}\cdot\mathbf{q}/\hbar}$, i.e., using $\mathrm{e}^{-\mathrm{i}\mathbf{u}\cdot\mathbf{q}/2\beta}$, so that

$$\psi' = \mathrm{e}^{-\mathrm{i}\mathbf{u}\cdot\mathbf{q}/2\beta}\psi .$$

Comparison with (8.9) shows that $\alpha = -2\beta$. We end up with one-particle *Bohmian mechanics* in the form

$$\frac{\mathrm{d}\mathbf{Q}}{\mathrm{d}t} = \mathbf{v}^\psi(\mathbf{Q},t) = \alpha \Im \frac{\nabla \psi}{\psi}(\mathbf{Q},t) , \tag{8.10}$$

where $\psi(\mathbf{q},t)$ is a solution of

$$\mathrm{i}\frac{\partial \psi(\mathbf{q},t)}{\partial t} = -\frac{\alpha}{2}\Delta \psi(\mathbf{q},t) . \tag{8.11}$$

The linearity of the equation blends well with the homogeneity of the velocity. Galilean invariance allows the addition of a Galilean real (because of time-reversal invariance) scalar function $G(\mathbf{q})$, so that the most general such equation reads

$$\mathrm{i}\frac{\partial \psi(\mathbf{q},t)}{\partial t} = -\frac{\alpha}{2}\Delta \psi(\mathbf{q},t) + G(\mathbf{q})\psi(\mathbf{q},t) . \tag{8.12}$$

The generalization to many particles is immediate by reading $\mathbf{q}$ as a configuration.

Before we do so, we would like to determine the proportionality constant α, by connecting the new fundamental theory with the particle mechanics we already know: Newtonian mechanics. We suspect that the connection is most easily made via the Hamilton–Jacobi formulation – at least for short times. In view of the velocity field (8.10), we write

$$\psi(\mathbf{q},t) = R(\mathbf{q},t)\mathrm{e}^{\mathrm{i}S(\mathbf{q},t)/\hbar} , \tag{8.13}$$

with real functions R and S. The factor $1/\hbar$ is nothing but a dimensional unit (and will later be recognized as Planck's constant) to ensure that S has the dimensions of an action. Then (8.10) becomes

$$\frac{\mathrm{d}\mathbf{Q}}{\mathrm{d}t} = \frac{\alpha}{\hbar}\nabla S\,. \tag{8.14}$$

Hence comparison with the first Hamilton–Jacobi equation suggests that $\alpha/\hbar = 1/m$ [see (2.35)], with m as the mass of the particle. With this identification we recognize (8.11) as the one-particle Schrödinger equation.

Of course, one would like to see whether this is consistent with an analogue of the second Hamilton–Jacobi equation. In fact, introducing (8.13) into (8.11), we obtain

$$\mathrm{i}\frac{\partial R}{\partial t} - \frac{1}{\hbar}R\frac{\partial S}{\partial t} = -\frac{\alpha}{2}\left[\Delta R + 2\mathrm{i}\nabla R\cdot\frac{1}{\hbar}\nabla S - R\left(\frac{1}{\hbar}\nabla S\right)^2 + \mathrm{i}R\frac{1}{\hbar}\Delta S\right] + GR\,.$$

Collecting the imaginary parts of that equation yields

$$\frac{\partial R}{\partial t} = -\frac{\alpha}{2}\left(2\nabla R\cdot\frac{1}{\hbar}\nabla S + R\frac{1}{\hbar}\Delta S\right)\,,$$

or

$$\frac{\partial R^2}{\partial t} = -\alpha\frac{1}{\hbar}\nabla\cdot\left(R^2\nabla S\right) = -\nabla\cdot\left(\mathbf{v}^{\Psi}R^2\right)\,, \tag{8.15}$$

while the real parts give

$$\frac{\partial S}{\partial t} - \frac{\alpha}{2}\hbar\frac{\Delta R}{R} + \hbar G + \frac{1}{2}\frac{\alpha}{\hbar}(\nabla S)^2 = 0\,. \tag{8.16}$$

Comparing (8.16) with (2.36), the identification $\alpha = \hbar/m$ is consistent, but observe the extra term $-\alpha\hbar\Delta R/2R$ which, compared with (2.36), "adds" to the potential $V \doteq \hbar G$. Bohm called this extra term the quantum potential. In any case, if that extra term is zero (or negligible), the Bohmian trajectories follow classical trajectories. The term is small, for example, when ψ is essentially a plane wave $\psi = \mathrm{e}^{\mathrm{i}\mathbf{k}\cdot\mathbf{q}}$. Then $S = \hbar\mathbf{k}\cdot\mathbf{q}$, and obviously

$$\mathbf{v}^{\psi} = \frac{\hbar\mathbf{k}}{m}\,.$$

That is, the Bohmian particle moves at least for a short time[3] along a straight line with velocity $\hbar\mathbf{k}/m$, and then as a classically moving particle we can assign to it a classical momentum $\mathbf{p} = m\hbar\mathbf{k}/m = \hbar\mathbf{k}$.

---

[3] For a definite assertion about classical motion we need to know how ψ changes in time, and we shall say more on this in the next chapter.

Only in special situations are the Bohmian trajectories Newtonian in character, namely those where the guiding wave can be approximated (locally, i.e., where the particle is) by an almost plane wave (where the wave number $\mathbf{k}$ may be a slowly varying function of position). Such situations may be described as classical. In many such scenarios these trajectories can be seen. A particle track in a cloud chamber, or the moon on its orbit are examples. But in so-called quantum mechanical situations, where the "wavy" character of the guiding wave plays a role (for example, through interference effects), the trajectories will be disturbed by the act of observation and therefore change under observation. Thus there may in general be a huge difference between "measured" trajectories and "unmeasured" trajectories. We shall give examples later.

Heisenberg concluded from this that there are no trajectories, and that one must not talk about a particle having a position. Many physicists found that conclusion logical and even banned the notion of particle altogether from physics. Here is further subject matter for historians, to find out how this could have happened. After all, classical physics is supposed to be incorporated in quantum mechanics, but when there is nothing, then there is nothing to incorporate, and hence no classical world.

The reader may wonder about equation (8.15). What is it good for? Well, looking at it once again one sees that this is the identity (7.11) which any solution of Schrödinger's equation fulfills. It will be the key to the statistical analysis of Bohmian mechanics. Having said that, we may add that the identity is for Bohmian mechanics what the Liouville theorem is for Hamiltonian mechanics. We shall say more about this in a moment.

Let us first note that generalizing to many particles leads to the defining equations (8.1)–(8.4), where $V \doteq \hbar G$ can now be identified as what will be, in classical situations, the "classical" potential.

Remark 8.1. Bohmian Mechanics Is Not Newtonian
David Bohm [5, 6] and many others, e.g., [7] present Bohmian mechanics as being Newtonian in appearance. This can easily be achieved by differentiating (8.3) with respect to t. Next to $-\nabla V$, the resulting right-hand side contains the extra term $-\nabla$(quantum potential), which has been called the quantum force. But this takes us off target. Differentiation of (8.3) with respect to t is redundant. Bohmian mechanics is a first order theory. The velocity is not a "phase" variable as in Newtonian mechanics, i.e., its initial value must be chosen (and can indeed be freely chosen). It becomes a phase variable in certain situations, which one may refer to as classical regimes. Casting Bohmian mechanics into a Newtonian mould is not helpful for understanding the behavior of the trajectories in quantum mechanical situations, because understanding means first explaining things on the basis of the equations which define the theory. Redundancies are more disturbing than helpful. Analogies may of course help, but they are secondary. ■

8.2 Bohmian Mechanics and Typicality

Now we come to the trajectories. We have a theory of particles in motion and we are now eager to compute them in situations of interest. Just as we compute the orbits of planets, we would like to compute the orbits of the Bohmian particles around the nucleus in an atom. Why should we be eager to do that? Because the theory allows us to do that? So the electron moves around the nucleus in some way. Is that exciting? Yes, you may say, because all kinds of exciting questions come up. For example, Heisenberg's uncertainty relation seems to contradict the existence of trajectories, so who is right?

Well, computing them would not help to decide on that. When the electron moves around the nucleus, why does it not radiate (since the electron will surely undergo accelerations) and fall into the nucleus? Compared to orthodox quantum mechanics, where there is nothing, nothing can fall into nothing? But joking aside, the mechanics here is Bohmian, not Newtonian, and one has to see what this new mechanics is like. Force and acceleration are not elements of the new theory, so any arguments based on that line of thought are off target. But the relevance for Heisenberg's uncertainty principle is something one needs to look into. In fact, the principle is a *consequence* of Bohmian mechanics. Let us see now how this works.

Boltzmann returns to the stage. What we have is a new mechanics, with the old statistical reasoning. The only difference is this. While in the Newtonian world non-equilibrium was the key to understanding why the world is the way we see it, equilibrium opens the door to understanding the microscopic world. Just as we describe a gas in a box by the equilibrium ensemble, we now describe a Bohmian particle in quantum equilibrium. While Boltzmann's reasoning was embarrassed in classical statistical mechanics by atypicality, which renders the justification of equilibrium ensembles almost impossible, we shall now meet a situation where Boltzmann's reasoning works right down the line.

Quantum equilibrium refers to the typical behavior of particle positions *given* the wave function. If you ask what happens to irreversibility, atypicality, and the arrow of time when everything is in equilibrium, then you must read the last sentence again. The particles are in equilibrium *given* the wave function. Non-equilibrium resides within the wave function. The following analogy may be helpful. The wave function generates a velocity vector field (on configuration space) which defines the Bohmian trajectories. This is the analogue of the idea that the Hamiltonian generates a vector field (on phase space) which defines classical trajectories. The Hamiltonian also defines the measure of typicality. Analogously, the wave function defines the measure of typicality. Does that not sound very much like what Born had in mind (see Chap. 7)?

Therefore, as in classical physics, we seek a measure of typicality, which is distinguished by the new physical law. As in the Boltzmann–Gibbs ensemble, we first seek such a measure with which we can form a statistical hypothesis about the typical empirical distribution over an ensemble of identical subsystems. The starting point for finding a measure of typicality is the continuity equation [see (2.3)] for the density $\rho(\mathbf{q},t)$ transported along the Bohmian flow of an N-particle system:

$$\Phi^{\psi}_{t,t_0} : \mathbb{R}^{3N} \longmapsto \mathbb{R}^{3N} , \quad t \in \mathbb{R}$$

$$\Phi^{\psi}_{t,t_0}\mathbf{q} = \mathbf{Q}(t,\mathbf{q}) = \text{solution of (8.2) with } \mathbf{Q}(t_0) = \mathbf{q} .$$

The equation reads

$$\frac{\partial \rho(\mathbf{q},t)}{\partial t} + \nabla \cdot \left[\mathbf{v}^{\psi}(\mathbf{q},t)\rho(\mathbf{q},t) \right] = 0 . \tag{8.17}$$

Since the guiding field $\psi(\mathbf{q},t)$ depends in general on time, the velocity field depends in general on time, and thus there is in general no stationary density. Nevertheless we require typicality, defined by the guiding field, to be time independent, so we look for a density which is a time-independent function of $\psi(\mathbf{q},t)$. This property, called *equivariance*, generalizes stationarity. That density was found by Born and Schrödinger, namely $\rho = |\psi|^2 = \psi^* \psi$. Putting this into (8.17), we obtain

$$\frac{\partial |\psi|^2}{\partial t} + \nabla \cdot \left(\mathbf{v}^{\psi} |\psi|^2 \right) = 0 , \tag{8.18}$$

which we know is satisfied from the previous chapter, because we compute

$$\mathbf{v}^{\psi} |\psi|^2 = \hbar m^{-1} \Im \frac{\nabla \psi}{\psi} |\psi|^2 = \hbar m^{-1} \Im \psi^* \nabla \psi = \mathbf{j}^{\psi} ,$$

so that (8.18) is identical to the quantum flux equation (7.11). We may view this result as analogous to the Liouville theorem. The physics gives us a distinguished measure of typicality.[4]

So what does all this mean? In fact it tells us that, if $\mathbf{Q}_0$ is distributed according to the density $|\psi(\cdot,0)|^2$ then $\Phi^{\psi}_t \mathbf{Q}_0 = \mathbf{Q}(\mathbf{Q}_0,t)$ is distributed according to $|\psi(\cdot,t)|^2$, where $\psi(\cdot,t)$ is the wave function of the system. In terms of expectation values,

$$\mathbb{E}^{\psi}\Big(f\big(\mathbf{Q}(t)\big) \Big) := \int f\big(\mathbf{Q}(t,\mathbf{q})\big) |\psi(\mathbf{q},0)|^2 \mathrm{d}^n q = \int f(\mathbf{q}) |\psi(\mathbf{q},t)|^2 \mathrm{d}^n q =: \mathbb{E}^{\psi_t}(f) . \tag{8.19}$$

But this is Born's statistical law, which says that $\rho^{\psi}(\mathbf{q},t) = |\psi(\mathbf{q},t)|^2$ is the probability density that the particle configuration is $\mathbf{Q} = \mathbf{q}$. Of course, in saying this we assume ψ to be *normalized*, i.e., $\int |\psi(\mathbf{q},t)|^2 \mathrm{d}^{3N} q = 1$. And in particular (as remarked in the last chapter) $\int |\psi(\mathbf{q},t)|^2 \mathrm{d}^{3N} q$ is independent of time. In the Boltzmannian way of thinking we are therefore led to formulate the following statistical hypothesis, which is the analogue of the Boltzmann–Gibbs hypothesis of thermal equilibrium in statistical mechanics:

---

[4] Is this density unique? Recall that in classical mechanics the stationarity requirement allows a great many densities, among which we discussed the microcanonical and canonical examples. In [8], it is shown that the $\rho = |\psi|^2$ density is the unique equivariant measure (under reasonable conditions of course).

Quantum Equilibrium Hypothesis. For an ensemble of identical systems, each having the wave function ψ, the typical empirical distribution of the configurations of the particles is given approximately by $\rho = |\psi|^2$. In short, Born's statistical law holds.

All this will be elaborated and the hypothesis justified in Chap. 11.

Remark 8.2. On the Existence and Uniqueness of Solutions of Bohmian Mechanics
In classical mechanics the question of existence and uniqueness of solutions of say gravitational motions is exceedingly hard. How about the new mechanics? Can it be that there is absolutely no problem with electrons moving around? After all, in most textbooks on quantum mechanics, it is asserted that particles cannot have a position and move. The equations are there, but perhaps, you might think, they do not have solutions. In fact (8.3) looks potentially dangerous, since the wave function is in the denominator, and the wave function can have nodes! It can and will be zero at various places! Trajectories can run into the nodes, and that finishes the mechanics. But typically they do not. One has existence and uniqueness of Bohmian trajectories for all times for almost all initial conditions (in the sense of Remarks 2.1 and 2.5), where "almost all" refers to the quantum equilibrium distribution [9, 10]. Note that in Bohmian mechanics the wave function must be differentiable, and therefore one needs the classical solutions of the Schrödinger equation we talked about in Remark 7.1. ∎

8.3 Electron Trajectories

Maybe the most exciting feature is the Bohmian trajectory of an "electron" guided by the ground state wave function of a hydrogen atom. To find that, one considers the one-particle stationary Schrödinger equation with $V(\mathbf{q})$ as the Coulomb potential for a fixed point charge nucleus, i.e., one solves the eigenvalue equation

$$-\frac{\hbar^2}{2m}\Delta\psi + V(\mathbf{q})\psi = E\psi\,.$$

The solutions ψ which are normalizable (i.e., square integrable) and bounded are called eigenstates. The ground state ψ_0 is the one which corresponds to the lowest eigenvalue. The eigenvalues E are commonly referred to as energy eigenvalues, and the lowest energy value is commonly referred to as the ground state energy. Given the underlying ideas, which we partly discussed in the lead-up to Schrödinger's equation, the name "energy" may seem appropriate. Of course, more elaboration would be needed to actually feel comfortable with calling objects of the new theory by names which have a clear connotation from classical physics. If we are to call the eigenvalues "energy", they had better bear some convincing relation to what we usually mean by energy.

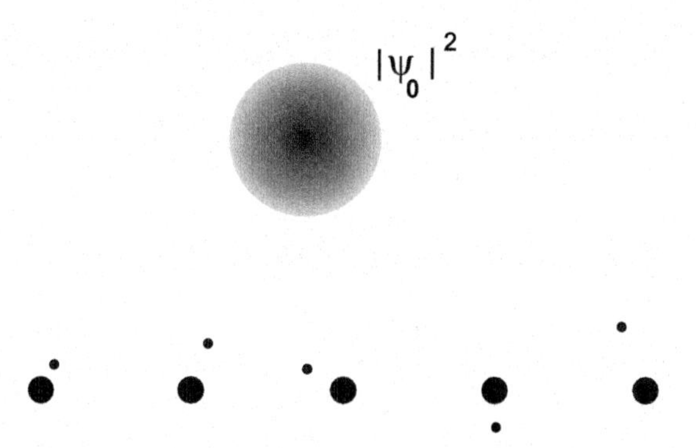

Fig. 8.1 Ground state wave function distribution (*dark regions* correspond to high probability) for an ensemble of hydrogen atoms in the ground state, with $|\psi_0|^2$ random electron positions

So let us consider the ground state. In principle, there could be many such states. But not for hydrogen. It is a mathematically rigorous statement that the ground state can always be chosen real (and everywhere positive). This is more than we need to know! If the guiding field is real, then the imaginary part is zero, so the velocity is zero, i.e., $\dot{\mathbf{Q}} = 0$, and $\mathbf{Q} = \mathbf{Q}_0$, so nothing moves! The reader should beware of the following idea, which is better considered as a bad joke: since the electron does not move it cannot radiate, and that explains why the electron does not fall into the nucleus.

But jokes aside, this finding is very much in contradiction with Bohr's naive model of electrons circling the nucleus. Moreover, it sharply contradicts Heisenberg's uncertainty principle. Although the position has some finite spread (we shall say what that means in a moment), the velocity, i.e., the momentum, is precisely zero! Is Heisenberg's uncertainty principle false after all? Of course, it is not! We must simply understand what it is telling us, and that means we must understand what the momentum spread refers to in quantum mechanics.

Clearly, Bohmian mechanics is a radical innovation, so different from what one naively thought! What else is new? What can we say about $\mathbf{Q}_0$? Only that it is $|\psi_0|^2$-distributed, by virtue of the quantum equilibrium hypothesis. This means that, in an ensemble of hydrogen atoms in the ground state (see Fig. 8.1), the empirical distribution of $\mathbf{Q}$ is typically close to $|\psi_0|^2$. That can be checked by experiment. The spread in position is then simply the variance of the $|\psi_0|^2$-distribution.

Stationary states belonging to higher energy values, called excited states, will be of the form $\psi_1 = f(r,\vartheta)\mathrm{e}^{\mathrm{i}\varphi}$ (which is due to the way the angle φ occurs in the Laplace operator), writing $\mathbf{q}$ in spherical coordinates (r,ϑ,φ). Then $S = \hbar\varphi$ and

$$\mathbf{v}^{\Psi}(\mathbf{q}) = \frac{1}{m}\nabla S = \frac{\hbar}{m}\frac{1}{r\sin\vartheta}\mathbf{e}_{\varphi}\frac{\partial}{\partial\varphi}\varphi = \frac{\hbar}{m}\frac{1}{r\sin\vartheta}\mathbf{e}_{\varphi}\,,$$

so that we find periodic orbits:

$$\mathbf{Q}(t) = \left(r = r_0 \,,\; \vartheta = \vartheta_0 \,,\; \varphi = \frac{\hbar}{m} \frac{1}{r_0 \sin \vartheta_0} t + \varphi_0 \right) . \tag{8.20}$$

But that should not excite anybody.

What happens to the motion of the particle when it is guided by a superposition of two real wave functions ψ_1 and ψ_2, for each of which $\mathbf{v} = 0$? It should be noted that

$$\psi = \psi_1 + \alpha \psi_2 \,, \qquad \alpha \in \mathbb{C} \,,$$

may produce a very complicated motion.

Is it in fact useful to compute Bohmian trajectories in various quantum mechanical situations we deem of interest? In general, it is not! But in certain asymptotic situations, when the trajectories are close to classical ones, that knowledge is useful and even crucial for understanding the microscopic physics. That is the case in scattering theory, which we shall deal with in a later chapter. But what about the above negative response? What we learn in general from the trajectories is that, at each moment of time t, the particles have positions which are distributed according to the quantum equilibrium hypothesis $|\psi(\mathbf{q},t)|^2$. Moreover, the theory is first order:

$$\dot{\mathbf{Q}} = \mathbf{v}^{\psi}(\mathbf{Q},t) \,.$$

From this it follows that the possible trajectories cannot cross in extended configuration space $\mathbb{R}^{3N} \times \mathbb{R}$, while this is possible in Newtonian mechanics.

With this observation, one can easily construct the trajectory picture in the famous double slit experiment for a stationary wave (see Figs. 8.2 and 8.3). In this experiment one sends particles one after the other (think of one particle per year, if you have enough time) towards a double slit. As any wave would do, after passage through the slits, the guiding wave ψ of every particle forms a diffraction pattern according to Huygens' principle. After the slit, this diffracted wave determines the

Fig. 8.2 Interference of spherical waves after a double slit

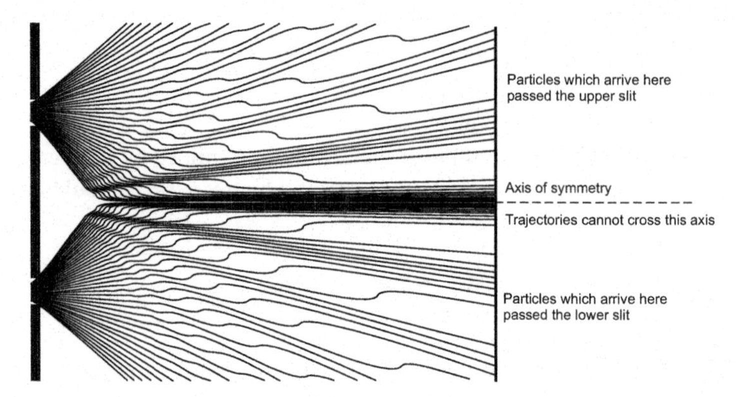

Fig. 8.3 Possible trajectories through the double slits

possible trajectories of the particle. The particles arrive one after the other (every year if you have the time) at the photographic plate and leave a black spot at their point of arrival.

If we wait long enough, the random black spots will eventually form a recognizable interference pattern, which is essentially the quantum equilibrium $|\psi|^2$ distribution[5] [11]. This is clear enough, because that is what the quantum equilibrium hypothesis says: in an ensemble of identical particles each having wave function ψ, the empirical distribution of positions is $|\psi|^2$-distributed. This is a rather dull observation, and yet the interference pattern of the double slit experiment is often taken in textbooks as proof that one cannot have moving particles in atomistic physics. We can say a bit more about the trajectories in the essentially two-dimensional experimental setup.

1. The trajectories cannot cross the axis of symmetry.
2. The trajectories move mostly along the maxima (hyperbola) of $|\psi|^2$ and spend only short times in the valleys of $|\psi|^2$ ($|\psi|^2 \approx 0$).
3. The trajectories cross the valleys since, right after the slits, the trajectories expand radially. This is turn happens because the guiding wave is a spherical wave close behind the slits (see Fig. 8.2), and while first feeding the nearest maxima, they must observe quantum equilibrium. Most trajectories will have to lie in the region of the main maximum (around the symmetry axis, which is the most clearly visible on the screen), i.e., trajectories must cross over from adjacent maxima to the main maximum.[6]
4. The arrival spot on the screen is random, in particular the slit through which each particle goes is random. The randomness is due to the random initial position $\mathbf{Q}_0$ of the particle with respect to the initial wave packet. By always preparing the same wave packet ψ, one prepares an ensemble of $|\psi|^2$-distributed positions.

[5] In fact it is the quantum flux across the surface of the photographic plate integrated over time (see Chap. 16).

[6] The double slit trajectory picture Fig. 8.3 can thus be more or less drawn by hand, or by numerical computation as done by Dewdney et al. [6].

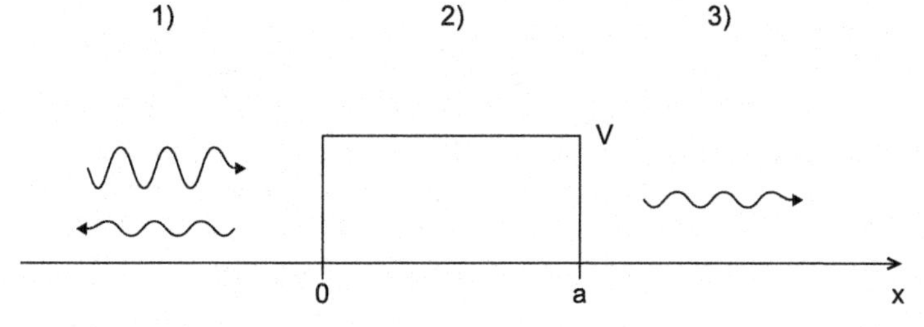

Fig. 8.4 Tunneling through a barrier

In many quantum mechanical situations one computes numbers from a stationary analysis, i.e., from a stationary wave picture. One must be clear about the meaning of such stationary computations. In the section on scattering theory, we shall examine the applicability of stationary pictures, but for now we shall only discuss the simple one-dimensional, textbook example of the stationary analysis of tunneling through a barrier. The latter is a potential of height V as shown in Fig. 8.4. One solves the time-independent Schrödinger equation in the three regions separately, with an incoming plane wave from the left:

$$
\begin{aligned}
&(1)\, x\leq 0: && -\frac{\hbar^2}{2m}\Delta\psi = E\psi\,, && \psi = \mathrm{e}^{\mathrm{i}kx} + A\mathrm{e}^{-\mathrm{i}kx}\,, \quad k = \frac{\sqrt{2mE}}{\hbar}\,,\\
&(2)\, 0<x<a: && -\frac{\hbar^2}{2m}\Delta\psi + V\psi = E\psi\,, && \psi = B\mathrm{e}^{lx} + C\mathrm{e}^{-lx}\,, \quad l = \frac{\sqrt{2m(V-E)}}{\hbar}\,,\\
&(3)\, a\leq x: && -\frac{\hbar^2}{2m}\Delta\psi = E\psi\,, && \psi = D\mathrm{e}^{\mathrm{i}kx}\,,
\end{aligned}
$$

where the constants A,B,C,D are determined by the requirement that the trajectories should be differentiable across the boundaries of the regions. This is commonly expressed by requiring $\mathbf{j}^{\psi}$ to be continuous.

In the region $x\leq 0$, we have a superposition of two wave functions, one incoming and one reflected, yielding a complicated current, while for $x\geq a$, we have a simple current. But the motion is one-dimensional and therefore particle trajectories cannot cross each other. Observing that the wave function on the left is in any case periodic, there cannot be two types of trajectory on the left side, moving towards and away from the barrier. By computation, one finds that all trajectories do in fact move towards the barrier, which seems to contradict the idea of back-scattering, since particles clearly get reflected.

The situation becomes more dramatic if the potential barrier becomes infinitely high, so that nothing can get through and everything is reflected (A becomes -1). Then on the left the wave is purely imaginary and nothing moves! Bohmian mechanics forces us to realize that the stationary picture is an idealization. This idealization is useful for computing the so-called transmission and reflection coefficients, which are computed from the ratios of the "incoming" and "outgoing" quantum fluxes, and are equal to $|D|^2$ and $|A|^2$. But it is only clear that A and D achieve this meaning when one considers the physically realistic setting where a wave packet $\psi_{\rm p}$ runs towards the barrier and part of the wave packet gets reflected, while part goes through.

If $\psi_{\rm p}$ represents the initial wave packet (the prepared one, which may look approximately like a plane wave), we can split the support of $\psi_{\rm p}$ into two intervals T and R. T is the area of initial positions giving rise to trajectories which reach $x \geq a$ (these are the positions in the front part of the wave packet), and R is the interval $(Q_0 \in R,\ \tilde{Q}_0 \in T \Longrightarrow Q_0 \leq \tilde{Q}_0)$ of initial points giving rise to trajectories which move towards $-\infty$. The probability that the particle starts in T is

$$\int_T |\psi_{\rm p}|^2 {\rm d}q = |D|^2 \,.$$

In the section on scattering theory, we shall say more about the meaning of the quantum flux, but we already have a pretty good understanding of how Bohmian mechanics works.

8.4 Spin

The reader who has come this far must eventually stumble across spin. Granted that particles have positions, that accounts for position, maybe even for momentum, once we understand what that refers to, but what is spin? How can a point have spin? What does rotation even mean for a point? Spin is indeed a truly quantum mechanical attribute. Wolfgang Pauli referred to it as nonclassical two-valuedness. Spin is in fact as quantum mechanical as the wave function, which is no longer simply complex-valued, but can be a vector with two, three, four, or any number of components.

While it is easy to handle and not at all strange or complicated, to explain from pure reasoning alone why spin arises would nevertheless go somewhat beyond the scope of this textbook. Spin plays a role only in connection with electromagnetic fields, and the ultimate reasoning must therefore invoke relativistic physics. We do not want to go that far here. So let us be pragmatic about it at this point, and simply consider the phenomenological facts.

A silver atom is electrically neutral and inherits a total magnetic moment from the spin of its one valence electron. If such an atom is sent through a Stern–Gerlach magnet, which produces a strongly inhomogeneous magnetic field, the guiding wave splits into two parts.[7] Each part follows a trajectory which resembles that of a magnetic dipole when it is sent through the Stern–Gerlach magnet, and when its orien-

[7] If one sent an electron through the Stern–Gerlach magnet, the splitting would not be as effective, due to strong disturbances from the Lorentz force which also acts on the electron. The reader might enjoy the following historical remark on this point. Bohr envisaged the impossibility of ever "seeing" the electron spin as a matter of principle: the spin being "purely quantum" and the results of experiments always being classically describable, the spin must not be observable!

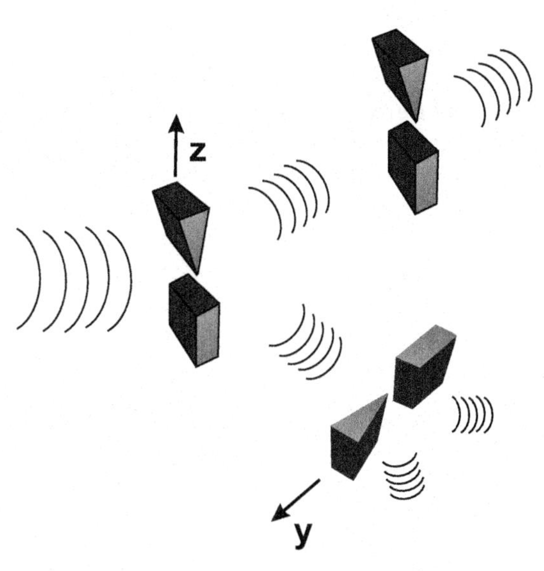

Fig. 8.5 In a Stern–Gerlach apparatus (strong inhomogeneous magnetic field), whose pole shoes are drawn schematically, a spinor wave function splits into two parts. If one of the parts goes through a Stern–Gerlach magnet of the same orientation, no further splitting occurs. If one of the parts goes through a magnet with a different orientation, a further splitting occurs

tation is either parallel (spin $+1/2$) or antiparallel (spin $-1/2$) to the gradient of the magnetic field.[8]

Suppose a Stern–Gerlach setup is oriented in the z-direction, and suppose the triangular pole shoe lies in the positive z-direction (see Fig. 8.5). This prepares a spin $+1/2$ and a spin $-1/2$ guiding wave, one moving towards positive z values (upwards), let us say in the direction of the triangular pole shoe, and one moving towards negative z values (downwards), i.e., in the direction of the rectangular pole shoe. Of course, when the wave packets are clearly separated, the particle will be in only one of the two wave packets. If we block let us say the downward-moving wave packet by a photographic screen, and if no black point appears, the particle must be in the other, upward-moving packet. In a manner of speaking that particle then has z-spin $+1/2$, and one might say that it has been prepared in the z-spin $+1/2$ state.

If one now sends that packet through a second Stern–Gerlach setup, also oriented in the z-direction, that packet does not split, but continues to move further upwards. But if the packet is sent through a Stern–Gerlach magnet which is oriented, say, in the y-direction (any direction orthogonal to the z-direction), then the wave splits again, and with probabilities of 1/2 the particle will be traveling in the positive or negative y-direction.

[8] The value 1/2 stands without explanation at this point. For the considerations in this book, we may as well replace 1/2 by unity. The reason why one chooses to talk about spin $\pm 1/2$ has to do with putting spin into analogy with rotation. This analogy is not so far-fetched, since spinors, as we shall see shortly, are acted on by a special representation of the rotation group.

Choosing a different non-orthogonal direction, the probabilities for the splitting vary with the angle. An angle of 90° yields probabilities of 1/2 for up and down, while another direction will give more weight to the spin axis closest to the z-direction. This has been observed experimentally.

The guiding field of the particle must obey the rules of the splitting and the simplest way to accomplish this is for the wave function itself to possess two degrees of freedom. It thus becomes a two-component wave function, i.e., a spinor wave function:

$$\psi(\mathbf{q}) = \begin{pmatrix} \psi_1(\mathbf{q}) \\ \psi_2(\mathbf{q}) \end{pmatrix} \in \mathbb{C}^2 .$$

Bohmian mechanics for spinor wave functions is simple. Start with one particle and write (8.1), viz.,

$$\mathbf{v}^{\psi} = \frac{\hbar}{m} \Im \frac{(\psi, \nabla\psi)}{(\psi, \psi)} , \tag{8.21}$$

where

$$(\psi, \varphi) = \begin{pmatrix} \psi_1^* \\ \psi_2^* \end{pmatrix} \cdot \begin{pmatrix} \varphi_1 \\ \varphi_2 \end{pmatrix} = \sum_{i=1,2} \psi_i^* \varphi_i ,$$

i.e., take the scalar product on the spinor degrees of freedom.

The Schrödinger equation is replaced by the so-called Pauli equation, which is an equation for the two-component wave function, in which the potential V is now a function with Hermitian matrix values. A Hermitian matrix can be decomposed in a special basis, comprising the Pauli matrices

$$\sigma_x = \begin{pmatrix} 0 & 1 \\ 1 & 0 \end{pmatrix} , \qquad \sigma_y = \begin{pmatrix} 0 & -\mathrm{i} \\ \mathrm{i} & 0 \end{pmatrix} , \qquad \sigma_z = \begin{pmatrix} 1 & 0 \\ 0 & -1 \end{pmatrix} ,$$

and the unit 2×2 matrix E_2:

$$V(\mathbf{q})\mathsf{E}_2 + \mathbf{B}(\mathbf{q}) \cdot \sigma = V(\mathbf{q})\mathsf{E}_2 + \sum_{k=1}^{3} B_k(\mathbf{q})\sigma_k ,$$

where $V \in \mathbb{R}$, $\mathbf{B} \in \mathbb{R}^3$, $\sigma := (\sigma_x, \sigma_y, \sigma_z)$. The Pauli equation for an uncharged, i.e., neutral, particle in a magnetic field $\mathbf{B}$ then reads:

$$\mathrm{i}\hbar \frac{\partial \psi}{\partial t}(\mathbf{q}, t) = -\frac{\hbar^2}{2m} \Delta\psi(\mathbf{q}, t) - \mu\sigma \cdot \mathbf{B}(\mathbf{q})\psi(\mathbf{q}, t) , \tag{8.22}$$

where μ is a dimensional constant called the gyromagnetic factor. This equation ensures that $\rho^\Psi = (\psi, \psi)$ is equivariant,[9] i.e., the four-divergence

$$(\partial_t, \nabla)\cdot(\rho^\Psi, \mathbf{v}^\Psi \rho^\Psi) = 0 \, ,$$

another way of expressing the continuity equation. So ρ^Ψ is the quantum equilibrium distribution.

We said at the beginning of the section that we do not want to engage in a discussion about the question of how spin arises from pure thought about nature. To do this, we would need to rework the arguments leading to the Dirac equation, of which (8.22) is a non-relativistic approximation for an uncharged particle like the silver atom sent through the Stern–Gerlach apparatus. However, it may be helpful to describe the mathematical reason for the appearance of spinors. The point here is that the rotation group $SO(3)$ has a double-valued representation, namely $SU(2)$, the group of unitary 2×2 matrices with unit determinant. This is called the spin 1/2 representation.

This becomes clear once one understands that $SO(3)$ is a Lie group, i.e., both a group and a manifold, and that, in the case of $SO(3)$, it is not simply connected. There is thus a topological reason for the double-valued covering. When we talk about "identical particles", we must return to topological considerations, but we shall not elaborate on this connection between $SO(3)$ and $SU(2)$. We shall limit ourselves to showing the double-valued representation at work.

Consider the rotation of a vector through an angle 2γ about the z-axis:

$$\begin{pmatrix} x' \\ y' \\ z' \end{pmatrix} = \begin{pmatrix} \cos 2\gamma & -\sin 2\gamma & 0 \\ \sin 2\gamma & \cos 2\gamma & 0 \\ 0 & 0 & 1 \end{pmatrix} \begin{pmatrix} x \\ y \\ z \end{pmatrix} .$$

Using the Pauli matrices, any vector $\mathbf{x}$ can be mapped to a Hermitian matrix

$$\mathbf{x} \cong x\sigma_x + y\sigma_y + z\sigma_z = \begin{pmatrix} z & x - \mathrm{i}y \\ x + \mathrm{i}y & -z \end{pmatrix} ,$$

a representation which originates from Hamilton's quaternions. It is easy to check that

$$\begin{pmatrix} z' & x' - \mathrm{i}y' \\ x' + \mathrm{i}y' & -z' \end{pmatrix} = \begin{pmatrix} \mathrm{e}^{-\mathrm{i}\gamma} & 0 \\ 0 & \mathrm{e}^{\mathrm{i}\gamma} \end{pmatrix} \begin{pmatrix} z & x - \mathrm{i}y \\ x + \mathrm{i}y & -z \end{pmatrix} \begin{pmatrix} \mathrm{e}^{\mathrm{i}\gamma} & 0 \\ 0 & \mathrm{e}^{-\mathrm{i}\gamma} \end{pmatrix} ,$$

where

$$U(\gamma) = \begin{pmatrix} \mathrm{e}^{\mathrm{i}\gamma} & 0 \\ 0 & \mathrm{e}^{-\mathrm{i}\gamma} \end{pmatrix} \in SU(2) \, .$$

[9] The potential matrix being Hermitian goes hand in hand with the Hamilton operator being Hermitian, which we shall discuss in detail later.

Spinors are now introduced as a kind of square root of the matrix, by writing it as sum of dyadic products of $\mathbb{C}^2$ vectors, which are called spinors in this context:

$$\begin{pmatrix} z & x-\mathrm{i}y \\ x+\mathrm{i}y & -z \end{pmatrix} = s_1 s_2^+ + s_2 s_1^+ ,$$

where $s = \begin{pmatrix} a \\ b \end{pmatrix} \in \mathbb{C}^2$ and $s^+ = (a^* \, b^*)$. For example,

$$\begin{pmatrix} 0 & -\mathrm{i}y \\ \mathrm{i}y & 0 \end{pmatrix} = \begin{pmatrix} 0 \\ y \end{pmatrix} (\mathrm{i}, 0) + \begin{pmatrix} -\mathrm{i} \\ 0 \end{pmatrix} (0, y) .$$

Then we can view $U(\gamma)$ as acting naturally by left and right (adjoint) action on the spinor factors, producing the rotation through 2γ in physical space. So while a rotation through 2π in physical space brings a vector back to its starting direction, the spinor rotates through only a half turn in its space. This explains why the spin is attributed the value of 1/2. We close this short interlude by observing that both $U(\gamma)$ and $-U(\gamma)$ represent the physical rotation through 2γ, which is why this is called a double-valued representation.

The Dirac equation is the simplest equation which relates the two Pauli spinors, now taken as functions on spacetime. In another representation of the Dirac equation, one collects the two Pauli spinors into a four-spinor, also called a Dirac spinor. Spinors are thus geometric objects on which the $SU(2)$ elements act naturally, and which carry a representation of the rotations.

We stress once again that this cannot be seen in the ad hoc Bohm–Pauli theory presented here. The vector character of the velocity (8.21) comes from the gradient, and the remaining spinor character is "traced out" by the scalar product. However, when one starts from the Dirac equation, one achieves the more fundamental view [12] we alluded to above. Then the velocity vector field can be constructed as a bilinear form from the Dirac spinors without invoking any gradient. A timelike vector j^μ can be written as a sesquilinear product of spinors $j^\mu = \psi^\dagger \gamma^0 \gamma^\mu \psi$, where γ^μ, $\mu = 0,1,2,3$, are the so-called γ matrices, 4×4 matrices containing the Pauli matrices as blocks.

One can also view the Dirac equation is the simplest equation which renders j^μ divergence free, i.e., $\partial_\mu j^\mu = 0$, whence j^μ can be viewed as a current, obeying the continuity equation. The Bohm–Pauli theory emerges as a non-relativistic approximation of the Bohm–Dirac theory.

But now we must move on. What is of more interest right now is to see how the splitting of the wave function emerges from the Pauli equation (8.22). Consider a Stern–Gerlach magnet oriented in the z-direction. For simplicity, the inhomogeneous magnetic field is assumed to be of the form $\mathbf{B} = \big(B_x(x,y), B_y(x,y), bz\big)$ (such that $\nabla \cdot \mathbf{B} = 0$), and the initial wave function is chosen to be

$$\Phi_0 = \varphi_0(z)\psi_0(x,y)\left[\alpha\begin{pmatrix}1\\0\end{pmatrix} + \beta\begin{pmatrix}0\\1\end{pmatrix}\right], \tag{8.23}$$

where $\binom{1}{0}$ and $\binom{0}{1}$ are eigenvectors of σ_z with eigenvalues[10] $1, -1$. (We have chosen the same function of position for both spinors merely for notational simplicity.) By virtue of the product structure of the function of position, the x,y evolution of the wave function separates from the z evolution, i.e., we obtain two Pauli equations, one for the x,y evolution and one for the z evolution. This is a straightforward consequence, whose demonstration is left to the reader.

We do not care about the x,y part, but focus instead on the z part of the evolution. We also assume that the velocity in the x-direction is rather large (the particle moves quickly through the magnetic field, which has only a small extension in the x,y-directions). So there is also a large inhomogeneity of the magnetic field in the x-direction, but since the particle is moving quickly, this should not matter much. In short, the x velocity essentially determines the amount of time τ the particle spends in the magnetic field.

Let us now focus on the z motion through the magnetic field. Due to the shortness of the time τ, we shall also ignore the spreading of the wave function. This amounts to ignoring the Laplace term in the equation. Then by linearity, we need only concentrate on the parts

$$\Phi^{(1)} = \varphi(z)\begin{pmatrix}1\\0\end{pmatrix}, \qquad \Phi^{(2)} = \varphi(z)\begin{pmatrix}0\\1\end{pmatrix}.$$

The wave packet

$$\varphi_0(z) = \int e^{ikz} f(k)\mathrm{d}k \tag{8.24}$$

is assumed to be such that $f(k)$ peaks around $k_0 = 0$. We have thus reduced the problem to an investigation of

$$\mathrm{i}\hbar\frac{\partial\Phi^{(n)}}{\partial t}(\mathbf{q},t) = \mu(-1)^n bz\Phi^{(n)}(\mathbf{q},t), \qquad n = 1,2.$$

Hence

$$\Phi^{(n)}(\tau) = \exp\left[-\mathrm{i}(-1)^n\frac{\mu b\tau}{\hbar}z\right]\Phi_0^{(n)},$$

and after leaving the magnetic field, so that they are once again in free space, the z waves of the wave packet (8.24) have wave numbers

$$\tilde{k} = k - (-1)^n\frac{\mu b\tau}{\hbar}.$$

[10] To avoid confusion, we talk about spin $\pm 1/2$, but the number 1/2 and the dimensional factor are of no importance for our considerations and absorbed in the factor μ.

As solutions of the free wave equation, they must therefore have frequency

$$\omega(\tilde{k}) = \frac{\hbar\tilde{k}^2}{2m} .$$

The group velocity of this wave is then (observing that $k_0 = 0$)

$$\frac{\partial\omega(\tilde{k})}{\partial\tilde{k}} = -(-1)^n \frac{\mu b \tau}{m} ,$$

whence the wave packet $\Phi^{(1)}$ moves in the positive z-direction, i.e., the direction of the gradient of $\mathbf{B}$ (*spin up*), while $\Phi^{(2)}$ moves in the negative z-direction (*spin down*). However, the superposition of both wave packets remains with the weights defined by the initial wave function (8.23), and hence, in quantum equilibrium, the particle will move upwards with probability $|\alpha|^2$. One then says that the particle has *spin up* with probability $|\alpha|^2$. And likewise, the particle moves downwards with probability $|\beta|^2$, and one says that it has *spin down* with probability $|\beta|^2$. Which spin the particle "ends up with" depends only on the initial position of the particle in the support of the initial wave function. That is were the "probability" comes from.

We close with an observation that will be used later. The "expectation value" of the spin is

$$\frac{1}{2}(|\alpha|^2 - |\beta|^2) .$$

This can be computed using the following scalar product, which will play an important role in later chapters:

$$\left\langle \Phi \middle| \frac{1}{2}\sigma_z \Phi \right\rangle = \frac{1}{2}\int \mathrm{d}x\mathrm{d}y\mathrm{d}z\, \Phi^* \cdot \sigma_z \Phi = \frac{1}{2}(|\alpha|^2 - |\beta|^2) . \tag{8.25}$$

This is easily verified.

The **a**-spin 1/2 spinors are defined as eigenvectors of the matrix $\mathbf{a}\cdot\sigma/2$, where the **a**-spin $+1/2$ spinor and the **a**-spin $-1/2$ spinor are orthogonal, and where **a** stands for some abitrary direction. If we choose

$$\begin{pmatrix} 1 \\ 0 \end{pmatrix} \quad \text{and} \quad \begin{pmatrix} 0 \\ 1 \end{pmatrix}$$

as above for the z spinor, then an arbitrary normalized wave function will look like

$$\begin{pmatrix} \psi_1(\mathbf{x}) \\ \psi_2(\mathbf{x}) \end{pmatrix} .$$

Sending this wave function through the Stern–Gerlach apparatus, the z-spin $+1/2$ will appear with probability

$$\int \mathrm{d}\mathbf{x} \left| \begin{pmatrix} \psi_1(\mathbf{x}) \\ \psi_2(\mathbf{x}) \end{pmatrix} \cdot \begin{pmatrix} 1 \\ 0 \end{pmatrix} \right|^2 = \int \mathrm{d}\mathbf{x} |\psi_1|^2(\mathbf{x}) .$$

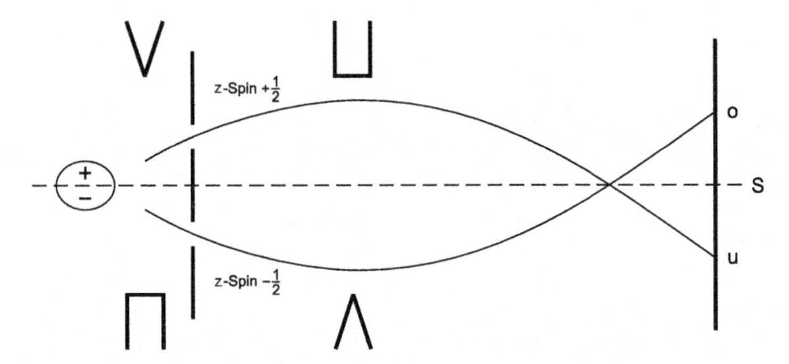

Fig. 8.6 Combination of two Stern–Gerlach experiments. *Curved lines* represent the paths of the wave packets

Finally, to crystallise our ideas on spin, let us consider a thought experiment devised by Tim Maudlin.[11] Send a y-spin $-1/2$ particle in the z-direction through a Stern–Gerlach apparatus. Then the wave function splits into two wave packets of equal weights belonging to a z-spin $+1/2$ particle and a z-spin $-1/2$ particle. Now let both wave packets pass through a Stern–Gerlach magnet which has opposite orientation (rotated through 180°), so that the two packets move towards each other, move through each other, and then hit a screen (see Fig. 8.6). If the particle hits the screen above (o) the symmetry axis S, then the particle must have z-spin $-1/2$, and if the particle hits the screen at u (below the axis), it must have z-spin $+1/2$. But the symmetry axis is still a topological barrier for the trajectories in a sufficiently symmetric setup. (As in the two slit experiment, we can even think of letting the wave go through a double slit.) This means that a particle that hits o was always above S, while a particle that hits u was always below S. What does this tell us? In fact it tells us that the particle changes its spin as a person changes his or her shirt. In other words, spin is *not* a property of a particle. It is dangerously misleading to speak of a particle with spin such and such. Spin is a property of the guiding wave function, which is a spinor. The particle itself has only one property: position.

In this thought experiment, the particle "has z-spin $+1/2$ once and z-spin $-1/2$ once", meaning that the particle is once guided by a z-spin $+1/2$ spinor and once by a z-spin $-1/2$ spinor. What spin the particle "has", i.e., in which packet the particle ends up, is decided by the initial position of the particle in the support of the y-spin initial wave function. In a totally symmetric setup (see Fig. 8.6), all initial positions in + will move so as to pass the first Stern–Gerlach apparatus upwards (z-spin $+1/2$), while all initial positions in – will move downwards (z-spin $-1/2$).

Let us push this idea a bit further. Returning to Fig. 8.5 and considering the lower outcome, imagine a series of Stern–Gerlach setups which split the wave packet further and further by having the magnets oriented as follows: z-direction, y-direction, z-direction, y-direction, for each outgoing wave packet. What we have then is obviously a quantum mechanical version of the Galton board. No matter how many

[11] Private communication.

magnets we put one after the other, the typical initial position will generate a typical run through this quantum mechanical Galton board which can be extended in principle ad infinitum.[12]

As a final remark, the wave function for many particles is an element of the *tensorial spin space*.[13] A typical N-particle wave function is a linear superposition of N-fold tensor products of one-particle spinor wave functions.

8.5 A Topological View of Indistinguishable Particles

In his talk about future physics entitled *Our Idea of Matter*,[14] Schrödinger speculates on what will remain of quantum mechanics, and whether a return to classical physics will be possible? Above all else, Schrödinger is firmly convinced that the return to discrete particles is impossible. By "discrete particles" he means particles in the sense we intend it in this book, that is, point particles which have a position. Now what is his reason for such a firm conviction?

To understand this we must consider the Bose–Fermi alternative. In many-particle quantum mechanics, the indistinguishability of particles is encoded in the symmetry (boson) or antisymmetry (fermion) of the wave function under exchange of particle coordinates. For example, a two-particle wave function of indistinguishable particles $\psi(x_1,x_2)$ is either symmetric (bosonic) $\psi(x_1,x_2) = \psi(x_2,x_1)$, or antisymmetric (fermionic) $\psi(x_1,x_2) = -\psi(x_2,x_1)$. According to Schrödinger, this cannot be if particles exist as discrete entities. Moreover many textbooks suggest that the quantum mechanical indistinguishability of particles forbids them from having positions. For if they had positions, then we could say that one particle is here and one particle is over there. And when they move, the one over here moves along here and the one over there moves along over there, so they are in fact perfectly distinguishable by the fact that one is here and one is over there.

The moral seems to be that quantum mechanical indistinguishability of particles is something fantastic, something new or even revolutionary, which absolutely necessitates the understanding that particles are not particles. But this is nonsense, and Schrödinger's conviction was unfounded. He was completely wrong, and so is any statement about the impossibility of a particle reality. Indeed, quite the opposite is true. Because one has particles, one can easily understand the symmetry properties of the wave function. To put it succinctly, the Bose–Fermi alternative is a straightforward prediction of Bohmian mechanics.

We wish to give an idea why this is so. What does it mean to have indistinguishable particles? One thing is clearly this: the labeling of particle positions, as we

[12] It may be interesting to note that, according to Wigner [13], von Neumann drew his intuition that "deterministic hidden variables" in quantum mechanics are not possible from this Gedankenexperiment. His intuition seemed to suggest that a "classical deterministic evolution" cannot produce an ideal Bernoulli sequence of "infinite" length.

[13] See Chap. 12 for the definition of tensor space.

[14] Unsere Vorstellung von der Materie [14].

did in this section when we wrote $\mathbf{Q}_i, i = 1,\ldots,N$, plays no physical role, i.e., it is purely for mathematical convenience. The particles are not physically distinguished, so it is better not to put any labels at all. But do we not then lose our nice configuration space picture $\mathbb{R}^{3N}$? The answer is that we do. The true configuration space of N indistinguishable particles is this. One has N positions in $\mathbb{R}^3$, and these N positions are N points of $\mathbb{R}^3$, so that we have a subset of $\mathbb{R}^3$ with N elements. Therefore the configuration space is

$$\mathscr{Q} = \left\{ q \subset \mathbb{R}^3 : |q| = N \,, \text{ i.e., } q = \{\mathbf{q}_1,\ldots,\mathbf{q}_N\} \,, \; \mathbf{q}_i \in \mathbb{R}^3 \right\} .$$

That space looks quite natural and simple, but topologically it is not. It is a far richer space than $\mathbb{R}^{3N}$.

For example, take a wave function defined on this space. Then it is a function depending on sets. Since the set has elements, it is a function also of elements, but as such it is symmetric, since exchanging the order of elements in a set does not change the set. Hence if we insist on writing the function $\psi(q)$ as a function of the elements $\mathbf{q}_i$, $i = 1,\ldots,N$, of the set q (this is like introducing coordinates on $\mathscr{Q}$), then $\psi(\mathbf{q}_1,\ldots,\mathbf{q}_N)$ will be symmetric under any permutation σ of the N indices:

$$\psi\big(\mathbf{q}_{\sigma(1)},\ldots,\mathbf{q}_{\sigma(N)}\big) = \psi(\mathbf{q}_1,\ldots,\mathbf{q}_N) \,.$$

This is pretty straightforward and implies that indistinguishable particles are bosons, i.e., they are guided by symmetric wave functions.

Interestingly, this is only half the truth. Since the discovery of Pauli's principle, usually phrased as saying that two or more electrons cannot occupy the same quantum state, together with the relativistic Dirac equation for electrons,[15] physicists have known that electrons are guided by *antisymmetric* wave functions. In the scalar case, these are wave functions for which

$$\psi\big(\mathbf{q}_{\sigma(1)},\ldots,\mathbf{q}_{\sigma(N)}\big) = \text{sign}(\sigma)\psi(\mathbf{q}_1,\ldots,\mathbf{q}_N) \,,$$

where $\text{sign}(\sigma)$ is the signature of the permutation, i.e., -1 if σ decomposes into an odd number of transpositions, and $+1$ if it decomposes into an even number of transpositions.

[15] Dirac's equation builds heuristically on a square root of the negative Laplacian, and contains a continuum of negative as well as positive energy states. To rule out unstable physics, Dirac invented the Dirac sea, which ensures that all negative energy states are occupied by particles. By Pauli's principle, no more than two particles can occupy one state, therefore particles of negative energy cannot radiate energy and thereby acquire an even more negative energy, because all negative energy states are already occupied. If that were not so, Dirac's equation would yield nonsensical physics, because all electrons would radiate endlessly, acquiring ever more negative energies. Dirac's sea is a way out, and it requires that electrons be described by antisymmetric wave functions (which is the mathematical formulation of Pauli's principle). Dirac's sea is reformulated in modern quantum field theory by using the notion of vacuum together with creation and annihilation operators which satisfy so-called anticommutation relations, and which describe the net balance of particles missing in the sea and particles "above" the sea.

But how can one understand that there are also antisymmetric wave functions? In fact, there is a principle that can be supported in the context of quantum field theory which says that half-integer-valued spinor representations of the rotation group for many particles must be antisymmetric wave functions, while integer-valued representations of the rotation group are to be described by symmetric wave functions. In the former case, the wave functions are said to be fermionic, and the particles guided by such wave functions are called fermions. In the latter case, they are described as bosonic, and the guided particles are called bosons. This principle is commonly referred to as the *spin–statistics theorem.*

However, we cannot go that far here, since a complete analysis would have to be relativistic. What we can do, however, is to explain a little further just where the antisymmetric wave functions are hiding when we look at things from a nonrelativistic standpoint (for an elaboration of topological effects in Bohmian mechanics, see [15]). To this end we observe a crucial topological characteristic of the configuration space $\mathcal{Q}$. This space can be expressed as

$$\mathcal{Q} \doteq (\mathbb{R}^{3N} \backslash \Delta^{3N})/S_N =: \mathbb{R}^{3N}_{\neq}/S_N \, .$$

The right-hand side means that we delete from $\mathbb{R}^{3N}$ the diagonal set

$$\Delta^{3N} = \left\{ (\mathbf{q}_1, \ldots, \mathbf{q}_N) \in \mathbb{R}^{3N} \middle| \, \mathbf{q}_i = \mathbf{q}_j \text{ for at least one } i \neq j \right\} ,$$

to yield the set we have denoted by $\mathbb{R}^{3N}_{\neq} := \mathbb{R}^{3N} \backslash \Delta^{3N}$, and than identify all those N-tuples in $\mathbb{R}^{3N}_{\neq}$ which are permutations of one another. The latter construction of equivalence classes is expressed as factorization by the permutation group S_N of N objects.

To understand what this factorization can do, let us consider a much simpler manifold, namely a circle, which arises from $[0,1]$ by identifying the points 0 and 1, or if one so wishes, from $\mathbb{R}$ by identifying the integers $n \in \mathbb{Z} \subset \mathbb{R}$ with zero. In terms of factorization, this is expressed by $\mathbb{R}/\sim$, where $x \sim y$ if and only if one has $x - y \in \mathbb{Z}$. $\mathbb{R}$ is topologically as simple as anything could be, but the circle is not. It contains closed curves winding around the circle, which cannot be homotopically deformed to a point. All closed curves on the circle fall into different equivalence classes (indexed by the winding number of the closed curve, which can be positive as well as negative). A class consists of all curves which can be homotopically deformed into one another. Closed curves with different winding numbers belong to different classes (a curve which winds around the circle 3 times clockwise and once counterclockwise belongs to class -2) and the set of classes can be made (with a rather straightforward definition of multiplication by joining curves) into a group called the fundamental group Π of the manifold. In the example of the circle, $\Pi = \mathbb{Z}$.

So the first lesson to learn from factorizing a topologically simple manifold (e.g., a simply connected manifold, which means that all closed curves can be deformed homotopically to a point) by an equivalence relation is that the resulting manifold will not in general be topologically simple (e.g., it may become *multiply connected*).

By analogy, $\mathbb{R}^{3N}_{\neq}/S_N$ will be multiply connected. The manifold will have "holes", so that classes of closed curves exist. We now switch to a different viewpoint. We wish to define a Bohmian vector field on such a strange manifold. The Bohmian vector field is the gradient of the phase of the wave function, and the phase is therefore the anti-derivative of the vector field. This parallels the integration of a vector field along a curve in the punctured plane, e.g., $1/z$ integrated over a circle around zero in the complex plane $\mathbb{C}\setminus\{0\}$. To construct the global anti-derivative, namely $\ln z$, Riemann sheets were invented. Quite analogously, one invents a *covering* of the circle by a winding staircase, so that the covering is $\mathbb{R}$ twisted into a spiral.

For the manifold $\mathscr{Q}=\mathbb{R}^{3N}_{\neq}/S_N$, the analogous construction leads to the universal covering space $\hat{\mathscr{Q}}=\mathbb{R}^{3N}_{\neq}$, which is a simply connected manifold covering the multiply connected manifold in the following sense. The covering space is mapped by a local diffeomorphism $\pi:\hat{\mathscr{Q}}\to\mathscr{Q}$ to the basis manifold, where the map is also referred to as a projection. This map is a coordinate map. What is local about it? Well, if one moves one flight up the staircase, one arrives at different (permuted) coordinates for the same point of the basis manifold. To any open neighborhood $U\subset\mathscr{Q}$ there corresponds a set of coordinate neighborhoods (leaves) $\pi^{-1}(U)$. All these different coordinates make up a covering fiber $\pi^{-1}(q)$, which is the set of all points in $\hat{\mathscr{Q}}=\mathbb{R}^{3N}_{\neq}$ which are projected down to q. If $q=\{\mathbf{q}_1,\dots,\mathbf{q}_N\}$, then

$$\pi^{-1}(q)=\Big\{\big(\mathbf{q}_{\sigma(1)},\dots,\mathbf{q}_{\sigma(N)}\big)\,,\ \sigma\in S_N\Big\}\,.$$

Now there is another group, the group of covering transformations. A covering transformation is an isomorphism which maps the covering space to itself, while preserving the covering fibers. The group of covering transformations is denoted by $Cov(\hat{\mathscr{Q}},\mathscr{Q})$. For two elements $\hat{q}$ and $\hat{r}$ in the same covering fiber, there is one element $\Sigma\in Cov(\hat{\mathscr{Q}},\mathscr{Q})$ such that $\hat{q}=\Sigma\hat{r}$. Since the fiber consists of permuted N-tuples, it is clear that $Cov(\hat{\mathscr{Q}},\mathscr{Q})$ is isomorphic to the permutation group S_N. It can be shown (and this may not come as a surprise) that the fundamental group $\Pi(\mathscr{Q})$ is isomorphic to $Cov(\hat{\mathscr{Q}},\mathscr{Q})$, whence the fundamental group is isomorphic to S_N.

We have now a good grasp of the connectedness of the true configuration space $\mathscr{Q}$, on which we wish to define a Bohmian vector field. We can thus look for wave functions $\hat{\psi}$ defined on the covering space. However, we must require such wave functions to behave properly under the projection along the fibers. Proper behavior means that the wave function must obey a certain periodicity condition. This is plain to see, otherwise the wave function cannot define a vector field on the configuration space. To begin with, $\hat{\psi}$ defines a Bohmian vector field $\hat{v}^{\hat{\psi}}$ on $\hat{\mathscr{Q}}$ in the usual way, and that vector field will have to be projected down to the configuration space. A vector field $\hat{v}$ on $\hat{\mathscr{Q}}$ is projectable[16] if and only if

$$\pi(\hat{q})=\pi(\hat{r}) \quad\Longrightarrow\quad \hat{v}(\hat{q})=\hat{v}(\hat{r})\,.$$

---

[16] The projection π generates a push forward $\pi*$ on vector fields which does the job of projecting the vector field to the manifold, but we do not want to formalize this any further.

The periodicity condition for the wave function is as follows. For $\Sigma \in Cov(\hat{\mathcal{Q}}, \mathcal{Q})$,

$$\hat{\psi}(\Sigma \hat{q}) = \gamma_\Sigma \hat{\psi}(\hat{q}) \, ,$$

where $\gamma_\Sigma \in \mathbb{C} \setminus \{0\}$. But this implies that $\hat{\psi}$ must be a an element of the representation space of the group of covering transformations represented by $\mathbb{C} \setminus \{0\}$, since

$$\hat{\psi}(\Sigma_2 \circ \Sigma_1 \hat{q}) = \gamma_{\Sigma_2} \gamma_{\Sigma_1} \hat{\psi}(\hat{q}) \, .$$

When we require that$|\gamma_\Sigma|^2 = 1$, i.e., $|\gamma_\Sigma| = 1$, we can project the equivariant evolution of the Bohmian trajectories on the covering space to the motion on the configuration space, where the probability density

$$|\hat{\psi}(\Sigma \hat{q})|^2 = |\gamma_\Sigma|^2 |\hat{\psi}(\hat{q})|^2 = |\hat{\psi}(\hat{q})|^2$$

projects to a function $|\psi(q)|^2$ on $\mathcal{Q}$. We thus obtain a unitary group representation, also called the character representation of the group. Translating this into coordinate language, we have

$$\hat{\psi}(\Sigma \hat{q}) = \hat{\psi}(\mathbf{q}_{\sigma(1)}, \dots, \mathbf{q}_{\sigma(N)}) = \gamma_\sigma \hat{\psi}(\mathbf{q}_1, \dots, \mathbf{q}_N) \, , \tag{8.26}$$

where we put the element from the group of covering transformations in correspondence with a permutation.

This is a formula that can be found in standard textbooks (perhaps without the hats), and we could have written it down immediately at the beginning of this section. Since we started the chapter with labeled particles and labeling is unphysical, we could have concluded that, because the $|\psi|^2$-distribution must not change under permutation of the arguments, the wave function should at best change by a phase factor when labels are permuted. But then we would have missed the truth behind indistinguishability, which lies in the topology of the configuration space.

But let us end the argument as briefly as possible. Consider (8.26) for σ a transposition τ. We have $\tau \circ \tau = \text{id}$, and hence $\gamma_\tau^2 = 1$, so that $\gamma_\tau = \pm 1$. So there we are. There exist two character representations of the permutation group, one with $\gamma_\tau = 1$, which takes us back to the bosonic wave functions, and one with $\gamma_\tau = -1$, which leads us to the fermionic wave functions. And moreover, we found that this is all there is. But as we pointed out earlier, this is not all there is to say. The whole construction should be done for spinor-valued wave functions. This is a bit more involved, but the principle is the same. On the basis of our present knowledge, the spin–statistics theorem seems not to be purely grounded on topology, and we shall not say more on this issue.

Here is a final remark. Let us return to $\mathbb{R}^{3N}_{\neq} := (\mathbb{R}^{3N} \setminus \Delta^{3N})$. Since the diagonal is taken out, this will produce holes in the space, and one may wonder whether this now allows for closed curves around those holes that cannot be deformed to zero curves. But the codimension of the diagonal set is the dimension of the physical space d. Consider therefore the following analogy. In a space of $d = 3$ dimensions, a set of codimension 3 is a point, and removing a point from $\mathbb{R}^3$ does not produce anything interesting, because all closed curves can still be deformed while avoiding the point. In $d = 2$, the diagonal has codimension two. That corresponds to a line in $\mathbb{R}^3$, and removing a line makes the space multiply connected!

The moral is that the configuration space of indistinguishable particles in two dimensions is again multiply connected, and its fundamental group is no longer isomorphic to the permutation group, but rather to a group which takes into account the extra obstacles arising from taking away the diagonal. This group is called the braid group. It allows for the continuum of characters of unit length, so any phase factor is allowed. The wave functions have therefore been called anyons.

References

1. J.S. Bell, *Speakable and Unspeakable in Quantum Mechanics* (Cambridge University Press, Cambridge, 1987)
2. D. Dürr, S. Goldstein, R. Tumulka, N. Zanghì: Phys. Rev. Lett. **93** (9), 090402, 4 (2004)
3. D. Dürr, S. Goldstein, R. Tumulka, N. Zanghì: J. Phys. A **38** (4), R1 (2005)
4. E. Deotto, G.C. Ghirardi: Found. Phys. **28** (1), 1 (1998)
5. D. Bohm: Physical Review **85**, 166 (1952)
6. D. Bohm, B.J. Hiley: *The Undivided Universe* (Routledge, London, 1995). An ontological interpretation of quantum theory
7. P.R. Holland: *The Quantum Theory of Motion* (Cambridge University Press, Cambridge, 1995)
8. S. Goldstein, W. Struyve: J. Stat. Phys. **128** (5), 1197 (2007)
9. K. Berndl, D. Dürr, S. Goldstein, G. Peruzzi, N. Zanghì: Commun. Math. Phys. **173** (3), 647 (1995)
10. S. Teufel, R. Tumulka: Commun. Math. Phys. **258** (2), 349 (2005)
11. A. Tonomura, J. Endo, T. Matsuda, T. Kawasaki, H. Ezawa: American Journal of Physics **57**, 117 (1989)
12. D. Dürr, S. Goldstein, K. Münch-Berndl, N. Zanghì: Physical Review A **60**, 2729 (1999)
13. E. Wigner: American Journal of Physics **38**, 1005 (1970)
14. E. Schrödinger: *Unsere Vorstellung von der Materie.* CD-ROM: Was ist Materie?, Original Tone Recording, 1952 (Suppose Verlag, Köln, 2007)
15. D. Dürr, S. Goldstein, J. Taylor, R. Tumulka, N. Zanghì: Ann. Henri Poincaré **7** (4), 791 (2006)

Chapter 9
The Macroscopic World

We need to cope with an unwanted heritage, namely the idea that physics is about measurements and nothing else. Bohmian mechanics is obviously not about measurements. However, since quantum mechanics is about measurements and at the same time plagued with the measurement problem, and since Bohmian mechanics is supposed to be a correct quantum mechanical description of nature, one may want to understand how Bohmian mechanics resolves the measurement problem. Measurements, the reading of apparatus states, belong to the non-quantum, i.e., classical world. Therefore the more general question, and a question of great interest on its own is this: How does the classical world arise from Bohmian mechanics? Concerning the latter we shall never, except for now, talk about $\hbar \to 0$. While this is a mathematically sensible limit procedure, it is physically meaningless because $\hbar$ is a constant that is not equal to zero. The physically well-posed question here is: Under what circumstances do Bohmian systems behave approximately like classical systems?

9.1 Pointer Positions

Since Bohmian mechanics is not about measurements but about ontology, namely particles, it has no measurement problem. However, since quantum mechanics does have the measurement problem, one may wonder whether Bohmian mechanics might not have another problem, namely that it is not a correct description of nature. As we shall argue, Bohmian mechanics is a correct description. To show this it may be quite helpful to phrase the measurement process in Bohmian terms.

First note that the very term "measurement process" suggests that we are concerned with a physical process during which a measurement takes place. But this in turn suggests that *something* is measured, and we are compelled to ask what this quantity is? In Bohmian mechanics two things spring to mind as quantities that could be measured: particle position and wave function. Apart from having a position, particles have no further properties. The wave function on the other hand does two things.

D. Dürr, S. Teufel, *Bohmian Mechanics*, DOI 10.1007/978-3-540-89344-8_9,

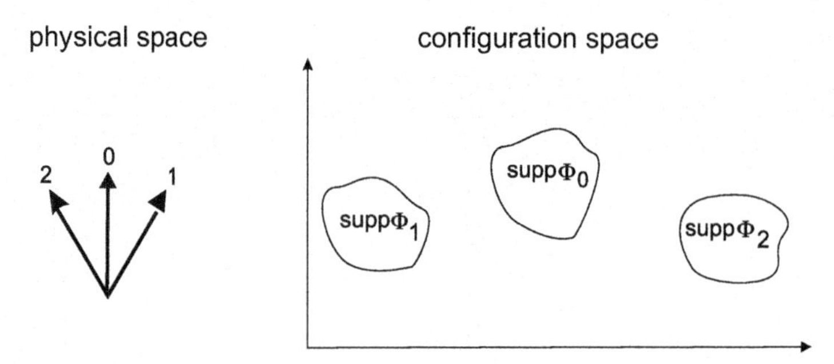

Fig. 9.1 Pointers in physical space and the supports of the pointer wave function in configuration space

Firstly, it guides the particle's motion, and secondly, it determines the statistical distribution of its position. The latter fact already suggests that at least the statistical distribution should be experimentally accessible, i.e., the particle's position should be measurable, for example by putting a photographic plate behind a double slit.

Even more famous than position measurement is "momentum measurement", since the Heisenberg uncertainty relation is about momentum and position measurements. But momentum is not a fundamental notion in Bohmian mechanics. On the other hand, particles do have velocities, so can they be measured? We need to say something about that, too. At the end of the day the moral to be drawn is simply this message from Bohr: in most cases a measurement is an experiment where nothing has been measured (in the sense the world is normally understood), but the experiment ends up with a classical pointer pointing to some value on a scale of values. In Bohmian terms one may put the situation succinctly by saying that most of what can be measured is not real and most of what is real cannot be measured, position being the exception.

So let us move on to the measurement problem. Consider a typical measurement experiment, i.e., an experiment in which the system wave function gets correlated with a pointer wave function. The latter is a macroscopic wave function, which can be imagined as a "random" superposition of macroscopically many ($\sim 10^{26}$) one-particle wave functions, with support[1] tightly concentrated around a region in configuration space (of $\sim 10^{26}$ particles) that makes up a pointer in physical space pointing in some direction, i.e., defining some pointer position. So different pointer positions belong to macroscopically disjoint wave functions, that is, wave functions whose supports are macroscopically separated in configuration space (see Fig. 9.1).

[1] The support of a function is the domain on which it is not equal to zero. The notions of support, separation of supports, and disjointness of supports have to be taken with a grain of salt. The support of a Schrödinger wave function is typically unbounded and consists of (nearly) the whole of configuration space. "Zero" has thus to be replaced by "appropriately small" (in the sense that the square norm over the region in question is negligible). Then, the precise requirement of macroscopic disjointness is that the overlap of the wave functions is extremely small in the square norm over any macroscopic region.

The coordinates of the system particles will be denoted by $\mathbf{X}$ in some m-dimensional configuration space, and for simplicity we assume that the system wave function is a superposition of the two wave functions $\psi_1(\mathbf{x})$ and $\psi_2(\mathbf{x})$ alone. The system interacts with an apparatus with particle coordinates $\mathbf{Y}$ in some n-dimensional configuration space. Possible wave functions of the apparatus will be a pointer "null" wave function $\Phi_0(\mathbf{y})$ ($Y \in \operatorname{supp}\Phi_0 \widehat{=} 0$) and two additional pointer position wave functions $\Phi_1(\mathbf{y})$ ($Y \in \operatorname{supp}\Phi_1 \widehat{=} 1$) and $\Phi_2(\mathbf{y})$ ($Y \in \operatorname{supp}\Phi_2 \widehat{=} 2$). We call an experiment a measurement experiment whenever the interaction and the corresponding Schrödinger evolution of the coupled system is constructed in such a way that

$$\psi_i \Phi_0 \overset{t \longrightarrow T}{\longrightarrow} \psi_i \Phi_i \,, \qquad i = 1,2 \,, \tag{9.1}$$

where T is the duration of the experiment and the arrow stands for the Schrödinger evolution. This means that the pointer positions correlate perfectly with the system wave functions.[2] We shall say that the pointer points to the outcomes 1 (resp. 2) if ψ_1 (resp. ψ_2) is the system wave function.

Concerning the initial wave function of the system and apparatus, we can make the following remark. Starting the measurement experiment with a product wave function $\Psi(\mathbf{q}) = \Psi(\mathbf{x},\mathbf{y}) = \psi_i(\mathbf{x})\Phi_0(\mathbf{y})$ expresses the fact that the system and apparatus are initially independent physical entities. The Schrödinger equation (8.4) for a product wave function separates into two independent Schrödinger equations, one for each factor, if the interaction potential $V(\mathbf{x},\mathbf{y}) \approx 0$. A warning is in order here: $V(\mathbf{x},\mathbf{y}) \approx 0$ on its own does not imply that the $\mathbf{x}$ and $\mathbf{y}$ parts develop independently, because the wave function $\Psi(\mathbf{x},\mathbf{y})$ need not be a product. To have physical independence, we also require the velocity field $\mathbf{v}^{\Psi}$ of the $\mathbf{X}$ system to be a function of $\mathbf{x}$ alone. In view of (8.1), we observe that

$$\frac{\nabla\Psi}{\Psi} = \nabla \ln \Psi \,,$$

so that

$$\begin{aligned} \nabla \ln \Psi(\mathbf{x},\mathbf{y}) &= \nabla \ln \big(\psi_i(\mathbf{x})\Phi_0(\mathbf{y})\big) \\ &= \nabla \ln \psi_i(\mathbf{x}) + \nabla \ln \Phi_0(\mathbf{y}) \\ &= \begin{pmatrix} \nabla_{\mathbf{x}} \ln \psi_i(\mathbf{x}) \\ \nabla_{\mathbf{y}} \ln \Phi_0(\mathbf{y}) \end{pmatrix} . \end{aligned}$$

Hence the particle coordinates $\mathbf{X}$ are indeed guided by the system wave function ψ_i if the combined system is guided by a product wave function. The arrow in (9.1) stands for a time evolution where the interaction is not zero, i.e., $V(\mathbf{x},\mathbf{y}) \neq 0$. How-

[2] The wave function at the end of the experiment need not have the idealized product structure $\psi_i\Phi_i$. It can be replaced by an entangled wave function $\Psi_i(\mathbf{x},\mathbf{y})$ without changing the following arguments.

ever, in this measurement experiment, these particular initial wave functions guarantee that, at the end of the experiment, we once again obtain a product wave function.

The measurement problem (see also Remark 9.1) is a trivial mathematical consequence of (9.1) and the linearity of the Schrödinger evolution. It comes about whenever the system wave function is a nontrivial superposition

$$\psi(\mathbf{x}) = \alpha_1 \psi_1(\mathbf{x}) + \alpha_2 \psi_2(\mathbf{x}) \,, \qquad |\alpha_1|^2 + |\alpha_2|^2 = 1 \,. \tag{9.2}$$

Then, by virtue of the linearity of the Schrödinger equation, (9.1) yields

$$\psi\Phi_0 = \sum_{i=1,2} \alpha_i \psi_i \Phi_0 \stackrel{t\longrightarrow T}{\longrightarrow} \sum_{i=1,2} \alpha_i \psi_i \Phi_i \,, \tag{9.3}$$

which is a *macroscopic superposition* of pointer wave functions. If one has nothing but wave functions, this is a bad thing, because such a macroscopic superposition has no counterpart in the macroscopic, i.e., classical world. What could this be? A pointer pointing simultaneously to 1 and 2? Did the apparatus become a mushy marshmallow?

In Bohmian mechanics on the other hand the pointer is *there* and it points at something definite. In Bohmian mechanics we have the evolution of the real state of affairs, namely the evolution of the coordinates given by (8.1). And by virtue of the quantum equilibrium hypothesis, we do not even need to worry about the detailed trajectories. If we are interested in the pointer position after the measurement experiment, i.e., in the actual configuration of the pointer particles $\mathbf{Y}(T)$ (see Fig. 9.2), we need only observe the following. Given the wave function on the right-hand side

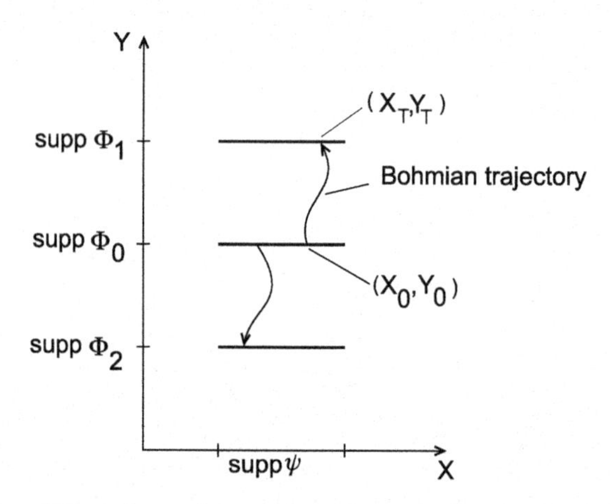

Fig. 9.2 Evolution of the system–pointer configuration in the measurement experiment. The Y-axis (the pointer axis) represents the configuration space of macroscopic dimensions, while the X-axis (the system axis) may be thought of as low-dimensional. Depending on the initial values (X_0, Y_0), at time T, the Bohmian trajectory ends up either in the upper or the lower of the two macroscopically distant subsets that the support is split into in the system–pointer configuration space

of (9.3), by the quantum equilibrium hypothesis, the probability that the pointer "points to k", i.e., the probability that its configuration is in the support of Φ_k, is given by

$$\begin{aligned}\mathbb{P}^{\Psi}\left(\mathbf{Y}(T) \mathrel{\widehat{=}} k\right) &= \mathbb{P}^{\Psi}\left(\mathbf{Y}(T) \in \operatorname{supp} \Phi_k\right) \\ &= \int_{\{(\mathbf{x},\mathbf{y})|\mathbf{y}\in\operatorname{supp}\Phi_k\}} \left| \sum_{i=1,2} \alpha_i \psi_i(\mathbf{x}) \Phi_i(\mathbf{y}) \right|^2 \mathrm{d}^m x \mathrm{d}^n y \\ &= \sum_{i=1,2} \int_{\{(\mathbf{x},\mathbf{y})|\mathbf{y}\in\operatorname{supp}\Phi_k\}} |\alpha_i|^2 \, |\psi_i(\mathbf{x})|^2 \, |\Phi_i(\mathbf{y})|^2 \, \mathrm{d}^m x \mathrm{d}^n y \\ &\quad +2\Re\left[\int_{\{(\mathbf{x},\mathbf{y})|\mathbf{y}\in\operatorname{supp}\Phi_k\}} \alpha_1 \alpha_2^* \psi_1(\mathbf{x}) \psi_2^*(\mathbf{x}) \Phi_1(\mathbf{y}) \Phi_2^*(\mathbf{y}) \mathrm{d}^m x \mathrm{d}^n y\right] \\ &\cong |\alpha_k|^2 \,.\end{aligned}$$

Note that $\operatorname{supp} \Phi_1 \cap \operatorname{supp} \Phi_2 \cong \emptyset$, and hence for $i \neq k$,

$$\int_{\operatorname{supp}\Phi_k} |\Phi_i(\mathbf{y})|^2 \mathrm{d}^n y \cong 0 \,,$$

and likewise for the mixed terms

$$\int_{\operatorname{supp}\Phi_k} \Phi_1(\mathbf{y}) \Phi_2(\mathbf{y}) \mathrm{d}^n y \cong 0 \,.$$

Therefore the pointer points to position 1 with probability $|\alpha_1|^2$ and to position 2 with probability $|\alpha_2|^2$. Which pointer position results depends solely on the initial coordinates $(\mathbf{X}_0, \mathbf{Y}_0)$ of the particles, as depicted in Fig. 9.2. Note that, while not necessary for measurement experiments, but possibly in line with one's intuition, it may often be the case that the pointer's final position is determined essentially by the initial positions $\mathbf{X}_0$ of the system particles alone and not by the coordinates $\mathbf{Y}_0$ of the pointer particles. This is the case in a "spin measurement" using a Stern–Gerlach apparatus (where the measurement is simply the detection of already split trajectories).

Remark 9.1. The Song and Dance about the Measurement Problem
Without Bohmian mechanics the result (9.3) has no physical meaning, unless one interprets the wave function as an instrument for computing the probabilities that this or that pointer position results. But this then means that there *is* a pointer, and this might as well be the Bohmian one, whence nothing more needs to be said. However, physicists have been convinced that the innovation of quantum mechanics is something like this: *the macroscopic world is real, but it cannot be described by microscopic constituents governed by a physical law.* On the other hand the pointer moves from 0 to 1 or 2, just as Schrödinger's cat either dies or stays alive, so some movement is going on. Why should that not be describable?

It seems therefore an inevitable consequence that one must deny altogether that there is anything real besides the wave function. This is a clean attitude, and one arrives at the clearest form of the measurement problem, because (9.3) is not what one sees. To overcome this defect the notion of "observer" was introduced. The observer makes everything alright, because *it/he/she collapses the pointer wave function to a definite outcome*. The observer brings things into being, so to speak, by observing, where "observing" is just a fancy way of saying "looking". As if the apparatus alone could not point to anything in particular, but rather – well what? As if the cat could not know whether it was alive, its brain being in a grotesque state of confusion, and it therefore had to wait to become definitely alive or dead until somebody looked at it.

But what qualifies an object to be an observer? Is a cat not an observer? Must the moon be observed in order to be there? Can only a subject with a PhD in physics act as an observer, as John Bell and Richard Feynman liked to ask with suitable irony? The observer is macroscopic alright, but how big is macroscopic? One must admit that these are all funny questions, but they have been earnestly discussed by physicists.

But let us get back to reason. An observer is simply another albeit huge pointer! This means that, when the observer looks, i.e., when she/he interacts with the system (the apparatus) in question, the disjointness of the wave function supports (now subsets of the system–apparatus–observer configuration space) becomes even greater, and we have the same line of argument again: the huge wave function (including the observer) splits into a macroscopic superposition and the measurement problem remains.

So now what? Well, nothing. Some say that decoherence comes to the rescue. When applied to the measurement problem, decoherence theory (which we discuss a little further below, in the context of effective collapse) attempts to explain or prove or argue the obvious: that it is impossible for all practical purposes to bring such macroscopically disjoint macroscopic wave functions into interference. In other words, for all practical purposes – fapp, to use John Bell's abbreviation – it is essentially impossible to bring the macroscopically disjoint supports of the wave functions together so that there is any significant overlap. Just as it is essentially impossible to have the wave functions of the dead and alive cat interfere. Just as it is essentially impossible for a gas which has left a bottle and filled a box to go back into the bottle. Of course, in principle it can be done, because one just has to reverse all the velocities at the same time and take care that there is no uncontrolled interaction with the walls of the box. So interference of macroscopically separated macroscopic wave packets is fapp impossible. Wave functions which consist of macroscopically separated macroscopic wave packets are fapp indistinguishable from statistical mixtures (which we also discuss further below) of two non-entangled wave functions. This is true. This is why Schrödinger needed to add in his cat example that there was a difference between a shaky or out-of-focus photograph and a snapshot of clouds and fog banks.[3]

Besides Bohmian mechanics, the only other serious approach to resolving the measurement problem is to make the evolution (9.3) theoretically impossible. This

[3] Es ist ein Unterschied zwischen einer verwackelten oder unscharf eingestellten Photographie und einer Aufnahme von Wolken und Nebelschwaden.

means that the *linear* Schrödinger evolution is not valid. It has to be replaced by one that does not allow macroscopic superpositions, i.e., wave packet reduction becomes part of the theoretical description. This is known as the GRW theory or dynamical reduction model, and good accounts can be found in [1–4]. ■

Remark 9.2. The Wave Function Is Not Measurable
One is sometimes told that the virtue of quantum mechanics is that it only talks about quantities that can be measured, and since by Heisenberg's uncertainty relation the momentum and position of a particle cannot be measured simultaneously, one must not introduce particles and their positions into the theory. We shall discuss this nonsense in more detail later, but first let us take the opportunity to point out a simple consequence of (9.3), namely that the wave function cannot be measured.

Here is the argument. Imagine a piece of apparatus which measures wave functions. This means that, for every wave function ψ a system can have, there is a pointer position such that, in view of (9.1), one has $\psi\Phi_0 \to \psi\Phi_\psi$, i.e., ψ is pointed at if ψ is the wave function the system happens to be in. Now such an apparatus does not exist. This is actually what (9.3) tells us! For suppose such an apparatus did point out either ψ_1 or ψ_2. Then if the system wave function is the superposition (9.2), the apparatus wave function must evolve by virtue of mathematical logic into (9.3), i.e., into a superposition of pointer positions, and not to $(\alpha_1\psi_1 + \alpha_2\psi_2)\Phi_{(\alpha_1\psi_1+\alpha_2\psi_2)}$. So it is easy to find a quantum mechanical quantity that cannot be measured. Fortunately, one has Bohmian mechanics and the particle positions, because they can be measured. And because of that, we can sometimes find out what the wave function is, using the Bohmian positions and *the theory*, of course. ■

9.2 Effective Collapse

Now let us return to (9.3) and more important issues. If the pointer points to 1, i.e., if

$$\mathbf{Y}(T) \in \operatorname{supp}\Phi_1 , \tag{9.4}$$

we can fapp forget the wave packet $\psi_2\Phi_2$ in the further physical description of the world. In view of (8.1), and because $\operatorname{supp}\mathbf{v}^{\Phi_1} \cap \operatorname{supp}\mathbf{v}^{\Phi_2} = \emptyset$, the velocity field on the combined configuration space for the system and apparatus induced by the wave function on the right-hand side of (9.3) becomes (the αs do not matter)

$$\mathbf{v}^{\psi_1\Phi_1+\psi_2\Phi_2} = \begin{cases} \mathbf{v}^{\psi_1\Phi_1} = \begin{pmatrix} \mathbf{v}_\mathbf{x}^{\psi_1} \\ \mathbf{v}_\mathbf{y}^{\Phi_1} \end{pmatrix} & \text{for } \mathbf{y} \in \operatorname{supp}\Phi_1 , \\ \mathbf{v}^{\psi_2\Phi_2} = \begin{pmatrix} \mathbf{v}_\mathbf{x}^{\psi_2} \\ \mathbf{v}_\mathbf{y}^{\Phi_2} \end{pmatrix} & \text{for } \mathbf{y} \in \operatorname{supp}\Phi_2 , \end{cases} \tag{9.5}$$

so that only $\psi_1\Phi_1$ "guides" if (9.4) holds, i.e., if the pointer position is 1. Fapp $\psi_2\Phi_2$ will not become effective again, as that would require $\psi_1\Phi_1$ and $\psi_2\Phi_2$ to interfere, either by controlling 10^{26} "random phases", an impossible task, or by Poincaré recurrence (see Sect. 9.5.2), which would take a ridiculously long lapse of time. Hence, as far as the further evolution of our world is concerned, we can forget about $\psi_2\Phi_2$, i.e., Bohmian mechanics provides us with a collapsed wave function. If $\mathbf{Y}(T) \in \operatorname{supp}\Phi_1$, the effective wave function is $\psi_1\Phi_1$ and that means, by the argument (9.2), that the effective wave function of the system is ψ_1. Put another way, we might say that, due to the measurement process ψ *collapses* to ψ_1.

This collapse is not a physical process, but an act of convenience. It is introduced because it would simply be uneconomical to keep the ineffective wave functions, and the price we pay for forgetting them amounts to nothing. In just the same way, we can safely forget the fact that the velocities of gas molecules that have escaped from a bottle into a (much larger) container are such that reversing them all at the same time would result in the gas rushing back into the bottle. The gas molecules will interact with the walls of the box (exchanging heat for example) and the walls of the box will further interact with the outside world. In the end, reversal of velocities will no longer be sufficient to get the gas back into the bottle, as the information that the gas was once in the bottle is now spread all over the place.

We have exactly the same effect in quantum mechanics, where it is called *decoherence*. The separation of waves happens more or less all the time, and more or less everywhere, because measurement experiments that measure position are ubiquitous. The pointer (as synonymous with the apparatus) looks at where the particle is, the air molecules in the lab look at where the pointer is (by bouncing off the pointer), the light passing through the lab looks at how the air molecules bounced off the pointer, and so it goes on.

Here is an example along these lines. A particle is initially in a superposition of two spatially separated wave packets $\psi_L(\mathbf{x}) + \psi_R(\mathbf{x})$, where ψ_L moves to the left and ψ_R moves to the right (see Fig. 9.3). To the right there is a photographic plate. If the particle hits the plate, it will blacken it at its point of arrival. So the photographic plate is a piece of apparatus and in this case the "pointer points" either at a black spot or nothing. If the particle is guided by ψ_R, then the plate will eventually feature a black spot. If no black spot shows up, then the particle travels with ψ_L. As in (9.3), Schrödinger evolution thus leads to

$$\left[\psi_L(\mathbf{x}) + \psi_R(\mathbf{x})\right]\Phi(\mathbf{y}) \overset{t \longrightarrow T}{\longrightarrow} \psi_L(\mathbf{x},T)\Phi(\mathbf{y}) + \Phi_B(\mathbf{y},\mathbf{x},T) \,, \tag{9.6}$$

where Φ_B stands for the blackened plate.

Can it happen that the position $\mathbf{X}$ is in the support of ψ_L and the plate nevertheless shows a black spot? In fact it cannot, because the supports of $\Phi(\mathbf{y})$ and $\Phi_B(\mathbf{y},\mathbf{x},T)$ are disjoint. Why is this? Because the black spot arises from a macroscopic chemical reaction at the particle's point of arrival (macroscopic because the black spot can be seen with the naked eye). Bohmian dynamics thus precludes the possibility of simultaneously having $\mathbf{X} \in \operatorname{supp}\psi_L(T)$ and $\mathbf{Y} \in \operatorname{supp}\Phi_B(T)$ (see Fig. 9.4).

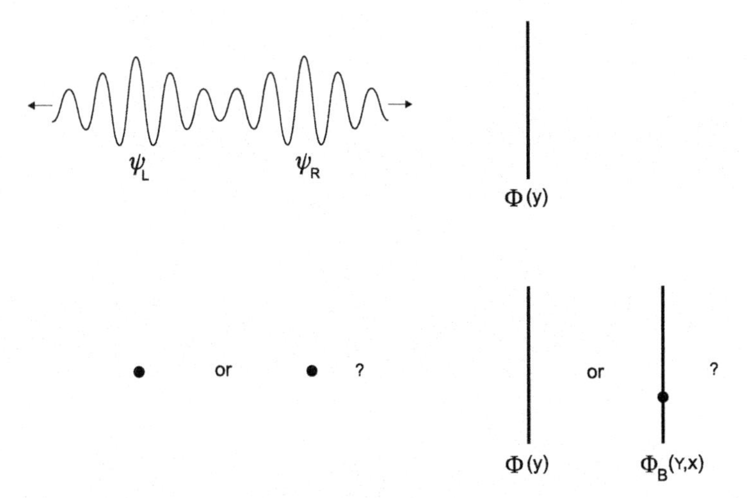

Fig. 9.3 The measurement process. Where is the particle?

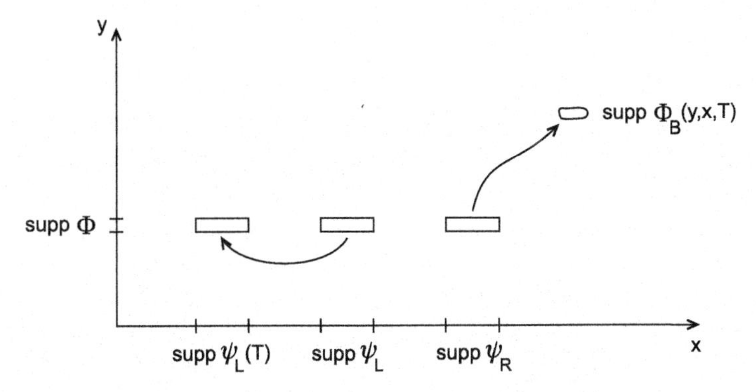

Fig. 9.4 Measurement of a particle's position, depicted in the configuration space of the photographic plate and particle

In the foregoing we described a position measurement with the help of a photographic plate. Is there anything special about the plate? Of course, there is nothing at all special about it! We could just as well have used light waves that scatter off the particle and measured its position in that way. Light rays scattered off a particle are different from undisturbed light rays. So just like a photographic plate, light rays produce decoherence. The particle wave function gets entangled with the wave function of the photons in the light ray. So once again we can effectively collapse the particle wave function to where the particle is. After all, since we wish to describe the world as we experience it, it is relevant to us where the particle is. So it is now the light waves that act as pointer (see Fig. 9.5).

But the procedure does not stop there. The light interacts with other things and so changes configurations in the environment (wherever the light goes, there will be changes). So the entanglement, the fapp impossibility of interference, continues

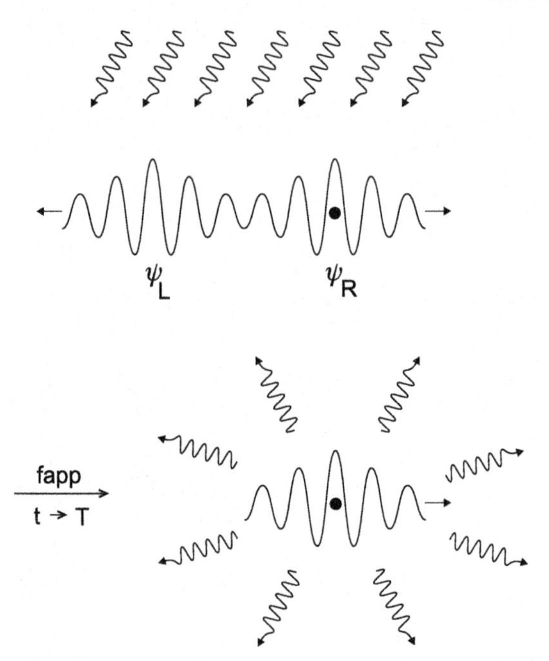

Fig. 9.5 Light waves as pointer

to grow. This is decoherence at work. But decoherence alone does not create facts. The facts are already there in the form of the Bohmian positions! Only the fapp description of that real state of affairs is a collapsed one. Schrödinger once thought that a cat was a big enough pointer to get that point across, and that some description of the real state of affairs was needed for a physical description.

The effective collapse becomes even more stable once the results of a measurement have been recorded. Interference of what was once the particle wave function now means that all the records must also be brought under control and made to interfere. That does seem hopeless. The collapse is stable for all times relevant to us. The moral is that interactions typically destroy coherence, since they "measure" the particle positions. Fapp collapse of wave functions is the rule, and that is essential for understanding how the classical macroscopic world emerges.

In Remark 9.1, we mentioned an alternative approach for solving the measurement problem, which involved introducing the collapse as a fundamental physical event. Such a fundamentally random theory, in which a random collapse is part of the theoretical description, will ensure that the collapse is effective only for large systems. How is it possible to distinguish experimentally between Bohmian mechanics and a collapse theory? The answer is, by experimentally achieving a macroscopic superposition, i.e., by forcing the wave functions of a dead and alive cat to interfere. But perhaps we had better not use a cat. Some less cruel and smaller sized experiment would do. However, one should realize the problematic nature of such experiments. For how can one hope to control the effects of decoherence? Without control over the ubiquitous decoherence, it does not prove much, if one cannot see interference of mesoscopic/macroscopic systems. On the other hand, if one could experimentally achieve interference, well then everything depends on whether the interfering system is big enough to be sure that it will be collapsed according to the collapse theory, which of course contains parameters that can be adjusted to cover quite a wide range. It will be a long time before we can hope for experimental help on deciding which of these theories, Bohmian mechanics or spontaneous localization, is physically correct.

9.3 Centered Wave packets

Let us follow the particle in Fig. 9.5 for a while. Suppose the particle is alone in the universe, "in" a wave packet of the form (6.15), viz.,

$$\psi(\mathbf{x},t) = \int \mathrm{e}^{\mathrm{i}[\mathbf{k}\cdot\mathbf{x}-\omega(\mathbf{k})t]} \hat{f}_{\mathbf{k}_0}(\mathbf{k})\mathrm{d}^3k \,,$$

with $\omega(\mathbf{k}) = \hbar k^2/2m$. This is the solution of the one-particle Schrödinger equation (8.4) with $V = 0$. The wave moves with group velocity $(\partial\omega/\partial\mathbf{k})(\mathbf{k}_0) = \hbar\mathbf{k}_0/m$, and according to the stationary phase argument, the width of the $\mathbf{k}$ distribution given by the exponential function is proportional to $1/t$ [see (6.18)]. The wave therefore spreads linearly with t. This spreading is a generic wave phenomenon for the Schrödinger evolution that results from the special dispersion relation. However, according to (6.18) (with $\gamma = \hbar/2m$), the larger the mass, the smaller the spreading rate, i.e., the wave function of a very massive particle stays localized (around its center) for quite a long time.

Now consider a wave packet over such a period of time that spreading can be neglected. Let us follow the evolution of the position $\mathbf{X}$ of a Bohmian particle. Due to quantum equilibrium, $\mathbf{X}(t)$ moves with the packet. But how does the packet move? According to Schrödinger's equation, of course:

$$\mathrm{i}\hbar\frac{\partial\psi}{\partial t}(\mathbf{x},t) = -\frac{\hbar^2}{2m}\Delta\psi(\mathbf{x},t) + V(\mathbf{x})\psi(\mathbf{x},t) = H\psi(\mathbf{x},t) \,, \tag{9.7}$$

with

$$H = -\frac{\hbar^2}{2m}\Delta + V(\mathbf{x}) \tag{9.8}$$

as Hamilton operator. For $\mathbf{X}(\mathbf{x},t)$ with initial value $\mathbf{x}$, we get

$$\dot{\mathbf{X}}(\mathbf{x},t) = \frac{\hbar}{m}\Im\frac{\nabla\psi}{\psi}(\mathbf{X}(\mathbf{x},t),t) \,.$$

Since the packet is well localized in position, instead of $\mathbf{X}(\mathbf{x},t)$, we consider its expectation value $\mathbb{E}^{\psi}\big(\mathbf{X}(t)\big)$. According to the quantum equilibrium hypothesis and equivariance, this is given by

$$\begin{aligned}\langle\mathbf{X}\rangle^{\psi}(t) &:= \mathbb{E}^{\psi}\big(\mathbf{X}(t)\big)\\ &= \int \mathbf{X}(\mathbf{x},t)|\psi(\mathbf{x},0)|^2\mathrm{d}^3x\\ &= \int \mathbf{x}|\psi(\mathbf{x},t)|^2\mathrm{d}^3x\,. \end{aligned}\tag{9.9}$$

For ease of notation we shall omit the superscript ψ from now on. Using (7.17), we obtain

$$\begin{aligned}\frac{\mathrm{d}}{\mathrm{d}t}\langle\mathbf{X}\rangle(t) &= \int \mathbf{x}\frac{\partial}{\partial t}|\psi(\mathbf{x},t)|^2\mathrm{d}^3x\\ &= -\int \mathbf{x}\nabla\cdot\mathbf{j}(\mathbf{x},t)\mathrm{d}^3x\\ &= \int \mathbf{j}(\mathbf{x},t)\mathrm{d}^3x\\ &= \big\langle \mathbf{v}^{\psi}\big(\mathbf{X}(t),t\big)\big\rangle\,,\end{aligned}\tag{9.10}$$

where we have used partial integration and set the boundary terms to zero. Intuitively, a Schrödinger wave that does not spread should move classically. Therefore we consider $\mathrm{d}^2\langle\mathbf{X}\rangle/\mathrm{d}t^2$. For this we need $\partial\mathbf{j}/\partial t$. Using (7.12), we get

$$\frac{\partial}{\partial t}\mathbf{j} = -\frac{\mathrm{i}\hbar}{2m}\frac{\partial}{\partial t}\left(\psi^*\nabla\psi-\psi\nabla\psi^*\right)\,,$$

and with (7.11) and (9.7),

$$\frac{\partial}{\partial t}\mathbf{j} = \frac{1}{2m}\Big[(H\psi^*)\nabla\psi-\psi^*\nabla(H\psi)+(H\psi)\nabla\psi^*-\psi\nabla(H\psi^*)\Big]\,.$$

Further, by partial integration, we have for wave packets ψ and φ

$$\int \psi^* H\varphi\mathrm{d}^3x = \int (H\psi^*)\varphi\mathrm{d}^3x\,,$$

a property which we shall later call symmetry (see Chap. 14). So

$$\begin{aligned}\frac{d^2}{dt^2}\langle \mathbf{X}\rangle(t) &= \int \frac{\partial}{\partial t}\mathbf{j}(\mathbf{x},t)d^3x \\ &= \frac{1}{2m}\int \left[\psi^* H\nabla\psi - \psi^*\nabla(H\psi) + \psi H\nabla\psi^* - \psi\nabla(H\psi^*)\right]d^3x \\ &= \frac{1}{2m}\int \left[\psi^* V\nabla\psi - \psi^*\nabla(V\psi) + \psi V\nabla\psi^* - \psi\nabla(V\psi^*)\right]d^3x \quad \text{[by (9.8)]} \\ &= \frac{1}{2m}\int \left[-(\nabla V)\psi^*\psi - (\nabla V)\psi\psi^*\right]d^3x \\ &= \frac{1}{m}\big\langle -\nabla V(\mathbf{X})\big\rangle(t)\,.\end{aligned}$$

Hence we arrive at Newtonian equations in the mean, so to speak, a version of the Ehrenfest theorem:

$$m\langle \ddot{\mathbf{X}}\rangle = \big\langle -\nabla V\big(\mathbf{X}(t)\big)\big\rangle\,, \tag{9.11}$$

and we would have the classical limit and the final identification of the parameter m with Newtonian mass if

$$\big\langle \nabla V\big(\mathbf{X}(t)\big)\big\rangle = \nabla V\big(\langle \mathbf{X}(t)\rangle\big)\,. \tag{9.12}$$

For this, however, we would have

$$\mathrm{Var}\,(\mathbf{X}) = \Big\langle (\mathbf{X} - \langle \mathbf{X}\rangle)^2\Big\rangle \approx 0\,,$$

which means that $\psi(\mathbf{x},t)$ would be a very well localized wave.

Let us make that a little more precise. Expand V up to third order around $\langle \mathbf{X}\rangle$:

$$V(\mathbf{X}) \approx V(\langle \mathbf{X}\rangle) + (\mathbf{X} - \langle \mathbf{X}\rangle)\nabla V(\langle \mathbf{X}\rangle) + \frac{1}{2}\Big[(\mathbf{X} - \langle \mathbf{X}\rangle)\cdot\nabla\Big]^2 V(\langle \mathbf{X}\rangle) + \frac{1}{3!}\Big[(\mathbf{X} - \langle \mathbf{X}\rangle)\cdot\nabla\Big]^3 V(\langle \mathbf{X}\rangle)\,. \tag{9.13}$$

So the expectation value of $\nabla V(\mathbf{X})$ is given by

$$\langle \nabla V(\mathbf{X})\rangle \approx \nabla V(\langle \mathbf{X}\rangle) + \frac{1}{2}\Big\langle \Big[(\mathbf{X} - \langle \mathbf{X}\rangle)\cdot\nabla\Big]^2\Big\rangle \nabla V(\langle \mathbf{X}\rangle)\,.$$

We thus establish as a rule of thumb that, in order to have classicality of the motion in a potential, the width of the wave function should obey

$$\mathrm{Var}(\mathbf{X}) \ll \sqrt{\frac{V'}{V'''}}\,. \tag{9.14}$$

So far so good. But now we have to admit that the spreading *will* eventually become effective. But the particle is not alone in the universe. It will interact with everything it can in its environment. If this interaction is strong enough, it may result in a measurement-like process (as we have discussed before), and an effective collapse will thus take place, countering the effect of spreading. So the collapsed wave packet remains localized and the above reasoning about classicality seems to be alright if we take into account the fact that the environment acts like a piece of position measuring apparatus. In other words, we do generically have (9.12), and hence with (9.11), classical physics.

However, this argument is faulty, because we have assumed the validity of the Schrödinger evolution. But why should that hold true for a wave function that collapses all the time due to interactions with the environment? Is the disturbance of the environment such that it only decoheres (i.e., counters the spreading of) the wave function, without disturbing the evolution of the center of the wave packet? The situation is reminiscent of Browian motion, where the effects of the environment lead to diffusion and friction (dissipation of energy). The question is therefore: Does the "reading" of the particle's position by the environment happen on a different (shorter) time scale than friction and dissipation, which are also generated by the interaction? The answer must be sought in the proper derivation of an appropriate new phenomenological equation which should describe the "reading" process in such a way that spreading is suppressed, i.e., (9.11) holds, only very little (or no) friction and diffusion is present, and (9.12) still holds. We shall say more about this in Sect. 9.4.

9.4 The Classical Limit of Bohmian Mechanics

In Bohmian mechanics the question of the classical limit is simply this: Under what physical conditions are the Bohmian trajectories close to Newtonian trajectories? In the previous section we discussed one possibility, in a rather hand-waving way, namely the evolution of narrow wave packets. If they move classically, the Bohmian trajectories do so, too, because they are dragged along, as it were. But now we would like to find a more general answer. At some point the narrow wave packets will certainly play a (technical) role, but they do not provide a fundamental answer. In fact the best hopes for a fundamental answer lie in exactly the opposite direction: a freely moving wave packet that spreads all over the place! In the long run the Bohmian trajectories of a freely moving wave packet become classical.

To see this we have to specify what we mean by "in the long run". We shall have to look at the wave function on a macroscopic scale (in time and space). Where else would we expect to see classical behavior? We have already encountered such a macroscopic scaling in our treatment of Brownian motion. However, the scaling here is a bit different, since we do expect ballistic rather than diffusive motion. Now, the macroscopic position of the Bohmian particle at a macroscopic time is

$$\mathbf{X}_{\varepsilon}(\mathbf{x},t) = \varepsilon \mathbf{X}\left(\frac{\mathbf{x}}{\varepsilon},\frac{t}{\varepsilon}\right) . \tag{9.15}$$

Here we think of ε as being very small, eventually even going to zero for precise limit statements. Note that $\dot{\mathbf{X}}_\varepsilon(\mathbf{x},t) = \mathbf{v}^\Psi(\mathbf{x}/\varepsilon, t/\varepsilon)$ is of order one. The wave function on this scale is

$$\psi_\varepsilon(\mathbf{x},t) = \varepsilon^{-3/2}\psi\left(\frac{\mathbf{x}}{\varepsilon},\frac{t}{\varepsilon}\right),$$

so the quantum equilibrium density (with norm one) becomes $|\psi_\varepsilon(\mathbf{x},t)|^2$. This is easily seen by computing any expectation value. A simple change of variables gives

$$\begin{aligned}\mathbb{E}^\psi\left[f(\mathbf{X}_\varepsilon(t))\right] &= \int\left|\psi\left(\mathbf{x},\frac{t}{\varepsilon}\right)\right|^2 f(\varepsilon\mathbf{x})\mathrm{d}^3x \\ &= \int\varepsilon^{-3}\left|\psi\left(\frac{\mathbf{x}}{\varepsilon},\frac{t}{\varepsilon}\right)\right|^2 f(\mathbf{x})\mathrm{d}^3x = \mathbb{E}^{\psi_\varepsilon}\left[f(\mathbf{X}(t))\right].\end{aligned}$$

The Schrödinger equation for the free evolution becomes after rescaling

$$\mathrm{i}\hbar\varepsilon\frac{\partial\psi_\varepsilon(\mathbf{x},t)}{\partial t} = -\frac{\hbar^2\varepsilon^2}{2m}\Delta_\mathbf{x}\psi_\varepsilon(\mathbf{x},t). \tag{9.16}$$

Now note that ε appears only in the combination $\varepsilon\hbar$. Thus the unphysical limit $\hbar\to 0$ can in fact be interpreted as the macroscopic limit $\varepsilon\to 0$.

If the Schrödinger equation contains a potential V, then (9.16) becomes

$$\mathrm{i}\hbar\varepsilon\frac{\partial\psi_\varepsilon(\mathbf{x},t)}{\partial t} = -\frac{\hbar^2\varepsilon^2}{2m}\Delta_\mathbf{x}\psi_\varepsilon(\mathbf{x},t) + V\left(\frac{\mathbf{x}}{\varepsilon}\right)\psi_\varepsilon(\mathbf{x},t).$$

We make this observation in order to establish the following point. For classical behavior, one will have to assume that the potential varies on the macroscopic scale, i.e., $V(\mathbf{x}/\varepsilon) = U(\mathbf{x})$. But the potential is given by the physical situation so, if anything, it is the potential that *defines* the macroscopic scale, i.e., it is the potential that defines the scaling parameter ε. More precisely, ε will be given by the ratio of two length scales, the width of the wave function and the characteristic length on which the potential varies [for the latter compare with (9.14)]. We do not wish to pursue this further here (but see [5]).

We continue with the free evolution. We wish to find an expression for $\psi_\varepsilon(\mathbf{x},t)$ when $\varepsilon\approx 0$. The solution of (9.16) is a superposition of plane waves and this is done with mathematical rigor in Remark 15.7. This involves nothing other than solving the equation via Fourier transformation (as we did for the heat equation). We should be able to do that with our eyes shut, never mind all the rigorous mathematics it may be shrouded in:

$$\begin{aligned}\psi_\varepsilon(\mathbf{x},t) &= (\varepsilon)^{-3/2}\psi\left(\frac{\mathbf{x}}{\varepsilon},\frac{t}{\varepsilon}\right) \\ &= (2\pi\varepsilon)^{-3/2}\int\exp\left[\mathrm{i}\left(\mathbf{k}\cdot\frac{\mathbf{x}}{\varepsilon} - \frac{\hbar k^2}{2m}\frac{t}{\varepsilon}\right)\right]\widehat{\psi}(\mathbf{k})\mathrm{d}^3k \\ &= (2\pi\varepsilon)^{-3/2}\int\exp\left[\frac{\mathrm{i}}{\varepsilon}\left(\mathbf{k}\cdot\mathbf{x} - \frac{\hbar k^2}{2m}t\right)\right]\widehat{\psi}(\mathbf{k})\mathrm{d}^3k,\end{aligned} \tag{9.17}$$

where $\widehat{\psi}(\mathbf{k})$ is the Fourier transform of the wave function ψ at time zero.

Now let us go on with the stationary phase argument. It says that for small ε the only contribution comes from that $\mathbf{k}$ value for which the phase

$$S(\mathbf{k}) = \mathbf{k} \cdot \mathbf{x} - \frac{\hbar k^2}{2m} t$$

is stationary. This value is given by $\mathbf{k}_0 = m\mathbf{x}/\hbar t$. Expanding around the stationary point,

$$\frac{1}{\varepsilon} S(\mathbf{k}) = \frac{mx^2}{2t\hbar\varepsilon} - \frac{\hbar t}{2m\varepsilon} (\mathbf{k} - \mathbf{k}_0)^2 ,$$

putting this into (9.17), and multiplying by the phase factor, this gives

$$\begin{aligned}
\exp\left(-\frac{\mathrm{i}}{\hbar\varepsilon}\frac{mx^2}{2t}\right)\psi_\varepsilon(\mathbf{x},t) &= (2\pi\varepsilon)^{-3/2} \exp\left(\frac{-\mathrm{i}}{\hbar\varepsilon}\frac{mx^2}{2t}\right) \int \exp\left[\frac{\mathrm{i}}{\varepsilon} S(\mathbf{k})\right] \widehat{\psi}(\mathbf{k}) \mathrm{d}^3 k \\
&= (2\pi\varepsilon)^{-3/2} \int \exp\left[-\frac{\mathrm{i}t}{2m\hbar\varepsilon}\hbar^2 (\mathbf{k}-\mathbf{k}_0)^2\right] \widehat{\psi}(\mathbf{k}) \mathrm{d}^3 k \\
&= (2\pi\varepsilon)^{-3/2} \int \exp\left[-\frac{\mathrm{i}t}{2m\hbar\varepsilon}\left(\mathbf{p} - m\frac{\mathbf{x}}{t}\right)^2\right] \widehat{\psi}\left(\frac{\mathbf{p}}{\hbar}\right) \frac{1}{\hbar^3} \mathrm{d}^3 p \\
&= (2\pi\varepsilon)^{-3/2} \int \exp\left(-\frac{\mathrm{i}t}{2m\hbar\varepsilon}\mathbf{u}^2\right) \widehat{\psi}\left(\frac{\mathbf{u}}{\hbar} + \frac{m}{\hbar}\frac{\mathbf{x}}{t}\right) \frac{1}{\hbar^3} \mathrm{d}^3 u \\
&= \left(\frac{m}{\pi\hbar t}\right)^{3/2} \int \exp\left(-\mathrm{i}\mathbf{u}^2\right) \widehat{\psi}\left(\frac{\sqrt{2m\varepsilon}}{\sqrt{\hbar t}}\mathbf{u} + \frac{m}{\hbar}\frac{\mathbf{x}}{t}\right) \mathrm{d}^3 u .
\end{aligned} \tag{9.18}$$

Next we let $\varepsilon \to 0$. (In Remark 15.8 we shall do these asymptotics in a rigorous manner, but for now we do not care about rigor.) The integral becomes a complex Gaussian integral,

$$\int \exp\left(-\mathrm{i}u^2\right) \mathrm{d}^3 u = \left(\frac{\pi}{\mathrm{i}}\right)^{3/2} . \tag{9.19}$$

Since the integrand is holomorphic, (9.19) is easily verified by a simple deformation of the integration path that changes the integral into a typical Gaussian integral [see (5.8)]. Thus

$$\exp\left(-\frac{\mathrm{i}}{\hbar\varepsilon}\frac{mx^2}{2t}\right)\psi_\varepsilon(\mathbf{x},t) \approx \left(\frac{m}{\mathrm{i}t\hbar}\right)^{3/2} \widehat{\psi}\left(\frac{m}{\hbar}\frac{\mathbf{x}}{t}\right) ,$$

or

$$\psi_\varepsilon(\mathbf{x},t) \approx \left(\frac{m}{\mathrm{i}t\hbar}\right)^{3/2} \exp\left(\frac{\mathrm{i}}{\hbar\varepsilon}\frac{mx^2}{2t}\right) \widehat{\psi}\left(\frac{m}{\hbar}\frac{\mathbf{x}}{t}\right) . \tag{9.20}$$

We introduce

$$S_{\text{class}}(\mathbf{x},t) := \frac{mx^2}{2t}, \qquad R(\mathbf{x},t) := \left(\frac{m}{\mathrm{i}t\hbar}\right)^{3/2} \widehat{\psi}\left(\frac{m}{\hbar}\frac{\mathbf{x}}{t}\right),$$

to get

$$\psi_\varepsilon(\mathbf{x},t) \approx R(\mathbf{x},t) \exp\left[\frac{\mathrm{i}}{\hbar\varepsilon} S_{\text{class}}(\mathbf{x},t)\right]. \tag{9.21}$$

Thus on the macroscopic scale the Bohmian velocity field is given by

$$\begin{aligned} \mathbf{v}^{\psi_\varepsilon}(\mathbf{x},t) : &= \mathbf{v}^{\psi}\left(\frac{\mathbf{x}}{\varepsilon},\frac{t}{\varepsilon}\right) = \frac{\hbar}{m}\Im\frac{\nabla_y \psi(\mathbf{y},t/\varepsilon)}{\psi(\mathbf{y},t/\varepsilon)}\bigg|_{\mathbf{y}=\mathbf{x}/\varepsilon} \\ &= \frac{\varepsilon\hbar}{m}\Im\frac{\nabla\psi_\varepsilon(\mathbf{x},t)}{\psi_\varepsilon(\mathbf{x},t)} \approx \frac{1}{m}\nabla S_{\text{class}}(\mathbf{x},t) = \frac{\mathbf{x}}{t}, \end{aligned} \tag{9.22}$$

since the contribution coming from R or $\widehat{\psi}$, which is complex and thus contributes to the derivative, is of order ε and thus negligible. Evaluated for macroscopic Bohmian trajectories, this gives

$$\dot{\mathbf{X}}_\varepsilon(\mathbf{x},t) = \mathbf{v}^{\psi_\varepsilon}\left(\mathbf{X}_\varepsilon(\mathbf{x},t),t\right) \approx \frac{\mathbf{X}_\varepsilon(\mathbf{x},t)}{t}, \tag{9.23}$$

so macroscopically the Bohmian trajectories will indeed become classical, in the sense that they become straight lines with the macroscopic velocities $\mathbf{X}_\varepsilon(\mathbf{x},t)/t$. We can reformulate this as a result holding for the long time asymptotics of Bohmian trajectories which are guided by a freely evolving wave function. In view of (9.15), we can write

$$\dot{\mathbf{X}}(\mathbf{x},t/\varepsilon) = \dot{\mathbf{X}}_\varepsilon(\varepsilon\mathbf{x},t) \approx \frac{\mathbf{X}_\varepsilon(\varepsilon\mathbf{x},t)}{t} = \frac{\mathbf{X}(\mathbf{x},t/\varepsilon)}{t/\varepsilon}. \tag{9.24}$$

This shows that, in the case of free motion in the long run, i.e., asymptotically ($t/\varepsilon \to \infty$ for $\varepsilon \to 0$), the Bohmian trajectories become straight lines with the asymptotic velocity $\mathbf{X}(\mathbf{x},t)/t$, for t big [6]. In particular we shall see in Chap. 16 on scattering that, after leaving the scattering center, a scattered particle will move asymptotically along a straight line, just as one observes in cloud chambers.

The moral of (9.20) is best understood when we forget ε (put it equal to unity) and consider x,t large! Equation (9.20) gives us the asymptotic form of the wave function for large times (or for large distances, depending on one's point of view):

$$\psi(\mathbf{x},t) \overset{t\,\text{large}}{\approx} \left(\frac{m}{\mathrm{i}t\hbar}\right)^{3/2} \exp\left(\frac{\mathrm{i}}{\hbar}\frac{mx^2}{2t}\right)\widehat{\psi}\left(\frac{m}{\hbar}\frac{\mathbf{x}}{t}\right). \tag{9.25}$$

This says that, for large times, the wave function will be localized at positions $\mathbf{x}$ such that the "momentum" $m\mathbf{x}/t$, or more precisely the wave vector $(m/\hbar)(\mathbf{x}/t)$, lies in the support of the Fourier transform $\widehat{\psi}$ of ψ. That is why the "momentum distribution" is given by $|\widehat{\psi}|^2$, as we shall now show in a little more detail.

For ε small, let us call

$$\mathbf{V}_\infty = \frac{\mathbf{X}_\varepsilon(\mathbf{x},t)}{t}$$

the asymptotic velocity. In (9.23), we identified this as the velocity of the (straight) macroscopic Bohmian trajectory. Using the feature (8.19) of the quantum equilibrium distribution and the asymptotic form (9.20), we show that $\mathbf{V}_\infty$ is distributed according to $|\widehat{\psi}|^2$. For any function f,

$$\begin{aligned}
\mathbb{E}^\psi\Big(f((V_\infty))\Big) &= \int f\left(\frac{\mathbf{X}_\varepsilon(\mathbf{x},t)}{t}\right)|\psi(\mathbf{x},0)|^2\mathrm{d}^3x \\
&= \int f\left(\frac{\mathbf{x}}{t}\right)|\psi_\varepsilon(\mathbf{x},t)|^2\mathrm{d}^3x \\
&\approx \int f\left(\frac{\mathbf{x}}{t}\right)\left|\widehat{\psi}\left(\frac{m}{\hbar}\frac{\mathbf{x}}{t}\right)\right|^2\left(\frac{m}{\hbar t}\right)^3\mathrm{d}^3x \\
&= \int f(\mathbf{v})\left|\widehat{\psi}\left(\frac{m}{\hbar}\mathbf{v}\right)\right|^2\left(\frac{m}{\hbar}\right)^3\mathrm{d}^3v\,,
\end{aligned} \tag{9.26}$$

where we have used the natural substitution $\mathbf{v} = \mathbf{x}/t$. We shall reconsider this in a rigorous manner in Chap. 15. To sum up, what we see on the macroscopic scale are straight trajectories starting at the origin, with velocity distribution $|\widehat{\psi}|^2$. In other words, the classical phase space ensemble $\delta(\mathbf{x})|\widehat{\psi}|^2(\mathbf{k})$ is transported along classical force free trajectories. If one wants to have another starting point $\mathbf{x}_0$, the initial wave function must be chosen to be supported around that point, $\psi(\cdot - \mathbf{x}_0/\varepsilon, 0)$.

All the above ultimately results from the dispersion of wave groups (waves of different wavelengths have different speeds), which we already alluded to in Chap. 6. We remark in passing that the distribution of the free asymptotic velocity [which according to (9.24) and (9.26) is given by $|\widehat{\psi}|^2$] has some meaning for Heisenberg's uncertainty relation. The latter relates this distribution to that of the initial position of the particle. We can already guess that, when all is said and done, Heisenberg's uncertainty relation will be recognized as a simple consequence of Bohmian mechanics and quantum equilibrium.

Let us go on with $S_{\text{class}}(\mathbf{x},t) := mx^2/2t$, the Hamilton–Jacobi function of a free particle of mass m. Let us call wave functions of the form (9.21) local plane wave packets. They consist of "local" plane waves, where each of them produces a straight line as trajectory. We can lift this picture to the situation where a wave moves in some potential V that varies on a macroscopic scale, meaning for example that a relation like (9.14) is satisfied. Then one expects the wave to obtain the form (9.21) once again, but now with the Hamilton–Jacobi function $S_{\text{class}}(V)$ that contains the potential V. Moreover, the local plane waves will be guided by the potential, changing the wavelengths in an analogous way to what happens when light passes through an optical medium. The classical phase space ensemble $\delta(\mathbf{x}-\mathbf{x}_0)|\widehat{\psi}|(\mathbf{k})^2$ will now be transported along the classical trajectories governed by $S_{\text{class}}(V)$.

In Sect. 2.3, we said that the Hamilton–Jacobi function was useless because typically it cannot be defined in a unique manner, at least, not if there are two separate points in configuration space that are, in a given time interval, connected by more than one (classical) trajectory, as we mentioned for the case of a ball reflected by a wall. Here we encounter this difficulty once more. If two different (classical) trajectories can pass from one point in configuration space to another, then this means that two local plane waves can originate from one (macroscopic) point and meet again in another, thereby interfering! The classical trajectories simply move through each other, while Bohmian trajectories cannot do that. Thus whenever the local plane waves meet again, we do not have classicality. But this is a typical situation! So have we uncovered a serious problem with Bohmian mechanics? Of course, we have not. We have simply ignored the fact that there is a world surrounding the particles we wish to look at. We have ignored the effects of decoherence, the "reading" of the environment.

So the full picture is as follows. Dissipation produces local plane waves that cannot interfere any more because of decoherence, each of them being multiplied by a wave function for the environment, as in the measurement experiments. We are thus left with one local plane wave, the one which guides the particle, which moves along a classical trajectory.

9.5 Some Further Observations

9.5.1 Dirac Formalism, Density Matrix, Reduced Density Matrix, and Decoherence

Effective wave functions, as we discussed them above, emerge from decoherence, which arises from interaction of the system in question with its environment. The environment may be a piece of apparatus (pointer) in a carefully designed experiment or simply the noisy environment that is always present. In the latter case, it is feasible that a random evolution for the effective wave function of the system might emerge from an analysis of the combined system and environment, in the same way as a Wiener process emerges from the analysis of the motion of a Brownian particle in its environment.

The types of evolution one would expect to emerge are those that have been suggested in the so-called spontaneous localization models [1]. These describe Brownian motion as a process on the space of wave functions, which, mathematically speaking, will be chosen to be the Hilbert space of square (Lebesgue) integrable functions. The wave functions thus follow diffusion-like paths through Hilbert space. The probability distribution of random wave functions is given by density matrices. But beware! A density matrix need not have the meaning of a probability distribution of random wave functions. We shall see this in a moment.

Phenomenological evolution equations for the so-called reduced density matrices (the analogue of the heat equation for Brownian motion) have been rather extensively studied and for a long time now. The evolution of a density matrix through such phenomenological equations, which in the course of time lead to a vanishing of the "off-diagonal" elements, has become a celebrated solution of the measurement problem. Celebrated because it is a solution from "within" quantum mechanics. However, the vanishing of the off-diagonal elements is nothing but a rephrasing of the fapp impossibility of bringing wave packets belonging to different pointer positions into interference. The sole purpose of the following section is to acquaint the reader with the basic technicalities so that she/he need not feel intimidated if she/he encounters the overwhelming technical arguments that are often claimed to be "solutions" to the measurement problem. We shall also quickly introduce the Dirac formalism, which is a wonderful formalism for vector spaces with a scalar product.

Remark 9.3. On the Dirac and Density Matrix Formalisms
The Dirac formalism is a symbolism which is well adapted to the computation of expectation values in quantum equilibrium, since they become expressions of scalar product type. In Sect. 15.2.1, we shall review the Dirac formalism with more mathematical background in hand.

A wave function ψ (i.e., an element of the Hilbert space) is represented by the symbol $|\psi\rangle$, and the projection (in the sense of scalar products) of φ onto ψ is denoted by

$$\langle\psi|\varphi\rangle := \int \psi^*(\mathbf{x})\varphi(\mathbf{x})\mathrm{d}^3x\,.$$

That is, $\langle\cdot|\cdot\rangle$ stands for the scalar product on the Hilbert space. $|\mathbf{x}\rangle$ symbolizes the wave function which is "localized" at $\mathbf{x}$, and one reads as follows:

$$\begin{aligned}
\sqrt{\langle\psi|\psi\rangle} = \|\psi\| &\equiv \text{ norm of } \psi\,,\\
\langle\mathbf{x}|\psi\rangle = \psi(\mathbf{x}) &= \text{value of } \psi \text{ at } \mathbf{x}\,,\\
\langle\psi|\mathbf{x}\rangle &= \psi^*(\mathbf{x})\,,\\
\langle\mathbf{x}'|\mathbf{x}\rangle &= \delta(\mathbf{x}'-\mathbf{x})\,,\\
|\mathbf{x}\rangle\langle\mathbf{x}| &= \text{orthogonal projection on } |\mathbf{x}\rangle\,.
\end{aligned}$$

The wave function ψ is the following superposition of the $|\mathbf{x}\rangle$:

$$|\psi\rangle = \int |\mathbf{x}\rangle\langle\mathbf{x}|\psi\rangle\mathrm{d}^3x\,, \tag{9.27}$$

which is simply the *coordinate representation* of the vector $|\psi\rangle$ in the basis $|\mathbf{x}\rangle$. In particular, for the identity I, we obtain

$$\int |\mathbf{x}\rangle\langle\mathbf{x}|\mathrm{d}^3x = \mathsf{I}\,, \tag{9.28}$$

and the scalar product is simply

$$\langle\psi|\psi\rangle = \int \langle\psi|\mathbf{x}\rangle\langle\mathbf{x}|\psi\rangle\mathrm{d}^3x\,.$$

We introduce the operator $\hat{\mathbf{x}}$ by defining its action on the basis $|\mathbf{x}\rangle$ as

$$\hat{\mathbf{x}}|\mathbf{x}\rangle = \mathbf{x}|\mathbf{x}\rangle\,, \tag{9.29}$$

or alternatively by its matrix elements

$$\langle\mathbf{x}'|\hat{\mathbf{x}}|\mathbf{x}\rangle = \mathbf{x}\delta(\mathbf{x}'-\mathbf{x})\,.$$

One finds (rather easily) that the matrix elements of the ∇ operator are

$$\langle\mathbf{x}'|\nabla|\mathbf{x}\rangle = \delta(\mathbf{x}'-\mathbf{x})\partial_{\mathbf{x}}\,. \tag{9.30}$$

Likewise for Δ

$$\langle\mathbf{x}'|\Delta|\mathbf{x}\rangle = \delta(\mathbf{x}'-\mathbf{x})\partial_{\mathbf{x}}^2\,. \tag{9.31}$$

The Schrödinger equation now reads

$$\mathrm{i}\hbar\frac{\partial}{\partial t}|\psi_t\rangle = H|\psi_t\rangle\,,$$

with H as Hamilton operator. As for any linear equation, the (formal) solution is given by

$$|\psi_t\rangle = \exp\left(-\frac{\mathrm{i}}{\hbar}Ht\right)|\psi_0\rangle\,.$$

Using the Dirac notation, the expectation value of the Bohmian position can be written as

$$\begin{aligned}
\langle\mathbf{X}\rangle^{\psi}(t) &= \mathbb{E}^{\psi}(\mathbf{X}(t)) \\
&= \int \mathbf{x}|\psi(\mathbf{x},t)|^2\mathrm{d}^3x \\
&= \langle\psi_t|\hat{\mathbf{x}}|\psi_t\rangle \\
&= \int \mathrm{d}^3x\langle\psi_t|\mathbf{x}\rangle\langle\mathbf{x}|\hat{\mathbf{x}}|\psi_t\rangle \\
&= \mathrm{tr}(\hat{\mathbf{x}}|\psi_t\rangle\langle\psi_t|) \\
&= \mathrm{tr}(\hat{\mathbf{x}}\rho_t)\,,
\end{aligned} \tag{9.32}$$

where $\rho_t = |\psi_t\rangle\langle\psi_t|$ is called the density matrix or statistical operator, and tr denotes the trace. Note that

$$\begin{aligned}\frac{\partial}{\partial t}\rho_t &= \frac{\partial}{\partial t}(|\psi_t\rangle\langle\psi_t|) \\ &= \left(\frac{\partial}{\partial t}|\psi_t\rangle\right)\langle\psi_t| + |\psi_t\rangle\left(\frac{\partial}{\partial t}\langle\psi_t|\right) \\ &= -\frac{\mathrm{i}}{\hbar}H|\psi_t\rangle\langle\psi_t| + |\psi_t\rangle\langle\psi_t|\left(\frac{\mathrm{i}}{\hbar}H\right) \\ &= -\frac{\mathrm{i}}{\hbar}[H,\rho_t]\,, \end{aligned} \tag{9.33}$$

where $[A,B] = AB - BA$ denotes the commutator of the two operators A and B. This is the quantum mechanical analogue of Liouville's equation. It is called the von Neumann equation.

Now to the point. The object $\rho = |\psi\rangle\langle\psi|$ has matrix elements

$$\rho(\mathbf{x},\mathbf{x}') = \langle\mathbf{x}|\psi\rangle\langle\psi|\mathbf{x}'\rangle = \psi(\mathbf{x})\psi^*(\mathbf{x}')\,, \tag{9.34}$$

and in view of (9.27), we may write ρ in the form

$$\rho = |\psi\rangle\langle\psi| = \iint \mathrm{d}^3x\mathrm{d}^3x'\,\psi(\mathbf{x})\psi^*(\mathbf{x}')|\mathbf{x}\rangle\langle\mathbf{x}'|\,. \tag{9.35}$$

This is often called a pure state.[4] In contrast, the diagonal density matrix

$$\rho = \int \mathrm{d}^3x|\psi(\mathbf{x})|^2|\mathbf{x}\rangle\langle\mathbf{x}| \tag{9.36}$$

can be read as a statistical mixture of the localized wave packets $|\mathbf{x}\rangle$, where each packet appears with probability $|\psi(\mathbf{x})|^2$. This is why the density matrix is also called the statistical operator.

For any kind of density matrice we can compute averages according to (9.32). But sometimes a density matrix is neither a pure state nor a mixture of states. Consider Fig. 9.5 and the scattered light waves, which are different depending on whether $\psi_\mathrm{L} = |l\rangle$ or $\psi_\mathrm{R} = |r\rangle$ is effective. We shall symbolize the light waves by "pointer" wave functions $\Phi_0 = |0\rangle$, $\Phi_\mathrm{L} = |L\rangle$ and $\Phi_\mathrm{R} = |R\rangle$. We can then describe the process as follows. Initially, we have $(\alpha_l|l\rangle + \alpha_r|r\rangle)|0\rangle$, where the environment (the light) is represented by the unscattered state $|0\rangle$. From this we obtain $\alpha_l|l\rangle|L\rangle + \alpha_r|r\rangle|R\rangle$. The wave function of the system is understood to be time dependent, so that $|l\rangle$ denotes a wave function moving to the left and $|r\rangle$ denotes a wave function moving to the right. The density matrix is a pure state of the form (9.35), and its change in time is given by

[4] For any pure state we have $\rho^2 = \rho$, which does not hold for a mixture (9.36).

$$\rho = |\alpha_l|^2 |l\rangle|0\rangle\langle l|\langle 0| + |\alpha_r|^2 |r\rangle|0\rangle\langle r|\langle 0| + \alpha_l\alpha_r{}^* |l\rangle|0\rangle\langle r|\langle 0| + \alpha_r\alpha_l{}^* |r\rangle|0\rangle\langle l|\langle 0|$$

$$\xrightarrow{t\longrightarrow T}$$

$$\rho_T = |\alpha_l|^2 |l\rangle|L\rangle\langle l|\langle L| + |\alpha_r|^2 |r\rangle|R\rangle\langle r|\langle R| + \alpha_l\alpha_r{}^* |l\rangle|L\rangle\langle r|\langle R| + \alpha_r\alpha_l{}^* |r\rangle|R\rangle\langle l|\langle L| \,. \tag{9.37}$$

This looks more complicated than it really is. Let us focus on the position of the particle, i.e., we are interested in expectation values of functions of the particle position alone. In that case we can "trace out" the environment. That is, in (9.32), we take the trace over all the states $|Y\rangle$ of the environment. This partial tracing yields the *reduced density matrix*

$$\begin{aligned}
\rho_T^{\text{red}}(x,x') &= [\mathrm{tr}_Y \rho_T](x,x') \\
&= \int \mathrm{d}Y \Big\langle Y \Big| \Big[|\alpha_l|^2 \langle x|l\rangle|L\rangle\langle l|x'\rangle\langle L| + |\alpha_r|^2 \langle x|r\rangle|R\rangle\langle r|x'\rangle\langle R| \\
&\qquad + \alpha_l\alpha_r^* \langle x|l\rangle|L\rangle\langle r|x'\rangle\langle R| + \alpha_r\alpha_l{}^* \langle x|r\rangle|R\rangle\langle l|x'\rangle\langle L| \Big] \Big| Y \Big\rangle \\
&= |\alpha_l|^2 \langle x|l\rangle\langle l|x'\rangle \int \mathrm{d}Y |\langle L|Y\rangle|^2 + |\alpha_r|^2 \langle x|r\rangle\langle r|x'\rangle \int \mathrm{d}Y |\langle R|Y\rangle|^2 \\
&\quad + \alpha_l\alpha_r^* \langle x|l\rangle\langle r|x'\rangle \int \mathrm{d}Y \langle Y|L\rangle\langle R|Y\rangle + \alpha_r\alpha_l{}^* \langle x|r\rangle\langle l|x'\rangle \int \mathrm{d}Y \langle Y|R\rangle\langle L|Y\rangle \,,
\end{aligned} \tag{9.38}$$

which can be used to compute all the relevant expectations.

Now we must think as follows. Let us picture the scattering states $|R\rangle$ and $|L\rangle$ as pointer states which have macroscopically disjoint supports in (light wave) configuration space. The light waves occupy different regions. Since $\langle Y|R\rangle$ contributes only with those Y that are in the support of $|R\rangle$, and $\langle Y|L\rangle$ contributes only with those Y in the support of $|L\rangle$, this then shows that, according to the very small overlap of $|L\rangle$ and $|R\rangle$, the last two integrals in (9.38) are very small. We thus arrive at a result which easily ranks among the most severely misinterpreted results in science: the reduced density matrix acquires almost diagonal form. Therefore it looks like the density matrix of the mixture of the wave functions $|l\rangle$ and $|r\rangle$ with weights $|\alpha_l|^2$ and $|\alpha_r|^2$, respectively, where we have used the normalization $\int \mathrm{d}Y |\langle Y|R\rangle|^2 = \int \mathrm{d}Y |\langle Y|L\rangle|^2 = 1$:

$$\rho_T^{\text{red}} = \mathrm{tr}_Y \rho_T \approx |\alpha_l|^2 |l\rangle\langle l| + |\alpha_r|^2 |r\rangle\langle r| \,. \tag{9.39}$$

Why is this result often misunderstood? Suppose we do not know the wave function of a system, but our ignorance about it can be expressed in terms of probabilities, viz., with probability $|\alpha_l|^2$ the wave function is $|l\rangle$ and with probability $|\alpha_r|^2$ it is $|r\rangle$. Then the corresponding density matrix would be exactly the right-hand side of (9.39). Therefore one may easily be trapped into thinking that the left-hand side of (9.39), which approximately equals the right-hand side, also submits to the ignorance interpretation. In short, decoherence seems to turn "and" into "or". But (9.39) is nothing but a rephrasing of (9.3). The only difference is the mathematical formulation.

Many physicists have nevertheless taken (9.39) as the solution to the measurement problem, as if Schrödinger had not been aware of this calculation. Of course, that is nonsense. Schrödinger *based* his cat story on the fapp impossibility of the interference of macroscopically disjoint wave packets. So let us repeat [7]:[5] There is a difference between a shaky or out-of-focus photograph and a snapshot of clouds and fog banks.

Everything else that needs to be said has already been pointed out in Remark 9.1. Only with Bohmian mechanics can (9.39) be interpreted as a mixture. ■

Remark 9.4. Collapse Equations

In Bohmian mechanics the result (9.39) means that just one of the wave functions will actually guide the particle. Moreover, the environment will continue to "read" the particle position, so the effective guiding wave will continue to be a localized wave packet. We wish to discuss briefly the type of equations that govern the corresponding reduced density matrix, which represents the probability distribution of the random wave function. That is, we wish to describe the time evolution of the reduced density matrix.

The evolution equation arises from the full quantum mechanical description of system and environment in an appropriate scaling, and it must satisfy two desiderata: the off-diagonal elements must go to zero and the Newtonian equations of motion should be satisfied in the mean. Otherwise we could not expect Newtonian behavior of Bohmian trajectories for highly localized wave functions. As a (mathematical) example we give the simplest such equation for ρ_t, namely,

$$\frac{\partial}{\partial t}\rho_t = -\frac{\mathrm{i}}{\hbar}[H,\rho_t] + \Sigma_t \,, \tag{9.40}$$

with ($\Lambda > 0$)

$$\langle \mathbf{x}|\Sigma_t|\mathbf{x}'\rangle = -\Lambda(\mathbf{x}-\mathbf{x}')^2\rho_t(\mathbf{x},\mathbf{x}') \,. \tag{9.41}$$

For $H = 0$, the solution is

$$\rho_t(\mathbf{x},\mathbf{x}') = \mathrm{e}^{-\Lambda t(\mathbf{x}-\mathbf{x}')^2}\rho_0(\mathbf{x},\mathbf{x}') \,,$$

and one sees that the off-diagonal elements vanish at the rate Λ. However, this is an unphysical model in the sense that the strength of decoherence saturates when the distance $\mathbf{x}-\mathbf{x}'$ has achieved a certain macroscopic value. Notwithstanding these shortcomings, equations of this type have been studied and "physical values" for Λ have been proposed [9]. However, we have seen that (9.40) and (9.41) describe at least roughly the emergence of a statistical mixture.

Next let us check whether the Newtonian equations of motion hold in the mean. We differentiate the last equality of (9.32), and with (9.40), we obtain

[5] Translation by John D. Trimmer in [8].

$$\begin{aligned}\frac{\mathrm{d}}{\mathrm{d}t}\langle \mathbf{X}(t)\rangle &= \frac{\mathrm{d}}{\mathrm{d}t}\mathrm{tr}(\hat{\mathbf{x}}\rho_t)\\ &= \mathrm{tr}\left(\hat{\mathbf{x}}\frac{\partial}{\partial t}\rho_t\right)\\ &= -\frac{\mathrm{i}}{\hbar}\mathrm{tr}\big(\hat{\mathbf{x}}[H,\rho_t]\big)+\mathrm{tr}(\hat{\mathbf{x}}\Sigma_t)\,.\end{aligned}$$

Differentiating once more yields[6]

$$\frac{\mathrm{d}^2}{\mathrm{d}t^2}\langle \mathbf{X}(t)\rangle = \frac{1}{m}\big\langle -\nabla V\big(\mathbf{X}(t)\big)\big\rangle - \frac{\mathrm{i}}{\hbar}\mathrm{tr}\big(\hat{\mathbf{x}}[H,\Sigma_t]\big)+\mathrm{tr}\left(\hat{\mathbf{x}}\frac{\partial}{\partial t}\Sigma_t\right)\,.$$

Note that we have already derived the first part of the equation in (9.11), without using abstract operator calculus. The result is consistent with classical motion if

$$\mathrm{tr}\left(\hat{\mathbf{x}}\frac{\partial}{\partial t}\Sigma_t\right)=0 \tag{9.42}$$

and

$$\mathrm{tr}\big(\hat{\mathbf{x}}[H,\Sigma_t]\big)=\mathrm{tr}(\hat{\mathbf{x}}H\Sigma_t)-\mathrm{tr}(\hat{\mathbf{x}}\Sigma_t H)=0\,. \tag{9.43}$$

For (9.42), it suffices that $\mathrm{tr}(\hat{\mathbf{x}}\Sigma_t)$ be constant. One requires

$$0=\mathrm{tr}(\hat{\mathbf{x}}\Sigma_t)=\int \mathrm{d}\mathbf{x}\,\mathbf{x}\,\langle \mathbf{x}|\Sigma_t|\mathbf{x}\rangle\,, \tag{9.44}$$

where in the last step we have computed the trace and used (9.29). The above certainly holds if the kernel $\langle \mathbf{x}'|\Sigma_t|\mathbf{x}\rangle$ vanishes on the diagonal. Concerning (9.43), note that the trace is invariant under cyclic permutations:

$$\mathrm{tr}(\hat{\mathbf{x}}\Sigma_t H)=\mathrm{tr}(H\hat{\mathbf{x}}\Sigma_t)\,, \tag{9.45}$$

so that from (9.43) we obtain the requirement

$$\mathrm{tr}\big(\hat{\mathbf{x}}[H,\Sigma_t]\big)=\mathrm{tr}\big([\hat{\mathbf{x}},H]\Sigma_t\big)=0\,. \tag{9.46}$$

Next we observe that, since V is a function of the position operator $\hat{\mathbf{x}}$, we have

$$[\hat{\mathbf{x}},V(\hat{\mathbf{x}})]=0\,,$$

and thus

[6] The untrained reader may have difficulty checking the computation. It is in fact straightforward, but one may need some practice. For example, one should first compute $[H,\hat{x}]$ for a Schrödinger Hamiltonian of the usual form $H=-(\hbar^2/2m)\Delta+V(\hat{x})$. Moreover, it should be understood that the trace remains unchanged under cyclic permutations of the arguments [see (9.45)].

$$\langle \mathbf{x}|[\hat{\mathbf{x}},H]|\mathbf{x}'\rangle = -\frac{\hbar^2}{2m}\langle \mathbf{x}|[\hat{\mathbf{x}},\Delta_{\mathbf{x}}]|\mathbf{x}'\rangle + \langle \mathbf{x}|[\hat{\mathbf{x}},V(\hat{\mathbf{x}})]|\mathbf{x}'\rangle = -\frac{\hbar^2}{2m}\langle \mathbf{x}|[\hat{\mathbf{x}},\Delta_{\mathbf{x}}]|\mathbf{x}'\rangle\,.$$

The first commutator can be handled using (9.31). Then, using (9.28), we may straightforwardly compute

$$\begin{aligned}\mathrm{tr}\left([\hat{\mathbf{x}},H]\Sigma_t\right) &= \int \mathrm{d}^3x \int \mathrm{d}^3x' \langle \mathbf{x}|[\hat{\mathbf{x}},H]|\mathbf{x}'\rangle\langle \mathbf{x}'|\Sigma_t|\mathbf{x}\rangle \\ &= \int \mathrm{d}^3x \int \mathrm{d}^3x' \delta\left(\mathbf{x}-\mathbf{x}'\right)\frac{\hbar^2}{m}\partial_{\mathbf{x}'}\langle \mathbf{x}'|\Sigma_t|\mathbf{x}\rangle\,.\end{aligned}$$

Thus (9.46) and hence (9.43) certainly hold if not only the kernel $\langle \mathbf{x}'|\Sigma_t|\mathbf{x}\rangle$ but also its differential (with respect to one of the variables) vanishes on the diagonal. Our choice (9.41) is one example that fulfills both these requirements.

Equation (9.40) with (9.41) is a special case of a general class of evolution equations for density matrices, the class of equations of Lindblad form [1, 9, 10]. Among other things, the Lindblad form guarantees that the solution ρ_t remains positive definite, so that its interpretation as a *statistical* operator remains consistent. The general form of Σ for a self-adjoint operator A (see Chap. 14 for the notion of self-adjointness) is (see, for instance, [1, 9])

$$\Sigma = A\rho A - \frac{1}{2}(A^2\rho + \rho A^2)\,, \tag{9.47}$$

The choice $A = \sqrt{2\Lambda}\hat{\mathbf{x}}$ yields

$$\begin{aligned}\langle \mathbf{x}|\Sigma|\mathbf{x}'\rangle &= \left\langle \mathbf{x}\left|A\rho A - \frac{1}{2}(A^2\rho + \rho A^2)\right|\mathbf{x}'\right\rangle \\ &= 2\Lambda\mathbf{x}\cdot\mathbf{x}'\langle \mathbf{x}|\rho|\mathbf{x}'\rangle - \Lambda x^2\langle \mathbf{x}|\rho|\mathbf{x}'\rangle - \Lambda x'^2\langle \mathbf{x}|\rho|\mathbf{x}'\rangle \\ &= -\Lambda(\mathbf{x}-\mathbf{x}')^2\langle \mathbf{x}|\rho|\mathbf{x}'\rangle\,.\end{aligned}$$

Of course, if we wish to keep the Newtonian equations in the mean, A must commute with $\hat{\mathbf{x}}$, i.e., it must be a function of $\hat{\mathbf{x}}$. However, this is not the place to pursue this further. ■

9.5.2 Poincaré Recurrence

Recurrence also appears in quantum mechanics, in the following sense [11]. Consider a system with discrete eigenvalues[7] E_n. Let ψ_0 be the system wave function at time 0, and $\varepsilon > 0$. Then there exists a time $T > 0$, such that the distance

$$\|\psi(T) - \psi_0\| < \varepsilon\,.$$

[7] This corresponds to the finite measure condition in the classical argument of Theorem 4.1.

Wave functions that are close in this sense yield almost the same statistics. Here is the argument for recurrence. Suppose

$$H|n\rangle = E_n|n\rangle \ ,$$

where $|n\rangle$, $n \in \mathbb{N}$, defines a basis of eigenfunctions. Then

$$\begin{aligned}|\psi_t\rangle &= \exp\left(-\frac{\mathrm{i}}{\hbar}Ht\right)|\psi_0\rangle \\ &= \sum_{n=0}^{\infty} \exp\left(-\frac{\mathrm{i}}{\hbar}Ht\right)|n\rangle\langle n|\psi_0\rangle \\ &= \sum_{n=0}^{\infty} \exp\left(-\frac{\mathrm{i}}{\hbar}E_n t\right)|n\rangle\langle n|\psi_0\rangle \ .\end{aligned}$$

Hence

$$|\psi_t\rangle - |\psi_0\rangle = \sum_{n=0}^{\infty}\left(\mathrm{e}^{-\mathrm{i}E_n t/\hbar} - 1\right)|n\rangle\langle n|\psi_0\rangle$$

and

$$\begin{aligned}\big\||\psi_t\rangle - |\psi_0\rangle\big\|^2 &= \sum_{n=0}^{\infty}\left|\mathrm{e}^{-\mathrm{i}E_n t/\hbar} - 1\right|^2 |\langle n|\psi_0\rangle|^2 \\ &= 2\sum_{n=0}^{\infty}\left[1 - \cos\left(\frac{E_n}{\hbar}t\right)\right]|\langle n|\psi_0\rangle|^2 \ .\end{aligned}$$

For appropriate N,

$$\sum_{n=N}^{\infty}\left[1 - \cos\left(\frac{E_n}{\hbar}t\right)\right]|\langle n|\psi_0\rangle|^2 \leq 2\sum_{n=N}^{\infty}|\langle n|\psi_0\rangle|^2$$

will be arbitrarily small, since

$$\sum_{n=0}^{\infty}\left|\langle n|\psi_0\rangle\right|^2 = \langle\psi_0|\psi_0\rangle = \|\psi_0\| = 1 \ .$$

Therefore we need only show that, for appropriately chosen T, the quantity

$$2\sum_{n=0}^{N-1}\left[1 - \cos\left(\frac{E_n}{\hbar}T\right)\right]|\langle n|\psi_0\rangle|^2$$

also becomes arbitrarily small. This can be done because any frequency can be approximated arbitrarily well by a rational frequency. Since T can be chosen as large as we wish, rational frequencies multiplied by T can be turned into integer multiples of 2π. This underlies the rigorous proof using almost periodic functions [12].

References

1. A. Bassi, G. Ghirardi: Phys. Rep. **379** (5–6), 257 (2003)
2. M. Bell, K. Gottfried, and M. Veltman (Eds.): *John S. Bell on the Foundations of Quantum Mechanics* (World Scientific Publishing Co. Inc., River Edge, NJ, 2001)
3. R. Tumulka: J. Stat. Phys. **125** (4), 825 (2006)
4. R. Tumulka: Proc. R. Soc. Lond. Ser. A Math. Phys. Eng. Sci. **462** (2070), 1897 (2006)
5. V. Allori, D. Dürr, S. Goldstein, N. Zanghì: Journal of Optics B **4**, 482 (2002)
6. S. Römer, D. Dürr, T. Moser: J. Phys. A **38**, 8421 (2005); math-ph/0505074
7. E. Schrödinger: Naturwissenschaften **23**, 807 (1935)
8. J.A. Wheeler, W.H. Zurek (Eds.): *Quantum Theory and Measurement*, Princeton Series in Physics (Princeton University Press, Princeton, NJ, 1983)
9. D. Giulini, E. Joos, C. Kiefer, J. Kumpsch, I.O. Stamatescu, H. Zeh: *Decoherence and the Appearance of a Classical World in Quantum Theory* (Springer-Verlag, Berlin, 1996)
10. G.C. Ghirardi, P. Pearle, A. Rimini: Phys. Rev. A **42** (1), 78 (1990)
11. P. Bocchieri, A. Loinger: Phys. Rev. **107**, 337 (1957)
12. H. Bohr: *Fastperiodische Funktionen* (Springer, 1932)

Chapter 10
Nonlocality

Bohmian mechanics is about particles guided by a wave. This is new, but not revolutionary physics. This chapter will now present a paradigm shift. It is about how nature is, or better, it is about how any theory which aims at a correct description of nature must be. Any such theory must be nonlocal. We do not attempt to define nonlocality (see [1] for a serious examination of the notion), but simply take it pragmatically as meaning that the theory contains action at a distance in the true meaning of the words, i.e., faster than light action between spacelike separated events. Since we shall exemplify the idea shortly, this should suffice for the moment.

We note that the action at a distance in question here is such that no information can be sent with superluminal speed, whence no inconsistency with special relativity arises. However, action at a distance does seem to be at odds with special relativity. Einstein held the view that nonlocality is unphysical, and referred to nonlocal actions as "ghost fields", a notion which expressed his contempt for nonlocal theories. But such theories are not unfamiliar in physics. For example, Newtonian mechanics, which is non-relativistic, is nonlocal. Bohmian mechanics is nonlocal, too. But there is a noteworthy difference between the nonlocality of Newtonian mechanics and that of Bohmian mechanics. For the latter is encoded in the wave function which lives on *configuration space* and is by its very nature a nonlocal agent. All particles are guided simultaneously by the wave function, and if the wave function is entangled, the nonlocal action does not get small with the spatial distance between the particles, in contrast to what happens in a Newtonian system with gravitational interaction.

In a two-particle system, with coordinates $\mathbf{X}_1(t)$ and $\mathbf{X}_2(t)$, we have

$$\dot{\mathbf{X}}_1(t) = \frac{\hbar}{m_1}\Im\left[\frac{\left.\dfrac{\partial}{\partial \mathbf{x}}\psi\big(\mathbf{x},\mathbf{X}_2(t)\big)\right|_{\mathbf{x}=\mathbf{X}_1(t)}}{\psi\big(\mathbf{X}_1(t),\mathbf{X}_2(t)\big)}\right],$$

whence the velocity of $\mathbf{X}_1$ at time t depends in general on $\mathbf{X}_2$ at time t, no matter how far apart the positions are. "In general" means here that the wave function $\psi(\mathbf{x},\mathbf{y})$ is entangled and not a product $\psi(\mathbf{x},\mathbf{y}) = \varphi(\mathbf{x})\Phi(\mathbf{y})$, for example. There is

D. Dürr, S. Teufel, *Bohmian Mechanics*, DOI 10.1007/978-3-540-89344-8_10,

no immediate reason why the wave function should become a product when **x** and **y** are far apart.[1] Therefore, if some local interaction changes the wave function at $\mathbf{X}_2(t)$, that will immediately affect the velocity of particle 1.

Bohmian mechanics is definitely nonlocal, because the wave function is a function on configuration space. In fact, since the wave function is an object of all quantum theories, Bohmian or not, quantum mechanics is nonlocal. One may dislike nonlocality. In that case one should ask: Can one do better? Is it possible to describe nature by a local theory? For Einstein, Podolsky, and Rosen (EPR) [2] the answer was unquestionably affirmative. They presented an argument based on their belief that all theories of nature must be local, which was supposed to prove that quantum mechanics is incomplete. Besides the (nonlocal) wave function (which Einstein considered as merely expressing "probability", which is naturally a function on phase space in classical physics, and hence also on configuration space), there are other – local – hidden variables that have been left out of the physical description.

We no longer need to argue that quantum mechanics is incomplete, but the EPR argument is nevertheless of interest, as it constitutes one part of Bell's proof of the nonlocality of nature. Bell's response to the question as to whether one can do better is Bell's theorem, and it answers in the negative: one cannot do better. Nature is nonlocal.

Bell's theorem has two parts. The first is the Einstein–Podolsky–Rosen argument [2] applied to the simplified version of the EPR Gedanken experiment considered by David Bohm [3], viz., the EPRB experiment. It is based on the fact that one can prepare a special pair (L, R) of spin 1/2 particles which fly apart in opposite directions (L to the left and R to the right), and which behave in the following well determined way. When both particles pass identically oriented Stern–Gerlach magnets, they get deflected in exactly opposite directions, i.e., if L moves up, R moves down, and vice versa. In quantum language, if L has **a**-spin $+1/2$, then R has **a**-spin $-1/2$, and vice versa, where **a** is the orientation of the magnets (see Fig. 10.1). This is true for all directions **a**. Moreover, the probability for (L up, R down) is 1/2. The two-particle wave function is called a singlet state [see (10.4)]. The total spin of this singlet state is zero. We shall give details later and simply note for now that such is the physical reality, and it is correctly described by quantum mechanics. We obtain opposite values for the spins when the particles move through **a**-directed magnets, for any **a**.

Now we come to the first part of the nonlocality argument. Measuring first the **a**-spin on L, we can predict with certainty the result of the measurement of the **a**-spin on R. This is true even if the measurement events at L and R are spacelike separated. Suppose therefore that the experiment is arranged in such a way that a light signal cannot communicate the L-result to the R-particle before the R-particle passes SGM-R. Suppose now that "locality" holds, meaning that the spin measurement on one side has no superluminal influence on the result of the spin measurement on the other side. Then we must conclude that the value we predict for the **a**-spin on R is preexisting. It cannot have been created by the result obtained on L, because we

[1] However, decoherence is always lurking there, awaiting an opportunity to destroy coherence, i.e., to produce an effective product structure.

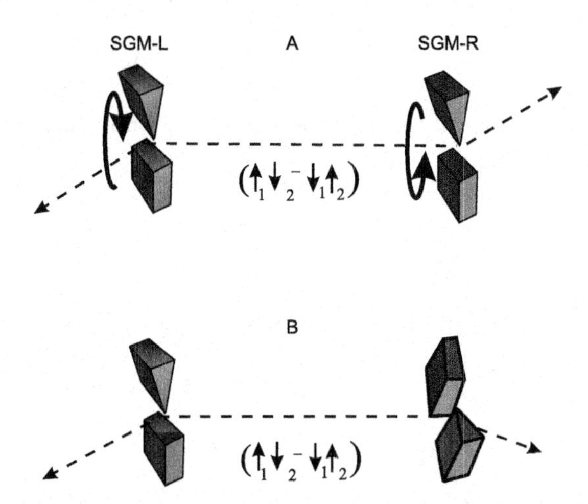

Fig. 10.1 The EPR experiment. Two particles in the singlet spin state (10.4) fly apart and move towards the Stern–Gerlach magnets SGM-L und SGM-R. In situation (**A**), the Stern–Gerlach magnets are parallel, while in situation (**B**), the directions of the magnets are rotated shortly before the particles arrive. The measurements are made in such a short time that no communication of results transmitted with the speed of light between the left and the right is possible within the duration of the measurement process

assume locality. Now let us reflect on that. If the value preexists, then that means that it exists even before the decision was taken in which direction **a** the spin on the left is to be measured. Hence the value preexists for any direction **a**. By symmetry this holds also for the values obtained on L. Therefore, by locality, we obtain the preexisting values of spins on either side in any direction. We collect the preexisting values in a family of variables $X_{\mathbf{a}}^{(\mathrm{L})}, X_{\mathbf{a}}^{(\mathrm{R})} \in \{-1,1\}$, with **a** indexing arbitrary directions and with $X_{\mathbf{a}}^{(\mathrm{L})} = -X_{\mathbf{a}}^{(\mathrm{R})}$.

The locality check is now simply to ask whether such preexisting values actually exist. In other words, do the preexisting values accommodate the measured correlations? This leads to the second part of Bell's proof, namely to show that the answer is negative. One might think that this would be a formidable task. One might think that because we make no assumptions about the nature of the variables. They can conspire in the most ingenious way. Contradicting Einstein's famous: "Subtle is the lord, but malicious He is not", we might say: "The lord could have been malicious" in correlating the variables in such an intricate manner that they do whatever they are supposed to do. Of course, the particles can no longer conspire when they are far apart, because that is forbidden by the locality assumption. But at the time when the particles are still together in the source, before they fly apart, they can conspire to form the wildest correlations.

But no, this part of the proof is trivial. There is no way the variables can reproduce the quantum mechanical (which are the Bohmian) correlations. Choose three directions, given by unit vectors $\mathbf{a},\mathbf{b},\mathbf{c}$, and consider the corresponding 6 variables $X_{\mathbf{y}}^{(\mathrm{L})}, X_{\mathbf{z}}^{(\mathrm{R})}$, $\mathbf{y},\mathbf{z} \in \{\mathbf{a},\mathbf{b},\mathbf{c}\}$. They must satisfy

$$\left(X_{\mathbf{a}}^{(\mathrm{L})}, X_{\mathbf{b}}^{(\mathrm{L})}, X_{\mathbf{c}}^{(\mathrm{L})}\right) = \left(-X_{\mathbf{a}}^{(\mathrm{R})}, -X_{\mathbf{b}}^{(\mathrm{R})}, -X_{\mathbf{c}}^{(\mathrm{R})}\right) . \tag{10.1}$$

We wish to reproduce the relative frequencies of the anti-correlation events

$$X_{\mathbf{a}}^{(\mathrm{L})} = -X_{\mathbf{b}}^{(\mathrm{R})} , \qquad X_{\mathbf{b}}^{(\mathrm{L})} = -X_{\mathbf{c}}^{(\mathrm{R})} , \qquad X_{\mathbf{c}}^{(\mathrm{L})} = -X_{\mathbf{a}}^{(\mathrm{R})} .$$

Adding the probabilities and using the rules of probability in the inequality, we get

$$\begin{aligned}
&\mathbb{P}\left(X_{\mathbf{a}}^{(\mathrm{L})} = -X_{\mathbf{b}}^{(\mathrm{R})}\right) + \mathbb{P}\left(X_{\mathbf{b}}^{(\mathrm{L})} = -X_{\mathbf{c}}^{(\mathrm{R})}\right) + \mathbb{P}\left(X_{\mathbf{c}}^{(\mathrm{L})} = -X_{\mathbf{a}}^{(\mathrm{R})}\right) \\
&\qquad = \mathbb{P}\left(X_{\mathbf{a}}^{(\mathrm{L})} = X_{\mathbf{b}}^{(\mathrm{L})}\right) + \mathbb{P}\left(X_{\mathbf{b}}^{(\mathrm{L})} = X_{\mathbf{c}}^{(\mathrm{L})}\right) + \mathbb{P}\left(X_{\mathbf{c}}^{(\mathrm{L})} = X_{\mathbf{a}}^{(\mathrm{L})}\right) \qquad [\text{by (10.1)}] \\
&\qquad \geq \mathbb{P}\left(X_{\mathbf{a}}^{(\mathrm{L})} = X_{\mathbf{b}}^{(\mathrm{L})} \text{ or } X_{\mathbf{b}}^{(\mathrm{L})} = X_{\mathbf{c}}^{(\mathrm{L})} \text{ or } X_{\mathbf{c}}^{(\mathrm{L})} = X_{\mathbf{a}}^{(\mathrm{L})}\right) \\
&\qquad = \mathbb{P}(\text{sure event}) = 1 ,
\end{aligned}$$

because $X_{\mathbf{y}}^{(i)}$, $i = \mathrm{L}, \mathrm{R}$, $y \in \{\mathbf{a}, \mathbf{b}, \mathbf{c}\}$, can only take two values.

This is thus one version of Bell's inequality:

$$\mathbb{P}\left(X_{\mathbf{a}}^{(\mathrm{L})} = -X_{\mathbf{b}}^{(\mathrm{R})}\right) + \mathbb{P}\left(X_{\mathbf{b}}^{(\mathrm{L})} = -X_{\mathbf{c}}^{(\mathrm{R})}\right) + \mathbb{P}\left(X_{\mathbf{c}}^{(\mathrm{L})} = -X_{\mathbf{a}}^{(\mathrm{R})}\right) \geq 1 . \tag{10.2}$$

We shall show in a moment that the quantum mechanical value (which is of course the Bohmian one) for the probability of perfect anti-correlations is 3/4 if the angles between $\mathbf{a}, \mathbf{b}, \mathbf{c}$ are each $120°$. Therefore quantum mechanics contradicts (10.2).

The logical structure of Bell's nonlocality argument is thus as follows [4]. Let P be the hypothesis of the existence of preexisting values $X_{\mathbf{a},\mathbf{b},\mathbf{c}}^{\mathrm{L,R}}$ for the spin components relevant to this EPRB experiment. Then

$$\begin{array}{ll}
\text{First part} & \text{quantum mechanics} + \text{locality} \implies \mathrm{P} , \\
\text{Second part} & \text{quantum mechanics} \implies \text{not } \mathrm{P} , \\
\text{Conclusion} & \text{quantum mechanics} \implies \text{not locality} .
\end{array} \tag{10.3}$$

To save locality, one could hope that the quantum mechanical value would be false. So let us forget about quantum mechanics and all other theories and simply take the experimental facts. They show quite convincingly[2] that (10.2) is violated by the observed relative frequencies [5]. There is no doubt that better experiments will corroborate the finding that (10.2) is violated. In (10.3) we can therefore replace "quantum mechanics" by "experimental facts". This then is independent of any theory and yields a conclusion about nature: nature is nonlocal. The practical meaning is that, if you devise a theory about nature, you had better make sure that it is nonlocal, otherwise it is irrelevant. Bohmian mechanics violates Bell's inequalities and

[2] The experiments done up until now contain so-called loopholes, which are based on detector deficiencies and the belief that nature behaves in a conspiratorial way. Since the experiments agree well with the quantum mechanical predictions, such beliefs seem unlikely to be well founded.

furthermore predicts the values measured in experiments. Concerning nonlocality, one cannot do better than Bohmian mechanics.

10.1 Singlet State and Probabilities for Anti-Correlations

The spin part of the singlet wave function is the antisymmetric vector

$$\psi_s = \frac{1}{\sqrt{2}}\left(|\uparrow\rangle_1|\downarrow\rangle_2 - |\downarrow\rangle_1|\uparrow\rangle_2\right) . \tag{10.4}$$

This factor multiplies the position dependent symmetric wave function

$$\psi(\mathbf{x}_1,\mathbf{x}_2) = \psi_L(\mathbf{x}_1)\psi_R(\mathbf{x}_2) + \psi_R(\mathbf{x}_1)\psi_L(\mathbf{x}_2) ,$$

which arranges for one particle to move to the left and one to move to the right. This spatial arrangement is not contained in the spin part (10.4). For the purpose of computing the probabilities, the spatial symmetrization is rather irrelevant. Furthermore, computing everything while maintaining the symmetrization is a bit demanding and, for the sake of simplicity, we focus only on the term $\psi_L(\mathbf{x}_1)\psi_R(\mathbf{x}_2)\psi_s$, thereby viewing the first factor in (10.4) as belonging to the particle on the left and the second factor as belonging to the particle on the right. Then, for the purpose of computing averages, we can forget about the position part altogether and focus on the spin part alone. But the reader should bear in mind that it is in fact the trajectory of each particle that determines the value of the spin.

Now for a few facts concerning (10.4). First $|\uparrow\rangle$ is the spinor for spin up $(+1/2)$ in an arbitrarily chosen direction, because the singlet is completely symmetric under change of basis. Suppose we express $|\uparrow\rangle_k, |\downarrow\rangle_k$ in another orthogonal basis $\mathbf{i}_k, \mathbf{j}_k$, viz.,

$$|\uparrow\rangle_k = \mathbf{i}_k \cos\alpha + \mathbf{j}_k \sin\alpha , \qquad |\downarrow\rangle_k = -\mathbf{i}_k \sin\alpha + \mathbf{j}_k \cos\alpha ,$$

then one readily computes

$$\psi_s = \frac{1}{\sqrt{2}}(\mathbf{i}_1\mathbf{j}_2 - \mathbf{j}_1\mathbf{i}_2) .$$

We note immediately that the total spin in the singlet state is zero. The following computation shows what is meant by this. One considers the spin operator for the combined system, i.e., for the two particles, which is given by

$$\mathbf{a}\cdot\sigma^{(1)} \otimes \mathsf{I} + \mathsf{I} \otimes \mathbf{a}\cdot\sigma^{(2)} ,$$

where $\mathbf{a}\cdot\sigma^{(1)} \otimes \mathsf{I}$ (and analogously the second summand) is to be read as follows. The operator describes the statistics of the measurement of spin in the direction $\mathbf{a}$ on SGM-L, and technically the first operator factor acts on the first spinor factor, while the second acts on the second spinor factor. The sum $\mathbf{a}\cdot\sigma^{(1)}\otimes \mathsf{I}+\mathsf{I}\otimes\mathbf{a}\cdot\sigma^{(2)}$ is thus the operator describing the statistics of measurements of spin on the left and on the right, which we call measurement of the total spin.

Now using (8.25) in Chap. 8, we compute the average value, in the singlet wave function, of the square of the total spin in an arbitrary direction, which we chose here as the $\mathbf{a}$ direction. We conveniently choose $|\uparrow\rangle$ and $|\downarrow\rangle$ in the singlet wave function as eigenvectors of the spin matrix $\mathbf{a}\cdot\sigma^{(k)}$. In the calculation, we use the fact that $\langle\uparrow|\mathsf{I}|\downarrow\rangle=0$ and $\langle\uparrow|\mathbf{a}\cdot\sigma^{(k)}|\downarrow\rangle=\pm\langle\uparrow|\downarrow\rangle=0$, and we suppress the indices $1,2$ on the spin factors. Then ignoring the dimension and scale factors $\hbar/2$ for the spin 1/2 particle, we compute

$$
\begin{aligned}
&\left\langle\psi_{\mathrm{s}}\left|\left(\mathbf{a}\cdot\sigma^{(1)}\otimes\mathsf{I}+\mathsf{I}\otimes\mathbf{a}\cdot\sigma^{(2)}\right)^2\right|\psi_{\mathrm{s}}\right\rangle\\
&\quad=\frac{1}{2}\left[\left\langle\uparrow\left|\left(\mathbf{a}\cdot\sigma^{(1)}\right)^2\right|\uparrow\right\rangle\langle\downarrow|\mathsf{I}^2|\downarrow\rangle+\left\langle\downarrow\left|\left(\mathbf{a}\cdot\sigma^{(1)}\right)^2\right|\downarrow\right\rangle\langle\uparrow|\mathsf{I}^2|\uparrow\rangle\right]\\
&\qquad+\frac{1}{2}\left[\langle\uparrow|\mathsf{I}^2|\uparrow\rangle\left\langle\downarrow\left|\left(\mathbf{a}\cdot\sigma^{(2)}\right)^2\right|\downarrow\right\rangle+\langle\downarrow|\mathsf{I}^2|\downarrow\rangle\left\langle\uparrow\left|\left(\mathbf{a}\cdot\sigma^{(2)}\right)^2\right|\uparrow\right\rangle\right]\\
&\qquad+\left\langle\uparrow\left|\left(\mathbf{a}\cdot\sigma^{(1)}\right)\right|\uparrow\right\rangle\left\langle\downarrow\left|\left(\mathbf{a}\cdot\sigma^{(2)}\right)\right|\downarrow\right\rangle+\left\langle\downarrow\left|\left(\mathbf{a}\cdot\sigma^{(1)}\right)\right|\downarrow\right\rangle\left\langle\uparrow\left|\left(\mathbf{a}\cdot\sigma^{(2)}\right)\right|\uparrow\right\rangle\\
&\quad=\frac{1}{2}\left[(-1)^2+1^2\right]+\frac{1}{2}\left[1^2+(-1)^2\right]-1-1\\
&\quad=0\,.
\end{aligned}
$$

Let us now turn to the correlations

$$
\mathbb{E}^{\psi_{\mathrm{s}}}_{\mathbf{a},\mathbf{b}}=\left\langle\psi_{\mathrm{s}}\left|\mathbf{a}\cdot\sigma^{(1)}\otimes\mathbf{b}\cdot\sigma^{(2)}\right|\psi_{\mathrm{s}}\right\rangle\,.
$$

This expression is bilinear in $\mathbf{a}$ and $\mathbf{b}$, and rotationally invariant. Therefore the expression must be a multiple of $\mathbf{a}\cdot\mathbf{b}$, i.e., $\mathbb{E}^{\psi_{\mathrm{s}}}_{\mathbf{a},\mathbf{b}}=\lambda\mathbf{a}\cdot\mathbf{b}$, where λ is determined by the value one gets for $\mathbf{a}=\mathbf{b}$. For the singlet, this is (the spin values are exactly opposite) $\mathbb{E}^{\psi_{\mathrm{s}}}_{\mathbf{a},\mathbf{a}}=-1$. Hence,

$$
\mathbb{E}^{\psi_{\mathrm{s}}}_{\mathbf{a},\mathbf{b}}=-\mathbf{a}\cdot\mathbf{b}\,. \tag{10.5}
$$

With this we can determine the quantum mechanical anti-correlation probability

$$
\mathbb{P}^{\psi_{\mathrm{s}}}_{\mathbf{a},\mathbf{b}}:=\mathbb{P}^{\psi_{\mathrm{s}}}\left(S^{(1)}_{\mathbf{a}}=-S^{(2)}_{\mathbf{b}}\right)\,,
$$

where $S^{(1)}_{\mathbf{a}},S^{(2)}_{\mathbf{b}}$ are the "spin values", without the need to introduce the quantum formalism of joint measurement statistics. According to the rules of probability,

$$
\mathbb{E}^{\psi_{\mathrm{s}}}_{\mathbf{a},\mathbf{b}}=-\mathbb{P}^{\psi_{\mathrm{s}}}_{\mathbf{a},\mathbf{b}}+\left(1-\mathbb{P}^{\psi_{\mathrm{s}}}_{\mathbf{a},\mathbf{b}}\right)=-2\mathbb{P}^{\psi_{\mathrm{s}}}_{\mathbf{a},\mathbf{b}}+1\,,
$$

that is,

$$\mathbb{P}^{\psi_s}_{\mathbf{a},\mathbf{b}} = \frac{1}{2} + \frac{1}{2}\mathbf{a}\cdot\mathbf{b}\,.$$

Choosing $120°$ for the angles between $\mathbf{a}, \mathbf{b}$, and $\mathbf{c}$, we get

$$\mathbb{P}^{\psi_s}_{\mathbf{a},\mathbf{b}} = \frac{1}{2} - \frac{1}{4} = \frac{1}{4}\,, \qquad \mathbb{P}^{\psi_s}_{\mathbf{a},\mathbf{c}} = \frac{1}{4}\,, \qquad \mathbb{P}^{\psi_s}_{\mathbf{b},\mathbf{c}} = \frac{1}{4}\,.$$

Hence for the sum of the probabilities on the left-hand side of (10.2), we obtain the value 3/4, as already claimed.

Let us rephrase the argument for later reference to hidden variables. The EPR argument yields, by the locality assumption, local hidden variables for spin, the ones we denoted by $X^{(\mathrm{L},\mathrm{R})}_{\mathbf{a},\mathbf{b},\mathbf{c}}$. The second part of Bell's theorem shows that the existence of spin hidden variables contradicts Bohmian mechanics (or quantum mechanics for that matter). In other words, local hidden variables cannot reproduce the experimental correlations. Expressed in yet another way, there exist no random variables $X^{(\mathrm{L},\mathrm{R})}_{\mathbf{a},\mathbf{b},\mathbf{c}}$ which have the quantum mechanical correlations (10.5).

The following terminology, which yields absolutely no new insights, has also been used to describe Bell's theorem. The spin measurement on the right side depends on the context in which the experiment is done, i.e., in the present case, it depends on what happens on the left, i.e., the hidden variable is, if it exists at all, "contextual". Hence Bell's theorem asserts that non-contextual hidden variables are not possible.

We conclude this section with some light entertainment:

- Bell's theorem has (quite often) been cited as proving that Bohmian mechanics is impossible. Why? Presumably because Bohmian mechanics was viewed as a hidden variable theory and the hearsay on Bell's theorem was that it proved that hidden variable theories conflict with quantum mechanics.
- It has also been said that quantum mechanics is local despite (10.3). How can that be? By forbidding or not believing that the steps in (10.3) are valid. How can that be? We do not know.
- It has also been said that Bell's nonlocality is nothing more than "learning at a distance". Wittgenstein's blue and brown book are wrapped in packages. Now suppose you and your friend each get a package without knowing which of the two books it contains. Your friend leaves for the moon with his package. When the spacecraft lands, you open your package and you unwrap the blue book. You know immediately that your friend on the moon has the brown book. Again one may wonder how anyone could reach the misunderstanding that this is a possible reading of Bell's nonlocality. Presumably because Bell's article was not actually read, and conclusions were drawn from hearsay about Bell's work. In his nice article entitled *Bertlmann's Socks and the Nature of Reality* [6], Bell elaborates on the distinction between this unspectacular effect of learning at a distance and the spectacular effect of nonlocality.

10.2 Faster Than Light Signals?

Bohmian mechanics is nonlocal. The wave function acts in a nonlocal way on the particles. Why can one not send signals faster than light? One cannot, because of the quantum equilibrium hypothesis. The action at a distance which the wave function mediates is randomized in such a way that it is unusable. If quantum equilibrium were false, superluminal signalling might perhaps be possible. Since there is no evidence that quantum equilibrium is false, there is no reason to speculate any further.[3]

Let us show for the sake of completeness how quantum equilibrium acts here. Let us take a general entangled two-particle state

$$\psi = a|\uparrow\rangle_1|\downarrow\rangle_2 + b|\downarrow\rangle_1|\uparrow\rangle_2 + c|\downarrow\rangle_1|\downarrow\rangle_2 + d|\uparrow\rangle_1|\uparrow\rangle_2 ,$$

with $|a|^2+|b|^2+|c|^2+|d|^2=1$. The probability of getting the spin value $|\uparrow\rangle_2$ at SGM-R is $|b|^2+|d|^2$. We now do a measurement, first on the left side in an arbitrarily chosen direction γ at SGM-L, where γ is the angle between the z-direction and the chosen direction. That is the freedom the experimenter has, and with which the experimenter can hope to affect the outcome on the right-hand side. Expressing the z-spin basis vectors in the corresponding γ-basis,

$$|\uparrow\rangle_1 = \mathbf{i}_1 \cos\gamma + \mathbf{j}_1 \sin\gamma , \qquad |\downarrow\rangle_1 = -\mathbf{i}_1 \sin\gamma + \mathbf{j}_1 \cos\gamma ,$$

we rewrite the above state as

$$\begin{aligned}\psi &= \mathbf{i}_1\Big[\cos\gamma\big(a|\downarrow\rangle_2 + d|\uparrow\rangle_2\big) - \sin\gamma\big(b|\uparrow\rangle_2 + c|\downarrow\rangle_2\big)\Big] \\ &\quad + \mathbf{j}_1\Big[\sin\gamma\big(a|\downarrow\rangle_2 + d|\uparrow\rangle_2\big) + \cos\gamma\big(b|\uparrow\rangle_2 + c|\downarrow\rangle_2\big)\Big] \\ &=: \psi_{\mathbf{i}_1} + \psi_{\mathbf{j}_1} ,\end{aligned}$$

from which we read off that $||\psi_{\mathbf{i}_1}||^2$ and $||\psi_{\mathbf{j}_1}||^2$ are the probabilities for the outcomes spin up or spin down when measuring first at SGM-L(γ). That measurement will "produce" a collapse of the entangled state. The collapse is the nonlocal effect which could be the source for nonlocal signalling. The collapsed wave function will be the one in the support of which the particle is located after leaving SGM-L(γ), i.e., it will be either $\psi_{\mathbf{i}_1}/||\psi_{\mathbf{i}_1}||$ or $\psi_{\mathbf{j}_1}/||\psi_{\mathbf{j}_1}||$, depending on the outcome at SGM-L(γ).

And what is now the effect of this measurement on the probability for the outcome at SGM-R? To find out, we now compute the quantum equilibrium probability that the particle, when going through SGM-R (oriented in the z-direction), is in the support of, let us say, the spin-up wave function $|\uparrow\rangle_2$. Repeating the argument now with the new wave function $\psi_{\mathbf{i}_1}/||\psi_{\mathbf{i}_1}||$, the probability will be

[3] That quantum equilibrium prevents superluminal signalling is taken, however, as a motivation for research on quantum non-equilibrium [7].

$$\frac{\left|\left|\mathbf{i}_1|\uparrow\rangle_2(d\cos\gamma - b\sin\gamma)\right|\right|^2}{||\psi_{\mathbf{i}_1}||^2},$$

while for $\psi_{\mathbf{j}_1}/||\psi_{\mathbf{j}_1}||$, it will be

$$\frac{\left|\left|\mathbf{j}_1|\uparrow\rangle_2(b\cos\gamma + d\sin\gamma)\right|\right|^2}{||\psi_{\mathbf{j}_1}||^2}.$$

From this, we obtain the probability for the outcome spin up on SGM-R(γ) by summing the "joint probabilities"

$$||\psi_{\mathbf{i}_1}||^2\frac{\left|\left|\mathbf{i}_1|\uparrow\rangle_2(d\cos\gamma - b\sin\gamma)\right|\right|^2}{||\psi_{\mathbf{i}_1}||^2} + ||\psi_{\mathbf{j}_1}||^2\frac{\left|\left|\mathbf{j}_1|\uparrow\rangle_2(b\cos\gamma + d\sin\gamma)\right|\right|^2}{||\psi_{\mathbf{j}_1}||^2},$$

which yields

$$|b|^2 + |d|^2 .$$

There is therefore no effect on the statistics of the outcomes on the right-hand side. They are the same, whether or not a measurement on SGM-L takes place first.

The key property we have used here is that we can infer from the "joint distribution" the probability for the outcome on the right by summing the joint probability over the possible values of the left outcome. We shall learn in Chap. 12 that the shorthand notation for this is that "observables" commute. The commutation of the spin observables on the left and on the right is in this sense an expression of the fact that one can perform "local operations" on the quantum system, i.e., that the pieces of apparatus SGM-L and SGM-R are decoupled, meaning that they function independently of one another.

References

1. T. Maudlin: *Quantum Non-Locality and Relativity: Metaphysical Intimations of Modern Physics*, Aristotelian Society Series, Vol. 13 (Oxford UK and Cambridge: Blackwell, 1994)
2. A. Einstein, B. Podolsky, N. Rosen: Physical Review **41**, 777 (1935)
3. D. Bohm: *Quantum Theory* (Prentice Hall, New York, 1951)
4. D. Dürr, S. Goldstein, R. Tumulka, N. Zanghì: In: *Encyclopedia of Philosophy*, ed. by D.M. Borchert (Macmillan Reference, USA, 2005)
5. A. Aspect, J. Dalibard, G. Roger: Phys. Rev. Lett. **49** (25), 1804 (1982)
6. J.S. Bell: *Speakable and Unspeakable in Quantum Mechanics* (Cambridge University Press, Cambridge, 1987)
7. A. Valentini: J. Phys. A **40** (12), 3285 (2007)

Chapter 11
The Wave Function and Quantum Equilibrium

After introducing Bohmian mechanics we described how macroscopic physics emerges from it, and how the microscopic wave function of a system determines probabilities for pointer positions. All this is based on the quantum equilibrium hypothesis. We used the notion "wave function of a system" rather loosely, without scrutinizing its meaning in any depth. In this chapter, we shall complete that description and justify the quantum equilibrium hypothesis. The basic idea as how to approach this justification has been presented in Chap. 4. We need to show that the empirical distribution of configurations is typically close to the quantum equilibrium distribution, which has been established experimentally to be empirically adequate. This can be done with surprising ease, and Bohmian mechanics thus recommends itself as the paradigm for Boltzmann's view of chance in physics.

11.1 Measure of Typicality

Equations (8.3) and (8.4) define Bohmian mechanics for an N-particle system which has no environment to interact with. It is an N-particle Bohmian universe. Following Boltzmann's understanding of chance in physics, the justification of the quantum equilibrium hypothesis must begin with a Bohmian universe, huge enough to allow for many subsystems, so that one can form an ensemble of subsystems. This allows for empirical distributions and statistical testing. We start with

$$(\mathscr{Q}, \Phi_t^{\Psi}, \mathbb{P}^{\Psi}) \tag{11.1}$$

as dynamical system with $\mathscr{Q}$ as configuration space, and Ψ as the wave function of the universe generating Φ_t^{Ψ}, the Bohmian flow on $\mathscr{Q}$:

$$\forall t \in \mathbb{R}\,, \qquad \Phi_t^{\Psi}(\mathbf{q}) = \mathbf{Q}(t,\mathbf{q}) = \text{solution of (8.3)}\,.$$

$\mathbb{P}^{\Psi}$ is the equivariant measure, the *quantum equilibrium measure*. This means the following. Let Ψ_t be the solution of (8.4) with initial condition Ψ. An *equivariant*

D. Dürr, S. Teufel, *Bohmian Mechanics*, DOI 10.1007/978-3-540-89344-8_11,

measure is a measure $\mathbb{P}^\Psi$, depending on Ψ in a particular way, viz.,

$$\mathbb{P}_t^\Psi(A) := \mathbb{P}^\Psi \circ (\Phi_t^\Psi)^{-1}(A) \equiv \mathbb{P}^\Psi\left((\Phi_t^\Psi)^{-1}(A)\right) = \mathbb{P}^{\Psi_t}(A)\,, \tag{11.2}$$

where the last equality expresses the condition for equivariance. In terms of expectation values of a function $f : \mathcal{Q} \to \mathbb{R}$,

$$\mathbb{E}^\Psi\left(f(\mathbf{Q}(t))\right) = \mathbb{E}^{\Psi_t}(f)\,. \tag{11.3}$$

This means that the mapping from Ψ to $\mathbb{P}^\Psi$ is invariant under the time evolution $\mathbb{P}_t^\Psi = \mathbb{P}^{\Psi_t}$. Diagrammatically,

$$\begin{array}{ccc} \Psi & \longrightarrow & \mathbb{P}^\Psi \\ U_t \downarrow & & \downarrow \circ(\Phi_t^\Psi)^{-1} \\ \Psi_t & \longrightarrow & \mathbb{P}^{\Psi_t} \end{array}$$

where U_t denotes the evolution $\Psi_t = U_t\Psi$ according to Schrödinger's equation and $\circ(\Phi_t^\Psi)^{-1}$ stands for the flow map defining the time evolution of the measure $\mathbb{P}_t^\Psi := \mathbb{P}^\Psi \circ (\Phi_t^\Psi)^{-1}$ along the Bohmian flow.

Equivariance generalizes stationarity and defines the quantum equilibrium measure, the measure singled out by the dynamical law itself, and which defines typicality. *The equivariance property ensures that typicality is time independent.* The generalization to equivariance is required, because stationarity has no meaning when the velocity field $v^\Psi(\mathbf{q},t)$ [see (8.1)] depends on time, since the wave function is time dependent. Finding an equivariant measure is potentially a hard task! In classical mechanics the stationary measure was easy to find because the divergence of the Hamiltonian vector field on phase space is zero (Liouville's theorem). No such property holds for the Bohmian vector field. Nevertheless, thanks to Born and Schrödinger, we already know the equivariant measure $\mathbb{P}^\Psi$ [see (7.11)– (7.14)], namely

$$\mathbb{P}^\Psi(A) = \int_A |\Psi(\mathbf{q})|^2 \mathrm{d}^n q\,, \tag{11.4}$$

normalized to unity,

$$\int |\Psi(\mathbf{q})|^2 \mathrm{d}^n q = 1\,. \tag{11.5}$$

Of course, we have no idea what the universal wave function looks like. Is it time dependent? One reason to think that the wave function of the universe is not stationary is macroscopic irreversibility, so that the wave function can be held responsible for the global non-equilibrium character of the universe. It is not unreasonable to think that the non-equilibrium character of the universe is encoded in a special initial wave function of the universe (for a discussion of typicality of wave functions, see [1]).

Assuming a special initial wave function allows one to separate the justification of the statistical hypothesis for the Bohmian particles from the issue of macroscopic irreversibility, which then resides solely in the wave function.

Note, however, that the assumption of a non-equilibrium universal wave function is plausible, but not necessary. To understand this better, suppose the universal wave function were stationary, not evolving in time. How is it possible then for things to move around? In orthodox quantum mechanics the world would be forever still. Not so in Bohmian mechanics. If the wave function contains a nontrivial phase, Bohmian particles move around and a world like ours is still possible. We give a simple example of a time-evolving Bohmian world with a stationary universal wave function in Remark 11.1. The irreversibility we experience in all macroscopic processes will then have to be explained by a special configuration of the Bohmian particles.

Readers who have absorbed the following sections may come back to this point and wonder how much of the following analysis remains valid when the wave function of the universe is stationary. Presumably the result will no longer be as easy to prove, but it might be worthwhile noting that there is no reason to think that the set of special initial configurations which yield a macroscopic non-equilibrium universe like ours is also the set of atypical Bohmian configurations for which the quantum equilibrium hypothesis does not hold. In other words, conditioning the quantum equilibrium measure on the set of initial conditions responsible for thermal non-equilibrium may imply that the quantum equilibrium hypothesis typically holds. Since we lack a good understanding of what the wave function of the universe looks like, these last remarks are a subject for future research, and we turn now to more modest and practical questions.

11.2 Conditional Wave Function

Given the Bohmian universe, how does one describe a subsystem? Asking the same question for a Newtonian universe, the answer is clear: just apply the Newtonian laws to the subsystem. Thinking about this for a moment, one understands that the answer is based on the possibility that influences from outside the system are negligible. If we throw a stone, it is not just the Newtonian laws of the arm giving momentum to the stone, but also the gravitational interaction between the stone and the earth which is relevant for its motion; but by all means forget the mass of the sun! It is too far away!

However, in Bohmian mechanics "too far away" has no obvious meaning. The universal wave function is a function on the configuration space of the universe. What does it mean to neglect "distant Bohmian bodies"? We need to analyze this. The last chapter already tells us that "far away" is not the essential feature on which an autonomous description of a subsystem can be based. It is rather a product structure of the wave function in conjunction with fapp-impossibility of interference.

We consider an m-dimensional subsystem (x-system) of particles given by their configuration $\mathbf{X}$, and denote the n-dimensional particle configuration of the rest of the universe by $\mathbf{Y}$, i.e., the total environment of the x-system is the y-system. The universal configuration therefore splits according to

$$\mathbf{Q} = (\mathbf{X}, \mathbf{Y}) \qquad \big(\mathbf{q} = (\mathbf{x}, \mathbf{y})\big) . \tag{11.6}$$

The x-system we should have in mind now is a physical system which one studies in a laboratory. That means that the experimenter is part of the environment. Everything the experimenter learns, writes down, or otherwise secures belongs to $\mathbf{Y}$. From a macroscopic point of view, we may say that we know the "relevant region" of configuration space in which $\mathbf{Y}$ lies quite well. What we do not know is the wave function of the universe Ψ. We need a concept to describe the x-system in Bohmian terms. Since we already have the Bohmian positions, all that is needed is the notion of the wave function for the system.

The Bohmian equation for the x-system is

$$\dot{\mathbf{X}}(t) = \mathbf{v}_x^{\Psi}\big(\mathbf{X}(t), \mathbf{Y}(t)\big) \sim \Im \frac{\nabla_x \Psi\big(\mathbf{x}, \mathbf{Y}(t)\big)}{\Psi\big(\mathbf{x}, \mathbf{Y}(t)\big)}\Bigg|_{\mathbf{x}=\mathbf{X}(t)} ,$$

suggesting the definition of a *conditional* wave function for the x-system:

$$\varphi^{\mathbf{Y}}(\mathbf{x}) = \frac{\Psi(\mathbf{x}, \mathbf{Y})}{\|\Psi(\mathbf{Y})\|} , \tag{11.7}$$

with the norm

$$\|\Psi(\mathbf{Y})\| = \left[\int |\Psi(\mathbf{x}, \mathbf{Y})|^2 \mathrm{d}^m x\right]^{1/2} .$$

Hence,

$$\dot{\mathbf{X}}(t) = \mathbf{v}^{\varphi^{\mathbf{Y}}}\big(\mathbf{X}(t)\big) .$$

We obtain the conditional wave function by replacing the y-part in the configuration coordinate $\mathbf{q} = (\mathbf{x}, \mathbf{y})$ of the universal wave function by the actual configuration $\mathbf{Y}$ and then normalizing. The conditional wave function is in general unknown and it will not generally evolve according to a Schrödinger equation for the x-system. However, look at the example in 11.1.

On the other hand, the conditional wave function connects directly with the conditional quantum equilibrium measure. Suppose we would like to make a typicality statement about the x-system. Which measure is relevant to that purpose? Since the environment of the x-system is macroscopically factual, i.e., the laboratory and the experimenter at work in it are facts, we must condition the quantum equilibrium measure on these facts.

The conditional measure $\mathbb{P}(A|B)$ of A given the event B is simply the measure restricted to B and appropriately renormalized:

$$\mathbb{P}(A|B) = \mathbb{P}(A \cap B)/\mathbb{P}(B) .$$

When we condition on a set b of measure zero, which will be the case relevant for us, since we wish to condition on the macroscopic facts encoded in the point $\mathbf{Y}$, we can consider the limit of $\mathbb{P}(A \cap B)/\mathbb{P}(B)$ as $\mathbb{P}(B) \to \mathbb{P}(b) = 0$. This exists when the measure has a density, as is the case for

$$\mathbb{P}^{\Psi}(\mathrm{d}^m x, \mathrm{d}^n y) = |\Psi(\mathbf{x},\mathbf{y})|^2 \mathrm{d}^m x \mathrm{d}^n y .$$

The conditional measure is then simply given by

$$\begin{aligned} \mathbb{P}^{\Psi}\Big(\{\mathbf{Q} = (\mathbf{X},\mathbf{Y}), \mathbf{X} \in \mathrm{d}^m x\}\big|\mathbf{Y}\Big) &= \frac{\big|\Psi\big((\mathbf{x},\mathbf{Y})\big)\big|^2 \mathrm{d}^m x}{\int \big|\Psi\big((\mathbf{x},\mathbf{Y})\big)\big|^2 \mathrm{d}^m x} \\ &= \big|\varphi^{\mathbf{Y}}(\mathbf{x})\big|^2 \mathrm{d}^m x . \end{aligned} \tag{11.8}$$

In (11.8) the specification of the environment to the configuration $\mathbf{Y}$ is much too specific for the formula to be applicable in relevant physical situations. We only know a few macroscopic facts about $\mathbf{Y}$, so the conditioning on $\mathbf{Y}$ seems ridiculous. However, we can gain a valuable formula from (11.8) by making the following observation. We can collect all $\mathbf{Q}$s which yield the same conditional wave function for the x-system into a set, say

$$\{\varphi^Y = \varphi\} := \Big\{(\mathbf{x},\mathbf{Y}) \in \mathscr{Q} \big| \varphi^Y(\mathbf{x}) = \varphi(\mathbf{x})\Big\} .$$

Then use the following simple property of conditional probabilities. Let $B = \bigcup B_i$ be a pairwise disjoint partition and let $\mathbb{P}(A|B_i) = a$ for all B_i. Then by the additivity of the measure, viz.,

$$\mathbb{P}(B)a = \sum_i \mathbb{P}(A|B_i)\mathbb{P}(B_i) = \sum_i \mathbb{P}(A \cap B_i) = \mathbb{P}(A \cap B) ,$$

and hence $\mathbb{P}(A|B) = a$. Therefore,

$$\mathbb{P}^{\Psi}\Big(\{\mathbf{Q} = (\mathbf{X},\mathbf{Y}), \mathbf{X} \in \mathrm{d}^m x\}\big|\{\varphi^{\mathbf{Y}} = \varphi\}\Big) = |\varphi|^2 \mathrm{d}^m x , \tag{11.9}$$

which for ease of notation we simply write as

$$\mathbb{P}^{\Psi}\big(\mathbf{X} \in \mathrm{d}^m x \big| \varphi^{\mathbf{Y}} = \varphi\big) = |\varphi|^2 \mathrm{d}^m x . \tag{11.10}$$

This formula is crucial for justifying the quantum equilibrium hypothesis. We wish to apply it to a situation where the conditional wave function of the x-system does not depend (at least for a certain amount of time) on $\mathbf{Y}$. That is what we believe to be the case in our world, i.e., that subsystems sometimes behave autonomously. This brings us then to the concept of a Bohmian subsystem, meaning that the subsystem has its own wave function, its own Schrödinger evolution, and hence an autonomous law for Bohmian trajectories.

11.3 Effective Wave function

If the universal wave function has a product structure, then in view of (9.5) the velocity field of the x-system is determined by ψ:

$$\Psi(\mathbf{x},\mathbf{y}) = \psi(\mathbf{x})\Phi(\mathbf{y}) \quad \Longrightarrow \quad \mathbf{v}_x^{\Psi} = \mathbf{v}_x^{\psi} . \tag{11.11}$$

But (11.11) is much too special, since any interaction between the x-system and the environment will destroy the product structure, leading to an entangled wave function as in the measurement process. It is unreasonable to assume that the universal wave function has product structure. Generically, it will be a superposition of products, i.e., a bona fide entangled wave function.

But we also know that for macroscopically disjoint wave packets when seen as functions of the macroscopic environment configuration $\mathbf{y}$ (see Fig. 9.2),

$$\Psi(\mathbf{x},\mathbf{y}) = \Psi_1(\mathbf{x},\mathbf{y}) + \Psi_2(\mathbf{x},\mathbf{y}) , \tag{11.12}$$

and only one of the packets will be effective in Bohmian mechanics, either Ψ_1 (if $\mathbf{Y} \in \operatorname{supp} \Psi_1$) or Ψ_2 (if $\mathbf{Y} \in \operatorname{supp} \Psi_2$) [see (9.5)], and we can fapp forget about the ineffective packet. The idea which leads to the relevant concept of the effective wave function of a subsystem comes from combining (11.11) and (11.12). As already remarked in Chap. 8, one must take the macroscopic disjointness of the packets in (11.12) with a pinch of salt. It will only be approximately satisfied, for example, in the sense of L^2, which means that

$$\Psi \approx \tilde{\Psi} \quad \Longleftrightarrow \quad \mathbb{P}^{\Psi} \approx \mathbb{P}^{\tilde{\Psi}} .$$

We introduce the concept of effective wave function for the x-system, which is the well-defined expression of the collapsed wave function of orthodox quantum mechanics. The effective wave function is the conditional wave function for a special physical situation. The x-system has an *effective wave function* φ if

$$\Psi(\mathbf{x},\mathbf{y}) = \varphi(\mathbf{x})\Phi(\mathbf{y}) + \Psi^{\perp}(\mathbf{x},\mathbf{y}) , \tag{11.13}$$

where Φ and $\Psi^{\perp}$ have *macroscopically disjoint* $\mathbf{y}$-supports and in addition

$$\mathbf{Y} \in \operatorname{supp} \Phi . \tag{11.14}$$

It is helpful to recall that the splitting (11.13) happens in the "measurement process". As in the discussion of the measurement experiment we can fapp forget the wave packet $\Psi^{\perp}$ if the environment is guided by Φ, i.e., if $\mathbf{Y} \in \operatorname{supp} \Phi$. We can forget it for as long as we wish, since interference is fapp impossible (until the universe comes to an end). In view of $\mathbf{v}^{\varphi\Phi}$, we see that the x-system will now be guided by φ. If the interaction term $V(\mathbf{x},\mathbf{y})\varphi(\mathbf{x})\Phi(\mathbf{y})$ in the Schrödinger equation is negligible (at least for a certain length of time), the x-system and the environment will be dynamically decoupled, and φ will obey a Schrödinger equation on its own. The x-system is then an isolated Bohmian system for that period of time. To repeat, the conditional wave function always exists, and it becomes an effective wave function when (11.13) holds with (11.14). We stress therefore that the effective wave function is a mathematically precise concept of the collapsed wave function of orthodox quantum theory.

Remark 11.1. Stationary Universal Wave Function with Random Conditional Wave Function and an Effective Wave Function with Nontrivial Time Dependence

We consider here a two-particle universe with "masses" $m_x = m$, $m_y = M$, and $\mathbf{q} = (x,y) \in \mathbb{R}^2$. Let the wave function be stationary:

$$\Psi(x,y) = R_1(x+y)R_2(y)\mathrm{e}^{\mathrm{i}k(x-y)} + R_3(x)R_4(y) ,$$

with real functions R_1, R_2, R_3, R_4 and $R_2(y) = 0$ for $y \geq 0$ and $R_4(y) = 0$ for $y < 0$. According to the equations of Bohmian mechanics, we have

$$Y_1(t) = Y_0 - \frac{\hbar k}{M}t ,$$

for $Y_0 < 0$, and

$$Y_2(t) = Y_0 ,$$

for $Y_0 \geq 0$. Suppose that $\int |R_1|^2 \mathrm{d}x = \int |R_3|^2 \mathrm{d}x = 1$. Then, with probability $p_1 = \int R_2^2(y)\mathrm{d}y$, the conditional wave function for the x-particle is (in the projective sense)

$$\begin{aligned}
\varphi_1(x,t) &= \frac{\Psi\big(x,Y_1(t)\big)}{\left[\int \mathrm{d}x R_1\big(x+Y_1(t)\big)^2 R_2\big(Y_1(t)\big)^2\right]^{1/2}} \\
&= c_1 R_1\left(x+Y_0-\frac{\hbar k}{M}t\right)\mathrm{e}^{\mathrm{i}k(x-Y_0+\hbar kt/M)} \\
&\mathrel{\widehat{=}} \tilde{c}_1 R_1\left(x+Y_0-\frac{\hbar k}{m}t\right)\mathrm{e}^{\mathrm{i}kx} ,
\end{aligned}$$

for $Y_0 < 0$, and

$$\varphi_2(x,t) = \frac{\Psi\big(x,Y_2(t)\big)}{\left[\int \mathrm{d}x R_3(x)^2 R_4(Y_0)^2\right]^{1/2}} = c_2 R_3(x) ,$$

for $Y_0 \geq 0$. We see that the conditional wave function is random.

Let us now slightly change the focus of the example and consider

$$\Psi(x,y) = \cos\left[k(x+y)\right]\mathrm{e}^{\mathrm{i}k(x-y)}$$

as "universal wave function". This happens to be a stationary solution of the Hamiltonian

$$H = -\frac{\hbar^2}{2m}\frac{\mathrm{d}^2}{\mathrm{d}x^2} - \frac{\hbar^2}{2M}\frac{\mathrm{d}^2}{\mathrm{d}y^2}\,.$$

The conditional wave function for the x-system is now

$$\varphi(x,t) = \cos\left(kx + kY_0 - \frac{\hbar k^2}{M}t\right)\mathrm{e}^{\mathrm{i}\left(kx - kY_0 + \hbar k^2 t/M\right)}\,.$$

Observing once again that the overall phase factors depending only on time are projective, and thus physically irrelevant, we obtain

$$\varphi(x,t) \hat{=} \tilde{\varphi}(x,t) := c\left[\mathrm{e}^{2\mathrm{i}(kx+2Y_0-\hbar k^2 t/M)} + 1\right]\,.$$

Observe next that the conditional wave function satisfies a Schrödinger equation on its own, so that we may consider it as an effective wave function, namely that of a free particle with mass M:

$$\mathrm{i}\hbar\frac{\partial}{\partial t}\tilde{\varphi} = -\frac{\hbar^2}{2M}\frac{\mathrm{d}^2}{\mathrm{d}x^2}\tilde{\varphi}\,.$$

The examples are interesting because the Bohmian positions and the effective wave function of a subsystem turn out to be non-trivial functions of time, even though the universal wave function is stationary (see also [2] for more on the meaning of wave functions). ■

11.4 Typical Empirical Distributions

We now justify the quantum equilibrium hypothesis as being the theoretical prediction for typical empirical distributions, which is known as Born's statistical interpretation of the wave function, and which we have already addressed in Chap. 8. The hypothesis reads as follows. *If a subsystem has effective wave function φ, then its particle coordinates are $|\varphi|^2$-distributed.* What does this mean? According to our understanding of Boltzmann's view, we ought to know by now! In an ensemble of similar subsystems, which all have effective wave function φ, the relative frequencies of the configuration coordinates will typically be close to the $|\varphi|^2$-distribution. We need to prove a law of large numbers!

Let us therefore consider the situation where the x-system consists of many similar microscopic subsystems $\mathbf{x}_1,\ldots,\mathbf{x}_N$, i.e., where $\mathbf{x} = (\mathbf{x}_1,\ldots,\mathbf{x}_N)$. Each of the x_i-systems is assumed to have (simultaneously) the effective wave function φ_i. If N

is not too large (not macroscopically large!), the x-system has effective wave function

$$\varphi(\mathbf{x}_1,\ldots,\mathbf{x}_N) = \prod_{i=1}^{N} \varphi_i(\mathbf{x}_i)\,. \tag{11.15}$$

This somewhat remarkable fact can be seen as follows. For each i we have, by virtue of (11.13) and (11.14),

$$\Psi(\mathbf{x},\mathbf{y}) = \varphi_i(\mathbf{x}_i)\Phi_i(\mathbf{y}_i) + \Psi_i^{\perp}(\mathbf{x}_i,\mathbf{y}_i)\,,$$

where Φ_i and $\Psi_i^{\perp}$ have macroscopically disjoint $\mathbf{y}_i$-supports and $\mathbf{Y}_i \in \text{supp}\,\Phi_i$. But the $\mathbf{x}_i$ are microscopically few coordinates and the number N of subsystems is not too large, so Φ_i and $\Psi_i^{\perp}$ must already have macroscopically disjoint $\mathbf{y}$-supports, where $\mathbf{q} = (\mathbf{x}_1,\ldots,\mathbf{x}_N,\mathbf{y})$. Furthermore we have

$$\mathbf{Y} \in \text{supp}\,\Phi_1 \cap \text{supp}\,\Phi_2 \cap \ldots \cap \text{supp}\,\Phi_N\,.$$

Therefore, for this $\mathbf{Y}$ and all i,

$$\Psi(\mathbf{x}_1,\ldots,\mathbf{x}_N,\mathbf{Y}) = \varphi_i(\mathbf{x}_i)\Phi_i(\mathbf{Y},\hat{\mathbf{x}}_i)\,, \tag{11.16}$$

with

$$\hat{\mathbf{x}}_i = (\mathbf{x}_1,\ldots,\mathbf{x}_{i-1},\mathbf{x}_{i+1},\ldots,\mathbf{x}_N)\,.$$

Hence let us write as an ansatz

$$\Psi(\mathbf{x}_1,\ldots,\mathbf{x}_N,\mathbf{Y}) = \prod_{i=1}^{N} \varphi_i(\mathbf{x}_i)\tilde{\Phi}(\mathbf{Y},\mathbf{x})\,.$$

Division by $\prod \varphi_i(\mathbf{x}_i)$ shows that, in view of (11.16),

$$\tilde{\Phi}(\mathbf{Y},\mathbf{x}) = \Phi(\mathbf{Y})\,.$$

So (11.15) is true.

Let us now move on to an ensemble of N subsystems which all have the same effective wave function φ. More precisely, we fix the same coordinate frame in all subsystems and φ is the effective wave function relative to that coordinate system. Each x_i-subsystem has coordinates $\mathbf{x}_i$, also relative to the chosen coordinate system. Then by virtue of (11.15), according to (11.8), we obtain for the distribution of the coordinates $\mathbf{x}_1,\ldots,\mathbf{x}_N$,

$$\begin{aligned}\mathbb{P}^{\mathbf{Y}}(\mathbf{X}_1 \in \mathrm{d}x_1,\ldots,\mathbf{X}_N \in \mathrm{d}x_N) &:= \mathbb{P}^{\Psi}(\mathbf{X}_1 \in \mathrm{d}x_1,\ldots,\mathbf{X}_N \in \mathrm{d}x_N|\mathbf{Y}) \\ &= \prod_{i=1}^{N} |\varphi(\mathbf{x}_i)|^2 \mathrm{d}x_i\,,\end{aligned} \tag{11.17}$$

where $\mathbf{Y}$ represents the environment of the $(x_1,\ldots,x_N)$-system. This formula enables us to predict the empirical distribution of the coordinates of what we might refer to as a $|\varphi|^2$-ensemble. Under the measure $\mathbb{P}^{\mathbf{Y}}$, according to (11.17), the coarse-graining functions $(\mathbf{X}_i)_{i=1,\ldots,N}$ defined on $\mathcal{Q}$ form a Bernoulli sequence of $|\varphi|^2$-distributed random variables. For such a sequence, we showed at the end of Chap. 4 that the law of large numbers (4.54) holds. Adjusting that assertion to the present setting, it implies that the relative frequencies of x-coordinates, i.e., the empirical distribution of the $X_1,\ldots,X_N$, is close to the $|\varphi|^2$-distribution for $\mathbb{P}^{\mathbf{Y}}$-typical configurations. This is exactly what the quantum equilibrium hypothesis says. Note that we acquire good information about the effective wave function, at least about its modulus squared, via the empirical statistics. The quantum equilibrium hypothesis, which is now no longer a hypothesis but a theorem, is the link (actually the only link) between theory and experience.

We formulate the theorem precisely as follows. Suppose that, say, at time t the x-system consists of N systems with coordinates $\mathbf{x}_1,\ldots,\mathbf{x}_N$ (relative to the same frame in each system), and that the configurations are $\mathbf{X}_t=(\mathbf{X}_1,\ldots,\mathbf{X}_N)$. Suppose the effective wave function is

$$\varphi_t(\mathbf{x})=\varphi(\mathbf{x}_1)\ldots\varphi(\mathbf{x}_N)\,,$$

and let $\mathbf{Y}_t=\mathbf{Y}$ be the environmental configuration at that time, in accordance with the fact that the effective wave function of the ensemble is $\varphi_t(\mathbf{x})$. Then

$$\begin{aligned}
\mathbb{P}^{\Psi_t}&\left(\left\{\mathbf{Q}\left|\left|\frac{1}{N}\sum_{i=1}^{N}f(\mathbf{X}_i)-\int f(\mathbf{x})|\varphi(\mathbf{x})|^2\mathrm{d}x\right|<\varepsilon\right.\right\}\Big|\mathbf{Y}_t=\mathbf{Y}\right)\\
&=\mathbb{P}^{\mathbf{Y}}_t\left(\left\{\mathbf{Q}\left|\left|\frac{1}{N}\sum_{i=1}^{N}f(\mathbf{X}_i)-\int f(\mathbf{x})|\varphi(\mathbf{x})|^2\mathrm{d}x\right|<\varepsilon\right.\right\}\right)\\
&=1-\delta(\varepsilon,f,N)\,,
\end{aligned}\tag{11.18}$$

and $\delta(\varepsilon,f,N)\longrightarrow 0$ for $N\longrightarrow\infty$.

Think of f as a characteristic function χ_A in (11.18). Then for a family χ_{A_α} defining the relative frequencies of the measured values ($\in A_\alpha$) and for N large enough, we have

$$\sum_\alpha \delta(\varepsilon,f_\alpha,N)\ll 1\,.$$

The bad set of initial configurations $\mathbf{Q}$, for which the empirical distribution is not close to the quantum equilibrium value, has very small $\mathbb{P}^{\mathbf{Y}}_t$-measure. There are so to speak only a few points $\mathbf{Q}\in\mathbf{Q}^{\mathbf{Y}}_t=\{\mathbf{Q}|\mathbf{Y}_t=\mathbf{Y}\}$ in the given environment which fail.

The reader may have many concerns with this assertion. We shall address two. The first may be this: What is special about the equivariance property of the universal measure $\mathbb{P}^{\Psi}$ defining typicality? Would another measure $\tilde{\mathbb{P}}$, say the one with density $|\Psi|^4$ (which is not equivariant), not yield typicality for the empirical distribution $|\varphi|^4$ by the same argument? In fact, it would, but only at *exactly* that moment of time where the measure has the $|\Psi|^4$ density. Since the measure is not equivariant, its density will soon change to something completely different. If $\tilde{\mathbb{P}}$ is supposed to be the "initial measure" of typicality at the "initial time" when the universe started, then who knows what the measure of typicality will look like today? That measure has no special significance. Why that measure and not some other? The equivariant measure of typicality on the other hand is special – as in Boltzmann's way of looking at statistical physics. Typicality defined by this measure does not depend on time. It is singled out by the physics itself. Time evolution does play a role, although so far we have only considered an ensemble at a single time (like tossing 10 000 coins at the same time). We shall say a bit more about repetitions of an experiment (ensemble in time, like tossing the same coin 10 000 times) later.

Another concern the reader may have is that the conditioning is much too strong and therefore irrelevant. Since we can never know what the exact environmental configuration $\mathbf{Y}$ is, we should condition on less, indeed condition only on the fact that an experiment of the kind we describe has been carried out and whatever else seems relevant for the experiment. But we have already observed in (11.10) that coarse-graining the conditioning to a set on which only the conditional (here effective) wave function is given makes no difference to the right-hand side. The conclusion holds just the same. Moreover, we need the assertion in the strongest possible form, which is the way we formulated it. Indeed, we must be sure that further knowledge of the environment, for example concerning the history of the ensemble system and whatever else we may deem relevant, does not affect the conditional distribution, given the effective wave function. Is it relevant that the experimenter chose a red tie that morning? Is it relevant that his car had a flat tire on the way to work? Who knows beforehand? The assertion (11.18) tells us that all those details are irrelevant.

It is crucial that the *conditional* measure $\mathbb{P}_t^{\mathbf{Y}}$ of the configurations for which the empirical statistics deviate from $|\varphi|^2$ should be small. Suppose we could only show that for the unconditional measure $\mathbb{P}^{\Psi}$. That would tell us nothing, because the set of environments which are in accordance with the experiments taking place may already have small $\mathbb{P}^{\Psi}$-measure, so this alone could be responsible for the smallness of the result. This is what happens in classical statistical mechanics. The set of initial conditions of the universe we happen to live in has extremely small equilibrium measure. Therefore, conditioning is crucial. Without it we would be empty-handed.

We could even condition – if that were necessary, for example, if the universal wave function were stationary – on such special environments as could explain irreversible evolution. But then, to justify the quantum equilibrium hypothesis, it must be the case that the set of special initial conditions $\mathscr{N} \subset \mathscr{Q}$ responsible for thermal non-equilibrium (in Remark 11.1 this could be taken as the set of positive $\mathbf{Y}_0$ values) is not the bad set of $\mathbf{Q}$s for which the quantum equilibrium hypothesis fails to be true.

We conclude this series of remarks with one more point. Suppose we forget all the metaphysics and say that we do not believe in all this talk about the physics determining its measure of typicality. But we must still prove a law of large numbers for empirical distributions, otherwise we have no link between the theory and experience. In this view the equivariant measure is a highly valuable technical tool, because this is the measure which allows us to prove the theorem! At time t, let us say today when we do the experiment, any other measure would look so odd (it would depend on Ψ_t in such an intricate way) that we would have no chance of proving anything! And if it did not look odd today, then it would look terribly odd tomorrow! The equivariant measure always looks the same and what we prove today about the empirical distribution will hold forever.

We also remark that any measure $\mathbb{P}$ which is absolutely continuous with respect to the equivariant measure $\mathbb{P}^{\Psi}$ ($\mathbb{P}$ and $\mathbb{P}^{\Psi}$ have a density with respect to each other) defines the same sense of typicality. The observation we have just made about the technical advantage of the equivariant measure applies here, too. To prove the law of large numbers with another measure from the equivalence class of measures of typicality would be an awkward thing to do, since it changes its form all the time, and would look so odd that we would have no chance of proving anything. In other words, equivariance is also technically crucial!

But now on to more practical concerns! So far our statistical analysis has been restricted to a spatially distributed ensemble $(\mathbf{x}_1, \dots, \mathbf{x}_N)$ at one time. But what really happens in experiments is that they are repeated. For example, in the two slit experiments, one sends a beam of particles through the slit and thereby creates an ensemble of independent subsystems. But it is an ensemble distributed over time (like letting balls drop through the Galton board). One can actually handle this too [3], but it is definitely more complicated. Here there is a subtlety that must be taken into account in the analysis of time ensembles. In the universe the times at which the experiments are done are also "random", i.e., functions of $\mathbf{Q}$, like the spatial locations of the systems.

For example, suppose an experimenter, eager to win the Nobel prize, starts an experiment which is supposed to measure the EPR correlations in a very fine way. Suppose the experiment shows in the first 100 runs that, when a particle is registered on the left, no particle is registered on the right. That makes the experimenter so upset that he destroys the laboratory in a fit of anger. No further experiment is done. Alternatively, the experimenter rethinks the experimental setup, finds a problem, and calls the repair man, only to find that he is away on holiday. The moral is that the times when the experiments are done are random! That must be taken into account in the mathematics.

And that is not all. In the single time ensemble, all we need to look at are the actual configurations. No measurement talk is needed. In the time ensemble, however, we must take into account the fact that measurements do take place. The position of the electron in the ground state does not change in time. For $\mathbf{X}(t_1), \mathbf{X}(t_2), \dots, \mathbf{X}(t_N)$ in the ground state φ, we have $\mathbf{X}(t_1) = \mathbf{X}(t_2) = \dots = \mathbf{X}(t_N)$. In other words, the random variables are not at all independent. By measuring the ground state distribution of a hydrogen atom, we disturb the state, e.g., we ionize the electron, let it fall back, ionize again, let it fall back and so on. It is important here that we ionize it and then let it fall back again. The positions after settling back in the ground state are independent random variables. It is not difficult to see why this is so. Imagine N pointer wave functions $\Phi_1, \ldots, \Phi_N$ with configurations $\mathbf{Z}_1, \ldots, \mathbf{Z}_N$, which indicate the measured values of the positions of the electron in the ground state. At the end of the day we have a product wave function of pointer positions $\prod \Phi_i$ for the measured values $\mathbf{X}_1, \ldots, \mathbf{X}_N$, i.e., the distribution of the measured values is again a product distribution, and so we have independence. Therefore the law of large numbers applies as before. The principle which underlies the "many-times analysis" should thus be reasonably clear, even though the precise analysis is, as mentioned, somewhat demanding [3].

We have understood and justified the quantum equilibrium hypothesis. It is in fact Born's statistical law for the wave function, sometimes called the Born rule, or Born interpretation of the wave function, and we understand that the wave function φ one talks about is the conditional or effective wave function. The hypothesis concerns the empirical distribution of coordinates of particles in an ensemble. The justification tells us that the quantum equilibrium distribution $\rho = |\varphi|^2$ is what we should always be experiencing. That is all.

But some more lessons can nevertheless be learned. When we know, say by measuring the position of a particle, that the particle is in some spatial region, then we know by the quantum equilibrium distribution that the effective wave function will have its support in that region. If that region happens to be tiny, then the effective wave function will be sharply localized. If the wave function is sharply localized, its Fourier decomposition into plane waves will involve a great spread of wave numbers k. Suppose the wave function now evolves freely. The plane wave packets will move apart due to the dispersion relation, and eventually separate. Depending on the exact initial position of the Bohmian particle, this particle will eventually be guided by one of the almost plane wave packets, i.e., it will eventually move along a straight line. We explained that in Sect. 9.4.

The initial randomness of the particle position translates into the randomness of the particle's asymptotic velocity, which is given by the modulus squared of the Fourier transform of the initial localized wave packet. That distribution is all the more spread out as the initial wave packet is sharply localised. This is Heisenberg's uncertainty relation. Obviously, the relation is a direct consequence of the quantum equilibrium distribution, i.e., Born's statistical law.

Can we by any clever tricks whatever know more about the particle position than that it is $|\varphi|^2$-distributed when the effective wave function is φ? The answer is that we cannot. That is what the quantum equilibrium hypothesis says, and what we have proven to be typical. Equilibrium, here quantum equilibrium, entails absolute uncertainty about the Bohmian positions, beyond the $|\varphi|^2$-distribution.

11.5 Misunderstandings

We have discussed the justification of the statistical hypothesis in classical statistical mechanics and in Bohmian mechanics. The hypothesis concerns the typical empirical distribution of the values in an ensemble. Typicality is defined via a measure of typicality. Can the measure of typicality be misunderstood as an empirical distribution? Hardly so, because the measure of typicality is a measure on the configuration space of the universe, and since we only have access to one universe, namely the one we live in, an ensemble of universes is meaningless for physics. It so happens that the quantum equilibrium measure of the universe defining typicality has the same form, namely $|\Psi|^2$, as the predicted empirical distribution of a system with effective wave function φ, namely $|\varphi|^2$. Although they look similar, they do not mean the same thing, since the effective wave function is a fundamentally different object from the wave function of the universe.

We recall that, in classical statistical mechanics, we have the microcanonical measure as a measure of typicality, and the empirical distributions of subsystems are typically canonical or grand canonical ensembles which look different from the microcanonical measure. In quantum equilibrium the situation is simpler. Bohmian mechanics is simpler than classical mechanics, and because of that we are able to justify the quantum equilibrium hypothesis with great ease. The price to pay is that we need to be careful not to treat things which are not the same as being the same.

11.6 Quantum Nonequilibrium

The second law of thermodynamics captures irreversibility, and at the same time points towards the *problem of irreversibility*, which is to justify the special atypical initial conditions on which, according to Boltzmann, the second law is based. Atypical initial conditions (which are synonymous with non-equilibrium) do of course exist. So do even grotesquely atypical initial conditions, as in the "Umkehreinwand". While typicality is a clear-cut concept which needs no further justification, atypicality is tricky, and we (humankind) should consider ourselves lucky that we have found the second law of thermodynamics. It tells us that we should not worry about grotesquely atypical initial conditions, and it tells us more or less how special the initial configurations are.

One could nevertheless spend one's time worrying about what very atypical initial conditions would produce. As a believer in strong atypicality, one could sit in front of a stone and wait for the stone to jump into the air, because in a very atypical world, that could happen, now, tomorrow, maybe the day after tomorrow. The second law tells us that there are better ways to spend one's time, but for some there may still be the bitter pill to swallow, that the second law is based on non-equilibrium. There is no question that we do need to worry about what justifies these special initial conditions, although we may safely say that this is a problem for future generations to handle.

The situation in a Bohmian universe is irrevocable. There is no need for a second law for the configurations in Bohmian mechanics. Quantum equilibrium, which like equilibrium needs no justification, is fortunately all we need to describe the empirical import of Bohmian mechanics.

References

1. S. Goldstein, J.L. Lebowitz, R. Tumulka, N. Zanghì: Phys. Rev. Lett. **96** (5), 050403, 3 (2006)
2. D. Dürr, S. Goldstein, N. Zanghì: In: *Experimental Metaphysics* (Boston, MA, 1994), Boston Stud. Philos. Sci., Vol. 193 (Kluwer Acad. Publ., Dordrecht, 1997) pp. 25–38
3. D. Dürr, S. Goldstein, N. Zanghì: Journal of Statistical Physics **67**, 843 (1992)

Chapter 12
From Physics to Mathematics

The foundations of quantum mechanics are concerned with Hilbert spaces, linear operators on Hilbert spaces, unitary operators, and self-adjoint operators and their spectra. The foundations of Bohmian mechanics contain nothing of that sort and nothing of that sort seems relevant. Of course, the Schrödinger equation is a partial differential equation and contains differential operators, but so does the Maxwell–Lorentz theory of electromagnetism, which one learns about without all those abstract notions. Why is quantum mechanics different? Why does it need to be based on such abstract mathematical notions?

The quantity which determines the empirical import of Bohmian mechanics is the effective wave function, the "collapsed" wave packet which guides the particles. Its modulus squared gives the statistical distribution of the particle configuration. That is all. It seems a meager content. But we shall explain in this chapter why the statistical import of Bohmian mechanics, which seems so meager, is in fact extremely rich. The quantum formalism in its most general formulation follows from it, and so does much more (see [1] for a detailed analysis). Readers who know quantum mechanics from textbooks will find this chapter to be a revelation. It prepares the insight needed for the abstract mathematics to be discussed in the next part of the book, the mathematics which is usually viewed as forming the foundations of quantum mechanics.

12.1 Observables. An Unhelpful Notion

It is natural to think that an observable is a variable that can be observed. In quantum mechanics, a self-adjoint operator is an observable. Sometimes it is said that one measures the observable, which would then mean that one measures the operator. For example, one can say that one measures the operator Δ. But this is clearly not intended to mean that one determines the area of the symbol Δ. So what is meant? In fact, something very abstract. And this is an abstraction that one should not be surprised about. After all, an operator on a Hilbert space is a very abstract object,

D. Dürr, S. Teufel, *Bohmian Mechanics*, DOI 10.1007/978-3-540-89344-8_12,

which has no obvious or direct relation to things going on in physical space. One should expect the bridge from the physics to the mathematics to be a bold construction. But in fact it is not. It is quite the opposite, in many ways rather boring: the operator observables of quantum mechanics are book-keeping devices for effective wave function statistics. Let us explain that.

Consider an experiment $\mathscr{E}$ in which a system (in our usual notation x, m-dimensional) and a piece of apparatus (y, n-dimensional) with discrete pointer states Φ_α become entangled. Under an appropriate Schrödinger evolution, the pointer wave functions get entangled with certain wave functions φ_α of the system:

$$\varphi_\alpha(\mathbf{x})\Phi(\mathbf{y}) \overset{\text{Schrödinger evolution}}{\longrightarrow} \varphi_\alpha(\mathbf{x})\Phi_\alpha(\mathbf{y}) , \tag{12.1}$$

and by linearity we obtain for

$$\varphi = \sum_\alpha c_\alpha \varphi_\alpha$$

the result

$$\varphi(\mathbf{x})\Phi(\mathbf{y}) \overset{\text{Schrödinger evolution}}{\longrightarrow} \sum_\alpha c_\alpha \varphi_\alpha(\mathbf{x})\Phi_\alpha(\mathbf{y}) , \tag{12.2}$$

which means that the initial effective wave function $\varphi = \sum c_\alpha \varphi_\alpha$ changes with probability $|c_\beta|^2$ to the effective wave function φ_β. Why is this? Because by virtue of the quantum equilibrium distribution we have $\mathbf{Y} \in \operatorname{supp} \Phi_\beta$ with probability

$$\begin{aligned} &\int_{\{\mathbf{x},\mathbf{y}|\mathbf{y}\in\operatorname{supp}\Phi_\beta\}} \left|\sum_\alpha c_\alpha \varphi_\alpha(\mathbf{x})\Phi_\alpha(\mathbf{y})\right|^2 \mathrm{d}^m x \mathrm{d}^n y \\ &\qquad = |c_\beta|^2 \int |\varphi_\beta(\mathbf{x})|^2 \mathrm{d}^m x \int |\Phi_\beta(\mathbf{y})|^2 \mathrm{d}^n y = |c_\beta|^2 , \end{aligned} \tag{12.3}$$

where we have used $\operatorname{supp}\Phi_\alpha \cap \operatorname{supp}\Phi_\beta \approx \emptyset$ for $\alpha \neq \beta$ and the fact that the wave functions are normalized to unity. According to our definition of the effective wave function and our understanding of the fapp collapse from previous chapters, φ_β is the new effective wave function of the system, and the wave parts involving Φ_α with $\alpha \neq \beta$ can be ignored fapp forever.

The wave functions φ_α and probabilities $|c_\alpha|^2$ are associated with the experiment $\mathscr{E}$ given by (12.1–12.2), and we wish to handle both these aspects in a comfortable way. To identify the right way, we recall (7.11) and the ensuing discussion, which says that the Schrödinger evolution preserves the norm $\| \;\|^2$ (and hence the norm itself), which is the integrated modulus squared of the wave function (assumed normalized to unity). The relation (12.2) then implies that

$$
\begin{aligned}
\|\varphi\Phi\|^2 &:= \int |\varphi(\mathbf{x})\Phi(\mathbf{y})|^2 \, \mathrm{d}^m x \, \mathrm{d}^n y \\
&= \int \left| \sum_\alpha c_\alpha \varphi_\alpha(\mathbf{x}) \Phi(\mathbf{y}) \right|^2 \mathrm{d}^m x \, \mathrm{d}^n y \\
&= \int \left| \sum_\alpha c_\alpha \varphi_\alpha(\mathbf{x}) \Phi_\alpha(\mathbf{y}) \right|^2 \mathrm{d}^m x \, \mathrm{d}^n y \\
&= \sum_\alpha |c_\alpha|^2 \int |\varphi_\alpha(\mathbf{x})|^2 \int |\Phi_\alpha(\mathbf{y})|^2 \, \mathrm{d}^m x \, \mathrm{d}^n y \\
&= \sum_\alpha |c_\alpha|^2 \,.
\end{aligned}
\tag{12.4}
$$

Factoring out and integrating the right-hand side of the second equality, we obtain by comparison

$$
\sum_{\alpha \neq \beta} c_\alpha^* c_\beta \int \varphi_\alpha^*(\mathbf{x}) \varphi_\beta(\mathbf{x}) \, \mathrm{d}^m x = 0 \,.
$$

Since the c_α can be chosen arbitrarily, we obtain

$$
\int \varphi_\alpha^*(\mathbf{x}) \varphi_\beta(\mathbf{x}) \, \mathrm{d}^m x = 0 \,,
$$

for $\alpha \neq \beta$. This is reminiscent of the notion of orthogonality when we view

$$
\langle \varphi | \psi \rangle := \int \varphi^*(\mathbf{x}) \psi(\mathbf{x}) \, \mathrm{d}^m x \tag{12.5}
$$

as a scalar product on the space of square-integrable wave functions. Hence,

$$
\langle \varphi_\alpha | \varphi_\beta \rangle = \delta_{\alpha,\beta} = \begin{cases} 0 \text{ for } \alpha \neq \beta \,, \\ 1 \text{ for } \alpha = \beta \,. \end{cases} \tag{12.6}
$$

Why do we find orthogonality for the φ_α? The answer is of course that Φ_α and Φ_β are macroscopically disjoint pointer positions! In other words, to actually have an evolution like (12.1–12.2), the φ_α must be orthogonal.

Now we have the power of unitary geometry at our disposal. We can compute with the scalar product

$$
c_\alpha = \int \varphi_\alpha^*(\mathbf{x}) \varphi(\mathbf{x}) \, \mathrm{d}^m x := \langle \varphi_\alpha | \varphi \rangle
$$

as the orthogonal projection of φ onto φ_α. Let P_{φ_α} denote the orthogonal projector onto that direction

$$
P_{\varphi_\alpha} \varphi = \varphi_\alpha \langle \varphi_\alpha | \varphi \rangle \,,
$$

and assume for the moment that the φ_α form a basis, in the sense that we can expand any φ in the form

$$\varphi = \sum_\alpha P_{\varphi_\alpha} \varphi \, .$$

Hence we can associate the family of projectors $(P_{\varphi_\alpha})_\alpha$ with the following properties with the experiment (12.1–12.2):

$$P_{\varphi_\alpha} \varphi = \varphi_\alpha \langle \varphi_\alpha | \varphi \rangle \, , \tag{12.7a}$$

$$P_{\varphi_\alpha} P_{\varphi_\beta} = 0 \, , \quad \text{for } \alpha \neq \beta \, , \tag{12.7b}$$

$$P^2_{\varphi_\alpha} = P_{\varphi_\alpha} P_{\varphi_\alpha} = P_{\varphi_\alpha} \, , \tag{12.7c}$$

$$\sum_\alpha P_{\varphi_\alpha} = \mathsf{I} \qquad \text{(unit matrix)} \, . \tag{12.7d}$$

These properties characterize $(P_{\varphi_\alpha})_\alpha$ as a *family of orthogonal projectors*. We remark in passing that these projectors are self-adjoint, i.e., $\langle P\varphi, \psi \rangle = \langle \varphi, P\psi \rangle$, denoted in this book by $P^* = P$.

Suppose now that the pointer points to values (numbers displayed by the apparatus) $\{\lambda_1, \ldots, \lambda_N\}$, and suppose that $Y \in \operatorname{supp} \Phi_\beta$ means that the value λ_β is pointed at. The experiment is thus also characterized by the displayed values $\lambda_\alpha \in \Lambda = \{\lambda_1, \ldots, \lambda_N\}$. The quantum equilibrium statistics translate to the statistical distribution of the λ values, and one may want to know the average value and variance of the displayed λ values in the long run (repeating the experiment many times). The answer is encoded in one operator, namely,

$$\hat{A} = \sum \lambda_\alpha P_{\varphi_\alpha} \, , \tag{12.8}$$

the quantum Swiss army knife, containing all that we need – fapp.

To understand this better we need some mathematical facts, and these will all be detailed in the coming chapters on mathematics. First $\hat{A}$ inherits self-adjointness from the projectors. The relation between a self-adjoint operator and the family of projectors is one-to-one and called the spectral theorem. This is trivial from right to left in (12.8), but from left to right one needs some linear algebra, and in general one needs the infinite-dimensional version of linear algebra called functional analysis. The λ_α are eigenvalues and the projector P_{φ_α} can be defined as the characteristic function of $\hat{A}$ for the value λ_α:

$$P_{\varphi_\alpha} = \chi_{\{\lambda_\alpha\}}(\hat{A}) \, .$$

We can now express "everything of interest" in terms of $\hat{A}$. The probability for the value λ_α, computed in (12.3), can be expressed in various ways using (12.7a–d):

$$\begin{aligned}\mathbb{P}_\varphi(\lambda_\alpha) = |c_\alpha|^2 &= |\langle\varphi_\alpha|\varphi\rangle|^2 \\ &= \langle\varphi|\varphi_\alpha\rangle\langle\varphi_\alpha|\varphi\rangle \\ &= \langle\varphi, P_{\varphi_\alpha}\varphi\rangle = \langle\varphi|\chi_{\{\lambda_\alpha\}}(\hat{A})\varphi\rangle \\ &= \langle P_{\varphi_\alpha}\varphi, P_{\varphi_\alpha}\varphi\rangle \\ &= \|P_{\varphi_\alpha}\varphi\|^2 .\end{aligned} \tag{12.9}$$

The probability of the sure event is unity:

$$\begin{aligned}\mathbb{P}_\varphi(\Lambda) = \sum_\alpha \mathbb{P}_\varphi(\lambda_\alpha) &= \sum_\alpha \langle\varphi|P_{\varphi_\alpha}\varphi\rangle \\ &= \left\langle\varphi\middle|\sum_\alpha P_{\varphi_\alpha}\varphi\right\rangle = 1 .\end{aligned} \tag{12.10}$$

The mean value of the λ values is

$$\begin{aligned}\mathbb{E}_\varphi(\lambda) &= \sum_\alpha \lambda_\alpha \mathbb{P}_\varphi(\lambda_\alpha) \\ &= \sum_\alpha \lambda_\alpha \langle\varphi|P_{\varphi_\alpha}\varphi\rangle \\ &= \left\langle\varphi\middle|\sum_\alpha \lambda_\alpha P_{\varphi_\alpha}\varphi\right\rangle \\ &= \langle\varphi|\hat{A}\varphi\rangle ,\end{aligned} \tag{12.11}$$

and the variance of the λ values is

$$\begin{aligned}\mathbb{E}_\varphi(\lambda^2) &= \sum_\alpha \lambda_\alpha^2 \mathbb{P}_\varphi(\lambda_\alpha) \\ &= \left\langle\varphi\middle|\sum_\alpha \lambda_\alpha^2 P_{\varphi_\alpha}\varphi\right\rangle \\ &= \left\langle\varphi\middle|\sum_\alpha \lambda_\alpha P_{\varphi_\alpha}\sum_\beta \lambda_\beta P_{\varphi_\beta}\varphi\right\rangle \qquad \text{[by (12.7b)]} \\ &= \langle\varphi|\hat{A}^2\varphi\rangle .\end{aligned} \tag{12.12}$$

We emphasize the use of (12.7b) in the computation of the variance. It makes the book-keeping operator $\hat{A}$ in (12.8) technically powerful, as a result of its self-adjointness, if one treats the operator as a priori. From the way in which $\hat{A}$ arises, it is automatically self-adjoint, because the display is made up of real numbers. If someone put imaginary units in front of the numbers in the display, the values would be imaginary and the book-keeping operator would no longer be self-adjoint.

Let us collect together these results. We may associate with the experiment $\mathscr{E}$ a family of orthogonal projectors P_{φ_α} and values $\lambda_\alpha \in \{\lambda_1, \ldots, \lambda_N\}$, both encoded in $\hat{A}$ through (12.8). In short $\mathscr{E} \Longrightarrow \hat{A}$, where the association denoted by the arrow means that we can use $\hat{A}$ to compute the statistics of the values when the experiment is repeated many times. Now we may understand why the idea of an operator observable has become such a dominant notion in quantum physics. In the experiment, pointers point to numbers and a situation where pointers point to numbers has, of course, measurement appeal.

Now, turning things upside-down, any self-adjoint operator $\hat{A}$ (think of it as a Hermitian matrix right now) uniquely defines a family of projectors P_{φ_α}, namely the projectors onto its eigenvectors, and a set of values, namely its eigenvalues. Now call the values "measured values", or again "measurable values", and call the operator "measurable". The experiment $\mathscr{E}$ with which $\hat{A}$ is associated can be referred to as the "measurement of $\hat{A}$". So there you have it. Confusion is programmed, since each $\hat{A}$ is now an "observable" and hence has a life of its own. But are all self-adjoint operators observables? If not, then which ones are? Is there a "classical" hidden variable behind the observable whose value is really measured? And in this way, many irrelevant questions arise.

An example for (12.1–12.2) and its association with an operator $\hat{A}$ is provided by our discussion of spin. Suppose the spinor wave function is

$$\begin{pmatrix} \psi_1(\mathbf{x}) \\ \psi_2(\mathbf{x}) \end{pmatrix} ,$$

in the eigenbasis of σ_z, and suppose the Stern–Gerlach magnet is oriented in the $\mathbf{a}$-direction. Then the wave function will split into two wave packets ϕ_+ and ϕ_-, the eigenfunctions of $\mathbf{a}\cdot\sigma$, as the example shows. The particle will be in the $\pm$ packet with probability

$$\int \mathrm{d}^3x \left| \phi_\pm \cdot \begin{pmatrix} \psi_1(\mathbf{x}) \\ \psi_2(\mathbf{x}) \end{pmatrix} \right|^2 = \left\| P^{\mathbf{a}}_\pm \begin{pmatrix} \psi_1 \\ \psi_2 \end{pmatrix} \right\|^2 ,$$

where $P^{\mathbf{a}}_\pm$ denotes the projector onto the corresponding spinor component. The associated operator is simply

$$\hat{A}_{\mathbf{a}} = +\frac{1}{2}P^{\mathbf{a}}_+ - \frac{1}{2}P^{\mathbf{a}}_- = \frac{1}{2}\mathbf{a}\cdot\sigma , \tag{12.13}$$

the spin operator in the direction of $\mathbf{a}$, $\|a\| = 1$.

This example reveals an interesting feature, namely that we did not have to talk about measurement apparatus in order to get the orthogonality of the possible effective wave functions. The Schrödinger evolution of the particle through the Stern–Gerlach magnet has already taken care of that. In the end one only needs to detect which of the wave packets the particle is riding along with. This adds something to the false intuition that there is a genuine quantity behind the observable "which is measured".

The moral of the foregoing is that one can associate operators with experiments as book-keeping devices for the statistics, where an "orthogonal" splitting of the wave function often comes for free. The apparatus is not even needed for that. In the end, the apparatus detects where the particle sits.

Let us summarize the mathematical structures that emerge from the statistics. The space of wave functions will be a linear space (reflecting the linearity of the Schrödinger evolution!) with a scalar product. A Hilbert space $\mathscr{H}$ will therefore be the most convenient setting. The space will in general be infinite-dimensional, as it contains the set of wave functions on configuration space. The set of values Λ which the pointer points at has so far been chosen as a discrete set (but that will change in the next section). For simplicity, we considered above the situation where one eigenvector φ_α corresponds to each value λ_α. In general this will not be so, because in general the orthogonal projector corresponding to a value λ, which we denote by P_λ, will not be one-dimensional, but rather will project onto a higher-, even infinite-dimensional subspace $\mathscr{H}_\lambda$ which contains all wave functions correlating appropriately with the pointer position, i.e., which lead to the displayed value λ. The $(P_\lambda)_{\lambda\in\Lambda}$ form a family of orthogonal (and hence self-adjoint) projections.

The statistical outcome of the experiment $\mathscr{E}$ can be encoded in the operator

$$\hat{A} = \sum_{\lambda\in\Lambda} \lambda P_\lambda \,, \qquad \text{with} \quad \sum_{\lambda\in\Lambda} P_\lambda = \mathrm{I} \,, \tag{12.14}$$

or in short

$$\mathscr{E} \quad \Longrightarrow \quad (\lambda, P_\lambda)_{\lambda\in\Lambda} \quad \Longleftrightarrow \quad \hat{A} = \sum_{\lambda\in\Lambda} \lambda P_\lambda \,. \tag{12.15}$$

We shall focus below on the first arrow. The second arrow is mathematics, the subject of the so-called spectral representation of a self-adjoint operator. The backwards arrow is mathematically somewhat difficult when the operator $\hat{A}$ is unbounded and has a continuous spectrum. Such operators will be discussed next.

12.2 Who Is Afraid of PVMs and POVMs?

We showed in the previous discussion of the experiment (12.1–12.2), where pointers display values, that the quantum equilibrium distribution for the probabilities of the values may be conveniently computed from a family of orthogonal projectors. The orthogonality of the projector arises in general from the orthogonality of pointer wave functions, since they have disjoint supports in configuration space. But the spin example shows that orthogonality may sometimes come without invoking pointers. The apparatus merely detects the position of the particle. In fact, from a Bohmian perspective, pointers are not needed to create facts or values. The Bohmian particle has a position, and one only needs to detect where it is. So sometimes the apparatus does not need to be mentioned at all, if all it does is to detect the actual state of affairs.

From a Bohmian perspective we should therefore formulate the experimental situation in much more general terms. A system (configuration x) which may get coupled to some apparatus (configuration y) defines (together with the apparatus, if the apparatus plays some role) the configuration space $\mathscr{Q}$. On the latter, one has a coarse-graining function $F : \mathscr{Q} \to \Lambda$ which maps configurations to the displayed values (e.g., pointer positions) in Λ. Quantum equilibrium determines the statistics of the values. Therefore the most general formalism encoding quantum equilibrium and the statistics of measurement experiments emerges from the sequence

$$\begin{array}{cl} \varphi(\mathbf{x}) & \\ \downarrow & \text{system couples (possibly) to apparatus} \\ \Psi(\mathbf{x},\mathbf{y}) = \varphi(\mathbf{x})\Phi(\mathbf{y}) & \\ \downarrow & \text{Schrödinger evolution} \\ \Psi_T(\mathbf{x},\mathbf{y}) & \qquad (12.16a) \\ \Downarrow & \text{quantum equilibrium distribution} \\ \rho^{\Psi_T} = |\Psi_T(\mathbf{x},\mathbf{y})|^2 & \qquad (12.16b) \\ \downarrow & \text{and we are only interested in} \\ \mathbb{P}_\varphi(A) := \mathbb{P}^{\Psi_T}\left(F^{-1}(A)\right) , & A \subset \Lambda , \qquad (12.16c) \end{array}$$

where single arrows denote linear maps and the double arrow denotes a *bilinear* or, more correctly, a sesquilinear (because of the complex conjugation of one of the factors) map. The "possibly" on the first arrow expresses the possibility that there need not be any coupling to a piece of apparatus, so that we focus only on the system configuration. As always, T is the duration of the experiment.

Before we become more abstract, let us discuss a few examples. In (12.1–12.2), we have

$$F(\mathbf{x},\mathbf{y}) = F(\mathbf{y}) \in \{\lambda_1,\ldots,\lambda_N\} = \Lambda ,$$

and in view of (12.9)

$$\mathbb{P}^{\Psi_T}\left(F^{-1}(\{\lambda_i\})\right) = \mathbb{P}_\varphi(\{\lambda_i\}) . \qquad (12.17)$$

In the Stern–Gerlach experiment,

$$F(\mathbf{x},\mathbf{y}) = F(\mathbf{x}) \in \left\{-\frac{1}{2},\frac{1}{2}\right\} .$$

The most direct and simple example, however, is the position of a particle $\mathbf{x} \in \mathbb{R}^3$ with effective wave function φ, without apparatus, and with $T = 0$, i.e., $\Psi_T = \varphi$. Taking $F = \mathrm{id}$ on $\Lambda = \mathbb{R}^3$, the sequences reduce to

$$\varphi(\mathbf{x}) \quad \Longrightarrow \quad |\varphi(\mathbf{x})|^2 , \qquad \text{or} \quad \mathbb{P}_\varphi(\mathrm{d}^3x) = |\varphi(\mathbf{x})|^2\mathrm{d}^3x . \qquad (12.18)$$

We would like now to shift the emphasis from the meaning of the family of projectors introduced in the previous section towards the idea of a *projection-valued measure* (PVM), a notion which comes naturally with (12.16a–c) and which combines discrete values with continuous values, as in (12.18). From the sequence, we read the bilinear map from "wave function space" to measures. In the case of (12.1–12.2), the measure is a discrete projection-valued measure (PVM) on subsets of the value space Λ. Using the rules of probability and the orthogonality of the projectors corresponding to different values, (12.17) implies

$$\begin{aligned}
\mathbb{P}_\varphi(\{\lambda_{\alpha_1},\dots,\lambda_{\alpha_n}\}) &= \mathbb{P}^{\Psi_T}\left(F^{-1}(\{\lambda_{\alpha_1},\dots,\lambda_{\alpha_n}\})\right)\\
&= \mathbb{P}^{\Psi_T}\Big(\{(\mathbf{x},\mathbf{y})\big|F(\mathbf{x},\mathbf{y})\in\{\lambda_{\alpha_1},\dots,\lambda_{\alpha_n}\}\}\Big)\\
&= \sum_i \mathbb{P}^{\Psi_T}\left(F^{-1}(\{\lambda_{\alpha_i}\})\right) = \sum_i \mathbb{P}^{\varphi}(\{\lambda_{\alpha_i}\})\\
&= \sum_i \|P_{\lambda_{\alpha_i}}\varphi\|^2 = \sum_i \langle\varphi|P_{\lambda_{\alpha_i}}\varphi\rangle\\
&= \Big\langle\varphi|\sum_i P_{\lambda_{\alpha_i}}\varphi\Big\rangle =: \langle\varphi|P_{\{\lambda_{\alpha_1},\dots,\lambda_{\alpha_n}\}}\varphi\rangle\,.
\end{aligned} \tag{12.19}$$

Hence to every subset of values $\{\lambda_{\alpha_1},\dots,\lambda_{\alpha_n}\}$ corresponds a projector

$$P_{\{\lambda_{\alpha_1},\dots,\lambda_{\alpha_n}\}} = \sum_i P_{\lambda_{\alpha_i}}\,,$$

indexed by that set. Hence we can view this as a family $(P_A)_{A\subset\Lambda}$ of projectors which acts like a (discrete) measure on the subsets of Λ : a PVM.

Since we are used to thinking of a measure on subsets of a continuum, the abstraction to the PVM structure becomes evident if the value set is continuous. So let us look at the position of the Bohmian particle (12.18), where the value space is now the continuum $\Lambda = \mathbb{R}^3$. In this case $F = \mathrm{id}$ and

$$\begin{aligned}
\mathbb{P}_\varphi(A) = \mathbb{P}^\varphi\left(F^{-1}(A)\right) &= \int \chi_A|\varphi|^2\mathrm{d}^3x\\
&= \langle\varphi|\chi_A\varphi\rangle\\
&= \int_A \langle\varphi|\chi_{\{\mathrm{d}^3x\}}\varphi\rangle\,.
\end{aligned} \tag{12.20}$$

What replaces the projectors P_λ? Obviously, $\chi_{\{\mathrm{d}^3x\}}$ takes on the role of P_λ. This suggests defining the continuous PVM O of (measurable) subsets of $\Lambda\,(=\mathbb{R}^3)$ taking values in the space of orthogonal (self-adjoint) projectors:

$$O_A : \varphi \longmapsto \chi_A\varphi(\mathbf{x}) = \begin{cases} \varphi(\mathbf{x})\,, & \mathbf{x}\in A\subset\mathbb{R}^3\,,\\ 0\,, & \text{otherwise}\,,\end{cases} \tag{12.21}$$

so that

$$\mathbb{P}_{\varphi}(A) = \langle \varphi | O_A \varphi \rangle = \int_A |\varphi(\mathbf{x})|^2 \mathrm{d}^3 x \tag{12.22}$$

comes out just right. The orthogonality

$$O_A O_B \varphi(\mathbf{x}) = \chi_A(\mathbf{x}) \chi_B(\mathbf{x}) \varphi(\mathbf{x}) = \chi_{A \cap B}(\mathbf{x}) \varphi(\mathbf{x}) = 0\,, \tag{12.23}$$

for $A \cap B = \emptyset$ is obvious, and so is the projector property

$$O_A^2 = O_A\,, \tag{12.24}$$

and the normalization

$$O_{\mathbb{R}^3} = \mathsf{I}\,, \tag{12.25}$$

the identity operator.

In Dirac's shorthand notation, we write

$$\langle \mathbf{x} | \psi \rangle = \psi(\mathbf{x}) \tag{12.26}$$

and

$$\mathrm{d}O_{\mathbf{x}} = |\mathbf{x}\rangle \langle \mathbf{x}| \mathrm{d}^3 x\,, \tag{12.27}$$

with

$$\langle \mathbf{x} | \mathbf{x}' \rangle = \delta(\mathbf{x} - \mathbf{x}') \tag{12.28}$$

representing the orthogonality of the projectors, while

$$\int |\mathbf{x}\rangle \langle \mathbf{x}| \mathrm{d}^3 x = \mathsf{I} \tag{12.29}$$

represents the normalization, often referred to as the completeness relation. We put

$$\hat{X} := \int \mathbf{x} \mathrm{d}O_{\mathbf{x}} =: \int \mathbf{x} |\mathbf{x}\rangle \langle \mathbf{x}| \mathrm{d}^3 x\,. \tag{12.30}$$

Therefore with the measurement of position (if one cares to detect where the particle is) one can associate a hatted position $\hat{X}$, an observable which encodes all the statistics. As in (12.12), we can compute the variance of the position distribution, viz.,

$$\mathbb{E}^{\varphi}(\mathbf{X}^2) - \mathbb{E}^{\varphi}(\mathbf{X})^2\,,$$

by virtue of (12.28), whence

$$\begin{aligned}
\mathbb{E}^{\varphi}(\mathbf{X}^2) - \mathbb{E}^{\varphi}(\mathbf{X})^2 &= \int \mathbf{x}^2 |\varphi(\mathbf{x})|^2 \mathrm{d}^3x - \left(\int \mathbf{x} |\varphi(\mathbf{x})|^2 \mathrm{d}^3x \right)^2 \\
&= \int \mathbf{x}^2 \langle \varphi | \mathrm{d}O_{\mathbf{x}} \varphi \rangle - \left(\int \mathbf{x} \langle \varphi | \mathrm{d}O_{\mathbf{x}} \varphi \rangle \right)^2 \\
&= \langle \varphi | \hat{X}^2 \varphi \rangle - \langle \varphi | \hat{X} \varphi \rangle^2 ,
\end{aligned} \tag{12.31}$$

using (12.28) to obtain the last equality.

A related and rather important example dealing with a continuum of values is provided by the asymptotic velocity and its distribution, which we calculated in (9.26):

$$\begin{aligned}
\mathbb{E}^{\varphi}(f(V_\infty)) &\approx \int f\left(\frac{\mathbf{x}}{t}\right) \left| \widehat{\varphi}\left(\frac{m}{\hbar}\frac{\mathbf{x}}{t}\right) \right|^2 \left(\frac{m}{\hbar t}\right)^3 \mathrm{d}^3x \\
&= \int f(\mathbf{v}) \left| \widehat{\varphi}\left(\frac{m}{\hbar}\mathbf{v}\right) \right|^2 \left(\frac{m}{\hbar}\right)^3 \mathrm{d}^3v \\
&= \int f\left(\frac{\hbar}{m}\mathbf{k}\right) |\widehat{\varphi}(\mathbf{k})|^2 \mathrm{d}^3k .
\end{aligned} \tag{12.32}$$

Taking $f = \chi_A$, we can rephrase this in the sense of our sequence (12.16a–c), choosing $T = t$, a time-dependent coarse-graining function

$$F_t(\mathbf{x}) = \frac{\mathbf{x}}{t} ,$$

and no apparatus, i.e., $\Psi^t = \varphi_t$. We then obtain

$$\begin{aligned}
\mathbb{E}^{\varphi}(\chi_A(V_\infty)) &= \lim_{t\to\infty} \mathbb{P}^{\varphi_t}(F^{-1}(A)) = \int \chi_A\left(\frac{\hbar}{m}\mathbf{k}\right) |\widehat{\varphi}(\mathbf{k})|^2 \mathrm{d}^3k \\
&= \langle \widehat{\varphi} | \chi_{mA/\hbar} \widehat{\varphi} \rangle = \langle \varphi | \mathscr{F}^{-1} \chi_{mA/\hbar} \mathscr{F} \varphi \rangle \\
&=: \langle \varphi | V_A \varphi \rangle ,
\end{aligned} \tag{12.33}$$

where V is also a PVM, since $\mathscr{F}$ acts isometrically on the Hilbert space of square-integrable functions (see Sect. 13.1.2).

Let us rephrase this in a less prosaic way. Let us recall what a measurement experiment for the asymptotic velocity might look like. Prepare a particle with a wave packet around here, let it evolve freely for quite some time, and catch the particle on a screen or with some other appropriate detector far away. Then take the ratio of the distance traveled to the time taken, and run the experiments many times to get the statistics. That is given by $|\widehat{\varphi}|^2$, as we computed from the quantum equilibrium distribution in Chap. 9, after (9.25).

To recall the simple explanation for that, let us remember what happens during the free evolution. The wave packet φ will spread according to the dispersion relation. Why is this? It is because the packet φ is composed of "plane wave packets" with wave numbers $\mathbf{k}$, and each such plane wave moves with velocity $\hbar\mathbf{k}/m$. That is the meaning of the dispersion relation. Hence the plane wave packets separate and after a long time occupy separate regions in configuration space. That is, they become orthogonal. Therefore the orthogonal plane wave parts $\mathrm{e}^{\mathrm{i}\mathbf{k}\cdot\mathbf{x}}$ will thus replace the φ_α in the discrete examples.

Dirac invented the notation $|\mathbf{k}\rangle\langle\mathbf{k}|$ [already used in (12.26)] for the projector onto the plane wave $\langle\mathbf{x}|\mathbf{k}\rangle = \mathrm{e}^{\mathrm{i}\mathbf{k}\cdot\mathbf{x}}$, meaning that $\langle\mathbf{k}|\varphi\rangle = \widehat{\varphi}(\mathbf{k})$, the Fourier transform of φ. As for position, a subset of $\Lambda = \mathbb{R}^3$ gets mapped to a projector. In fact, $A \subset \mathbb{R}^3$ is mapped to the orthogonal projection given by

$$V_A = \int \chi_A\left(\frac{\hbar}{m}\mathbf{k}\right)|\mathbf{k}\rangle\langle\mathbf{k}|\mathrm{d}^3k = \int_{mA/\hbar}|\mathbf{k}\rangle\langle\mathbf{k}|\mathrm{d}^3k\,. \tag{12.34}$$

The probability for the asymptotic velocity to lie within A when the wave function is φ is

$$\mathbb{P}^\varphi(V_\infty \in A) = \langle\varphi|V_A\varphi\rangle = \int_{mA/\hbar}\langle\varphi|\mathbf{k}\rangle\langle\mathbf{k}|\varphi\rangle\mathrm{d}^3k = \int_{mA/\hbar}|\langle\varphi|\mathbf{k}\rangle|^2\mathrm{d}^3k\,.$$

When $A = \mathbb{R}^3$, we have $V_{\mathbb{R}^3} = \mathsf{I}$, the identity operator. In terms of the sandwich with a wave packet, this is simply Plancherel's identity

$$\langle\varphi|V_{\mathbb{R}^3}\varphi\rangle = \int_{\mathbb{R}^3}|\langle\varphi|\mathbf{k}\rangle|^2\mathrm{d}^3k = \int_{\mathbb{R}^3}|\varphi(\mathbf{x})|^2\mathrm{d}^3x = 1\,.$$

The orthogonality is again captured by

$$\langle\mathbf{k}|\mathbf{k}'\rangle = \delta(\mathbf{k}-\mathbf{k}')\,,$$

and one has $V_{\mathbb{R}^3} = \mathsf{I}$, the so-called completeness relation:

$$\int_{\mathbb{R}^3}|\mathbf{k}\rangle\langle\mathbf{k}|\mathrm{d}^3\mathbf{k} = \mathsf{I}\,.$$

We find that the asymptotic velocity can also be associated with a hatted observable, namely

$$\hat{V}_\infty := \int\frac{\hbar}{m}\mathbf{k}|\mathbf{k}\rangle\langle\mathbf{k}|\mathrm{d}^3k\,. \tag{12.35}$$

Multiplying this by m, the mass of the particle, we get a hatted momentum, usually called the momentum operator.

Mathematically, this is nothing but a kind of diagonal representation. Indeed, a PVM is (in functional analytical terms) the spectral resolution of a self-adjoint operator. There is a one-to-one correspondence between PVMs $\mathrm{d}P_\lambda$ and self-adjoint operators $\hat{A}$. The former defines a self-adjoint operator $\hat{A} = \int\lambda\mathrm{d}P_\lambda$ and the latter defines a projection-valued measure via its spectral representation. All this, and all the examples above, will be put on a mathematically rigorous basis in Chap. 15.

Let us make a historical remark. The move from classical variables to operator observables was Heisenberg's invention to explain the discreteness of atomic spectra. But he did so without Schrödinger's equation and quantum equilibrium. He postulated that this abstract kind of observable is the new measurable quantity. Many physicists hoped that some "ordinary" quantities would underlie these abstract observables, and that whilst these quantities remained hidden, they could be held responsible for the outcome in a measurement.

Bohr on the other hand held the clear and correct view that, in an experiment which "measures an observable", nothing is actually being measured. We can understand why that is correct. The role of the observables is just a book-keeping role for the statistics of an experiment. More generally, we may rewrite (12.15) using the notion of PVM, which can be a discrete measure as in (12.15) or a continuous measure like Lebesgue measure, to highlight once again the association of an operator with the experiment (12.16a):

$$\mathscr{E} \quad \Longrightarrow \quad (\Lambda, \mathrm{d}P_\lambda) \quad \Longleftrightarrow \quad \hat{A} = \int_\Lambda \lambda \mathrm{d}P_\lambda \,. \tag{12.36}$$

But sometimes something is measured: the Bohmian positions.

Note in passing that the position observable and the momentum observable are non-commuting! This is easily seen, since $|\mathbf{k}\rangle\langle\mathbf{k}|$ does not commute with $|\mathbf{x}\rangle\langle\mathbf{x}|$. But this is no world-shattering discovery, for there is absolutely no reason why it should be the case. And that the variance in a position measurement of a particle with effective wave function φ is roughly inverse to the variance of the asymptotic velocity of the particle is also rather unexciting. Or better, it is trivially so in quantum equilibrium, once one understands the dispersion of waves. In short, Heisenberg's uncertainty relation is wholly unsurprising once quantum equilibrium has been understood.

Remark 12.1. On the Dirac Formalism
In the Dirac formalism one uses bra $\langle\cdot|$ and ket $|\cdot\rangle$ symbols to denote dual vectors and vectors, as in (12.5). But the symbolism becomes technically powerful when the same symbols are used in the PVMs, as in (12.27). One must be aware, however, that $|\varphi\rangle$ denotes a vector in Hilbert space which has a finite length $\sqrt{\langle\varphi|\varphi\rangle}$, and that $|\mathbf{x}\rangle$ has no such meaning since $\langle\mathbf{x}|\mathbf{x}'\rangle = \delta(\mathbf{x}-\mathbf{x}')$. But then, once one has that clear in one's mind, it does not hurt to think heuristically of $|\mathbf{x}\rangle$ as a wave packet which is "highly" peaked at $\mathbf{x}$. ■

Now we come to POVMs. What is the general abstract structure emerging from (12.16a–c)? At the end of (12.16c) stands a positive measure (a probability measure) on Λ, and that is built from a bilinear map acting on wave functions. In other words, wave functions are mapped (sesquilinearly) to positive measures. Without the wave function sandwich, the measure is operator-valued. Nothing says that the operators must be projectors, but they must be positive operators. A positive operator is one for which the sandwich with a wave function is always a positive number. We denote the POVM simply by $\mathrm{d}P_\lambda$, as a measure on Λ. PVMs are special cases of POVMs.

Thus the most general structure one can infer from the experiment (12.16a) is

$$\mathscr{E} \Longrightarrow (\Lambda, \mathrm{d}P_\lambda) , \tag{12.37}$$

without the further double arrow to a book-keeping operator. We shall see from the following example why there is no point in introducing a book-keeping operator when the POVM is not a PVM. Imagine a detection of position. Realistically, such a detection will come with a measurement error. How does quantum mechanics handle that with the position observable?

We know how to deal with this trivially in Bohmian mechanics. Let $p(\mathbf{x})$ be the probability density on $\mathbb{R}^3$ which describes the error due to the apparatus. The measured position $\tilde{\mathbf{X}}$ is therefore the sum of two random variables $\mathbf{X}+\mathbf{Y}$, with $\mathbf{X}$ distributed according to $|\varphi|^2$ and $\mathbf{Y}$ distributed according to p. It is reasonable to assume that $\mathbf{X}$ and $\mathbf{Y}$ are independent, which implies that the distribution of $\tilde{\mathbf{X}}$ is the convolution[1]

$$\tilde{\rho}(\mathbf{x}) = \int p(\mathbf{x}-\mathbf{y})|\varphi(\mathbf{y})|^2 \mathrm{d}^n y .$$

The probability is therefore

$$\begin{aligned}
\mathbb{P}^\varphi(\tilde{\mathbf{X}} \in A) &= \int_A \tilde{\rho}(\mathbf{x})\,\mathrm{d}^n x = \int_A \int p(\mathbf{x}-\mathbf{y})|\varphi(\mathbf{y})|^2 \mathrm{d}^n y \,\mathrm{d}^n x \\
&= \int \left[\int \chi_A(\mathbf{x}) p(\mathbf{x}-\mathbf{y})\,\mathrm{d}^n x\right] |\varphi(\mathbf{y})|^2 \mathrm{d}^n y \\
&=: \langle \varphi | \tilde{O}_A \varphi \rangle ,
\end{aligned} \tag{12.38}$$

where we have introduced the POVM

$$\tilde{O}_A = \int p(\mathbf{x}-\mathbf{y}) \chi_A(\mathbf{x})\,\mathrm{d}^n x ,$$

$$\tilde{O}_A : \varphi \longmapsto \int_A p(\mathbf{x}-\mathbf{y})\,\mathrm{d}^n x \; \varphi(\mathbf{y}) . \tag{12.39}$$

In general,

$$\tilde{O}_A^2 \neq \tilde{O}_A ,$$

where equality holds only if $p(\mathbf{x}) = \delta(\mathbf{x})$, i.e., when the POVM is a PVM.

---

[1] Consider the Fourier transform

$$\hat{\tilde{\rho}} = \mathbb{E}\left(\mathrm{e}^{\mathrm{i}\lambda\cdot\tilde{\mathbf{X}}}\right) = \mathbb{E}\left(\mathrm{e}^{\mathrm{i}\lambda\cdot(\mathbf{X}+\mathbf{Y})}\right) = \mathbb{E}\left(\mathrm{e}^{\mathrm{i}\lambda\cdot\mathbf{X}}\right)\mathbb{E}\left(\mathrm{e}^{\mathrm{i}\lambda\cdot\mathbf{Y}}\right) = \widehat{|\varphi|^2}\hat{p} ,$$

where we use the independence of X and Y to obtain the third equality. Now recall that the Fourier transform of a product is a convolution.

Suppose we want to know the variance of $\tilde{\mathbf{X}}$ and compare that with (12.31). Then computing the second moment, we obtain

$$\begin{aligned}\mathbb{E}(\tilde{\mathbf{X}}^2) &= \int \mathbf{x}^2 \tilde{\rho}(\mathbf{x})\,\mathrm{d}^n x = \int \mathbf{x}^2 \langle \varphi | \mathrm{d}\tilde{O}_{\mathbf{x}} \varphi \rangle \\ &= \iint \mathbf{x}^2 p(\mathbf{x}-\mathbf{y}) |\varphi(\mathbf{y})|^2 \,\mathrm{d}^n y\,\mathrm{d}^n x\,, \end{aligned} \tag{12.40}$$

and that is all there is to it. Compare this with (12.31) to understand why the introduction of an operator observable serves no purpose.

12.2.1 The Theory Decides What Is Measurable

It has been said that a theory must only be about measurable quantities. But that is not a very intelligent thing to say. How does one know beforehand, before the theory is built, what is measurable and what is not? If the theory does not contain an electric field, the electric field is not measurable. If it contains an electric field then obviously *the theory will have to tell us whether and how the electric field is measurable.* It is the theory which reveals the world to us in notions that are particular to the theory, and which make known to us the correct elements of the world, if the theory has the appeal of beauty and elegance. Recall the discussion on Wheeler–Feynman electromagnetism. In this theory, there is no electromagnetic field, so nothing of that kind could ever be measured. In Maxwell–Lorentz theory, there is. It is true that in these examples the variables entering the theories turn out to be measurable, but measurability was not the key consideration when constructing the theories. It turned out that way, accidentally as it were. But it need not be the case. Here is an example.

Bohmian mechanics contains variables which are not measurable in the sense of the experiment (12.16a). There is no apparatus measuring wave functions and no apparatus measuring the Bohmian velocity. The fact that there is no apparatus measuring wave functions has already been said, and the fact that there is no apparatus measuring the Bohmian velocity (and the trajectory) is of course related to this. But beware! Any Newtonian motion, i.e., classical motion which we see and measure, for example an apple falling from a tree or an electron in a cloud chamber, is a Bohmian motion, albeit one where the particle moves with the classically moving localized wave packet. Some care is therefore in order with such statements, and it is important to be clear about what is meant.

The question intended is as follows: In a situation where quantum mechanical interference acts, can one measure the actual velocity of the particle? Here is a quick and very general argument based on POVMs which answers in the negative. The distribution of the values of the velocity must be given by the sesquilinear form (quantum equilibrium distribution)

$$\mathbb{P}^{\psi}(A) = \text{sesquilinear form}\,(\psi)(A)\,,$$

on the subsets $A \subset \mathbb{R}^3$. For sesquilinear forms, we have the binomial formula estimate

$$\mathbb{P}^{\psi_1+\alpha\psi_2}(A) \leq 2\mathbb{P}^{\psi_1}(A) + 2\mathbb{P}^{\psi_2}(A) , \tag{12.41}$$

where $|\alpha| = 1$. Now take two real wave functions ψ_1, ψ_2, and α a complex phase factor. Then $\psi_1 + \alpha\psi_2$ will generate some velocity field that is not everywhere zero, while on the right-hand side the probabilities are concentrated on zero. So for a set A not containing 0, we get a contradiction since the right-hand side is zero.

What should we conclude? That Bohmian mechanics is not a physical theory, just because of the rather foolish and unsubstantiated claim that a theory must only be about measurable quantities. Of course, we could not conclude in that way. Furthermore, one should not go too far with all this formalism. Suffice it to say that we sometimes know the velocity even without measuring it.[2] In the hydrogen ground state, the electron is at rest. Measuring its position we know where it was and what velocity it had. And if the reader now points out that we only know the velocity because the theory tells us what it is, then she or he has understood. It is always the theory that tells us what exists, whether what exists is measurable, and if it is measurable, how to measure it. The reader should revisit the double slit experiment in Chap. 8 in order to understand that this provides another example of the maxim that *it is the theory that decides what is measurable.*

12.2.2 Joint Probabilities

We shall shortly discuss a sequence of measurements, performed one after the other, and we do so in the case of a discrete PVM. Suppose we have two pieces of measurement apparatus with pointers Φ_α, $\alpha \in I$, and Ψ_β, $\beta \in J$, which point to values $\lambda_\alpha \in \Lambda$ and $\mu_\beta \in \Pi$. How is this described? As always, of course. We simply have a slightly more complicated Schrödinger evolution (12.2), with the only difference that it ends in

$$\sum_{\alpha\in I,\beta\in J} \varphi_{\alpha,\beta}\Phi_\alpha\Psi_\beta , \tag{12.42}$$

where we keep the new effective wave functions $\varphi_{\alpha,\beta} := P_\beta P_\alpha \varphi$ non-normalized for reasons of notational simplicity. As a consequence, no $c_{\alpha,\beta}$ appear.

Repeating the computation (12.3), measuring first with the α apparatus, we see that the values $\lambda_\alpha, \mu_\beta$ come with "joint probability"

$$\mathbb{P}_\varphi(\lambda_\alpha, \mu_\beta) = \|P_\beta P_\alpha \varphi\|^2 . \tag{12.43}$$

[2] One can, however, measure the Bohmian velocity in a so-called weak measurement [2].

This is all very simple, but there is something malicious going on, as indicated by the quotation marks on "joint probability". Let us now bring this to light.

When we wish to interpret the left-hand side as the probability distribution of two random variables, it should be the case that if we ignore the outcome of one of the measurements, i.e., if we sum over all values of one variable, then we should get the probability of the other variable:

$$\sum_{\alpha \in I} \mathbb{P}_\varphi(\lambda_\alpha, \mu_\beta) = \mathbb{P}_\varphi(\mu_\beta)\,, \qquad \sum_{\beta \in J} \mathbb{P}_\varphi(\lambda_\alpha, \mu_\beta) = \mathbb{P}_\varphi(\lambda_\alpha)\,. \tag{12.44}$$

This is pure logic applied to the joint probability.[3] Summing over all values of one random variable amounts to ignoring it, and we are only left with the probability distribution of the other random variable. But is this true? Well, it is true if the following computation can be carried out:

$$\begin{aligned} \|P_\beta P_\alpha \varphi\|^2 = \langle P_\beta P_\alpha \varphi, P_\beta P_\alpha \varphi \rangle &= \langle \varphi, (P_\beta P_\alpha)^* P_\beta P_\alpha \varphi \rangle \\ &= \langle \varphi, {P_\alpha}^* {P_\beta}^* P_\beta P_\alpha \varphi \rangle \\ &= \langle \varphi, P_\alpha P_\beta P_\beta P_\alpha \varphi \rangle \\ &= \langle \varphi, P_\alpha P_\beta P_\alpha \varphi \rangle \\ &= \langle \varphi, P_\beta P_\alpha P_\alpha \varphi \rangle = \langle \varphi, P_\beta P_\alpha \varphi \rangle \,. \end{aligned}$$

This computation uses self-adjointness of the projectors $P^* = P$, the projector property $P^2 = P$, *and* the commutativity

$$[P_\alpha, P_\beta] := P_\alpha P_\beta - P_\beta P_\alpha = 0\,. \tag{12.45}$$

Now we can sum over either of the values in (12.44), using (12.7a). But we must note that the requirement (12.45) is very special and by no means natural. Think of the position and asymptotic velocity PVMs. We have already remarked that they do not commute.

For a discrete example, we may take the "spin measurement" in some direction $\mathbf{a}$, with which we may associate the operator $\hat{A}_\mathbf{a}$. One easily checks that $[\hat{A}_\mathbf{x}, \hat{A}_\mathbf{y}] \neq 0$, whence their PVMs do not commute. We note, however, that the spin observables discussed in the EPR–Bohm setup $\mathbf{a} \cdot \sigma^{(1)} \times \mathsf{I}$ and $\mathsf{I} \times \mathbf{b} \cdot \sigma^{(2)}$ do commute, a fact which is basic to the insight that, due to quantum equilibrium, Bohmian nonlocality cannot be used for faster than light signalling. If the PVMs do not commute, then (12.43) is all there is, and

$$\mathbb{P}_\varphi(\lambda_\alpha, \mu_\beta) = \|P_\beta P_\alpha \varphi\|^2 \neq \langle \varphi, P_\beta P_\alpha \varphi \rangle \quad \text{in general}\,. \tag{12.46}$$

Summing now over α, (12.44) does not hold in general.

---

[3] For more than two entries, one speaks of a consistent family of joint distributions, where the $(n-k)$-point distribution (with $n-k$ entries) of the family is given by the sum over k entries of the n-point distribution of the family. For two random outcomes this is (12.44).

Suppose the observables corresponding to the PVMs are $\hat{A}$, $\hat{B}$, and suppose we first "measure" (the reader hopefully understands by now what the notion "measuring an observable" means) $\hat{A} = \sum_\alpha \lambda_\alpha P_\alpha$ and then $\hat{B} = \sum_\beta \mu_\beta P_\beta$. Recall now a basic fact of linear algebra, where one learns that two Hermitian matrices have a common eigenbasis if and only if they commute. The same goes here, i.e., $\hat{A}$ and $\hat{B}$ commute if and only if their PVMs commute. If $\hat{A}$ and $\hat{B}$ do not commute, the sequential measurement of $\hat{A}$, $\hat{B}$ cannot be described by a PVM and thus not by an observable. But it is a measurement! A measurement which is not a measurement of an observable? Disaster? Of course not. It is described by a POVM.

The moral is this. In general, i.e., in the case of non-commuting observables, the probability formula (12.43) does not define a joint probability. Only commuting observables have joint probabilities. Is there anything deep in all this? Well, no, there is not. Any measurement experiment (12.2) channels the system's wave function into orthogonal pieces, thereby leading to a new effective wave function that will be the input for the next experiment. Then summing over the possible values of the first experiment does not undo the physical change which the systems underwent during that experiment.

12.2.3 Naive Realism about Operators

The notion of observable and the wording "measurement of an observable" have lured quite a few physicists into thinking that a measurement of an observable which results in pointing out a value reveals the actual value of a variable, a variable which for some reason is, however, not yet part of the theoretical description, i.e., a hidden variable, which is merely represented by the observable. We referred to this in Chap. 10 in the context of the EPR argument as measurement of a preexisting value. Naively, the observable describes a factual property of the system which is actually being measured in an experiment. The following question then arose: Could hidden variables be responsible for the outcome of measurements?

How could one phrase this question in mathematical terms? One way might be to ask: Is there a map from observables to random variables which is such that the joint statistics of any family of commuting observables are preserved, i.e., the corresponding family of random variables has the same joint statistics? The answer here is negative. There is no such map. This is sometimes dramatically referred to as a no-go theorem. The name "theorem" suggests that it might involve heavy mathematical machinery. However, it does not. Actually, the nonlocality proof in Chap. 10 is an example. We shall give two arguments which make it clear why the no-go theorem is nothing but a simple fact. The first is technical and throws light on the strategy of proof, while the second is obvious and shows that one should not waste time trying to prove the obvious.

Take observables $\hat{A}$, $\hat{B}$, and $\hat{C}$, and assume that $\hat{A}$ commutes with $\hat{B}$ and $\hat{B}$ with $\hat{C}$, but that $\hat{A}$ does not commute with $\hat{C}$. Then we have joint probabilities for $\hat{A}$ and $\hat{B}$ as well as for $\hat{B}$ and $\hat{C}$, but not for $\hat{A}$ and $\hat{C}$. Do such observables exist? Yes they do.

An example is the spin observable $\mathbf{a}\cdot\sigma$ in the $\mathbf{a}$-direction, which does not generally commute with $\mathbf{b}\cdot\sigma$, the spin observable in the $\mathbf{b}$-direction. Then as in an EPR situation, we consider the two-particle spin observables $\mathbf{a}\cdot\sigma^{(1)}\otimes\mathsf{I}$ and $\mathsf{I}\otimes\mathbf{b}\cdot\sigma^{(2)}$. These commute trivially. But $\mathbf{c}\cdot\sigma^{(1)}\otimes\mathsf{I}$ does not in general commute with $\mathbf{a}\cdot\sigma^{(1)}\otimes\mathsf{I}$. Now random variables X_A, X_B, and X_C always have joint probabilities, no matter what. That is the trivial conflict on which the no-go theorems are based [1, 3].

Trivial it is, but food for mysticism nevertheless. One measures $\hat{B}$ jointly with $\hat{A}$, which can be done since the PVMs commute, or one measures $\hat{B}$ simultaneously with $\hat{C}$, and one says that the properties which these observables represent are "contextual", that is, they depend on the context in which they are measured. On the one side observables, contextual properties, "non-classical" logic, complementarity, wave–particle duality, uncertainty, intrinsic probability, cat paradox, no-go mysticism. And on the other side, the two equations defining Bohmian mechanics, governing the whole of the (non-relativistic) world. Which side should physics be on?

We promised a second argument which makes all thoughts about hidden variables obsolete. It goes as follows. The observable is a book-keeping device for the quantum equilibrium statistics in experiments. The observable is associated with the experiment, i.e., there is a map

$$\mathscr{E} \Longrightarrow \hat{A}\,.$$

The experiment is the "real thing", even for a quantum hardliner. That, if anything, must be real. There is a map from reality to the observable. Can this map be inverted? Well, no, of course not, because the map is many-to-one. How would one ever come up with the idea that the association of the book-keeping device with the experiment could be a one to one correspondence? Only if one thinks that the experiment is truly a measurement of the observable. But who would be so naive?

Remark 12.2. Measuring the Position Operator
Measuring an operator means doing an experiment, the values and the statistics of which are encoded in $\hat{A}$, and in particular in the PVM: $\mathscr{E}\Longrightarrow\hat{A}$. We gave as example the trivial position PVM of a particle and the corresponding position operator $\hat{X}$. Now "measure" $\hat{X}$. Does that mean that one measures the position of the particle? No, not by any means. Why should it mean that? It simply means that we carry out an experiment whose values and statistics are those determined by $\hat{X}$. We give an example in Remark 15.4. ■

12.3 Schrödinger's Equation Revisited

Should we talk about existence and uniqueness of solutions of the Schrödinger equation? The answer is that we should not, unless something really catastrophic might happen. But the Schrödinger equation is linear, hence rather boring (see Remark 7.1 on that). What would be bad in this context? Well, we would be in difficulty if the quantum equilibrium distribution $\rho = |\psi|^2$ were to be called into question. For recall the continuity equation (7.17),

$$\frac{\partial |\psi|^2}{\partial t} = -\nabla \cdot \mathbf{j}^\psi = \frac{-\hbar}{2\mathrm{i}m}\nabla \cdot (\psi^* \nabla \psi - \psi \nabla \psi^*) .$$

The integral form of this guarantees that the probability always remains normalized because, by Gauss' theorem and assuming that the wave function goes to zero at spatial infinity, we have

$$\int \frac{\partial |\psi|^2}{\partial t} \mathrm{d}^n x = -\int \nabla \cdot \mathbf{j}^\psi d^n x = -\int \mathbf{j}^\psi \cdot \mathrm{d}\sigma = 0 . \tag{12.47}$$

But does (12.47) hold without further requirements? Let us consider a simple one-dimensional example, namely a "free" particle moving along the half-line $x > 0$. This means that we consider the Schrödinger equation on the half-line:

$$\mathrm{i}\hbar \frac{\partial}{\partial t} \psi = -\frac{\hbar^2}{2m} \frac{\mathrm{d}^2}{\mathrm{d}x^2} \psi , \qquad x \in (0, \infty) , \tag{12.48}$$

and we read (12.47) on the half-line. For notational simplicity, we now put $\hbar/m = 1$. The solution of (12.48) reads [see (5.9) with $D = 1$ and $\mathrm{i}t$ instead of t]

$$\psi(x,t) = \int \mathrm{d}y \frac{1}{\sqrt{2\pi \mathrm{i}t}} \exp\left[\mathrm{i}\frac{(x-y)^2}{2t}\right] \psi_0(y) . \tag{12.49}$$

We start with $\psi(y,0) = \psi_0(y)$ as a function which has compact support on the positive real line $\mathbb{R}^+$, so that $\psi_0(y) = 0$ for $y \leq 0$. But the solution (12.49) will not be zero for $x \leq 0$ (actually as soon as $t > 0$). This is easily seen for large t by recalling (9.25), viz.,

$$\psi(x,t) \approx \frac{1}{\sqrt{2\pi \mathrm{i}t}} \mathrm{e}^{\mathrm{i}x^2/2t} \hat{\psi}_0\left(\frac{x}{t}\right) ,$$

where $\hat{\psi}_0$ is the Fourier transform of ψ_0. This will not generally be zero for $x < 0$, since the Fourier transform of a compactly supported function is analytic.

We can also compute the current $j^\psi(0,t)$ through the origin for t large. We see then that, for some time,

$$\int_0^\infty |\psi(x,t)|^2 \mathrm{d}x < \int_0^\infty |\psi_0(x)|^2 \mathrm{d}x = 1 .$$

This means that (12.47) is false, i.e., $j^\psi(0,t) \neq 0$. The lesson is that equation (12.48) does not suffice to capture the idea of a free particle moving along the half-line $\mathbb{R}^*$.

What is missing are the *boundary conditions* at $x = 0$. In this simple example the boundary conditions are immediate. The particle trajectory must not cross the origin, i.e.,

$$v \sim \mathrm{Im}\frac{\nabla\psi}{\psi}(x,t) = 0\,, \qquad \text{at } x = 0\,,$$

which means that

$$\frac{\nabla\psi}{\psi}(0,t) \in \mathbb{R}\,,$$

or

$$\nabla\psi(0,t) = a\psi(0,t), \qquad a \in \mathbb{R}\,, \tag{12.50}$$

seems "good" as boundary condition. For each choice of a the condition is linear, so that it holds for superpositions of wave functions, each of which satisfies the boundary condition. This is a good point, but one still needs to check that the time evolution respects the chosen boundary condition. Given $a \in \mathbb{R}$, we require that, at $t = 0$,

$$\nabla\psi_0(0) = a\psi_0(0)\,.$$

We then show that, for $\psi(x,t)$, a solution of (12.48) with $\psi(x,0) = \psi_0(x)$ satisfies

$$\nabla\psi(0,t) = a\psi(0,t)\,.$$

More important than this, once we understand that boundary conditions are needed, we understand that the solution of (12.48) with a given initial wave function is not unique. When the support of $\psi_0(y)$ is away from $y = 0$, $\psi_0(y)$ does not feel the presence of the boundary. Then there are arbitrarily many solutions $\psi(y,t)$ of (12.48) with $\psi(y,0) = \psi_0(y)$. Here is another, different from (12.49):

$$\psi(x,t) = \int \mathrm{d}y \frac{1}{\sqrt{2\pi i t}} \left\{ \exp\left[-\mathrm{i}\frac{(x-y)^2}{2t}\right] - \exp\left[-\mathrm{i}\frac{(x+y)^2}{2t}\right] \right\} \psi_0(y)\,, \qquad x \geq 0\,,$$

for which we actually have $\psi(0,t) = 0$ for all $t \geq 0$.

Putting this together we see something nice emerging. The problem of boundary conditions, which we need for the quantum equilibrium hypothesis to hold true for all times, goes hand in hand with the existence and uniqueness of solutions of the Schrödinger equation. Mathematically, the quantum equilibrium hypothesis requires wave functions to remain normalized during the time evolution. The norm of the wave function is, according to the geometrical picture we have developed so far, the unitary one:

$$\|\psi_t\| := \sqrt{\langle\psi_t|\psi_t\rangle} := \sqrt{\int \psi_t(q)^*\psi_t(q)\,\mathrm{d}^n q}\,.$$

The time evolution must therefore be given by a group of unitary operators, generated by the Schrödinger equation. This is all captured in the self-adjointness of the differential operator on the right-hand side of the Schrödinger equation, the so-called Hamilton operator.

A related remark is this. We shall see that each choice of $a \in \mathbb{R}$ is a good boundary condition. This means that for each choice $a \in \mathbb{R}$, we have the physics of the free particle moving on the half-line. In other words, the phrase "free particle on the half-line" does not describe a unique physical system. Only the boundary conditions define the physical theory uniquely.

The reader may well get the impression that we are insisting on an artificial scenario of a particle moving on a half-line, something which will never occur like that in nature, since one needs a potential to constrain a particle, and a boundary condition is merely an idealization. This is correct, but the example is very informative and only prepares the ground for the more realistic physics of atoms. The Schrödinger equation for a particle in a Coulomb potential $V(x) = -e_1 e_2/\|\mathbf{x}\|$ reads

$$i\hbar\frac{\partial}{\partial t}\psi = \left(-\frac{\hbar^2}{2m}\Delta - \frac{e_1 e_2}{\|\mathbf{x}\|}\right)\psi\,, \qquad x \neq 0\,,$$

i.e., it is only defined for $\mathbf{x} \neq 0$. There is nothing one can do about that. The origin is a no-go point. We do not want the particle ever to reach the origin, otherwise it might vanish there, and that would be the end of quantum mechanics as we know it. The singular point $\mathbf{x} = 0$ is a boundary point of the physical theory, and boundary conditions are therefore needed. In three dimensions, we can consider the radial motion of the particle, and we do not want the radial trajectory to hit zero. We see that our half-line example is not so artificial after all.

12.4 What Comes Next?

We are ready to go on with the mathematics. We need to give a proper description of the vector space of functions which contains the physically relevant wave functions. We shall introduce the scalar product and with it a norm, and for technical reasons we shall want a complete vector space. This is the mathematics of Hilbert spaces. We also need to give a proper description of the (boundary) conditions under which the Hamilton operator generates a unitary evolution, so that quantum equilibrium will be valid. We shall give a more precise introduction of POVMs and PVMs, which describe the empirical import of quantum equilibrium, describing the relations between PVMs and observables so that we may link them to the textbook quantum formalism.

The one-to-one correspondence is called the spectral theorem. This is technically very important because it incorporates the diagonalization of the operators and allows one to compute formulas. We will then have a complete grasp of the free Schrödinger evolution and understand the structure of the spectrum of Schrödinger Hamiltonians when an interaction potential is present. A particularly relevant combination of interaction and free evolution forms the mathematical basis of scattering theory. In Chap. 16, we shall return to physics to discuss scattering theory (and more) from the standpoint of Bohmian mechanics.

References

1. D. Dürr, S. Goldstein, N. Zanghi: J. Stat. Phys. **116** (1–4), 959 (2004)
2. D. Dürr, S. Goldstein, N. Zanghì: preprint (2008)
3. J.S. Bell: *Speakable and Unspeakable in Quantum Mechanics* (Cambridge University Press, Cambridge, 1987)

Chapter 13
Hilbert Space

The Schrödinger equation is a linear equation: linear superpositions of solutions are again solutions. We also need square integrability of the solutions. Thus one is naturally led to a *vector space* of *square integrable* functions for the space of wave functions. It turns out that this space has additional mathematical structure, namely the structure provided by an inner product.

Definition 13.1. An inner product (scalar product) on a complex vector space $\mathscr{H}$ is a positive definite sesquilinear form, i.e., a map

$$\langle\cdot|\cdot\rangle : \mathscr{H}\times\mathscr{H}\to\mathbb{C}$$

with the following properties. Let $\varphi,\psi\in\mathscr{H}$ and $\alpha\in\mathbb{C}$, then:

(i) $\langle\varphi|\varphi\rangle\geq 0$ and $\langle\varphi|\varphi\rangle=0 \iff \varphi=0$,
(ii) $\langle\varphi|\varphi+\psi\rangle=\langle\varphi|\varphi\rangle+\langle\varphi|\psi\rangle$,
(iii) $\langle\varphi|\alpha\psi\rangle=\alpha\langle\varphi|\psi\rangle$,
(iv) $\langle\varphi|\psi\rangle=\langle\psi|\varphi\rangle^*$.

Property (i) is called positive definiteness, and properties (ii)–(iv) define sesquilinearity.

Note that (iii) and (iv) imply antilinearity in the first argument, i.e.,

$$\langle\alpha\varphi|\psi\rangle=\alpha^*\langle\varphi|\psi\rangle\,.$$

As first examples, we have $\mathbb{C}^n$ with the inner product

$$\langle z|w\rangle=\sum_{i=1}^{n} z_i^* w_i\,,\qquad z_i,w_i\in\mathbb{C}\,,$$

and $C([a,b])$, the space of continuous complex-valued functions on the interval $[a,b]$, with

$$\langle f|g\rangle=\int_a^b f^*(x)g(x)\,\mathrm{d}x\,.$$

D. Dürr, S. Teufel, *Bohmian Mechanics*, DOI 10.1007/978-3-540-89344-8_13,

The role of the inner product is to introduce the notion of orthogonality of vectors. Two vectors φ and ψ in $\mathscr{H}$ are said to be orthogonal if $\langle\varphi|\psi\rangle = 0$. A sequence $(\varphi_i)_{i\in\mathbb{N}} \in \mathscr{H}$ is called an orthonormal sequence if its elements are pairwise orthogonal and normalized, i.e., $\langle\varphi_i|\varphi_j\rangle = \delta_{ij}$.

Next we consider orthogonal decompositions. Let $(\varphi_i)_{i\in\mathbb{N}}$ be an arbitrary orthonormal sequence. Then, for all $\varphi \in \mathscr{H}$ and each $N \in \mathbb{N}$, the decomposition

$$\varphi = \underbrace{\sum_{i=1}^{N}\langle\varphi_i|\varphi\rangle\varphi_i}_{\psi_N} + \underbrace{\varphi - \sum_{i=1}^{N}\langle\varphi_i|\varphi\rangle\varphi_i}_{\psi_N^\perp}, \tag{13.1}$$

is orthogonal, meaning that ψ_N and $\psi_N^\perp$ are orthogonal. Hence

$$\|\varphi\|^2 := \langle\varphi|\varphi\rangle = \sum_{i=1}^{N}|\langle\varphi_i|\varphi\rangle|^2 + \left\|\varphi - \sum_{i=1}^{N}\langle\varphi_i|\varphi\rangle\varphi_i\right\|^2,$$

and (i) implies

$$\|\varphi\|^2 \geq \sum_{i=1}^{N}|\langle\varphi_i|\varphi\rangle|^2 \qquad \text{(Bessel inequality)}. \tag{13.2}$$

For $\varphi_1 = \psi/\|\psi\|$ and $N = 1$, we obtain as a special case the important Schwarz inequality:

$$|\langle\varphi|\psi\rangle| \leq \|\varphi\|\|\psi\|, \qquad \forall\varphi,\psi \in \mathscr{H}. \tag{13.3}$$

We can now easily show that $\|\cdot\| = \sqrt{\langle\cdot|\cdot\rangle}$ defines a *norm*,[1] which turns $\mathscr{H}$ into a normed linear space. It is positive definite and homogeneous by Definition 13.1 (i), (iii), and (iv), and the triangle inequality

$$\|\varphi + \psi\| \leq \|\varphi\| + \|\psi\|$$

follows from the Schwarz inequality as an easy exercise.

Such a normed linear space with inner product is very similar to $\mathbb{R}^n$ (or $\mathbb{C}^n$). What is missing? Analysis in $\mathbb{R}$ is based on *completeness*. Having completeness means not having to worry about infinite sequences and limits. However, there is a price to pay. In the case of wave functions, the price of the comfort of having completeness – or better for considering the norm closure of the space of physical wave functions – is rather high. As we shall see, most elements in the completion of the vector space of wave functions are irrelevant and abstract. They cannot be considered as wave functions, i.e., as physical states. They are not differentiable, nor even continuous.

---

[1] Recall that a norm $\|\cdot\|$ on a vector space V is a map $\|\cdot\| : V \to [0,\infty)$ such that (i) $\|v\| = 0 \Leftrightarrow v = 0$, (ii) $\|\alpha v\| = |\alpha|\|v\|$ for all $\alpha \in \mathbb{C}$, $v \in V$, and (iii) $\|v+w\| \leq \|v\| + \|w\|$ for all $v, w \in V$.

13.1 The Hilbert Space L^2

A *complete* normed space is called a Hilbert space if its norm is given by an inner product, $\|\cdot\| = \sqrt{\langle\cdot|\cdot\rangle}$. Completeness means that every Cauchy sequence $(\varphi_n)_{n\in\mathbb{N}} \in \mathscr{H}$ with respect to the norm $\|\cdot\|$ in $\mathscr{H}$ converges in $\mathscr{H}$. Obvious examples are $\mathbb{R}^n$ and $\mathbb{C}^n$. But we are interested in completeness of the space of wave functions with respect to the integral norm

$$\|\psi\|^2 := \int_{\mathbb{R}^n} \psi^*(\mathbf{x})\psi(\mathbf{x})\mathrm{d}^n x\,.$$

Here the integral must be understood as a Lebesgue integral, otherwise the resulting normed space of square integrable functions would not be complete, i.e., there would exist Cauchy sequences that did not converge to a Riemann square integrable function. But with the Lebesgue integral we obtain (as will be shown) a Hilbert space

$$L^2(\mathbb{R}^n, \mathrm{d}^n x) = \left\{ \varphi : \mathbb{R}^n \to \mathbb{C} \text{ measurable with } \|\varphi\| = \sqrt{\int |\varphi|^2 \mathrm{d}^n x} < \infty \right\},$$

with inner product given by

$$\langle\varphi|\psi\rangle = \int \varphi^*(\mathbf{x})\psi(\mathbf{x})\mathrm{d}^n x\,.$$

Note that the Schwarz inequality (13.3) implies $\langle\varphi|\psi\rangle < \infty$ for all $\varphi, \psi \in L^2$.

Remark 13.1. About Equality in L^2
The notion of equality of elements in L^2 is a rather special one. The relation $\varphi = \psi$ in L^2 means that $\|\varphi - \psi\| = 0$, which in turn means that $\varphi = \psi$ almost surely,[2] i.e., the equality $\varphi(\mathbf{x}) = \psi(\mathbf{x})$ might be violated for a Lebesgue null set of $\mathbf{x}$ values. The elements of L^2 are thus actually equivalence classes of functions. Each equivalence class consists of functions which are equal almost everywhere. But this need not worry us most of the time. We just go on talking about "functions" $f \in L^2$, although they are actually equivalence classes of functions. Most of the time the distinction between functions and equivalence classes is irrelevant. It is only if one needs pointwise evaluations of functions that one has to be careful. In general $f(\mathbf{x})$ makes no sense for an element of L^2, but of course $\int |f(\mathbf{x})|^2 \mathrm{d}x$ does. If an equivalence class contains a continuous or even differentiable representative (which is then unique), one naturally associates the class with this special function and pointwise evaluation becomes meaningful again. ■

The fact that L^2 is complete and thus a Hilbert space, and indeed the completeness of any L^p space,

[2] "Almost surely" is synonymous with "almost everywhere".

$$L^p(\mathbb{R}^n, \mathrm{d}^n x) = \left\{ \varphi : \mathbb{R}^n \to \mathbb{C} \text{ measurable with } \|\varphi\| = \left(\int |\varphi|^p \mathrm{d}^n x \right)^{1/p} < \infty \right\},$$

is the content of the Riesz–Fischer theorem:

Theorem 13.1. *The space $L^2(\mathbb{R}^n, \mathrm{d}^n x)$, and indeed every space $L^p(\mathbb{R}^n, \mathrm{d}^n x)$ with $1 \leq p \leq \infty$, is complete, i.e., every Cauchy sequence with respect to the norm in L^2 converges in the norm to an element of L^2.*

Only when $p = 2$ is the norm given by an inner product, which makes this value special. The idea of the proof is simple. Convergence of a subsequence of a Cauchy sequence implies convergence of the whole sequence. Thus we can pick a subsequence $(\varphi_k)_{k\in\mathbb{N}} := (\tilde{\varphi}_{n_k})_{k\in\mathbb{N}}$ of a given Cauchy sequence $(\tilde{\varphi}_n)_{n\in\mathbb{N}}$, such that

$$\|\varphi_k - \varphi_{k+1}\| < 2^{-k}. \tag{13.4}$$

Then chose any sequence of representatives $\varphi_k(\mathbf{x})$ and put

$$\psi_m(\mathbf{x}) = \sum_{k=1}^{m-1} |\varphi_k(\mathbf{x}) - \varphi_{k+1}(\mathbf{x})|$$

and $\psi_\infty(\mathbf{x}) = \lim_{m\to\infty} \psi_m(\mathbf{x})$, which may be infinite. [Note that $\big(\psi_m(\mathbf{x})\big)_{m\in\mathbb{N}}$ is monotonically increasing.] But because of (13.4), we have $\|\psi_m\| < 1$, i.e.,

$$\int |\psi_m|^2 \mathrm{d}^n x \leq 1 .$$

Lebesgue's theorem of monotone convergence implies that $|\psi_\infty|^2$ is integrable, i.e., that $\psi_\infty \in L^2$. Hence $|\psi_\infty|^2$ can be infinite only on a Lebesgue null set, and we have pointwise convergence of

$$\varphi_m = \varphi_1 + \sum_{k=1}^{m-1} (\varphi_{k+1} - \varphi_k)$$

almost everywhere to a function φ. Since

$$|\varphi_m|^2 \leq 2\big(|\varphi_1|^2 + |\psi_m|^2\big) \leq 2\big(|\varphi_1|^2 + |\psi_\infty|^2\big) \in L^1 , \tag{13.5}$$

we can apply Lebesgue's theorem of dominated convergence to conclude that $|\varphi|^2 \in L^1$, and hence $\varphi \in L^2$. The sequence $h_m := \varphi_m - \varphi$ converges almost surely to zero. From (13.5), it follows that

$$|h_m|^2 \leq 2\big(|\varphi_m|^2 + |\varphi|^2\big) \leq 2\Big[2\big(|\varphi_1|^2 + |\psi_\infty|^2\big) + |\varphi|^2\Big] \in L^1 ,$$

and a second application of dominated convergence gives $|h_m|^2 \longrightarrow 0$ in L^1, and finally

$$h_m \to 0 \text{ in } L^2 \quad \Longrightarrow \quad \varphi_m \to \varphi \text{ in } L^2 .$$

13.1.1 The Coordinate Space ℓ^2

While completeness allows us to do analysis, the inner product structure allows us to transfer our geometric intuition from finite-dimensional spaces to infinite-dimensional ones. For us (and for almost everybody else), only so called *separable* Hilbert spaces are interesting, because they admit *countable* orthonormal bases. An orthonormal sequence $(\varphi_n)_{n\in\mathbb{N}} \in \mathscr{H}$ is an orthonormal basis if every $\varphi \in \mathscr{H}$ can be represented through an orthonormal decomposition with respect to $(\varphi_n)_{n\in\mathbb{N}} \in \mathscr{H}$:

$$\varphi = \sum_{k\in\mathbb{N}} \langle \varphi_k | \varphi \rangle \varphi_k \,. \tag{13.6}$$

Here the equality means that the series on the right-hand side converges in the norm of $\mathscr{H}$ to φ.

Remark 13.2. The Notion of Separability
A topological space is called separable if it contains a countable dense subset. Hence $\mathscr{H}$ is separable if there is a sequence $(\psi_n)_{n\in\mathbb{N}}$ such that, for every element $\phi \in \mathscr{H}$ and $\varepsilon > 0$, there is an index $n \in \mathbb{N}$ with $\|\psi_n - \phi\| < \varepsilon$. But this is clearly equivalent to the existence of a countable orthonormal basis: by inductively removing those elements from $(\psi_n)_{n\in\mathbb{N}}$ which are finite linear combinations of the preceding ones, we can construct a linearly independent subsequence $(\psi_{n_k})_{k\in\mathbb{N}}$ such that $(\psi_n)_{n\in\mathbb{N}}$ can be recovered from finite linear combinations of elements in $(\psi_{n_k})_{k\in\mathbb{N}}$, i.e., such that $\operatorname{span}(\psi_{n_k})_{k\in\mathbb{N}} = \operatorname{span}(\psi_n)_{n\in\mathbb{N}}$. The Gram–Schmidt orthonormalization procedure applied to $(\psi_{n_k})_{k\in\mathbb{N}}$ finally yields an orthonormal basis $(\varphi_k)_{k\in\mathbb{N}}$. On the other hand we can construct a countable dense set from an orthonormal basis $(\varphi_k)_{k\in\mathbb{N}}$ by approximating the complex coefficients $\langle \varphi_k | \varphi \rangle$ by rational (complex) coefficients. (Recall that countable unions of countable sets are countable.) ■

It is in general not easy to check whether (13.6) holds for a given orthonormal sequence. However, using the fact that $\varphi - \sum \langle \varphi_k | \varphi \rangle \varphi_k$ must be zero gives the following more practical reformulation of (13.6).

Remark 13.3. On Orthonormal Bases
An orthonormal sequence $(\varphi_k)_{k\in\mathbb{N}}$ is an orthonormal basis if and only if

$$\langle \varphi_k | \varphi \rangle = 0\,, \quad \text{for all } k \in \mathbb{N} \qquad \Longrightarrow \qquad \varphi = 0\,. \tag{13.7}$$

To see that (13.7) implies (13.6), note that (13.2) implies, for arbitrary N,

$$\infty > \|\varphi\|^2 \geq \sum_{k=1}^{N} |\langle \varphi_k | \varphi \rangle|^2 \,.$$

Hence $(\tilde{\varphi}_N)_{N\in\mathbb{N}}$ with

$$\tilde{\varphi}_N = \sum_{k=1}^{N} \langle \varphi_k | \varphi \rangle \varphi_k$$

is a Cauchy sequence which converges to some $\tilde{\varphi}$. But for all k, we have

$$\begin{aligned}\langle \tilde{\varphi} - \varphi | \varphi_k \rangle &= \lim_{N \to \infty} \langle \tilde{\varphi}_N - \varphi | \varphi_k \rangle = \lim_{N \to \infty} \left\langle \sum_{l=1}^{N} \langle \varphi_l | \varphi \rangle \varphi_l - \varphi \middle| \varphi_k \right\rangle \\ &= \langle \varphi_k | \varphi \rangle^* - \langle \varphi | \varphi_k \rangle = 0 \,.\end{aligned}$$

With (13.7), we may thus conclude that $\tilde{\varphi} = \varphi$. ■

From (13.6) it follows that, for an orthonormal basis, the Bessel inequality (13.2) turns into an equality, called the Pythagorean theorem, or in this context, the Parseval equality:

$$\|\varphi\|^2 = \sum_{k=1}^{\infty} |\langle \varphi_k | \varphi \rangle|^2 \qquad \text{(Parseval equality)} \,. \tag{13.8}$$

This motivates the introduction of a further important Hilbert space, which is the coordinate representation of any separable Hilbert space: ℓ^2 is the space of all square-integrable sequences, viz.,

$$\ell^2 = \left\{ (x_n)_{n \in \mathbb{N}} : x_n \in \mathbb{C}, \sum_{n=1}^{\infty} |x_n|^2 < \infty \right\} ,$$

with the inner product

$$\langle x | y \rangle = \sum_{n=1}^{\infty} x_n^* y_n \,.$$

It is not a very difficult exercise in analysis to show that this space is complete, and we skip the proof here.

From (13.6) and (13.8), we conclude that any separable infinite-dimensional Hilbert space $\mathscr{H}$ is *isomorphic* to ℓ^2, $\mathscr{H} \cong \ell^2$, i.e., there exists a bijective linear isometry $U : \mathscr{H} \longrightarrow \ell^2$. Such an operator U is said to be *unitary*, and is characterized by the property that U is surjective and isometric, i.e., for all $\varphi \in \mathscr{H}$,

$$\|U\varphi\|_{\ell^2} = \|\varphi\|_{\mathscr{H}} \,.$$

Note that isometries are always injective, and it follows from the *polarization identity*,

$$\langle \varphi | \psi \rangle = \frac{1}{4} \left[\left(\|\varphi + \psi\|^2 - \|\varphi - \psi\|^2 \right) - \mathrm{i} \left(\|\varphi + \mathrm{i}\psi\|^2 - \|\varphi - \mathrm{i}\psi\|^2 \right) \right] , \tag{13.9}$$

that any isometry, and hence any unitary operator, satisfies

$$\langle U\varphi | U\psi \rangle_{\ell^2} = \langle \varphi | \psi \rangle_{\mathscr{H}} \,, \qquad \forall \varphi, \psi \in \mathscr{H} \,. \tag{13.10}$$

For any orthonormal basis $(\varphi_k)_{k \in \mathbb{N}}$ of $\mathscr{H}$, the coordinate map

$$U : \mathscr{H} \to \ell^2 , \quad \varphi \mapsto U\varphi = (\langle \varphi_k | \varphi \rangle)_{k \in \mathbb{N}}$$

is such a unitary operator. It is surjective by definition and an isometry by Parseval (13.8). The Fourier series of a function in $L^2([0,2\pi])$ is a prominent example of such a coordinate representation. The elements of the orthonormal basis are

$$\frac{1}{\sqrt{2\pi}} \mathrm{e}^{\mathrm{i}kx} , \qquad k \in \mathbb{Z} ,$$

and the coefficients of the series are the coordinate sequence in ℓ^2.

As we will try to understand in the following, $L^2(\mathbb{R}^n, \mathrm{d}^n x)$ is also separable. This might be surprising at first sight, since the elements $\varphi(\mathbf{x})$, $\mathbf{x} \in \mathbb{R}^n$, could be naively interpreted as "uncountable vectors" $\varphi_{\mathbf{x}}$ (with $\mathbf{x}$ as index like the i in y_i for $\mathbf{y} \in \mathbb{R}^n$, $i = 1, \ldots, n$). From this analogy one would expect uncountable bases for $L^2(\mathbb{R}^n, \mathrm{d}^n x)$. [Think also of (9.27), where we wrote $\psi(x) = \int |x\rangle \langle x | \varphi \rangle \mathrm{d}x$.] Of course, this intuition would also apply to $L^2([0,2\pi])$, where we have already seen that it is wrong. Indeed, the arbitrary assignment of a value φ_x to every x would typically yield very irregular functions, and functions in L^2 are not that irregular. However, note that the space of essentially bounded functions $L^\infty([0,2\pi])$ is *not* separable, although $L^\infty([0,2\pi]) \subset L^2([0,2\pi])$. So separability really depends on the norm, i.e., on how we measure the distance between functions.

We now show that $L^2(\mathbb{R}, \mathrm{d}x)$ is separable by explicitly constructing a countable orthonormal basis. Then separability of $L^2(\mathbb{R}^n, \mathrm{d}^n x)$ follows because of the *tensor product structure*, which we discuss later on. We will see that

$$\overline{\mathrm{span}\left\{ \left(x^n \mathrm{e}^{-x^2/2} \right)_{n \in \mathbb{N}_0} \right\}} = L^2(\mathbb{R}, \mathrm{d}x) , \tag{13.11}$$

which is a natural guess, since the monomials x^n are linear independent and $\mathrm{e}^{-x^2/2}$ makes them square integrable. The Gram–Schmidt orthogonalization procedure turns this linearly independent sequence into an orthonormal sequence

$$H_n(x) = P_n(x) \mathrm{e}^{-x^2/2} , \quad n \in \mathbb{N} ,$$

where the H_n are called Hermite functions and the P_n are polynomials of degree n, the Hermite polynomials. According to Remark 13.3, for $(H_n)_{n \in \mathbb{N}_0}$ to be a basis as claimed, we must have

$$\forall n \in \mathbb{N}_0 , \quad \langle H_n | \varphi \rangle = 0 \quad \Longrightarrow \quad \varphi = 0 . \tag{13.12}$$

But since the P_n are polynomials of degree n, (13.12) is equivalent to the requirement that the original system is a basis, i.e., that

$$\forall n \in \mathbb{N}_0 , \quad \left\langle x^n \mathrm{e}^{-x^2/2} | \varphi \right\rangle = 0 \quad \Longrightarrow \quad \varphi = 0 , \tag{13.13}$$

and we now show (13.13). Assuming that

$$\forall n \in \mathbb{N}_0 , \quad \langle x^n \mathrm{e}^{-x^2/2} | \varphi \rangle = 0 ,$$

then also

$$\forall k \in \mathbb{R} , \quad \langle \mathrm{e}^{\mathrm{i}kx} \mathrm{e}^{-x^2/2} | \varphi \rangle = 0 ,$$

since

$$\mathrm{e}^{\mathrm{i}kx} \mathrm{e}^{-x^2/2} = \sum_{m=0}^{\infty} \frac{(\mathrm{i}k)^m}{m!} x^m \mathrm{e}^{-x^2/2} , \tag{13.14}$$

in the L^2 sense. So the Fourier transform of $\mathrm{e}^{-x^2/2}\varphi$ must be identically zero:

$$\frac{1}{\sqrt{2\pi}} \langle \mathrm{e}^{\mathrm{i}kx} \mathrm{e}^{-x^2/2} | \varphi \rangle = \mathscr{F}\left(\mathrm{e}^{-x^2/2}\varphi\right)(k) = 0 .$$

As we shall see in the next section, the Fourier transformation $\mathscr{F} : L^2 \longrightarrow L^2$ is a unitary map, and hence $\varphi = 0$. So $L^2(\mathbb{R})$ is separable.

Remark 13.4. Proof of Convergence
Since it is not completely obvious that (13.14) really converges in L^2, we give the argument in this remark. The sequence of partial sums

$$\sum_{m=0}^{N} \frac{(\mathrm{i}k)^m}{m!} x^m \mathrm{e}^{-x^2/2}$$

is a Cauchy sequence, because for each $k \in \mathbb{R}$ we have

$$\begin{aligned} \left\| \frac{(\mathrm{i}k)^m}{m!} x^m \mathrm{e}^{-x^2/2} \right\|^2 &= \frac{k^{2m}}{(m!)^2} \int_{-\infty}^{\infty} x^{2m} \mathrm{e}^{-x^2} \mathrm{d}x \\ &= \frac{k^{2m}}{m!} 2^m \int_{-\infty}^{\infty} \left(\frac{x^2}{2}\right)^m \frac{1}{m!} \mathrm{e}^{-x^2} \mathrm{d}x \\ &\leq \frac{k^{2m}}{m!} 2^m \int_{-\infty}^{\infty} \mathrm{e}^{-x^2/2} \mathrm{d}x = \frac{(2k^2)^m}{m!} \sqrt{2\pi} . \end{aligned}$$

So it converges, not only pointwise, but also in L^2. ■

13.1.2 Fourier Transformation on L^2

On $L^2([0, 2\pi])$, the Fourier transformation is just the orthogonal decomposition with respect to the orthonormal basis $(\mathrm{e}^{\mathrm{i}kx})_{k \in \mathbb{Z}}$:

$$\mathscr{F} : L^2([0,2\pi]) \longrightarrow \ell^2 ,$$
$$f \longmapsto (\widehat{f}_k)_{k\in\mathbb{Z}} = (\langle \mathrm{e}^{ikx}, f\rangle)_{k\in\mathbb{Z}} .$$

But the plane waves e^{ikx} are not square integrable over $\mathbb{R}$ and thus, in particular, do not form a basis of $L^2(\mathbb{R})$. The Fourier transformation is now a unitary map

$$\mathscr{F} : L^2(\mathbb{R}^n) \longmapsto L^2(\mathbb{R}^n) ,$$
$$f \longmapsto \widehat{f} ,$$

with

$$\langle g|f\rangle = \langle \widehat{g}|\widehat{f}\rangle \qquad \text{(Plancherel equality)}, \tag{13.15}$$

where the Fourier transform of integrable functions is defined as usual by

$$\mathscr{F} f = \widehat{f}(\mathbf{k}) = \left(\frac{1}{2\pi}\right)^{n/2} \int_{\mathbb{R}^n} \mathrm{e}^{-\mathrm{i}\mathbf{k}\cdot\mathbf{x}} f(\mathbf{x})\,\mathrm{d}^n x \tag{13.16}$$

and

$$\mathscr{F}^* \widehat{f} = \mathscr{F}^{-1}\widehat{f} = f(\mathbf{x}) = \left(\frac{1}{2\pi}\right)^{n/2} \int_{\mathbb{R}^n} \mathrm{e}^{\mathrm{i}\mathbf{k}\cdot\mathbf{x}} \widehat{f}(\mathbf{k})\,\mathrm{d}^n k . \tag{13.17}$$

Integrable functions (f or $\widehat{f}$) are L^1-functions. While on bounded domains, L^2 is contained in L^1, this no longer holds on unbounded domains. Thus for some L^2-functions, we cannot define the Fourier transform by (13.16). In order to define Fourier transformation on all of $L^2(\mathbb{R}^n)$, one first analyses the behavior of "nice" functions under the Fourier transformation, as defined by the above integrals. For "not so nice" L^2-functions, we define the transformation by approximating with nice functions, which must therefore be dense.

It is convenient to begin by analyzing the Fourier transformation on "very nice" functions. The Schwartz space of rapidly decaying smooth functions $\mathscr{S}(\mathbb{R}^n) \subset L^2(\mathbb{R}^n)$ is the space of all C^∞-functions (functions for which all partial derivatives of any order exist and are continuous), which decay faster than any inverse polynomial as $|\mathbf{x}| \longrightarrow \infty$, and for which the same holds for all their partial derivatives. Since $C_0^\infty(\mathbb{R}^n)$, the space of smooth compactly supported functions is contained in $\mathscr{S}(\mathbb{R}^n)$ and dense in $L^2(\mathbb{R}^n)$, so $\mathscr{S}(\mathbb{R}^n)$ is dense in $L^2(\mathbb{R}^n)$.

Schwartz functions are certainly integrable, and $\mathscr{S}(\mathbb{R}^n)$ is invariant under the Fourier transformation [1–3]:[3]

$$\mathscr{F} : \mathscr{S} \longrightarrow \mathscr{S} , \qquad \mathscr{F}^{-1} : \mathscr{S} \longrightarrow \mathscr{S} . \tag{13.18}$$

This follows in a rather straightforward way from the definition. To get (13.18), we must first show that arbitrary partial derivatives $D_k^\beta \widehat{f}(\mathbf{k})$ of $\widehat{f}(\mathbf{k})$ multiplied by any

[3] The reference [1] is highly recommended reading.

polynomial k^α remain bounded. Here we use the multi-index notation

$$\alpha \in \mathbb{N}_0^n\,, \quad k^\alpha := k_1^{\alpha_1} \dots k_n^{\alpha_n}\,, \quad D_y^\alpha := \frac{\partial^{|\alpha|}}{\partial y_1^{\alpha_1} \dots \partial y_n^{\alpha_n}}\,, \quad |\alpha| = \alpha_1 + \dots + \alpha_n\,.$$

Referring to dominated convergence, we can differentiate the integral by differentiating the integrand to obtain

$$\begin{aligned} k^\alpha D_k^\beta \widehat{f}(\mathbf{k}) &= \left(\frac{1}{2\pi}\right)^{n/2} \int k^\alpha (-\mathrm{i}x)^\beta \mathrm{e}^{-\mathrm{i}\mathbf{k}\cdot\mathbf{x}} f(\mathbf{x}) \mathrm{d}^n x \\ &= \left(\frac{1}{2\pi}\right)^{n/2} \int \mathrm{i}^{|\alpha|} \left[D_x^\alpha \left(\mathrm{e}^{-\mathrm{i}\mathbf{k}\cdot\mathbf{x}}\right)\right] (-\mathrm{i}x)^\beta f(\mathbf{x}) \mathrm{d}^n x \\ &= \left(\frac{1}{2\pi}\right)^{n/2} (-\mathrm{i})^{|\alpha|} \int \mathrm{e}^{-\mathrm{i}\mathbf{k}\cdot\mathbf{x}} D_x^\alpha \left[(-\mathrm{i}x)^\beta f(\mathbf{x})\right] \mathrm{d}^n x\,, \end{aligned}$$

integrating $|\alpha|$-times by parts to get the last equality. Hence

$$\sup_{\mathbf{k}} \left|k^\alpha D_k^\beta \widehat{f}(\mathbf{k})\right| \leq \left(\frac{1}{2\pi}\right)^{n/2} \int \left|D_x^\alpha \left(x^\beta f(\mathbf{x})\right)\right| \mathrm{d}^n x\,.$$

From this one easily concludes that $\mathscr{F}, \mathscr{F}^* : \mathscr{S} \to \mathscr{S}$. We still need to show that $\mathscr{F}^* = \mathscr{F}^{-1}$. To get this, we consider compactly supported smooth functions, i.e., in C_0^∞, because we can continue such a function periodically and consider its Fourier series. The Fourier series of a smooth function converges uniformly to the function. Thus the idea is to approximate functions f from $\mathscr{S}$ by sequences (f_k) from C_0^∞. However, to conclude this density argument, we need continuity of $\mathscr{F}$ as a map from $\mathscr{S}$ to $\mathscr{S}$ in a suitable sense.

In order to define a suitable notion of convergence of sequences in $\mathscr{S}$, we continue the above estimate and introduce the following family of seminorms ("semi" means that $\|f\| = 0$ does not necessarily imply $f = 0$):

$$\|f\|_{\alpha,\beta} := \sup_{\mathbf{y}} \left|y^\alpha D_y^\beta f(\mathbf{y})\right|\,, \qquad \text{for arbitrary multi-indices } \alpha, \beta\,.$$

Then the previous estimate yields (one should remember the following trick for later use)

$$
\begin{aligned}
\|\mathscr{F} f\|_{\alpha,\beta} = \|\widehat{f}\|_{\alpha,\beta} &\leq \left(\frac{1}{2\pi}\right)^{n/2} \int \left| D_x^\alpha \left(x^\beta f(\mathbf{x})\right)\right| \mathrm{d}^n x \\
&\leq \left(\frac{1}{2\pi}\right)^{n/2} \int \frac{1+x^{2n}}{1+x^{2n}} \left| D_x^\alpha \left(x^\beta f(\mathbf{x})\right)\right| \mathrm{d}^n x \\
&\leq \left(\frac{1}{2\pi}\right)^{n/2} \sup_{\mathbf{x}} \left| \left(1+x^{2n}\right) D_x^\alpha \left(x^\beta f(\mathbf{x})\right)\right| \int \frac{1}{1+x^{2n}} \mathrm{d}^n x \\
&\leq C(\alpha,\beta) \sum_{|\gamma|\leq|\beta|+2n, |\delta|\leq|\alpha|} \|f\|_{\gamma,\delta} \,. \qquad (13.19)
\end{aligned}
$$

The same holds for $\mathscr{F}^*$. (This family of seminorms defines a metric[4] on $\mathscr{S}$. Convergence of sequences in $\mathscr{S}$ with respect to this metric is equivalent to convergence with respect to $\|\cdot\|_{\alpha,\beta}$ for all α and β. For details see, for example, [2, 3]. However, this is not really important for the following.) For us it is important that we can approximate $f \in \mathscr{S}$ by $f_0 \in C_0^\infty$ in such a way that $\mathscr{F}(f-f_0)$ is also small. More precisely, we want to show that

$$
\mathscr{F}^* \mathscr{F} f = f \,, \quad \text{or } \mathscr{F}^* = \mathscr{F}^{-1} \,, \quad \text{on } \mathscr{S} \,.
$$

We will show that

$$
\mathscr{F}^* \widehat{f_0} = f_0 \,, \qquad f_0 \in C_0^\infty \,.
$$

Then a simple triangulation together with (13.19) (for $\mathscr{F}^*$ and $\mathscr{F}$) yields

$$
\begin{aligned}
\left\|\mathscr{F}^* \widehat{f} - f\right\|_{0,0} &\leq \left\|\mathscr{F}^* \widehat{f} - \mathscr{F}^* \widehat{f_0}\right\|_{0,0} + \left\|\mathscr{F}^* \widehat{f_0} - f_0\right\|_{0,0} + \left\|f - f_0\right\|_{0,0} \\
&\leq C \sum_{|\gamma|\leq 2n} \left\|f - f_0\right\|_{\gamma,0} \,.
\end{aligned}
$$

Hence we need to approximate f by f_0 with respect to the seminorms $\|\cdot\|_{\gamma,0}$, $|\gamma| \leq 2n$. To do this we first define a C_0^∞-function Φ that is a "smooth version" of the characteristic function of the unit ball: $\Phi(x) = 1$ for $|x| \leq 1$, $\Phi(x) = 0$ for $|x| \geq 2$, and Φ is smooth and monotonic for $1 \leq |x| \leq 2$. In dimension $n = 1$, this looks like a snake that swallowed an elephant. It is easy to construct such a function Φ using, e.g.,

$$
\exp\left(-\frac{1}{1-y^2} + 1\right) \,, \quad 0 < y < 1 \,,
$$

which smoothly interpolates the values 1 for $y \leq 0$ and 0 for $y \geq 1$. Then the functions $\Phi_m(x) := \Phi(x/m)$ are 1 on the ball of radius m and have derivatives of order $1/m$. Hence the sequence $f\Phi_m$ in C_0^∞ approximates f with respect to all the seminorms $\|\cdot\|_{\alpha,\beta}$.

[4] We will refer to this metric later on when we discuss the dual space of $\mathscr{S}$.

Now we go on with $f_0 \in C_0^\infty$. Let L be such that $\operatorname{supp} f_0 \subset [-L,L]^n$, the centered cube in $\mathbb{R}^n$ with edges of length $2L$. We can consider f_0 as a smooth periodic function over the cube decomposition of $\mathbb{R}^n$. With the orthonormal basis

$$\phi_\mathbf{l} = \left(\frac{1}{2L}\right)^{n/2} \exp\left(\mathrm{i}\frac{\pi}{L}\mathbf{l}\cdot\mathbf{x}\right), \qquad \mathbf{l} \in \mathbb{Z}^n ,$$

the Fourier coefficients are

$$c_\mathbf{l} = \langle \phi_\mathbf{l} | f_0 \rangle = \left(\frac{1}{2L}\right)^{n/2} \int_{[-L,L]^n} \exp\left(-\mathrm{i}\frac{\pi}{L}\mathbf{l}\cdot\mathbf{x}\right) f_0(\mathbf{x}) \mathrm{d}^n x . \tag{13.20}$$

From standard analysis, we know that the Fourier series of f_0 converges uniformly to f_0, and we now apply this. To relate to the Fourier transform of f_0, we note that (13.20) implies

$$\widehat{f_0}\left(\frac{\pi}{L}\mathbf{l}\right) = \left(\frac{L}{\pi}\right)^{n/2} c_\mathbf{l} ,$$

which suggests taking the limit $L \to \infty$ of the L-Fourier series of f_0. The latter converges uniformly and is given by

$$\begin{aligned} f_0(\mathbf{x}) &= \sum_\mathbf{l} \phi_\mathbf{l}(\mathbf{x}) c_\mathbf{l} \\ &= \sum_\mathbf{l} \left(\frac{1}{2L}\right)^{n/2} \exp\left(\mathrm{i}\frac{\pi}{L}\mathbf{l}\cdot\mathbf{x}\right) \left(\frac{2\pi}{2L}\right)^{n/2} \widehat{f_0}\left(\frac{\pi}{L}\mathbf{l}\right) \\ &= \left(\frac{1}{2\pi}\right)^{n/2} \left(\frac{\pi}{L}\right)^n \sum_\mathbf{l} \exp\left(\mathrm{i}\frac{\pi}{L}\mathbf{l}\cdot\mathbf{x}\right) \widehat{f_0}\left(\frac{\pi}{L}\mathbf{l}\right) . \end{aligned}$$

But the right-hand side is just a Riemann sum for the integral $\mathscr{F}^*\widehat{f_0}$, and it converges to exactly this integral for $L \to \infty$, since $\widehat{f_0} \in \mathscr{S}$. Hence $f_0(\mathbf{x}) = \mathscr{F}^*\widehat{f_0}$, and we can conclude that $\mathscr{F}^* = \mathscr{F}^{-1}$, i.e., (13.18).[5] From this we easily get (13.15) for $g, f \in \mathscr{S}$:

[5] Note that our computation was just a mathematical way of expressing the fact that

$$\frac{1}{\sqrt{2\pi}} \int \mathrm{e}^{\mathrm{i}kx} \mathrm{d}k = \delta(x) .$$

We will comment on this formula later on.

$$\begin{aligned}\langle g|f\rangle &= \int g^*(\mathbf{x})f(\mathbf{x})\,\mathrm{d}^n x\\ &= \int g^*(\mathbf{x})\left[\left(\frac{1}{2\pi}\right)^{n/2}\int \mathrm{e}^{\mathrm{i}\mathbf{k}\cdot\mathbf{x}}\widehat{f}(\mathbf{k})\,\mathrm{d}^n k\right]\mathrm{d}^n x\\ &= \int \widehat{f}(\mathbf{k})\left[\left(\frac{1}{2\pi}\right)^{n/2}\int \mathrm{e}^{\mathrm{i}\mathbf{k}\cdot\mathbf{x}}g^*(\mathbf{x})\,\mathrm{d}^n x\right]\mathrm{d}^n k\\ &= \int \widehat{g}^*(\mathbf{k})\widehat{f}(\mathbf{k})\,\mathrm{d}^n k\\ &= \langle\widehat{g}|\widehat{f}\rangle\,,\end{aligned}$$

where interchanging the order of integration is unproblematic because the integrals converge absolutely. In addition, this immediately implies

$$\|f\|_{L^2} = \|\widehat{f}\|_{L^2}\,, \tag{13.21}$$

for $f \in \mathscr{S}$, and thus the continuity of the map $\mathscr{F} : \mathscr{S} \to \mathscr{S}$ with respect to the L^2-norm.

Another useful property of the Fourier transformation is that the Fourier transform of a product of two functions is given by the convolution of the Fourier transforms of the two functions.

Remark 13.5. Convolution
For $f, g \in \mathscr{S}$

$$\mathscr{F}(fg)(\mathbf{k}) = \left(\frac{1}{2\pi}\right)^{n/2}\int \widehat{f}(\mathbf{k}-\mathbf{k}')\widehat{g}(\mathbf{k}')\,\mathrm{d}^n k' =: \left(\frac{1}{2\pi}\right)^{n/2}\left(\widehat{f}*\widehat{g}\right)(\mathbf{k})\,,$$

and

$$\begin{aligned}\mathscr{F}(f*g)(\mathbf{k}) &= \left(\frac{1}{2\pi}\right)^{n/2}\int \mathrm{e}^{-\mathrm{i}\mathbf{k}\cdot\mathbf{x}}\int f(\mathbf{x}-\mathbf{y})g(\mathbf{y})\,\mathrm{d}^n y\,\mathrm{d}^n x\\ &= \left(\frac{1}{2\pi}\right)^{n/2}\int\int \mathrm{e}^{-\mathrm{i}\mathbf{k}\cdot(\mathbf{z}+\mathbf{y})}f(\mathbf{z})g(\mathbf{y})\,\mathrm{d}^n z\,\mathrm{d}^n y\\ &= (2\pi)^{n/2}\widehat{f}(\mathbf{k})\,\widehat{g}(\mathbf{k})\,.\end{aligned}$$

Note that the convolution integral $f*g$ is also well defined for $f, g \in L^2$:

$$(f*g)(\mathbf{x}) = \langle\tilde{f}(\cdot-\mathbf{x}), g\rangle_{L^2}\,,$$

where $\tilde{f}(\mathbf{y}) := \overline{f}(-\mathbf{y})$. ■

In the next step, we use a density argument again, in order to extend the Fourier transformation to a unitary map on L^2. One way of doing this is to say that, with

(13.21), the map $\mathscr{F}$ defines a continuous (with respect to the L^2-norm) linear operator on the dense subspace $\mathscr{S} \subset L^2$, and thus uniquely extends to a continuous linear operator on the whole space L^2. But since we have not introduced these notions yet, we shall work out the argument in more detail.

Let $f \in L^2$ and let f_n be a sequence in $\mathscr{S}$ such that f_n converges to f in the L^2-norm. Such a sequence always exists, since $\mathscr{S}$ is dense in L^2. Then f_n is a Cauchy sequence in L^2, and with (13.15) $\widehat{f_n}$ is also a Cauchy sequence in L^2. By completeness it therefore has a limit $\phi \in L^2$. This limit is independent of the choice of the original sequence f_n. For let f'_n be another Cauchy sequence converging to f. Then $(f_1, f'_1, f_2, f'_2, \ldots)$ converges to f as well, and so is Cauchy. But then by the same reasoning as before, $(\widehat{f_1}, \widehat{f'_1}, \widehat{f_2}, \widehat{f'_2}, \ldots)$ is Cauchy and converges to ϕ, since the subsequence f_n converges to ϕ. So every other subsequence, and in particular f'_n, converges to ϕ.

Consequently, we can define the Fourier transform of $f \in L^2$ to be given by the unique limit ϕ,

$$\widehat{f} = \mathscr{F} f := \phi \,.$$

It remains to show that (13.15) (which we showed for $f, g \in \mathscr{S}$) extends to $f, g \in L^2$. But this follows immediately from the norm continuity of the inner product $\langle \cdot, \cdot \rangle$. Let $(f_n), (g_n) \subset \mathscr{S}$ converge to $f, g \in L^2$ in norm. Then $\langle f_n, g_n \rangle$ also converges to $\langle f, g \rangle$. This follows from the Schwarz inequality (13.3):

$$\begin{aligned} \big|\langle f_n, g_n\rangle - \langle f, g\rangle\big| &\le \big|\langle f_n - f, g\rangle\big| + \big|\langle f_n, g - g_n\rangle\big| \\ &\le \|f_n - f\|\,\|g\| + \|f_n\|\,\|g_n - g\| \,. \end{aligned}$$

Hence $\mathscr{F}$ is a unitary operator on L^2.

If one wants to compute $\mathscr{F} f$ in the case where f is indeed only an L^2-function, then one can approximate f for example by the $L^1 \cap L^2$-function $f\chi_{[-l,l]^n}$, and approximate $\widehat{f}$ as in

$$\widehat{f} = \left(\frac{1}{2\pi}\right)^{n/2} L^2\text{-}\lim_{l\to\infty} \int \chi_{[-l,l]^n}(\mathbf{x}) \mathrm{e}^{-\mathrm{i}\mathbf{k}\cdot\mathbf{x}} f(\mathbf{x})\, \mathrm{d}^n x \,.$$

Remark 13.6. Distributions
We extended the Fourier transformation from $\mathscr{S}$ to L^2 by using the L^2 continuity of $\mathscr{F} : \mathscr{S} \to \mathscr{S}$. But as we have seen, $\mathscr{F} : \mathscr{S} \to \mathscr{S}$ is continuous in a much stronger sense, namely with respect to the family of seminorms $\|\cdot\|_{\alpha,\beta}$. This allows one to extend $\mathscr{F}$ to a much larger class of "functions", called tempered distributions. A tempered distribution $\varphi \in \mathscr{S}'$ is a continuous linear functional on $\mathscr{S}$, i.e., a linear map $\varphi : \mathscr{S} \to \mathbb{C}$ such that $\|f_n - f\|_{\alpha,\beta} \to 0$ for all $\alpha, \beta \in \mathbb{N}_0^n$ implies $\varphi(f_n) \to \varphi(f)$ in $\mathbb{C}$. The space of linear continuous functionals on a topological vector space $\mathscr{T}$ is called its dual space, denoted by $\mathscr{T}'$. So the space of tempered distributions $\mathscr{S}'$ is

just the dual space of $\mathscr{S}$ with respect to the topology on $\mathscr{S}$ induced by the family of seminorms $\|\cdot\|_{\alpha,\beta}$.[6]

For example, every function in $\mathscr{S}$ itself defines a tempered distribution as follows. For $\varphi \in \mathscr{S}$ let

$$T_\varphi(f) := \int \varphi(\mathbf{x}) f(\mathbf{x}) \mathrm{d}^n x = \langle \varphi^* | f \rangle_{L^2} , \qquad \forall f \in \mathscr{S} . \tag{13.22}$$

Motivated by the "similarity" with L^2, one often extends the inner product notation to the natural pairing of linear functionals with vectors, e.g., in the present case, for $T \in \mathscr{S}'$ and $f \in \mathscr{S}$, one writes

$$T(f) =: (T|f)_{\mathscr{S}',\mathscr{S}} . \tag{13.23}$$

For any continuous linear map $A : \mathscr{S} \to \mathscr{S}$, one can now define the adjoint map $A^* : \mathscr{S}' \to \mathscr{S}'$ through

$$(A^* \varphi)(f) := \varphi(Af) .$$

This simple trick allows one to extend the Fourier transform (and, of course, also its inverse) from $\mathscr{S}$ to $\mathscr{S}'$. Let $\varphi \in \mathscr{S}'$. Then $\mathscr{F}\varphi \in \mathscr{S}'$ is defined by

$$\mathscr{F}\varphi(f) := \varphi(\mathscr{F}f) , \qquad \forall f \in \mathscr{S} , \tag{13.24}$$

or symbolically

$$\widehat{\varphi}(f) := \varphi(\widehat{f}) , \qquad \forall f \in \mathscr{S} .$$

This is indeed an extension of the Fourier transformation on $\mathscr{S}$ in the sense of (13.22). For $\varphi, f \in \mathscr{S}$, we have

$$\begin{aligned}
\widehat{T_\varphi}(f) &\overset{(13.24)}{=} T_\varphi(\widehat{f}) \overset{(13.22)}{=} \int \varphi(\mathbf{x}) \widehat{f}(\mathbf{x}) \mathrm{d}^n x \\
&= \langle \varphi^* | \widehat{f} \rangle \overset{(13.15)}{=} \langle \widehat{\varphi}^* | f \rangle \\
&= \int \widehat{\varphi}(\mathbf{x}) f(\mathbf{x}) \mathrm{d}^n x \\
&= T_{\widehat{\varphi}}(f) .
\end{aligned}$$

However, the identification of a function φ with a distribution T_φ as in (13.22) makes sense for a much larger class of functions than $\mathscr{S}$. A sufficient condition would be, for example, that φ is measurable and polynomially bounded or, alternatively, that $\varphi \in L^2$. In both cases, T_φ defines a tempered distribution. An interesting example of

[6] One can carry out a similar construction replacing $\mathscr{S}$ by $\mathscr{D} := C_0^\infty$. The dual space $\mathscr{D}'$ of $\mathscr{D}$ is called the space of distributions. For the purposes of Fourier transformation, $\mathscr{S}$ and $\mathscr{S}'$ are advantageous, because, as we have already seen, $\mathscr{S}$, and as a consequence also $\mathscr{S}'$, is invariant under $\mathscr{F}$.

this kind is the constant function

$$\varphi = \left(\frac{1}{2\pi}\right)^{n/2},$$

i.e., the distribution

$$T_\varphi(f) = \left(\frac{1}{2\pi}\right)^{n/2} \int f(\mathbf{x})\mathrm{d}^n x\,.$$

Its Fourier transform is given by

$$\widehat{T_\varphi}(f) = T_\varphi(\widehat{f}) = \left(\frac{1}{2\pi}\right)^{n/2} \int \widehat{f}(\mathbf{k})\,\mathrm{d}^n k = f(0) \qquad \left[= \int \delta(\mathbf{x}) f(\mathbf{x})\mathrm{d}^n x \right],$$

so that, heuristically (hence the inverted commas),

$$\text{“}\,\widehat{\varphi} = \left(\frac{1}{2\pi}\right)^{n} \int \mathrm{d}^n k\, \mathrm{e}^{-\mathrm{i}\mathbf{k}\cdot\mathbf{x}} = \delta(\mathbf{x})\,\text{”}\,.$$

This is the heuristic formula $\mathscr{F}1 = \delta/(2\pi)^{n/2}$, or more simply,

$$\int \mathrm{d}^n k \mathrm{e}^{-\mathrm{i}\mathbf{k}\cdot\mathbf{x}} = (2\pi)^n \delta(\mathbf{x})\,.$$

It provides a powerful heuristic computational tool. For example, one obtains the inversion formula simply through

$$\begin{aligned}
\left(\mathscr{F}^* \widehat{f}\right)(\mathbf{x}) &= \left(\frac{1}{2\pi}\right)^{n/2} \int \mathrm{e}^{\mathrm{i}\mathbf{k}\cdot\mathbf{x}} \widehat{f}(\mathbf{k})\,\mathrm{d}^n k \\
&= \left(\frac{1}{2\pi}\right)^{n} \iint \mathrm{e}^{\mathrm{i}\mathbf{k}\cdot\mathbf{x}} \mathrm{e}^{-\mathrm{i}\mathbf{k}\cdot\mathbf{x}'} f(\mathbf{x}')\,\mathrm{d}^n k\,\mathrm{d}^n x' \\
&= \left(\frac{1}{2\pi}\right)^{n} \int f(\mathbf{x}')\,\mathrm{d}^n x' \int \mathrm{e}^{\mathrm{i}\mathbf{k}\cdot(\mathbf{x}-\mathbf{x}')}\mathrm{d}^n k \\
&= \int \mathrm{d}^n x'\, f(\mathbf{x}')\,\delta(\mathbf{x}-\mathbf{x}') \\
&= f(\mathbf{x})\,.
\end{aligned}$$

In the same way as for the Fourier transformation, one can extend other continuous linear mappings from $\mathscr{S}$ to $\mathscr{S}$ to mappings from $\mathscr{S}'$ to $\mathscr{S}'$. An important example are the partial derivatives $\partial_{x_j} : \mathscr{S}(\mathbb{R}^n) \to \mathscr{S}(\mathbb{R}^n)$, where one defines the distributional derivative for $\varphi \in \mathscr{S}'$ by

$$\left(\frac{\partial}{\partial x_j}\varphi\right)(f) := \varphi\left(-\frac{\partial}{\partial x_j} f\right), \qquad \forall f \in \mathscr{S}\,.$$

Here the minus sign on the right-hand side ensures that the distributional derivative does indeed extend the usual derivative, in the sense that, for $\varphi \in \mathscr{S}$, one has

$$\left(\partial_{x_j} T_\varphi\right)(f) = T_{\partial_{x_j}\varphi}(f)\,, \qquad \forall f \in \mathscr{S}\,.$$

In this way one can, for example, understand the Laplace operator Δ_x appearing in the Schrödinger equation as an operator acting on distributions, and thus in particular also as an operator acting on L^2-functions. One can read Schrödinger's equation

$$\mathrm{i}\frac{\mathrm{d}}{\mathrm{d}t}\psi(t) = -\frac{1}{2}\Delta_x\psi(t)$$

as an equality for distributions, and would say that $\psi(t)$ solves the Schrödinger equation in the distributional sense. What we really mean then is that, for every test function $f \in \mathscr{S}$, the time-dependent distribution $\psi(t) \in \mathscr{S}'$ satisfies [and here the alternative notation (13.23) is convenient]

$$\mathrm{i}\frac{\mathrm{d}}{\mathrm{d}t}\left(\psi(t)|f\right)_{\mathscr{S}',\mathscr{S}} = \left(-\frac{1}{2}\Delta_x\psi(t)\Big|f\right)_{\mathscr{S}',\mathscr{S}} := \left(\psi(t)\Big|-\frac{1}{2}\Delta_x f\right)_{\mathscr{S}',\mathscr{S}}\,,$$

which is now an equality in $\mathbb{C}$.

Another example is convolution, where for $\psi \in \mathscr{S}$ the map

$$\begin{aligned}\psi * : \mathscr{S} &\longrightarrow \mathscr{S}\,,\\ \varphi &\longmapsto \psi * \varphi\,,\end{aligned}$$

can be extended to a map

$$\begin{aligned}\psi * : \mathscr{S}' &\longrightarrow \mathscr{S}'\,,\\ \varphi &\longmapsto \psi * \varphi\,.\end{aligned}$$

■

Remark 13.7. Uniqueness of the Extensions

Both extensions of the Fourier transform, the one to L^2 and the one to $\mathscr{S}'$, are unique. This is because in both cases the extensions are continuous maps which are uniquely defined by their values on the dense set $\mathscr{S}$. [Although we did not show that $\mathscr{S}$ with the identification (13.22) is dense in $\mathscr{S}'$, this is in fact true.] Now we saw that L^2-functions can also be identified with distributions by the identification (13.22), i.e., $L^2 \subset \mathscr{S}'$. It is also true, and easily seen, that the restriction of $\mathscr{F} : \mathscr{S}' \to \mathscr{S}'$ to $L^2 \subset \mathscr{S}'$ agrees with the Fourier transform which was directly defined on L^2 previously. ■

Remark 13.8. On the Decay of L^2-Functions

One might naively expect (square) integrable functions to decay at ∞. The example $\mathrm{e}^{-x^2(\sin x)^2}$ shows that this is not true in general. However, for $L^2(\mathbb{R})$, the following shows what happens. If, not only φ, but also φ' is square integrable, i.e., $\varphi, \varphi' \in L^2$, then $\varphi(x) \longrightarrow 0$ for $x \longrightarrow \pm\infty$. Here $\varphi' \in \mathscr{S}'$ is first defined as the distributional derivative of $\varphi \in L^2$, but we say $\varphi' \in L^2$ if the distribution φ' is indeed given as the distribution associated with the L^2-function φ'. It turns out that, for $\varphi, \varphi' \in L^2(\mathbb{R})$, one has the usual theorem of calculus

$$\varphi(x) = \int^x \varphi'(y)\mathrm{d}y + C\,,$$

where φ is an absolutely continuous function of x [4]. Hence we can do integration by parts and find that

$$\begin{aligned}\frac{1}{2}\left|\varphi^2(b) - \varphi^2(a)\right| &= \left|\int_a^b \varphi(x)\varphi'(x)\mathrm{d}x\right| \\ &\leq \left[\int_a^b |\varphi(x)|^2\mathrm{d}x \int_a^b |\varphi'(x)|^2\mathrm{d}x\right]^{1/2} \qquad \text{[by (13.3)]} \\ &\longrightarrow 0\,,\end{aligned}$$

for $a, b \longrightarrow \infty$, since both integrands are integrable. Hence $\varphi^2(x)$ converges, since every subsequence $\varphi(x_n)$ is Cauchy and therefore convergent. But the limit must be zero, otherwise $|\varphi|^2$ could not be integrable. ■

13.2 Bilinear Forms and Bounded Linear Operators

We now return from L^2 to the abstract Hilbert space setting. Our aim is to understand that the symmetric bilinear (more precisely sesquilinear) forms are in one-to-one correspondence with the bounded symmetric operators. This is completely analogous to the same statement in linear algebra and rooted in the self-duality of Hilbert spaces.

A bounded linear functional ℓ on $\mathscr{H}$ is a linear map $\ell : \mathscr{H} \to \mathbb{C}$, for which there is a constant $c < \infty$ such that

$$|\ell(\varphi)| \leq c\|\varphi\|\,, \qquad \forall \varphi \in \mathscr{H}\,.$$

Clearly, every $\psi \in \mathscr{H}$ defines a bounded linear functional via the inner product, i.e.,

$$\begin{aligned}\psi : \mathscr{H} &\longrightarrow \mathbb{C}\,, \\ \varphi &\longmapsto \langle\psi|\varphi\rangle\,.\end{aligned}$$

But are these *all* the bounded linear functionals on $\mathscr{H}$? The answer is affirmative:

Theorem 13.2. *Let ℓ be a bounded linear functional on $\mathscr{H}$. Then there is a unique vector $\Psi_\ell \in \mathscr{H}$ such that $\ell(\varphi) = \langle\Psi_\ell|\varphi\rangle$ for all $\varphi \in \mathscr{H}$.*

This means that, to each linear functional ℓ, there corresponds a unique vector onto which ℓ projects. As mentioned before, the space of bounded linear functionals on

$\mathscr{H}$ is called the *dual space* of $\mathscr{H}$ and is denoted by $\mathscr{H}^*$. The unique identification of $\ell \in \mathscr{H}^*$ with $\psi \in \mathscr{H}$ claimed in the theorem yields $\mathscr{H}^* \cong \mathscr{H}$, i.e., $\mathscr{H}$ is dual to itself. Before we come to the simple but not completely trivial proof of the theorem, we need to understand some general geometric properties of Hilbert spaces.

Let $\mathscr{M} \subset \mathscr{H}$ be some arbitrary subset of $\mathscr{H}$. Then one defines the orthogonal complement of $\mathscr{M}$ in $\mathscr{H}$ as

$$\mathscr{M}^\perp := \left\{ \psi \in \mathscr{H} : \langle \varphi, \psi \rangle = 0, \quad \text{for all}\, \varphi \in \mathscr{M} \right\} . \tag{13.25}$$

The linearity and continuity of the inner product imply immediately that $\mathscr{M}^\perp$ is a closed subspace of $\mathscr{H}$. The following theorem also holds for non-separable Hilbert spaces, but we prove it only for separable ones.

Theorem 13.3. *Let $\mathscr{M} \subset \mathscr{H}$ be a closed subspace of a separable Hilbert space $\mathscr{H}$. Then one has $\mathscr{H} = \mathscr{M} \oplus \mathscr{M}^\perp$, i.e., each vector $\varphi \in \mathscr{H}$ can be uniquely decomposed as $\varphi = \psi + \psi^\perp$, with $\psi \in \mathscr{M}$ and $\psi^\perp \in \mathscr{M}^\perp$.*

To prove this, we first note that, as closed subspaces of a separable Hilbert space, $\mathscr{M}$ and $\mathscr{M}^\perp$ are by themselves separable Hilbert spaces, and as such allow for orthonormal bases $(\phi_n)_{n\in\mathbb{N}}$ and $(\phi_m^\perp)_{m\in\mathbb{N}}$, respectively. We will show that the orthonormal sequence $(\phi_n)_{n\in\mathbb{N}} \cup (\phi_m^\perp)_{m\in\mathbb{N}}$ is an orthonormal basis for $\mathscr{H}$, using the criterion (13.7). Assume that

$$\langle \phi_n, \varphi \rangle = 0 = \langle \phi_m^\perp, \varphi \rangle , \tag{13.26}$$

for all $n, m \in \mathbb{N}$. The first equality implies that $\langle \phi, \varphi \rangle = 0$, for all $\phi \in \mathscr{M}$, and therefore $\varphi \in \mathscr{M}^\perp$. Since $(\phi_m^\perp)_{m\in\mathbb{N}}$ is by assumption an orthonormal basis of $\mathscr{M}^\perp$, the second equality in (13.26) implies that $\varphi = 0$. Hence by (13.7), we see that $(\phi_n)_{n\in\mathbb{N}} \cup (\phi_m^\perp)_{m\in\mathbb{N}}$ is an orthonormal basis of $\mathscr{H}$. Moreover, for any $\varphi \in \mathscr{H}$, we have $\varphi = \psi + \psi^\perp$, with

$$\psi = \sum_{n=1}^{\infty} \langle \phi_n, \varphi \rangle \phi_n \in \mathscr{M} , \qquad \psi^\perp = \sum_{m=1}^{\infty} \langle \phi_m^\perp, \varphi \rangle \phi_m^\perp \in \mathscr{M}^\perp .$$

Uniqueness of the decomposition is a very easy exercise.

We now come to the proof of Theorem 13.2. Pick any bounded linear functional ℓ on $\mathscr{H}$. We are looking for a corresponding vector $\psi \in \mathscr{H}$ on which ℓ projects. In particular, all vectors orthogonal to ψ are mapped to zero by ℓ. Hence we look at the null space of ℓ, viz.,

$$\mathscr{M} = \left\{ \varphi \in \mathscr{H} \,|\, \ell(\varphi) = 0 \right\} .$$

Note that $\mathscr{M}$ is a closed subspace of $\mathscr{H}$, since ℓ is bounded and therefore continuous. If $\mathscr{M} = \mathscr{H}$, we could pick $\psi = 0$ and we would be done. Otherwise, by Theorem 13.3, the orthogonal complement $\mathscr{M}^\perp$ is at least one-dimensional. Indeed, $\mathscr{M}^\perp$ is exactly one-dimensional, which is basically the statement of the theorem. To see this, let $\psi_0, \psi_1 \in \mathscr{M}^\perp \setminus \{0\}$. Then

$$\ell(\psi_0 - \alpha\psi_1) = \ell(\psi_0) - \alpha\,\ell(\psi_1) = 0\,,$$

for $\alpha = \ell(\psi_0)/\ell(\psi_1)$, and thus

$$\psi_0 - \frac{\ell(\psi_0)}{\ell(\psi_1)}\psi_1 \in \mathscr{M} \cap \mathscr{M}^\perp = \{0\}\,.$$

Hence $\mathscr{M}^\perp$ is one-dimensional, and with a normalized $\psi_0 \in \mathscr{M}^\perp$, we can use Theorem 13.3 (and its proof) to uniquely decompose any $\varphi \in \mathscr{H}$ as

$$\varphi = \psi + \psi^\perp = \psi + \langle\psi_0|\varphi\rangle\,\psi_0\,,$$

with $\psi \in \mathscr{M}$. But then we find the desired result

$$\ell(\varphi) = \ell(\psi + \psi^\perp) = \ell(\psi) + \ell(\langle\psi_0|\varphi\rangle\,\psi_0) = \ell(\psi_0)\langle\psi_0|\varphi\rangle = \langle\Psi_\ell|\varphi\rangle\,,$$

with $\Psi_\ell = \ell(\psi_0)^*\psi_0$, where we used antilinearity in the first argument of the inner product.

Now we come to the equivalence of bounded linear operators and bounded bilinear, or more precisely sesquilinear, forms. A linear operator $\hat{A} : \mathscr{H} \to \mathscr{H}$ is bounded if there is a constant $C < \infty$ such that

$$\|\hat{A}\varphi\| \le C\|\varphi\|\,, \qquad \text{for all}\, \varphi \in \mathscr{H}\,.$$

The norm of $\hat{A}$ is the smallest such C,

$$\|\hat{A}\| := \sup_{\varphi} \frac{\|\hat{A}\varphi\|}{\|\varphi\|} = \sup_{\|\varphi\|=1} \|\hat{A}\varphi\|\,, \tag{13.27}$$

and it is quite easy to see that this definition turns the space of bounded linear operators on $\mathscr{H}$ into a complete normed space. Note that a bounded operator is obviously continuous, but the converse is true as well: every continuous linear operator is bounded.

Theorem 13.4. *Let $B(\,\cdot\,,\,\cdot\,)$ be a map from $\mathscr{H} \times \mathscr{H}$ to $\mathbb{C}$ with the following properties. For all $\varphi, \psi, \chi \in \mathscr{H}$ and $\alpha, \beta \in \mathbb{C}$ one has*

$$\text{(i)}\ B(\varphi, \alpha\psi + \beta\chi) = \alpha B(\varphi, \psi) + \beta B(\varphi, \chi)\,,$$
$$\text{(ii)}\ B(\varphi, \psi) = B(\psi, \varphi)^*\,,$$
$$\text{(iii)}\ |B(\varphi, \psi)| \le C\|\varphi\|\|\psi\|\,.$$

Then there exists a unique symmetric bounded linear operator $\widehat{A}$ on $\mathscr{H}$ such that

$$B(\varphi, \psi) = \langle\hat{A}\varphi|\psi\rangle\,, \qquad \textit{for all}\, \varphi, \psi \in \mathscr{H}\,.$$

This is an immediate consequence of Theorem 13.2, since $B(\varphi,\,\cdot\,)$ is obviously a bounded linear functional. Hence the action of $B(\varphi,\,\cdot\,)$ is given by projecting onto a unique vector $\tilde{\varphi}$, viz.,

$$B(\varphi, \psi) = \langle \tilde{\varphi} | \psi \rangle \ .$$

The mapping $\varphi \longrightarrow \tilde{\varphi}$ defines an operator $\hat{A}$ through $\hat{A}\varphi = \tilde{\varphi}$ with the property that $\langle \hat{A}\varphi | \psi \rangle = B(\varphi, \psi)$. But the properties (i)–(iii) of B imply immediately that $\hat{A}$ is linear and bounded. Linearity is obvious and boundedness is

$$||\hat{A}\varphi||^2 = \langle \hat{A}\varphi | \hat{A}\varphi \rangle = B(\varphi, \hat{A}\varphi) \leq C||\varphi|| \, ||\hat{A}\varphi|| \ ,$$

whence

$$||\hat{A}\varphi|| \leq C||\varphi|| \ .$$

Finally (ii) implies symmetry of $\hat{A}$, which means that for all $\varphi, \psi \in \mathscr{H}$,

$$\langle \hat{A}\varphi | \psi \rangle = B(\varphi, \psi) = B(\psi, \varphi)^* = \langle \hat{A}\psi | \varphi \rangle^* = \langle \varphi | \hat{A}\psi \rangle \ . \tag{13.28}$$

13.3 Tensor Product Spaces

Now we come to the question of how to describe entanglement of wave functions for N particles. Let us proceed in a purely axiomatic way for the moment. For one particle, the state space is the Hilbert space of what we shall call wave functions $L^2(\mathbb{R}^3, \mathrm{d}^3x)$. For two particles[7] we have several possibilities, if we proceed axiomatically, without taking the physical theory into account. For example, one could take the *direct* sum of the one-particle spaces $\mathscr{H}_1$ and $\mathscr{H}_2$:

$$\mathscr{H}_1 \oplus \mathscr{H}_2 = \left\{ \begin{pmatrix} \varphi_1 \\ \varphi_2 \end{pmatrix}, \varphi_1 \in \mathscr{H}_1, \varphi_2 \in \mathscr{H}_2 \right\} , \tag{13.29}$$

with the inner product

$$\left\langle \begin{pmatrix} \varphi_1 \\ \varphi_2 \end{pmatrix} \middle| \begin{pmatrix} \psi_1 \\ \psi_2 \end{pmatrix} \right\rangle = \langle \varphi_1 | \psi_1 \rangle_{\mathscr{H}_1} + \langle \varphi_2 | \psi_2 \rangle_{\mathscr{H}_2} \ .$$

Physically, this amounts to de Broglie's conception of having one wave per particle (the dimensions of the spaces add up). However, we know already that the wave function lives on configuration space, i.e., it lies in $L^2(\mathbb{R}^6, \mathrm{d}^6x)$. And this is not the direct sum of the one particle spaces $L^2(\mathbb{R}^3, \mathrm{d}^3x)$, but the *tensor product*, i.e., the space of linear superpositions of products $\varphi_1(\mathbf{x}_1)\varphi_2(\mathbf{x}_2)$.

Formally, one can form the tensor product of two given spaces $\mathscr{H}_1$ and $\mathscr{H}_2$, by just taking linear combinations of "formal products" $\phi_1 \otimes \phi_2$, $\phi_i \in \mathscr{H}_i$. One then defines an inner product on that vector space of sums of formal products in a natural

[7] We leave it to the reader to formulate the results of this section for N particles. To simplify the notation, we exemplify with 2 particles only.

way, in order to turn it, after completion, into a Hilbert space once again. However, this is so formal that we should hold on for a second and ask ourselves why we should be interested in getting the many-particle spaces in such an abstract manner. The reason is that in many models one would like to couple different systems with different kinds of degrees of freedom. With the notion of tensor product, one obtains the full Hilbert space of such coupled systems in a straightforward and unambiguous way, by just taking the tensor product of the single-particle Hilbert spaces. For example one can now easily combine spatial degrees of freedom with spin degrees of freedom.

We will discuss this later in more detail, but the moral is that we are not so much concerned about the many-particle L^2-space, which is given by physics anyway. Instead we want a convenient way to include new degrees of freedom into our models. Still we do not wish to proceed too abstractly, and choose a slightly more concrete road to introduce the tensor product. For $\varphi_1 \in \mathscr{H}_1$ and $\varphi_2 \in \mathscr{H}_2$, we define the bilinear (for N factors an N-linear) map

$$\varphi_1 \otimes \varphi_2 : \mathscr{H}_1 \times \mathscr{H}_2 \longrightarrow \mathbb{C} ,$$

$$\varphi_1 \otimes \varphi_2(\psi_1, \psi_2) := \langle \varphi_1 | \psi_1 \rangle_{\mathscr{H}_1} \langle \varphi_2 | \psi_2 \rangle_{\mathscr{H}_2} . \tag{13.30}$$

Then we take all finite linear combinations with coefficient in $\mathbb{C}$, i.e.,

$$\mathrm{span}^{\mathbb{C}}(\otimes) := \mathrm{span}^{\mathbb{C}}(\varphi_1 \otimes \varphi_2, \varphi_1 \in \mathscr{H}_1, \varphi_2 \in \mathscr{H}_2) ,$$

and on that space define the inner product as the linear extension of

$$\langle \varphi_1 \otimes \varphi_2 | \psi_1 \otimes \psi_2 \rangle_{\otimes} = \langle \varphi_1 | \psi_1 \rangle_{\mathscr{H}_1} \langle \varphi_2 | \psi_2 \rangle_{\mathscr{H}_2} . \tag{13.31}$$

What is still missing for $\mathrm{span}^{\mathbb{C}}(\otimes)$ to be a Hilbert space? Completeness, i.e., closedness under Cauchy convergence! We therefore define $\mathscr{H}_{\otimes} = \mathscr{H}_1 \otimes \mathscr{H}_2$ as the completion of $\mathrm{span}(\otimes)$ under the norm $\|\cdot\|_{\otimes} := \sqrt{\langle \cdot | \cdot \rangle_{\otimes}}$.

Remark 13.9. About Completion
We should say a few words about the general idea of completion. Completing a unitary space $\mathscr{M}$ means finding a complete unitary space $\mathscr{H}$, i.e., a Hilbert space, in which $\mathscr{M}$ can be isometrically and densely embedded. The canonical way to construct $\mathscr{H}$ is to consider equivalence classes of Cauchy sequences. Two Cauchy sequences $f_n, g_n \in \mathscr{M}$ are equivalent, if $\lim_{n\to\infty} \|f_n - g_n\| = 0$. This obviously defines an equivalence relation and $\mathscr{H}$ can be defined as the space of equivalence classes of Cauchy sequences. One can now define an inner product on $\mathscr{H}$ through $\lim_{n\to\infty} \langle f_n | g_n \rangle$, which is independent of the chosen representatives, due to continuity of the inner product. Finally, one can prove completeness of $\mathscr{H}$ along the following lines. From a given Cauchy sequence $F_n \in \mathscr{H}$ (a sequence of sequences in $\mathscr{M}$) extract the diagonal sequence $f = (F_{n,n})_n$, which is again a Cauchy sequence in $\mathscr{M}$ (use the triangle inequality here), i.e., $f \in \mathscr{H}$. Finally convince yourself that f is the limit of the Cauchy sequence F_n. One can now recover $\mathscr{M}$ in $\mathscr{H}$ by identi-

fying with $f \in \mathscr{M}$ the constant Cauchy sequence $f_n = f$ in $\mathscr{H}$. By construction the constant sequences are dense in $\mathscr{H}$, and isometry is evident. ■

However, we were a bit hasty when we talked about extending (13.31) by linearity to span($\otimes$) (resp., $\mathscr{H}_\otimes$), since it is not a priori clear whether this procedure is unique. More precisely, if for $\Phi, \Psi \in \text{span}(\otimes)$ we define $\langle \Phi | \Psi \rangle_\otimes$ through linear extension of (13.31), the result must be independent of the representation of Φ and Ψ. In other words, if Φ is the zero form, then $\langle \Phi | \Psi \rangle_\otimes = 0$ must hold for all $\Psi \in \text{span}(\otimes)$. To see this let

$$\Psi = \sum_{k=1}^{N} \alpha_k \varphi_k \otimes \eta_k \,.$$

Then, because of linearity,

$$\left\langle \Phi \,\middle|\, \sum_{k=1}^{N} \alpha_k \, \varphi_k \otimes \eta_k \right\rangle_\otimes = \sum_{k=1}^{N} \alpha_k^* \langle \Phi | \varphi_k \otimes \eta_k \rangle_\otimes = \sum_{k=1}^{N} \alpha_k^* \Phi(\varphi_k, \eta_k) = 0 \,,$$

since Φ is the zero form.

Moreover, the inner product must be positive, i.e., $\langle \Psi | \Psi \rangle_\otimes > 0$ for $\Psi \neq 0$. To check this, let

$$\Psi = \sum_{k=1}^{N} \alpha_k \varphi_k \otimes \eta_k \,.$$

By decomposing the vectors $(\varphi_k)_{k=1,\dots,N}$ and $(\eta_k)_{k=1,\dots,N}$ with respect to orthogonal bases $(\tilde{\varphi}_k)$ and $(\tilde{\eta}_k)$ of the corresponding subspaces $\text{span}(\varphi_k) \subset \mathscr{H}_1$ and $\text{span}(\eta_k) \subset \mathscr{H}_2$, we can write

$$\Psi = \sum_{k,l=1}^{N} \alpha_{kl} \tilde{\varphi}_k \otimes \tilde{\eta}_l \,,$$

whence we find that

$$\begin{aligned}
\langle \Psi | \Psi \rangle_\otimes &= \left\langle \sum_{k,l=1}^{N} \alpha_{kl} \tilde{\varphi}_k \otimes \tilde{\eta}_l \,\middle|\, \sum_{n,m=1}^{N} \alpha_{nm} \, \tilde{\varphi}_n \otimes \tilde{\eta}_m \right\rangle_\otimes \\
&= \sum_{k,l,m,n=1}^{N} \alpha_{kl}^* \alpha_{nm} \langle \tilde{\varphi}_k \otimes \tilde{\eta}_l | \tilde{\varphi}_n \otimes \tilde{\eta}_m \rangle_\otimes \\
&= \sum_{k,l,m,n=1}^{N} \alpha_{kl}^* \alpha_{nm} \langle \tilde{\varphi}_k | \tilde{\varphi}_n \rangle_{\mathscr{H}_1} \langle \tilde{\eta}_l | \tilde{\eta}_m \rangle_{\mathscr{H}_2} \\
&= \sum_{k,l,m,n=1}^{N} \alpha_{kl}^* \alpha_{nm} \delta_{kn} \delta_{lm} = \sum_{k,l=1}^{N} \alpha_{kl}^* \alpha_{kl} = \sum_{k,l=1}^{N} |\alpha_{kl}|^2 > 0 \,.
\end{aligned}$$

Remark 13.10. Warning: Product Functions Are Not Typical
Generically, $\Psi \in \mathscr{H}_{\otimes}$ is not of product form $\Psi = \varphi_1 \otimes \varphi_2$, but of the form

$$\Psi = \sum_{i,j=1}^{\infty} \alpha_{ij} \varphi_i \otimes \varphi_j \, .$$

Note also that the dimension of the tensor product space is the product of the dimensions of the single spaces. This is a clear statement for finite-dimensional spaces, and it also holds true for infinite-dimensional spaces in the following sense.

Theorem 13.5. *If* (φ_k) *and* (ψ_l) *are orthonormal bases of* $\mathscr{H}_1$ *and* $\mathscr{H}_2$*, then the set* $(\varphi_k \otimes \psi_l)_{kl}$ *is an orthonormal basis of* $\mathscr{H}_1 \otimes \mathscr{H}_2$*.*

Clearly, the $(\varphi_k \otimes \psi_l)_{kl}$ are an orthonormal system. One way to see that they do indeed form a basis is to show that the closure S of $\text{span}^{\mathbb{C}}\big((\varphi_k \otimes \psi_l)_{kl}\big)$ contains $\text{span}^{\mathbb{C}}(\otimes)$ and therefore also its closure $\mathscr{H}_1 \otimes \mathscr{H}_2$. But this follows if we just show that $\varphi \otimes \psi \in S$, for all $\varphi \in \mathscr{H}_1$ and $\psi \in \mathscr{H}_2$. Now let

$$\varphi = \sum_{k=1}^{\infty} \alpha_k \varphi_k \, , \qquad \psi = \sum_{l=1}^{\infty} \beta_l \psi_l \, ,$$

and define

$$\varphi(N) = \sum_{k=1}^{N} \alpha_k \varphi_k \, , \qquad \psi(N) = \sum_{l=1}^{N} \beta_l \psi_l \, .$$

Then $\varphi(N) \otimes \psi(N) = \sum_{k,l=1}^{N} \alpha_k^* \beta_l \varphi_k \otimes \psi_l$, and we see that the difference

$$\big\| \varphi \otimes \psi - \varphi(N) \otimes \psi(N) \big\|$$

goes to zero as $N \to \infty$:

$$\begin{aligned} \big\| \varphi \otimes \psi - \varphi(N) \otimes \psi(N) \big\| &= \Big\| \big[\varphi - \varphi(N)\big] \otimes \psi - \varphi(N) \otimes \big[\psi(N) - \psi\big] \Big\| \\ &\leq \|\varphi - \varphi(N)\| \|\psi\| + \|\varphi(N)\| \|\psi(N) - \psi\| \\ &\leq \|\varphi - \varphi(N)\| + \|\psi(N) - \psi\| \longrightarrow 0 \, . \end{aligned}$$

■

After these abstract considerations, let us look at a concrete example. The tensor product $L^2(\mathbb{R}, \mathrm{d}x) \otimes L^2(\mathbb{R}, \mathrm{d}y)$ is naturally isomorphic to $L^2(\mathbb{R}^2, \mathrm{d}x\,\mathrm{d}y)$. If we identify an element Ψ of $L^2(\mathbb{R}, \mathrm{d}x) \otimes L^2(\mathbb{R}, \mathrm{d}y)$ with a linear combination of products of functions,

$$\Psi = \sum_{k,l} \alpha_{kl} \varphi_k \otimes \psi_l \mathrel{\widehat{=}} \sum_{k,l} \alpha_{kl} \varphi_k(x) \psi_l(y) = \Psi(x,y) \, ,$$

then the function $\Psi(x,y)$ is indeed an element of $L^2(\mathbb{R}^2, \mathrm{d}x\,\mathrm{d}y)$, since

$$\iint |\Psi(x,y)|^2 \mathrm{d}x\,\mathrm{d}y = \sum_{k,l,m,n} \alpha_{kl}^* \alpha_{mn} \left[\int \varphi_k^*(x)\varphi_m(x)\mathrm{d}x\right]\left[\int \psi_l^*(y)\psi_n(y)\mathrm{d}y\right]$$
$$= \sum_{k,l} |\alpha_{kl}|^2 < \infty .$$

Here and in the following, $(\varphi_k)_k$ and $(\psi_l)_l$ are orthonormal bases of $L^2(\mathbb{R}, \mathrm{d}x)$. To see that this map from $L^2(\mathbb{R}, \mathrm{d}x) \otimes L^2(\mathbb{R}, \mathrm{d}y)$ to $L^2(\mathbb{R}^2, \mathrm{d}x\,\mathrm{d}y)$ is indeed an isomorphism, we show that products $(\varphi_k(x)\psi_l(y))_{kl}$ form an orthonormal basis of $L^2(\mathbb{R}^2, \mathrm{d}x\,\mathrm{d}y)$. We see this by once again using our criterion (13.6) for orthonormal bases. Let $\Psi \in L^2(\mathbb{R}^2, \mathrm{d}x\,\mathrm{d}y)$ be such that

$$\iint \Psi^*(x,y)\varphi_k(x)\psi_l(y)\mathrm{d}x\,\mathrm{d}y = 0 , \qquad \forall k,l .$$

We show that the only vector orthogonal to all $(\varphi_k(x)\psi_l(y))_{kl}$ is $\Psi = 0$. This follows from Fubini's theorem:

$$\int \left[\int \Psi^*(x,y)\varphi_k(x)\mathrm{d}x\right] \psi_l(y)\mathrm{d}y = 0 , \quad \forall\, l ,$$

implies that the function

$$g_k(y) = \int \Psi^*(x,y)\varphi_k(x)\mathrm{d}x \in L^2(\mathbb{R}, \mathrm{d}y)$$

vanishes outside a null set N_k. Hence, for $y \notin \bigcup N_k$, we have

$$\int \Psi^*(x,y)\varphi_k(x)\mathrm{d}x = 0 , \quad \forall k .$$

But then $\Psi^*(x,y)$ vanishes almost everywhere (with respect to "dx") and $\Psi(x,y)$ is zero almost everywhere with respect to "dx dy". Thus $(\varphi_k(x)\psi_l(y))_{kl}$ is an orthonormal basis of $L^2(\mathbb{R}^2, \mathrm{d}x\,\mathrm{d}y)$.

As a consequence, the mapping

$$U : \varphi_k \otimes \psi_l \longrightarrow \varphi_k(x)\psi_l(y)$$

maps the orthogonal basis $L^2(\mathbb{R}, \mathrm{d}x) \otimes L^2(\mathbb{R}, \mathrm{d}y)$ onto an orthogonal basis of the space $L^2(\mathbb{R}^2, \mathrm{d}x\,\mathrm{d}y)$, and we can extend it to a unitary operator

$$U : L^2(\mathbb{R}, \mathrm{d}x) \otimes L^2(\mathbb{R}, \mathrm{d}y) \longrightarrow L^2(\mathbb{R}^2, \mathrm{d}x\,\mathrm{d}y)$$

by linearity.

In this sense $L^2(\mathbb{R}, \mathrm{d}x) \otimes L^2(\mathbb{R}, \mathrm{d}y)$ is canonically isomorphic to $L^2(\mathbb{R}^2, \mathrm{d}x\,\mathrm{d}y)$, and in general

$$\bigotimes_{i=1}^{3N} L^2(\mathbb{R}, \mathrm{d}x_i) \cong L^2(\mathbb{R}^{3N}, \mathrm{d}^{3N}x) . \tag{13.32}$$

In Bohmian mechanics product functions $\varphi(x)\psi(y)$ (or $\varphi \otimes \psi$) in the tensor space describe statistical and metaphysical independence of the x- and y-systems. In general, however, the wave function evolves, through interaction potentials in the Schrödinger equation, into a typical element of the tensor space that cannot be written as a product, but is of the form

$$\sum_{k,l} \alpha_{kl} \varphi_k \otimes \psi_l \mathrel{\widehat{=}} \sum_{k,l} \alpha_{kl} \varphi_k(x) \psi_l(y) = \Psi(x,y) .$$

Remark 13.11. On Spinor Wave Functions
For the space $L^2(\mathbb{R}^3, \mathrm{d}^3x; \mathbb{C}^2)$ of spinor-valued wave functions

$$\begin{pmatrix} \varphi_1 \\ \varphi_2 \end{pmatrix} ,$$

we also find a natural isomorphism with the tensor space $L^2(\mathbb{R}^3, \mathrm{d}^3x) \otimes \mathbb{C}^2$, viz.,

$$L^2(\mathbb{R}^3, \mathrm{d}^3x; \mathbb{C}^2) \cong L^2(\mathbb{R}^3, \mathrm{d}^3x) \otimes \mathbb{C}^2 ,$$

through the identification

$$U : \varphi \otimes v \longrightarrow \varphi(\mathbf{x})v ,$$

for $\varphi \in L^2(\mathbb{R}^3, \mathrm{d}^3x)$ and $v \in \mathbb{C}^2$ and its linear extension. One can replace $\mathbb{C}^2$ by an arbitrary Hilbert space. In particular, for the wave function space of N particles with spin,

$$\begin{aligned} \bigotimes_{k=1}^{N} L^2(\mathbb{R}^3, \mathbb{C}^2) &= \bigotimes_{k=1}^{N} \left[L^2(\mathbb{R}^3) \otimes \mathbb{C}^2 \right] = \left[\bigotimes_{k=1}^{N} L^2(\mathbb{R}^3) \right] \otimes \left(\bigotimes_{k=1}^{N} \mathbb{C}^2 \right) \\ &= L^2(\mathbb{R}^{3N}) \otimes \mathbb{C}^{2^N} . \end{aligned}$$

In many applications, one reduces spin-related problems to involve only the "spin degrees of freedom". This is possible if the full wave function $\Psi \in L^2(\mathbb{R}^{3N}) \otimes \mathbb{C}^{2^N}$ has a product form $\Psi = \psi \otimes \phi$ with $\psi \in L^2(\mathbb{R}^{3N})$ and $\phi \in \mathbb{C}^{2^N}$, and if this product structure is (approximately) conserved under the time evolution, i.e., if there is no coupling between translational motion and spin dynamics. However, even if the decoupling condition is not satisfied, e.g., when discussing the EPR experiment, one often only explicitly considers the spin factor $\phi \in \mathbb{C}^{2^N}$ and its dynamics. In these cases it is very important to bear the full picture in mind. ■

Remark 13.12. On the Schmidt Basis in Tensor Spaces
A given function $\psi(\mathbf{x}, \mathbf{y})$ can always be represented in a bi-orthogonal basis with non-negative coefficients, i.e.,

$$\psi(\mathbf{x},\mathbf{y}) = \sum_n \alpha_n \tilde{\varphi}_n(\mathbf{x}) \tilde{\psi}_n(\mathbf{y}) , \tag{13.33}$$

where $(\tilde{\varphi}_n)$ and $(\tilde{\psi}_n)$ are orthonormal bases and $\alpha_n \geq 0$. This is emphasized, for example, in the context of the wave function of the universe, because (13.33) has a formal similarity with the superposition emerging from a measurement process. Mathematically, this observation goes back to work by Schmidt [5]. Although the physical implications of (13.33) are overrated, it is often convenient to have the representation (13.33) for other reasons, so we shall explain briefly how to get it.

In general, we showed that, for orthonormal bases (φ_n) and (ψ_n), one has

$$\psi(\mathbf{x},\mathbf{y}) = \sum_{n,m} \alpha_{nm} \varphi_n(\mathbf{x}) \psi_m(\mathbf{y}) .$$

For simplicity, we consider only finite sums and do some linear algebra with

$$\psi(\mathbf{x},\mathbf{y}) = \sum_{n,m=1}^{N} \alpha_{nm} \varphi_n(\mathbf{x}) \psi_m(\mathbf{y}) .$$

Under unitary transformations S and T, i.e., $S^*S = T^*T = \mathsf{I}$, the orthonormal bases (φ_n) and (ψ_n) are mapped to orthonormal bases:

$$S^* \varphi_n = \tilde{\varphi}_n , \qquad T^* \psi_n = \tilde{\psi}_n .$$

With $A = (\alpha_{nm})$, we write

$$\psi(\mathbf{x},\mathbf{y}) = \sum_{n,m=1}^{N} (SAT)_{nm} \tilde{\varphi}_n(\mathbf{x}) \tilde{\psi}_m(\mathbf{y}) ,$$

and hence (13.33) follows if we can show that there exist S and T that diagonalize A, i.e.,

$$SAT = D , \tag{13.34}$$

for some diagonal matrix D. To see that such matrices exist, let $(\mathbf{e}_k)$ denote the canonical basis of $\mathbb{R}^N$. With $T\mathbf{e}_k =: \mathbf{t}_k$ and $S\mathbf{s}_k := \mathbf{e}_k$, we can write (13.34) as

$$SAT\mathbf{e}_k = SA\mathbf{t}_k = S\delta_k\mathbf{s}_k = \delta_k\mathbf{e}_k ,$$

whenever

$$A\mathbf{t}_k = \delta_k\mathbf{s}_k . \tag{13.35}$$

Then $D = (\delta_k)$. So the question now is whether there are bases $\mathbf{s}_k$ and $\mathbf{t}_k$ such that (13.35) holds, and the answer is affirmative: they do indeed exist. With the adjoint matrix A^*, we have that A^*A and AA^* are positive self-adjoint matrices and hence diagonalizable with positive eigenvalues α_k^2 and $\tilde{\alpha}_n^2$, and an orthonormal eigenbasis

$\mathbf{t}_k$ and $\mathbf{s}_n$. But

$$A^*A\mathbf{t}_k = \alpha_k^2\mathbf{t}_k$$

implies

$$AA^*A\mathbf{t}_k = \alpha_k^2 A\mathbf{t}_k \, ,$$

i.e., $A\mathbf{t}_k$ is an eigenvector of AA^*, whose eigenvalues are $\tilde{\alpha}_n^2$ with eigenvectors $\mathbf{s}_n$. After renumbering where necessary, we have $\tilde{\alpha}_k^2 = \alpha_k^2$ and $A\mathbf{t}_k = \delta_k s_k$, with

$$\delta_k^2 = (\delta_k\mathbf{s}_k, \delta_k\mathbf{s}_k) = (A\mathbf{t}_k, A\mathbf{t}_k) = (A^*A\mathbf{t}_k, \mathbf{t}_k) = \alpha_k^2(\mathbf{t}_k, \mathbf{t}_k) = \alpha_k^2 \, ,$$

i.e., $\delta_k = \alpha_k$. ■

From now on, we shall say that the wave function is an element of a Hilbert space $\mathscr{H}$, and we shall usually mean

$$\mathscr{H} = L^2(\mathbb{R}^{3N}, \mathrm{d}^{3N}x) \, .$$

We know that only the nice smooth functions in that space, the real wave functions, are physically relevant.

References

1. H. Dym, H.P. McKean: *Fourier Series and Integrals*, Probability and Mathematical Statistics, No. 14 (Academic Press, New York, 1972)
2. M. Reed, B. Simon: *Methods of Modern Mathematical Physics I: Functional Analysis*, revised and enlarged edn. (Academic Press, San Diego, 1980)
3. M. Reed, B. Simon: *Methods of Modern Mathematical Physics. II. Fourier Analysis, Self-Adjointness* (Academic Press [Harcourt Brace Jovanovich Publishers], New York, 1975)
4. W. Rudin: *Principles of Mathematical Analysis*, 3rd edn., International Series in Pure and Applied Mathematics (McGraw-Hill Book Co., New York, 1976)
5. E. Schmidt: Math. Annalen **64** (1907)

Chapter 14
The Schrödinger Operator

In Bohmian mechanics, the dynamics of the wave function is determined by the Schrödinger equation (8.4) and the dynamics of the particle positions is determined by the guiding equation (8.3). In Remarks 7.1 and 8.2, we noted that in Bohmian mechanics the wave function must be differentiable, i.e., Bohmian mechanics is based on classical solutions of Schrödinger's equation. We will not be concerned with classical solutions in the present chapter, which develops the point of view already initiated in Chap. 12. We discuss here a new notion of solution in the sense that the Schrödinger equation gives rise to a unitary time evolution on Hilbert space.

14.1 Unitary Groups and Their Generators

As in every physical theory, the mathematical equations should be specified in such a way that, for suitable initial data, the solutions are uniquely determined for all times. In this chapter we discuss in particular the mathematical problem of setting up the Schrödinger equation in such a way that the solutions are uniquely determined for all times by the initial wave function at some arbitrary initial time. Moreover, the solution $\psi(t)$ of the Schrödinger equation also enters the guiding equation for the particles, for which we also expect the existence of solutions at all times. But if (almost) all trajectories exist for all times, then equivariance of the $|\psi|^2$-distribution leads to a further minimal requirement on $\psi(t)$, namely the conservation of the total probability $\int |\psi(t)|^2 \mathrm{d}^n x = 1$, i.e., conservation of the L^2-norm.

However, for singular potentials, like the physically relevant Coulomb potential, or for configuration spaces with boundary, the mere fact of specifying the potential does not lead to either uniqueness or, in general, conservation of norm for the solution of Schrödinger's equation. Indeed, in these cases, for any given initial data ψ_0, there exist many solutions of Schrödinger's equation, some with growing or decreasing norm. As we understood already in Chap. 12, one needs additional *boundary conditions* in order to make the physical situation described by the equations unique, and thus to select the correct physical solution. As we shall explain, using

D. Dürr, S. Teufel, *Bohmian Mechanics*, DOI 10.1007/978-3-540-89344-8_14,

the concept of self-adjointness one can at the same time enforce uniqueness of the solution and conservation of norm. Then one is left to select the physically correct self-adjoint version of the equation in order to get unique solutions.

We treat this problem as a mathematical problem, whence physical constants and dimensions will be irrelevant, and we put $\hbar = m = 1$ so that the Schrödinger equation becomes

$$\mathrm{i}\frac{\partial \psi(x,t)}{\partial t} = -\tfrac{1}{2}\Delta_x \psi(x,t) + V(x)\psi(x,t) =: H(\Delta, V)\psi(x,t) , \tag{14.1}$$

with $\psi(t=0) = \psi_0 \in L^2$ as initial condition.[1] Thinking of $\psi(t,x)$ as a vector-valued function $\psi : \mathbb{R} \to L^2, t \mapsto \psi(t)$, we can also understand (14.1) as an ordinary linear differential equation.

$$\mathrm{i}\frac{\mathrm{d}}{\mathrm{d}t}\psi(t) = H\psi(t) .$$

Then at least formally we obtain $\psi(t) = \mathrm{e}^{-\mathrm{i}tH}\psi_0$ as the solution and, since H is "real", we have, with the i in the exponent, that $U(t) = \mathrm{e}^{-\mathrm{i}tH}$ is a bounded linear operator on the Hilbert space $\mathscr{H}$. Indeed, formally we even expect $U(t) = \mathrm{e}^{-\mathrm{i}tH}$ to be unitary, so that $\|\psi(t)\| = \|U(t)\psi(0)\| = \|\psi(0)\|$. Our formal expectations motivate the following definition:

Definition 14.1. A (strongly continuous) unitary one-parameter group

$$U(t) : \mathscr{H} \longrightarrow \mathscr{H}$$

is a family of linear operators (one operator for each $t \in \mathbb{R}$) on a Hilbert space $\mathscr{H}$, such that:

(i) $t \mapsto U(t)\psi$ is continuous for each $\psi \in \mathscr{H}$,
(ii) $U(t+s) = U(t)U(s), U(0) = \mathrm{I}$,
(iii) $\|U(t)\psi\| = \|\psi\|$ for all $t \in \mathbb{R}$ and all $\psi \in \mathscr{H}$.

Condition (i) is called strong continuity, and it means that $\|U(t)\psi - U(s)\psi\| \to 0$ for $t \to s$. From (ii), it follows that $U(t)$ is invertible with inverse $U(-t)$. Hence, together with (iii), it follows that $U(t)$ is indeed unitary [see also the paragraph above (13.10)]. Note also that, for unitary groups, strong continuity is equivalent to weak continuity, since

$$\lim_{t\to 0}\left\| \left[U(t) - \mathrm{I}\right]\psi \right\|^2 = 2\|\psi\|^2 - 2\lim_{t\to 0}\mathrm{Re}\langle\psi|U(t)\psi\rangle = 0 .$$

This definition captures all our requirements for the Schrödinger flow on the space of wave functions: existence, uniqueness, and conservation of norm. The only thing

[1] In the last chapter, we saw that derivatives can be defined in a weak or distributional sense, so we know how Δ_x acts on an L^2-function, even if the latter is not differentiable.

missing is the connection with the Schrödinger equation, namely the requirement that

$$\mathrm{i}\frac{\mathrm{d}}{\mathrm{d}t}U(t) = HU(t)$$

should hold with H as the Hamilton operator, or for short, the Hamiltonian, on $\mathscr{H}$. At this point we must be careful to distinguish between $H(\Delta, V)$ in (14.1) and the Hamilton operator H. Why is this?

In fact, $H(\Delta, V)$ is just a *differential operator* defined on a subset of differentiable L^2-functions, e.g., on C_0^∞. In particular, while densely defined, $H(\Delta, V)$ is certainly not an operator defined on all of L^2. And if we define it on all of L^2 in the distributional sense, then it does not map L^2 into L^2. Hence we have to specify on precisely which set we want H to act,[2] i.e., we have to specify its *domain* $\mathscr{D}(H)$.

Recall the simple example of the particle on the half-line. There the prescription for "solving" the Schrödinger equation only for $x > 0$ and for initial data ψ_0 with $\operatorname{supp}\psi_0 \in (0,\infty)$ does not yield a unique solution. We had to specify *boundary conditions* – but this just means that, for each boundary condition $a \in \mathbb{R}$ that we impose [see (12.50)], we pick a domain of definition $\mathscr{D}(H_a)$ for H. Hence we actually talk about different operators H_a that all act like $-\Delta/2$ on the functions in their domain, but the domains $\mathscr{D}(H_a)$ differ. And this is exactly the subtlety we need to consider when introducing the *Schrödinger operator* H together with its domain $\mathscr{D}(H)$.

But what exactly are good boundary conditions? Just specifying some domain $\mathscr{D}(H)$ will not do anything for us, as the simple example of the half-line shows. By just putting

$$\mathscr{H} = L^2\big((0,\infty), \mathrm{d}x\big)\,, \qquad H = -\frac{1}{2}\frac{\mathrm{d}^2}{\mathrm{d}x^2}\,, \qquad \mathscr{D}(H) = \Big\{\varphi \in C_0^\infty\big((0,\infty)\big)\Big\}\,,$$

everything is precisely defined, but our problem is not solved. For an initial $\varphi \in \mathscr{D}(H)$, there is *no* solution with $\varphi(t) \in \mathscr{D}(H)$ for $t > 0$, but *many different* solutions with $\varphi(t) \notin \mathscr{D}(H)$.

Hence "good" boundary conditions, or more generally "good" domains $\mathscr{D}(H)$, should be such that, for any initial state $\varphi(0)$ in the domain, there is a unique solution $\varphi(t)$ of (14.1) which remains in the domain for all times. In other words the unitary group $U(t)$ corresponding to H with domain $\mathscr{D}(H)$ should leave the domain invariant:

$$U(t)\mathscr{D}(H) = \mathscr{D}(H)\,, \qquad \text{for all } t \in \mathbb{R}\,.$$

We now turn the desired connection between the unitary solution group $U(t)$ and its generator H with domain $\mathscr{D}(H)$ into a definition. By doing this, we adopt a rather

---

[2] Linear operators that cannot be defined on the whole Hilbert space, but only on a dense subspace, are called densely defined unbounded operators. If a densely defined operator were bounded, it could be uniquely extended to the whole space by continuity.

non-standard approach, turning everything upside down, as it may seem. But, at least for the present purpose of understanding the time evolution in Bohmian (and thus in quantum) mechanics, this is indeed the proper way to proceed.

To recapitulate, we want to identify operators H and domains $\mathscr{D}(H)$ such that the solutions of Schrödinger's equation are given by a unique unitary group $U(t)$, i.e., the unique solution to

$$\mathrm{i}\frac{\mathrm{d}}{\mathrm{d}t}U(t) = HU(t) \,.$$

As we shall see, all requirements are captured by the following definition.

Definition 14.2. A densely defined operator H is called the *generator* of a unitary group $U(t)$ if the following holds:

(i) $\mathscr{D}(H) = \left\{\varphi \in \mathscr{H} \,|\, t \mapsto U(t)\varphi \text{ is differentiable}\right\}$,

(ii) $\mathrm{i}\frac{\mathrm{d}}{\mathrm{d}t}U(t)\varphi = HU(t)\varphi$, for all $\varphi \in \mathscr{D}(H)$.

Note. The statement that $t \mapsto U(t)\varphi$ is differentiable means (using the group property) that there is an element $\psi \in \mathscr{H}$ such that

$$\lim_{t\to 0}\left\|\frac{U(t)\varphi - \varphi}{t} - \psi\right\| = 0 \,.$$

It then follows from Definition 14.2 (ii) that this ψ is actually given by $-\mathrm{i}H\varphi$. This motivates the notation $U(t) = \mathrm{e}^{-\mathrm{i}tH}$ (see also Remark 14.3). In particular, we can also differentiate within the inner product (actually within any bounded linear functional):

$$\begin{aligned}
\frac{\mathrm{d}}{\mathrm{d}t}\langle\psi\,|\,U(t)\varphi\rangle\Big|_{t=0} &= \lim_{t\to 0}\left\langle\psi\,\Big|\,\frac{U(t)\varphi-\varphi}{t}\right\rangle \\
&= \lim_{t\to 0}\left\langle\psi\,\Big|\,\frac{U(t)\varphi-\varphi}{t} + \mathrm{i}H\varphi\right\rangle - \langle\psi\,|\,\mathrm{i}H\varphi\rangle \\
&= -\langle\psi\,|\,\mathrm{i}H\varphi\rangle \,,
\end{aligned}$$

where the first term is zero by continuity of the inner product.

The definition is very compact. For example, Definition 14.2 (i) together with the group property of $U(t)$ implies the invariance of the domain $\mathscr{D}(H)$, i.e., for all $t \in \mathbb{R}$, one has $U(t)\mathscr{D}(H) = \mathscr{D}(H)$. Moreover, it follows that $U(t)H\varphi = HU(t)\varphi$ for all $t \in \mathbb{R}$, since

$$U(t)H\varphi = U(t)\mathrm{i}\,\frac{\mathrm{d}}{\mathrm{d}s}U(s)\varphi\Big|_{s=0} = \mathrm{i}\,\frac{\mathrm{d}}{\mathrm{d}s}U(s)U(t)\varphi\Big|_{s=0} = HU(t)\varphi \,.$$

Hence, $\|HU(t)\varphi\| = \|H\varphi\|$.

It is also easy to see that the group $U(t)$ is uniquely determined by H. Let $\tilde{U}(t)$ be a unitary group, also generated by H, and consider

$$\begin{aligned}\frac{\mathrm{d}}{\mathrm{d}t}\left\|\left[U(t)-\tilde{U}(t)\right]\varphi\right\|^2 &= 2\frac{\mathrm{d}}{\mathrm{d}t}\left[\|\varphi\|^2-\Re\langle U(t)\varphi|\tilde{U}(t)\varphi\rangle\right]\\ &= -2\Re\left[\langle -\mathrm{i}HU(t)\varphi|\tilde{U}(t)\varphi\rangle+\langle U(t)\varphi|-\mathrm{i}H\tilde{U}(t)\varphi\rangle\right]\\ &= -2\Re\left[\mathrm{i}\langle HU(t)\varphi|\tilde{U}(t)\varphi\rangle-\mathrm{i}\langle U(t)\varphi|H\tilde{U}(t)\varphi\rangle\right].\end{aligned}$$

This is zero if H is *symmetric* and then we would have uniqueness, since

$$\left\|\left[U(t)-\tilde{U}(t)\right]\varphi\right\| = \left\|\left[U(0)-\tilde{U}(0)\right]\varphi\right\| = 0\,,$$

by (ii) in Definition 14.1.

Definition 14.3. An operator H is called symmetric (or Hermitian), if $\langle\varphi|H\psi\rangle = \langle H\varphi|\psi\rangle$ holds for all $\varphi,\psi\in\mathscr{D}(H)$.

And indeed we have that

$$H \text{ generator} \implies H \text{ symmetric}\,, \tag{14.2}$$

since, for $\varphi,\psi\in\mathscr{D}(H)$, it follows that

$$\begin{aligned}0 = \frac{\mathrm{d}}{\mathrm{d}t}\langle\varphi|\psi\rangle &= \frac{\mathrm{d}}{\mathrm{d}t}\langle U(t)\varphi|U(t)\psi\rangle\\ &= \langle -\mathrm{i}HU(t)\varphi|U(t)\psi\rangle+\langle U(t)\varphi|-\mathrm{i}HU(t)\psi\rangle\\ &= \mathrm{i}\langle U(t)H\varphi|U(t)\psi\rangle-\mathrm{i}\langle U(t)\varphi|U(t)H\psi\rangle\\ &= \mathrm{i}\big(\langle H\varphi|\psi\rangle-\langle\varphi|H\psi\rangle\big)\,.\end{aligned}$$

In conclusion, if the Schrödinger operator is the generator of a unitary group, then we have all we were asking for: existence, uniqueness, and conservation of norm. Hence all we are missing is a good *criterion* to actually determine whether a given operator generates a unitary group.

To this end, first recall (12.47), the vanishing of the flux integral that yields the conservation of probability:

$$\begin{aligned}\int\frac{\partial|\psi(t)|^2}{\partial t}\mathrm{d}x &= -\mathrm{i}\int\left[\psi(t)^*H(\Delta,V)\psi(t)-\psi(t)H(\Delta,V)\psi(t)^*\right]\mathrm{d}x\\ &= -\int\nabla\cdot\mathbf{j}^{\psi(t)}\mathrm{d}x = -\int\mathbf{j}^{\psi(t)}\cdot\mathrm{d}\sigma = 0\,.\end{aligned} \tag{14.3}$$

Hence, we find the criterion

$$\int\psi^*H(\Delta,V)\psi\mathrm{d}x = \int\psi H(\Delta,V)\psi^*\mathrm{d}x\,, \tag{14.4}$$

which is just the symmetry we already have. It is clear that this cannot be the end of the story. In our example,

$$H_0 = -\frac{1}{2}\frac{\mathrm{d}^2}{\mathrm{d}x^2} \quad \text{on } \mathscr{D}(H_0) = C_0^\infty\big((0,\infty)\big)$$

is symmetric [see (14.11)]. But H cannot be a generator because the time evolution is not uniquely determined.

How can (14.4) hold, when (14.3) does not? In the example, we saw that the left-hand side of (14.3) can be negative, since there are solutions where the particle leaves the half-line at the origin and vanishes. But such solutions are not zero at the origin, $\psi(t,x=0) \neq 0$, and hence $\psi(t) \notin \mathscr{D}(H_0)$. Our notation in (14.4) was sloppy, since (14.4) must hold for $\psi(t)$. And in general it can happen that $\psi(0) \in \mathscr{D}(H)$ but $\psi(t) \notin \mathscr{D}(H)$, and then (14.4) is not defined. We thus need to pick the domain $\mathscr{D}(H)$ of H so that it remains invariant under the time evolution. The domain $\mathscr{D}(H_0) = C_0^\infty\big((0,\infty)\big)$ for $H_0 = -(1/2)\mathrm{d}^2/\mathrm{d}x^2$ is certainly not invariant under the time evolution, since sooner or later (actually sooner) the solution will reach the origin.

14.2 Self-Adjoint Operators

So far so good. The invariance of the domain was already part of Definition 14.2, but we have come a good way towards uncovering the difference between "symmetric" and "generating". Let us repeat what we have understood so far. If the domain is too small, there may be no solutions that stay within the domain. On the other hand, if the domain is too big, there may be more than one solution. So exactly how big should the domain of a generator be? Let us try to answer that question heuristically to begin with.

Consider a symmetric operator H_0 with a – possibly too small – domain $\mathscr{D}(H_0)$. Let $H_{\max}$ be the "same operator" on a domain $\mathscr{D}(H_{\max}) \supset \mathscr{D}(H_0)$ on which it is maximally symmetric. More precisely, we assume that $H_{\max}|_{\mathscr{D}(H_0)} = H_0$, that $H_{\max}$ is still symmetric, but that there is no larger domain to which $H_{\max}$ can be extended and still be a symmetric operator. Now let us assume that, at least for small times $|t| < t_0$ and $\psi(0) \in \mathscr{D}(H_0)$, the Schrödinger equation

$$\mathrm{i}\frac{\mathrm{d}}{\mathrm{d}t}\psi(t) = H_{\max}\,\psi(t)$$

has a solution $\psi(t)$ with $\psi(t) \in \mathscr{D}(H_{\max})$, but not necessarily $\psi(t) \in \mathscr{D}(H_0)$. But where can $\psi(t)$ go when it leaves $\mathscr{D}(H_0)$? Because of the symmetry of $H_{\max}$, and because $H_{\max}$ extends H_0, i.e., $\mathscr{D}(H_{\max}) \supset \mathscr{D}(H_0)$ and $H_{\max}|_{\mathscr{D}(H_0)} = H_0$, we have, for all $\varphi \in \mathscr{D}(H_0)$,

$$\big\langle \varphi | H_{\max}\psi(t)\big\rangle = \big\langle H_{\max}\varphi|\,\psi(t)\big\rangle = \big\langle H_0\varphi|\,\psi(t)\big\rangle\,. \tag{14.5}$$

Hence $\psi(t)$ always ends up in the domain of the adjoint operator H_0^*, which we now define with (14.5) in mind.

Definition 14.4. Let H be a densely defined linear operator on $\mathscr{H}$. Let $\mathscr{D}(H^*)$ be the set of all $\psi \in \mathscr{H}$ for which there exists an $\eta \in \mathscr{H}$ such that

$$\langle \psi | H\varphi \rangle = \langle \eta | \varphi \rangle , \qquad \text{for all } \varphi \in \mathscr{D}(H) . \tag{14.6}$$

For each $\psi \in \mathscr{D}(H^*)$, we define $H^*\psi = \eta$ and call H^* the adjoint operator to H. Hence we have $\langle \psi | H\varphi \rangle = \langle H^*\psi | \varphi \rangle$ for all $\varphi \in \mathscr{D}(H)$ and $\psi \in \mathscr{D}(H^*)$.

Since $\mathscr{D}(H)$ is dense, η in (14.6) is unique if it exists, and the operator H^* is thus also uniquely defined. The star notation is reminiscent of complex conjugation. This is intended, since the adjoint of a complex number c, or more precisely of the operator $c\mathrm{l}$, is just the complex conjugate number c^*. While $\mathscr{D}(H^*)$ need not be dense in general, for symmetric operators H, the adjoint H^* is again a densely defined operator. This is because H is symmetric (see Definition 14.3) if and only if $\mathscr{D}(H) \subset \mathscr{D}(H^*)$ and $H\varphi = H^*\varphi$ for all $\varphi \in \mathscr{D}(H)$. We now formulate the result of our heuristic argument (14.5) as a theorem.

Theorem 14.1. *H is the generator of a strongly continuous unitary group if and only if $H = H^*$, i.e., H is symmetric and $\mathscr{D}(H) = \mathscr{D}(H^*)$. Such an operator H is said to be self-adjoint.*

This is the main content of Stone's theorem [1]. Stone's theorem states in addition that any strongly continuous unitary group has a generator, which by the above statement must be self-adjoint. Self-adjointness is a property of operators that is of general mathematical interest. As such, it is usually considered independently of the property of being a generator, and this is why we did not introduce only the terminology "self-adjoint". However, in the following, we will use it synonymously with "generator".

The following argument shows once again why we expect any generator H of a unitary group to be self-adjoint. Let $\psi \in \mathscr{D}(H^*)$. Then we have, for all $\varphi \in \mathscr{D}(H)$,

$$\langle \psi | H\varphi \rangle = \left\langle \psi \middle| \mathrm{i}\frac{\mathrm{d}}{\mathrm{d}t} U(t)\varphi \right\rangle \Big|_{t=0} = \mathrm{i}\frac{\mathrm{d}}{\mathrm{d}t} \langle \psi | U(t)\varphi \rangle \Big|_{t=0} = \mathrm{i}\frac{\mathrm{d}}{\mathrm{d}t} \langle U(-t)\psi | \varphi \rangle \Big|_{t=0} .$$

Since $\psi \in \mathscr{D}(H^*)$, there exists $\eta \in \mathscr{H}$ such that $\langle \psi | H\varphi \rangle = \langle \eta | \varphi \rangle$ and thus

$$\mathrm{i}\frac{\mathrm{d}}{\mathrm{d}t} \langle U(-t)\psi | \varphi \rangle \Big|_{t=0} = \langle \eta | \varphi \rangle ,$$

i.e., $U(t)\psi$ is weakly differentiable at $t = 0$. If we assume for the moment that this implies also strong differentiability of $U(-t)\psi$, then it would follow that $\psi \in \mathscr{D}(H)$, and therefore $\mathscr{D}(H) = \mathscr{D}(H^*)$, i.e., $H^* = H$.

In order to close the gap in the above argument and to understand that the converse is also true, i.e., H self-adjoint $\Longrightarrow$ H generates a unitary group, we need to develop the idea of self-adjointness. In particular, we need to find a convenient criterion for self-adjointness, since explicitly determining the domain of the adjoint operator is not practicable in many cases.

The way we would like to proceed for the Schrödinger operator is the following. We start with a simple set of "nice" functions, e.g., $C_0^\infty(\mathbb{R}^n)$, on which the operator $H(\Delta, V)$ is symmetric. Then we extend this set further by adding, e.g., all twice differentiable functions φ with $H(\Delta, V)\varphi \in L^2$, and so on. But we have already seen that, if we add too much, the operator may no longer be symmetric. On the other hand, if the domain $\mathscr{D}(H)$ is too small, then $\mathscr{D}(H^*)$ is too big, i.e., $\mathscr{D}(H^*) \supsetneq \mathscr{D}(H)$. One can study the question of "balancing" the domains most conveniently by considering the graph of an operator.

Definition 14.5. The graph of an operator H is the linear subspace

$$\Gamma(H) = \{(\varphi, H\varphi) | \varphi \in \mathscr{D}(H)\} \subset \mathscr{H} \oplus \mathscr{H} .$$

We say that H is closed, if $\Gamma(H)$ is a closed subset of $\mathscr{H} \oplus \mathscr{H}$.

Clearly two operators are the same if and only if their graphs are the same.

Definition 14.6. Let H_1 and H be operators on $\mathscr{H}$. If $\Gamma(H) \subset \Gamma(H_1)$, then H_1 is an extension of H, in short $H \subset H_1$. An operator H is called closable if it has a closed extension. The smallest closed extension of a closable operator H is called the closure of H and denoted by $\overline{H}$. Clearly, $\overline{\Gamma(H)} = \Gamma(\overline{H})$.

For symmetric operators H, we have seen that the adjoint H^* is an extension of H, so in a certain sense $\mathscr{D}(H^*)$ contains all that was missing in $\mathscr{D}(H)$. It is not therefore too surprising to find that the following theorem holds.

Theorem 14.2. *Let H be a densely defined operator on $\mathscr{H}$. Then H^* is closed.*

The proof uses concepts that turn out to be useful for the following. According to Definition 14.4, we have

$$\begin{aligned}(\psi, \eta) \in \Gamma(H^*) &\Longleftrightarrow \langle \psi | H\varphi \rangle = \langle \eta | \varphi \rangle , \quad \text{for all } \varphi \in \mathscr{D}(H) , \\ &\Longleftrightarrow \langle \psi | H\varphi \rangle - \langle \eta | \varphi \rangle = 0 , \quad \text{for all } \varphi \in \mathscr{D}(H) ,\end{aligned}$$

or equivalently, with the inner product on $\mathscr{H} \oplus \mathscr{H}$ [see (13.29)],

$$\big\langle (\psi, \eta) | (-H\varphi, \varphi) \big\rangle = 0 , \quad \text{for all } \varphi \in \mathscr{D}(H) . \tag{14.7}$$

Geometrically, this means that (ψ, η) is orthogonal to $(-H\varphi, \varphi)$, which is essentially the graph of H. Introducing the unitary map

$$\begin{aligned} W : \mathscr{H} \oplus \mathscr{H} &\longrightarrow \mathscr{H} \oplus \mathscr{H} , \\ (\varphi, \psi) &\longmapsto W(\varphi, \psi) = (-\psi, \varphi) ,\end{aligned}$$

equation (14.7) says that $(\psi, \eta) \in \Gamma(H^*)$ if and only if $(\psi, \eta) \in W\big(\Gamma(H)\big)^\perp$, i.e.,

$$\Gamma(H^*) = W\big(\Gamma(H)\big)^{\perp} . \tag{14.8}$$

Since orthogonal complements are always closed [see (13.25)], $\Gamma(H^*)$ is closed.

Now, if H is symmetric, H is closable since $H \subset H^*$. But H^* need not be the closure $\overline{H}$ of H, and H is not necessarily the adjoint of H^*.

Corollary 14.1. *Let H be densely defined and closable and assume that H^* is also densely defined. Then $\overline{H} = H^{**} := (H^*)^*$ and $(\overline{H})^* = H^{***} = H^*$.*

Proof. It is easy to see that, for arbitrary subspaces $\mathscr{M} \subset \mathscr{H} \oplus \mathscr{H}$, one has $W(\mathscr{M}^{\perp}) = W(\mathscr{M})^{\perp}$. From (14.8), we conclude that

$$\begin{aligned}\Gamma(H^{**}) &= W\big(\Gamma(H^*)\big)^{\perp} = W\Big(W\big(\Gamma(H)\big)^{\perp}\Big)^{\perp} = W\left(\Big(W\big(\Gamma(H)\big)^{\perp}\Big)^{\perp}\right) \\ &= W\Big(W\big(\Gamma(H)^{\perp}\big)^{\perp}\Big) = W\Big(W\Big(\big(\Gamma(H)^{\perp}\big)^{\perp}\Big)\Big) .\end{aligned}$$

For general subspaces $\mathscr{M}$, we have

$$(\mathscr{M}^{\perp})^{\perp} = \overline{\mathscr{M}}$$

(not $\mathscr{M}$ but its closure, since orthogonal complements are always closed). Hence with $W^2 = -\mathbb{I}$, we finally get

$$\Gamma(H^{**}) = \overline{\Gamma(H)} = \Gamma(\overline{H}) .$$

By the same reasoning,

$$\Gamma(H^{***}) = \overline{\Gamma(H^*)} = \Gamma(H^*) ,$$

since H^* is closed.

For symmetric H, we thus have

$$\begin{aligned} &\text{(i)} \quad H \subset H^{**} = \overline{H} \subset H^* , \\ &\text{(ii)} \quad \overline{H}^* = H^* , \end{aligned} \tag{14.9}$$

and, if H is self-adjoint,

$$H = H^{**} = H^* . \tag{14.10}$$

From (14.9) (i) we see immediately that, if H^* is symmetric, i.e., $H^* \subset H^{**}$, then

$$H^{**} = \overline{H} = H^* ,$$

and thus $H^* = \overline{H}$ is self-adjoint. This motivates the following definition.

Definition 14.7. A symmetric operator H is essentially self-adjoint if its closure $\overline{H}$ is self-adjoint.

Corollary 14.2. *A symmetric operator H is essentially self-adjoint if and only if H^* is symmetric. Then $\overline{H} = H^*$.*

Proof. We have already shown that H^* symmetric $\Longrightarrow$ H essentially self-adjoint and $\overline{H} = H^*$. For the other direction, let H be essentially self-adjoint, i.e., $\overline{H} = \overline{H}^*$. With (14.9) (ii), we have $\overline{H}^* = H^*$, and with the equality $\overline{H} = H^{**}$ in (i), it follows that $H^* = H^{**}$. Thus H^* is self-adjoint and in particular symmetric.

Essential self-adjointness is sufficient for characterizing a self-adjoint operator uniquely. We only need to know a domain of essential self-adjointness or a *core* of a self-adjoint operator H, where $\mathscr{D} \subset \mathscr{D}(H)$ is a core for H if

$$\overline{H|_{\mathscr{D}}} = H \,,$$

i.e., if the closure of the restriction of H to $\mathscr{D}$ is again H itself.

Let us examine the concept of self-adjointness for two simple but very instructive examples. The free Hamiltonian

$$H_0 = -\frac{\mathrm{d}^2}{\mathrm{d}x^2} \,, \quad \text{with domain } \mathscr{D}(H_0) = C_0^\infty(\mathbb{R}) \,, \tag{14.11}$$

is essentially self-adjoint. For $\varphi \in C_0^\infty$ and $\psi \in L^2$ such that $\psi' \in L^2$, we have

$$\begin{aligned} \langle \psi | H_0 \varphi \rangle &= \int_{-\infty}^{\infty} \psi^* H_0 \varphi \,\mathrm{d}x = -\int_{-\infty}^{\infty} \psi^* \varphi'' \mathrm{d}x = \int_{-\infty}^{\infty} \psi^{*\prime} \varphi' \mathrm{d}x - \psi^* \varphi' \Big|_{-\infty}^{\infty} \\ &= -\int_{-\infty}^{\infty} \psi^{*\prime\prime} \varphi \,\mathrm{d}x + \psi^{*\prime} \varphi \Big|_{-\infty}^{\infty} - \psi^* \varphi' \Big|_{-\infty}^{\infty} \,, \end{aligned} \tag{14.12}$$

where the derivatives on ψ^* are taken in the distributional sense. From this we read off the following facts:

1. H_0 is symmetric on $C_0^\infty(\mathbb{R})$, since for $\varphi \in C_0^\infty$ the boundary terms vanish, and thus for $\psi, \varphi \in C_0^\infty$, the right-hand side equals $\langle H_0 \psi \,|\, \varphi \rangle_{L^2}$.
2. The domain of the adjoint operator H_0^* is

$$\mathscr{D}(H_0^*) = \left\{ \psi \in L^2 \middle| \psi'' \in L^2 \right\} = \left\{ \psi \in L^2 \middle| \, |k|^2 \widehat{\psi} \in L^2 \right\} ,$$

   since exactly for $\psi \in \mathscr{D}(H_0^*)$, we have

$$-\int_{-\infty}^{\infty} \psi^{*\prime\prime} \varphi \,\mathrm{d}x = \langle -\psi'' | \varphi \rangle_{L^2} \,.$$

3. H_0^* is symmetric and thus, according to Corollary 14.2, $H_0^* = \overline{H}$ is self-adjoint. Symmetry of H_0^* follows most directly in the Fourier representation,

$$\langle\psi|H_0^*\varphi\rangle = \langle\widehat{\psi}|\widehat{H_0^*\varphi}\rangle = \langle\widehat{\psi}|\,k^2\widehat{\varphi}\rangle = \langle k^2\widehat{\psi}|\widehat{\varphi}\rangle = \langle H_0^*\psi|\varphi\rangle\,.$$

Alternatively, we can show symmetry of H_0^* along the lines of (14.12). From Remark 13.8, we take

$$\psi' \in L^2 \iff k\widehat{\psi} \in L^2\,,$$

and since $k^2\widehat{\psi} \in L^2$ and $\widehat{\psi} \in L^2$, we have

$$\langle\widehat{\psi}|k^2\widehat{\psi}\rangle = \langle k\widehat{\psi}|k\widehat{\psi}\rangle = \|k\widehat{\psi}\|^2\,,$$

i.e., $k\widehat{\psi} \in L^2$ and therefore $\psi' \in L^2$. From this we get that $\psi, \psi', \psi'' \in L^2$ as well as $\psi(x) \longrightarrow 0$ and $\psi'(x) \longrightarrow 0$, for $|x| \longrightarrow \infty$. Hence, the boundary terms also vanish for $\psi \in \mathscr{D}(H_0^*)$.

On the other hand,

$$H_0 = -\frac{\mathrm{d}^2}{\mathrm{d}x^2}\,, \quad \text{with domain } \mathscr{D}(H_0) = C_0^\infty\big((0,\infty)\big) \subset L^2(\mathbb{R}^+,\mathrm{d}x) \tag{14.13}$$

is not essentially self-adjoint. For $\varphi \in C_0^\infty\big((0,\infty)\big)$, one has

$$\begin{aligned}\int_0^\infty \psi^* H_0\varphi\,\mathrm{d}x &= -\int_0^\infty \psi^*\varphi''\mathrm{d}x \\ &= -\int_0^\infty \psi^{*\prime\prime}\varphi\,\mathrm{d}x + \psi^{*\prime}\varphi\Big|_0^\infty - \psi^*\varphi'\Big|_0^\infty\,.\end{aligned} \tag{14.14}$$

As in the previous example one reads off directly that:

1. H_0 is symmetric on $C_0^\infty\big((0,\infty)\big)$.
2. The domain of H_0^* is $\mathscr{D}(H_0^*) = \Big\{\psi \in L^2(\mathbb{R}^+,\mathrm{d}x)\,\big|\,\psi'' \in L^2(\mathbb{R}^+,\mathrm{d}x)\Big\}$.
3. H_0^* is not symmetric on $\mathscr{D}(H_0^*)$, since the boundary terms at 0 no longer vanish. (The boundary terms at ∞ vanish by the same argument as in the previous example.) Hence H_0 is not essentially self-adjoint on $C_0^\infty\big((0,\infty)\big)$.

The problem in this example is clearly that $\mathscr{D}(H_0^*)$ is too big for $-\mathrm{d}^2/\mathrm{d}x^2$ to still be symmetric. By inspecting (14.14), we see that we can shrink $\mathscr{D}(H_0^*)$ by enlarging $\mathscr{D}(H_0)$. But we have to do it carefully in order not to destroy the symmetry of H_0. In this simple example, the solution is easy to guess: if the boundary terms in (14.14) at 0 do not vanish individually, they must at least cancel out. This happens if, for $a \in \mathbb{R}$, we define the operator

$$H_{0,a} = -\frac{\mathrm{d}^2}{\mathrm{d}x^2}$$

with domain

$$\mathscr{D}(H_{0,a}) = \Big\{\varphi \in L^2(\mathbb{R}^+,\mathrm{d}x)\,\big|\,\varphi'' \in L^2(\mathbb{R}^+,\mathrm{d}x), \varphi'(0) = a\,\varphi(0)\Big\}\,.$$

And indeed, $H_{0,a}$ is self-adjoint [see also (12.50)], since, for $\varphi \in \mathscr{D}(H_{0,a})$, we have

$$\int_0^\infty \psi^* H_{0,a}\varphi \,\mathrm{d}x = -\int_0^\infty \psi^{*\prime\prime}\varphi \,\mathrm{d}x - \psi^{*\prime}(0)\varphi(0) + \psi^*(0)\varphi'(0)$$
$$= -\int_0^\infty \psi^{*\prime\prime}\varphi \,\mathrm{d}x - \psi^{*\prime}(0)\varphi(0) + a\,\psi^*(0)\varphi(0)\,.$$

Now $\psi \in \mathscr{D}(H_{0,a}^*)$ if and only if the boundary terms vanish and $\psi'' \in L^2(\mathbb{R}^+, \mathrm{d}x)$. However, the boundary terms vanish if and only if $\psi'(0) = a\psi(0)$, and therefore $\mathscr{D}(H_{0,a}^*) = \mathscr{D}(H_{0,a})$.

In general, however, one cannot explicitly compute the domain of the adjoint operator, and what we need is an abstract and at the same time accessible criterion for self-adjointness. To recapitulate, in order to have $H = H^*$ for a symmetric operator H, necessary and sufficient conditions are that H is closed (since adjoints are always closed) and that H^* is symmetric. Since symmetric operators have real eigenvalues, a necessary condition for H^* to be symmetric is that neither i nor $-$i are eigenvalues. The following theorem states that this last condition is indeed sufficient to ensure self-adjointness of H.

Theorem 14.3. *Let H be a densely defined symmetric operator on a Hilbert space $\mathscr{H}$. Then the following assertions are equivalent:*[3]

(i) H *is self-adjoint.*
(ii) H *is closed and* $\mathrm{Ker}(H^* \pm \mathrm{i}) = \{0\}$.
(iii) $\mathrm{Ran}(H \pm \mathrm{i}) = \mathscr{H}$.

Proof. (i) $\Longrightarrow$ (ii). As argued before, if H is self-adjoint then $H^* = H$ is symmetric and thus only has real eigenvalues. The argument for this is the same as in linear algebra. Let A be symmetric and $A\varphi = \lambda\varphi$ for $\varphi \in \mathscr{D}(A) \setminus \{0\}$. Then,

$$\lambda\langle\varphi|\varphi\rangle = \langle\varphi|\lambda\varphi\rangle = \langle\varphi|A\varphi\rangle = \langle A\varphi|\varphi\rangle = \langle\lambda\varphi|\varphi\rangle$$
$$= \lambda^*\langle\varphi|\varphi\rangle\,.$$

So $\mathrm{Ker}(H^* - \mathrm{i}) = \{0\}$. Closedness of H^* and thus of H was shown in Theorem 14.2.

(ii) $\Longleftrightarrow$ (iii). We formulate as a lemma.

Lemma 14.1. *Let $T : \mathscr{H} \supset D(T) \to \mathscr{H}$ be densely defined. Then,*

1. $\mathrm{Ker}(T^* \mp \mathrm{i}) = \mathrm{Ran}(T \pm \mathrm{i})^\perp$ *and hence, in particular,*

$$\mathrm{Ker}(T^* \mp \mathrm{i}) = \{0\} \quad \Longleftrightarrow \quad \mathrm{Ran}(T \pm \mathrm{i}) \textit{ is dense in } \mathscr{H}\,.$$

2. If T is closed and symmetric, then $\mathrm{Ran}(T \pm \mathrm{i})$ *is closed.*

[3] We recall the definitions of the kernel and the range of a linear operator, which are the subspaces $\mathrm{Ker}(T) := \{\psi \in D(T)\,|\,T\psi = 0\}$ and $\mathrm{Ran}(T) := \{\varphi \in \mathscr{H}\,|\,\varphi = T\psi \text{ for some } \psi \in D(T)\}$.

For (1), note that $(T+\mathrm{i})^* = T^* - \mathrm{i}$ and therefore

$$\begin{aligned}\psi \in \mathrm{Ran}(T+\mathrm{i})^\perp &\Longleftrightarrow \langle \psi | (T+\mathrm{i})\varphi \rangle = 0\,, \quad \forall \varphi \in D(T) \\ &\Longleftrightarrow \psi \in D(T^*) \text{ and } (T^* - \mathrm{i})\psi = 0 \\ &\Longleftrightarrow \psi \in \mathrm{Ker}(T^* - \mathrm{i})\,.\end{aligned}$$

The other sign is treated analogously.

For (2), one uses the fact that, for symmetric T and $\varphi \in D(T)$, one always has $\langle \varphi, T\varphi \rangle \in \mathbb{R}$. Thus

$$\begin{aligned}\|(T+\mathrm{i})\varphi\|^2 &= \|T\varphi\|^2 + \|\varphi\|^2 + 2\mathrm{Re}\langle \mathrm{i}\varphi, T\varphi \rangle \\ &= \|T\varphi\|^2 + \|\varphi\|^2 \geq \|\varphi\|^2\,,\end{aligned} \tag{14.15}$$

which implies that $T+\mathrm{i}$ is injective and $(T+\mathrm{i})^{-1} : \mathrm{Ran}(T+\mathrm{i}) \to D(T)$ is bounded. Now let ψ_n be a sequence in $\mathrm{Ran}(T+\mathrm{i})$ with $\psi_n \to \psi$. Then $\varphi_n := (T+\mathrm{i})^{-1}\psi_n$ converges to $\varphi = (T+\mathrm{i})^{-1}\psi$. Since the graph of T is closed, the graph of $T+\mathrm{i}$ is also closed, and it follows that $(\varphi, \psi) \in \Gamma(T+\mathrm{i})$, and thus $\psi \in \mathrm{Ran}(T+\mathrm{i})$. So $\mathrm{Ran}(T+\mathrm{i})$ is closed.

It remains to show (iii) $\Longrightarrow$ (i). We have already shown that $H \subset H^*$ for symmetric operators [see (14.9)], and it remains to show $\mathscr{D}(H^*) \subset \mathscr{D}(H)$. So let $\psi \in \mathscr{D}(H^*)$. Since $\mathrm{Ran}(H - \mathrm{i}) = \mathscr{H}$, there exists $\varphi \in \mathscr{D}(H)$ such that

$$(H - \mathrm{i})\varphi = (H^* - \mathrm{i})\psi\,.$$

But from $H \subset H^*$, we get

$$(H^* - \mathrm{i})\varphi = (H^* - \mathrm{i})\psi \text{ or } (H^* - \mathrm{i})(\varphi - \psi) = 0\,.$$

By Lemma 14.1, it follows that $\psi = \varphi \in \mathscr{D}(H)$.

Since a self-adjoint operator is usually characterized by providing a core, the following corollary is very useful.

Corollary 14.3. *Let H be a densely defined symmetric operator on a Hilbert space $\mathscr{H}$. Then the following assertions are equivalent:*

(i) *H is essentially self-adjoint.*
(ii) $\mathrm{Ker}(H^* \pm \mathrm{i}) = \{0\}$.
(iii) $\mathrm{Ran}(H \pm \mathrm{i})$ *is dense.*

While (ii)$\Longleftrightarrow$(iii) is again Lemma 14.1, the equivalence (i)$\Longleftrightarrow$(ii) follows from:

$$\begin{aligned}H \text{ essentially self-adjoint} \quad &\Longleftrightarrow \quad \overline{H} = H^{**} \text{ self-adjoint} \\ &\Longleftrightarrow \quad H^{**} \text{ closed and } \mathrm{Ker}(H^{***} \pm \mathrm{i}) = \{0\} \\ &\Longleftrightarrow \quad \mathrm{Ker}(H^* \pm \mathrm{i}) = \{0\}\,.\end{aligned}$$

Remark 14.1. Instead of only looking at the points $\pm \mathrm{i}$ in Theorem 14.3 (ii) and (iii) and Corollary 14.3 (ii) and (iii), we could refer to all $\lambda \in \mathbb{C} \setminus \mathbb{R}$, i.e., $\mathrm{Im}\lambda \neq 0$. One can see this from the proof, or just convince oneself that

$$H \text{ self-adjoint} \quad \Longleftrightarrow \quad aH + b \text{ self-adjoint for } a, b \in \mathbb{R} \text{ on } \mathscr{D}(H)\,.$$

■

If H is a differential operator, as in the Schrödinger case, i.e., $H = -\Delta + V$, then with Corollary 14.3 (ii), we can show that an operator is not essentially self-adjoint by looking for solutions of the corresponding stationary Schrödinger equation

$$\left[-\Delta_x + V(x)\right]\psi(x) = \mathrm{i}\psi(x) \text{ or } \left[-\Delta_x + V(x)\right]\psi(x) = -\mathrm{i}\psi(x)$$

within the domain of the adjoint operator. Of course, we must allow not only for smooth solutions, but also for solutions in the distributional sense. However, in concrete applications, the potentials are smooth away from singular points and the corresponding distributional solutions are also smooth away from the singular points. In example (14.11), we can either conclude from the symmetry of H_0^* that H_0 is self-adjoint, or we can use Corollary 14.3, since

$$\mathscr{D}(H_0^*) = \left\{\varphi \in L^2 \middle| \, |k|^2 \widehat{\varphi} \in L^2\right\},$$

and a linear ordinary differential equation $-\varphi'' = \mathrm{i}\varphi$ has only the two linearly independent solutions

$$\exp\left(\pm \frac{1-\mathrm{i}}{\sqrt{2}} x\right),$$

which are not in $L^2(\mathbb{R}, \mathrm{d}x)$. The same holds for the solutions of $-\varphi'' = -\mathrm{i}\varphi$, and hence $\mathrm{Ker}(H_0^* \pm \mathrm{i}) = \{0\}$.

On the other hand, in (14.13) the solutions of $-\varphi'' = \mathrm{i}\varphi$ are again

$$\exp\left(\pm \frac{1-\mathrm{i}}{\sqrt{2}} x\right), \qquad x \in (0, \infty)\,,$$

where now, on the half-line,

$$\exp\left(-\frac{1-\mathrm{i}}{\sqrt{2}} x\right) \in L^2\left(\mathbb{R}^+, \mathrm{d}x\right),$$

that is,

$$\exp\left(-\frac{1-\mathrm{i}}{\sqrt{2}} x\right) \in \mathrm{Ker}\left(H_0^* - \mathrm{i}\right).$$

Therefore, H_0 is not essentially self-adjoint, but we expect $\dim \mathrm{Ker}(H_0^* - \mathrm{i}) = \dim \mathrm{Ker}(H_0^* + \mathrm{i})$.

Remark 14.2. Deficiency Index
The dimensions dim $\text{Ker}(H^* \pm \text{i}) = n_\pm$ of the kernels are called deficiency indices of the operator H. When does a symmetric operator have self-adjoint extensions? If and only if the deficiency indices are the same, i.e., $n_+ = n_-$. If $n_+ = n_- \neq 0$, then H does indeed have infinitely many self-adjoint extensions, more precisely a family of self-adjoint extensions parameterized by n_+ real parameters. This is covered by von Neumann's theory of self-adjoint extension (see, for example, [2]). Like most things, this is not surprising, since for $n_+ = n_-$ one can balance the mass-loss solutions $(+\text{i})$ with the mass-gain solutions $(-\text{i})$. It is also clear that this is connected to time-reversal invariance. If H commutes with complex conjugation, i.e., if the Schrödinger equation is invariant under time reversal, then one has $n_+ = n_-$. ■

Remark 14.3. On Self-Adjointness and Generators
As announced previously, we would like to make it at least plausible that the reverse direction of Theorem 14.1 should hold, i.e., that self-adjoint operators should generate unitary groups. First of all, in the case of a bounded self-adjoint operator H, one can define the corresponding unitary group directly through the convergent exponential series

$$U(t) := \text{e}^{-\text{i}Ht} := \sum_{n=0}^{\infty} \frac{(-\text{i}Ht)^n}{n!} ,$$

and just check by explicit computations that this really defines a unitary group in the sense of Definition 14.1.

In order to understand why, for unbounded operators H, the more technical notion of self-adjointness becomes relevant, we explain how Theorem 14.3 and, in particular, $\text{Ran}(H \pm \text{i}) = \mathscr{H}$ [which is as good as $\text{Ran}(H \pm \text{i}\lambda) = \mathscr{H}$ with $0 \neq \lambda \in \mathbb{R}$] enters into the construction of the unitary group $U(t) = \text{e}^{-\text{i}tH}$. In the case of unbounded H, we can no longer define the exponential $\text{e}^{-\text{i}tH}$ through the series. However, an alternative approach which is closer to the idea of a group is

$$\text{e}^{-\text{i}tH} = \lim_{n \to \infty} \left(1 + \frac{\text{i}tH}{n}\right)^{-n} = \lim_{n \to \infty} \text{i}^{-n} \left(\frac{tH}{n} - \text{i}\right)^{-n} , \tag{14.16}$$

where the thinking behind the negative exponent $-n$ will become clear in a moment. With Theorem 14.3, we have $\text{Ran}(\alpha H - \text{i}) = \mathscr{H}$ and $\text{Ker}(\alpha H - \text{i}) = \{0\}$ for $\alpha \in \mathbb{R}$. Hence $(\alpha H - \text{i})$ is invertible on $\mathscr{H}$ and

$$\left\|(\alpha H - \text{i})\varphi\right\|^2 = \alpha^2 \|H\varphi\|^2 + \|\varphi\|^2 \geq \|\varphi\|^2 ,$$

i.e., for $\varphi = (\alpha H - \text{i})^{-1}\psi$, one has

$$\left\|(\alpha H - \text{i})^{-1}\psi\right\| \leq \|\psi\| .$$

With $\text{Ran}(\alpha H - \text{i}) = \mathscr{H}$, it follows that

$$(\alpha H - \text{i})^{-1} : \mathscr{H} \longrightarrow \mathscr{D}(H) ,$$

and therefore $(tH/n - \mathrm{i})^{-n} : \mathscr{H} \longrightarrow \mathscr{D}(H)$ is bounded for any $n \in \mathbb{N}$. Hence, the product on the right-hand side of (14.16) defines a bounded operator with norm bounded by one for any $n \in \mathbb{N}$, and we can expect the limit to exist. ■

14.3 The Atomistic Schrödinger Operator

Let us now apply these ideas to the atomistic Hamiltonian with Coulomb potential. We consider

$$H_C = -\Delta - \frac{1}{r}\,, \quad \text{on } \mathscr{D}(H_C) = C_0^\infty\left(\mathbb{R}^3 \setminus \{0\}\right)\,,$$

where $|\mathbf{x}| = r$ and we put the electric charge equal to 1. Why $C_0^\infty(\mathbb{R}^3 \setminus \{0\})$? The reason is that $r = 0$ is clearly a singular point, and the Schrödinger equation can only hold for $r > 0$. Recall the example of the particle on the half-line. There the origin could not be crossed. But this was in one dimension, and here we have three dimensions, so taking out a single point could be less problematic. However, this is true only from dimension four upwards, and we must still be careful here.

One can understand the situation by looking at the simpler problem $H_0 = -\Delta$ with $\mathscr{D}_0 = C_0^\infty\left(\mathbb{R}^3 \setminus \{0\}\right)$. Even the free Laplacian without any singular potential is not essentially self-adjoint on a domain with a missing point. To see this, according to Corollary 14.3, we need to find solutions of $H_0^* \psi_\pm = \mp \mathrm{i} \psi_\pm$. We know that, on a suitable domain, we have $H_0^* = -\Delta$, and thus we first look for solutions of

$$-\Delta_x \psi_\pm(x) = \mp \mathrm{i} \psi_\pm(x)\,, \tag{14.17}$$

for which we will show later that $\psi_\pm \in D(H_0^*)$. Equation (14.17) is most conveniently solved in spherical coordinates, and seeking spherically symmetric solutions, one arrives at

$$\left(-\frac{\mathrm{d}^2}{\mathrm{d}r^2} - \frac{2}{r}\frac{\mathrm{d}}{\mathrm{d}r}\right)\psi_\pm(r) = \mp \mathrm{i} \psi_\pm(r)\,. \tag{14.18}$$

This is an ordinary differential equation with two linearly independent solutions

$$\psi_\pm(r) = \frac{\mathrm{e}^{-(1\pm\mathrm{i})r}}{r}\,.$$

Clearly, $\psi_\pm \in L^2(\mathbb{R}^3)$, since the $1/|x|$ singularity is square integrable in three space dimensions. However, we still need to show that $\psi_\pm$ are in the domain of H_0^*. This follows from the usual computation using integration by parts. Let $\varphi \in D(H_0)$. Then $\varphi(0) = 0$ and the boundary terms vanish:

$$\begin{aligned}
\langle \psi_\pm | H_0 \varphi \rangle_{L^2(\mathbb{R}^3)} &= -\int_{\mathbb{R}^3} \mathrm{d}x\, \psi_\pm(x)^* \Delta_x \varphi(x) \\
&= -\int_{S^2} \mathrm{d}\Omega \int_0^\infty r^2 \mathrm{d}r\, \psi_\pm(r)^* \left(\frac{\mathrm{d}^2}{\mathrm{d}r^2} + \frac{2}{r}\frac{\mathrm{d}}{\mathrm{d}r} + \frac{1}{r^2}\Delta_\omega \right) \varphi(r,\omega) \\
&= -\int_{S^2} \mathrm{d}\Omega \int_0^\infty r^2 \mathrm{d}r \left[\left(\frac{\mathrm{d}^2}{\mathrm{d}r^2} + \frac{2}{r}\frac{\mathrm{d}}{\mathrm{d}r} \right) \psi_\pm(r) \right]^* \varphi(r,\omega) \\
&= \langle \mp \mathrm{i} \psi_\pm | \varphi \rangle_{L^2(\mathbb{R}^3)} ,
\end{aligned}$$

whence $\psi_\pm \in D(H_0^*)$ and $H_0^* \psi_\pm = \mp \mathrm{i} \psi_\pm$. Thus not even $H_0 = -\Delta$ is essentially self-adjoint on the domain $\mathscr{D}_0 = C_0^\infty(\mathbb{R}^3 \setminus \{0\})$. It has many different self-adjoint extensions.[4] There is of course no problem with that. It is a matter of physics to select the correct "physical" extension. We would have come to exactly the same conclusion with the Coulomb potential added to H_0. Hence the atomistic Schrödinger operator H_C with $\mathscr{D}(H_\mathrm{C}) = C_0^\infty(\mathbb{R}^3 \setminus \{0\})$ is not essentially self-adjoint, but has many different self-adjoint extensions. But the different self-adjoint extensions have different eigenvalues and generate different time evolutions, i.e., they correspond to different physics.

Remark 14.4. About Self-Adjoint Extensions
What should one do now? For the Schrödinger operator with the Coulomb potential, we have the same problem as in the simple example (14.13), where we were able to select self-adjoint extensions by posing proper boundary conditions. In principle, we should be able to do the same in the case of the Coulomb potential. However, this will be more complicated, since the potential is singular at the boundary. The origin is a singular point of the potential, where the wave function need not and will not be differentiable. In spherical coordinates, the eigenvalue equation for the Coulomb problem reads [see (14.18)]

$$\left(-\frac{\mathrm{d}^2}{\mathrm{d}r^2} - \frac{2}{r}\frac{\mathrm{d}}{\mathrm{d}r} + \frac{1}{r} \right) \psi = E \psi . \tag{14.19}$$

The "physical" ground state eigenfunction of the hydrogen atom corresponding to the ground state energy $E_0 = -1/4$ is

$$\psi_0(\mathbf{x}) = \exp\left(-\frac{|\mathbf{x}|}{2} \right) .$$

While ψ_0 is not differentiable at the origin, it is bounded. And being bounded is also a sort of boundary condition. To see that there are also unbounded solutions of (14.19), note that ψ solves (14.19) if and only if $f = r\psi$ solves

[4] One of these extensions with $\mathscr{D}(H_0) = \{ \varphi \in L^2 |\ |k|^2 \widehat{\varphi} \in L^2 \}$ is the generator of the free time evolution. The other extensions correspond to so called δ-interactions, i.e., to Dirac delta potentials of various strengths at the origin.

$$-f'' - \frac{1}{r} f = Ef \,. \tag{14.20}$$

With each solution of this equation,

$$F(f) = f(r) \int_0^r \frac{1}{[f(s)]^2} \mathrm{d}s$$

is also a solution of (14.20). In particular, $f_1 = F(f_0)$ yields a solution f_1/r of (14.19) which diverges at the origin as $\psi \sim 1/r$, but is still square integrable at the origin. However, since $f_1 \sim \mathrm{e}^r$ for $r \to \infty$, one needs to go to $F(f_1)$ to obtain a solution with good decay properties at infinity, but which is still nonzero at the origin. The corresponding singular but square integrable ψ-functions are usually excluded as being unphysical in the physics textbooks, because they are not defined at the origin (like the Coulomb Hamiltonian itself!). Note that, being unbounded, they too cannot be contained in the domain $\mathscr{D}(H_0)$. As the following clever computation shows, each $\varphi \in \mathscr{D}(H_0) = \{\varphi \in L^2 | \, |\mathbf{k}|^2 \hat{\varphi} \in L^2\}$ is bounded. With

$$\varphi(x) = (2\pi)^{-3/2} \int \mathrm{e}^{\mathrm{i}\mathbf{k}\cdot\mathbf{x}} \hat{\varphi}(\mathbf{k}) \mathrm{d}^3 k \,,$$

the usual trick yields

$$\begin{aligned}
\|\varphi\|_\infty &\le (2\pi)^{-3/2} \int |\hat{\varphi}(\mathbf{k})| \mathrm{d}^3 k \\
&= (2\pi)^{-3/2} \int \frac{1}{1+k^2} |\hat{\varphi}(\mathbf{k})| (1+k^2) \mathrm{d}^3 k \\
&\overset{(*)}{\le} C \left[\int |\hat{\varphi}(\mathbf{k})|^2 (1+k^2)^2 \mathrm{d}^3 k \right]^{1/2} \\
&\le \sqrt{2} C \left[\int |\hat{\varphi}(\mathbf{k})|^2 (1+k^4) \mathrm{d}^3 k \right]^{1/2} \\
&= \tilde{C} \left(\|\hat{\varphi}\|^2 + \|\widehat{H_0 \varphi}\|^2 \right)^{1/2} \\
&\overset{(**)}{\le} \tilde{C} (\|\varphi\| + \|H_0 \varphi\|) \,.
\end{aligned} \tag{14.21}$$

In $(*)$, we used the Cauchy–Schwarz inequality with $\int \mathrm{d}^3 k/(1+k^2)^2 < \infty$, and in $(**)$, the Plancherel equality (13.15). This shows that the self-adjoint extensions of the Coulomb Hamiltonian with such unbounded eigenfunctions cannot be defined on the natural domain $\mathscr{D}(H_0) \cap \mathscr{D}(V)$. [We will soon see that this is actually equal to $\mathscr{D}(H_0)$.] ■

Back to the general question. It was Kato's idea to ignore the problem of boundary conditions and to focus on the domain. The natural domain is clearly $\mathscr{D}(H_0) \cap \mathscr{D}(V)$, and on that domain the "physical" Coulomb Hamiltonian should be self-adjoint. We first convince ourselves that $\mathscr{D}(V) \subset \mathscr{D}(H_0)$.

To this end, we split the potential into $V = c/r = V_1 + V_2$, where $\|V_1\| \leq \varepsilon$ and

$$\|V_2\|_\infty = \sup |V_2(x)| = a < \infty ,$$

that is,

$$V_1 = -\frac{c}{r}\chi_{\{r<\varepsilon/4\pi c^2\}} , \qquad V_2 = -\frac{c}{r}\chi_{\{r\geq\varepsilon/4\pi c^2\}} .$$

With this splitting and for $\varphi \in \mathscr{D}(H_0)$ with (14.21), we obtain

$$\begin{aligned} \|V\varphi\| &\leq \|V_1\varphi\| + \|V_2\varphi\| \leq \|V_1\|\|\varphi\|_\infty + \|V_2\|_\infty\|\varphi\| \\ &\leq \varepsilon\tilde{C}\big(\|\varphi\| + \|H_0\varphi\|\big) + a\|\varphi\| . \end{aligned} \tag{14.22}$$

We can now finish the argument. We try to interpret $V(r)$ as a "perturbation" of the free Hamiltonian $H_0 = -\Delta$, where the domain $\mathscr{D}(H_0)$ of the self-adjoint operator H_0 should determine the "correct" self-adjoint version of $H = H_0 + V$. Since the domain $\mathscr{D}(V)$ of V (as a multiplication operator) contains $\mathscr{D}(H_0)$, one can define the operator $H = H_0 + V$ on $\mathscr{D}(H_0)$. We now continue abstractly. For $H = H_0 + V$ to be self-adjoint, according to Theorem 14.3, we must have

$$\mathrm{Ran}(H_0 + V + \lambda) = \mathscr{H} ,$$

for $\lambda \in \mathbb{C} \setminus \mathbb{R}$. This follows if $(H_0 + V + \lambda)^{-1}$ exists as a bounded operator on $\mathscr{H}$. But,

$$\begin{aligned} (H_0 + V + \lambda)^{-1} &= \frac{1}{1 + V(H_0+\lambda)^{-1}}\frac{1}{H_0+\lambda} \\ &= \left[1 + V(H_0+\lambda)^{-1}\right]^{-1}(H_0+\lambda)^{-1} , \end{aligned}$$

where $(H_0 + \lambda)^{-1}$ is a well defined bounded operator, since H_0 is self-adjoint and therefore

$$\mathrm{Ran}(H_0+\lambda)^{-1} = \mathscr{H} , \qquad \mathrm{Ker}(H_0+\lambda) = \{0\} .$$

Bearing in mind that

$$\frac{1}{1-x} = \sum_{n=0}^{\infty} x^n ,$$

$\left[1 + V(H_0+\lambda)^{-1}\right]^{-1}$ can be represented as a geometric series if

$$\left\|V(H_0 + \mathrm{i}\lambda)^{-1}\right\| < 1 ,$$

that is, if

$$\left\|V(H_0+\lambda)^{-1}\varphi\right\| < \|\varphi\|$$

holds for all $\varphi \in \mathscr{H}$. (We will discuss this in more detail in the next chapter.) Now put $\psi = (H_0+\lambda)^{-1}\varphi$, so that

$$\|V\psi\| < \left\|(H_0+\lambda)\psi\right\| .$$

Since

$$\left\|(H_0+\lambda)\psi\right\|^2 = \|H_0\psi\|^2 + \lambda^2\|\psi\|^2 ,$$

we can equivalently require that

$$\|V\psi\|^2 \le \tilde{\alpha}^2\|H_0\psi\|^2 + \tilde{\beta}^2\|\psi\|^2 , \tag{14.23}$$

for all $\psi \in \mathscr{D}(H_0)$ with $\tilde{\alpha} < 1$. Finally, observe that (14.23) is equivalent to

$$\|V\psi\| \le \alpha\|H_0\psi\| + \beta\|\psi\| , \tag{14.24}$$

for all $\psi \in \mathscr{D}(H_0)$. The fact that (14.23) $\Longrightarrow$ (14.24) is obvious, and (14.24) $\Longrightarrow$ (14.23) can be seen with $\tilde{\alpha}^2 = (1+\varepsilon)\alpha^2$ and $\tilde{\beta}^2 = (1+1/\varepsilon)\beta^2$ for arbitrary $\varepsilon > 0$.

Definition 14.8. Let H_0 with domain $\mathscr{D}(H_0)$ be self-adjoint, and let V with domain $\mathscr{D}(V)$ be symmetric. One says that V is relatively bounded with respect to H_0 (or more briefly, H_0-bounded), if $\mathscr{D}(H_0) \subset \mathscr{D}(V)$ and if (14.24) or equivalently (14.23) holds for some $\alpha, \beta \in \mathbb{R}^+$, or $\tilde{\alpha}, \tilde{\beta} \in \mathbb{R}^+$. The infimum of all admissible α is called the relative bound and agrees with the infimum of all admissible $\tilde{\alpha}$.

From the above argument we obtain the following theorem due to Kato.

Theorem 14.4. *Let H_0 on $\mathscr{D}(H_0)$ be self-adjoint. Let V be H_0-bounded with relative bound $\alpha < 1$. Then $H = H_0 + V$ is self-adjoint on $\mathscr{D}(H) = \mathscr{D}(H_0)$.*

According to (14.22), the Coulomb potential is H_0-bounded with relative bound equal to zero and if we refer to the Coulomb Hamiltonian we mean exactly the one defined on $\mathscr{D}(H_0)$ as a self-adjoint operator in Theorem 14.4. And if physicists compute the spectrum of the Coulomb Hamiltonian, they only take the eigenvalues with bounded eigenfunctions, i.e., they consider H on $\mathscr{D}(H) = \mathscr{D}(H_0)$.

References

1. M. Reed, B. Simon: *Methods of Modern Mathematical Physics I: Functional Analysis*, revised and enlarged edn. (Academic Press, San Diego, 1980)
2. M. Reed, B. Simon: *Methods of Modern Mathematical Physics II. Fourier Analysis, Self-Adjointness* (Academic Press [Harcourt Brace Jovanovich Publishers], New York, 1975)

Chapter 15
Measures and Operators

We now discuss the operator calculus anticipated in Chap. 12. This calculus allows us to do computations within quantum equilibrium in a very efficient and concise way. The core element is the spectral theorem for self-adjoint operators, which establishes a precise connection between operator-valued measures and self-adjoint operators.

In (12.16a), we reduced the quantum equilibrium statistics to the map

$$\begin{aligned} \psi \mapsto \mathbb{P}^{\Psi_T}\left(F^{-1}(A)\right) &= \int_{F^{-1}(A)} |\Psi_T(x,y)|^2 \mathrm{d}x\,\mathrm{d}y \\ &= \left\langle \psi \otimes \Phi \,\middle|\, U(T)^* \chi_{F^{-1}(A)} U(T) \psi \otimes \Phi \right\rangle \\ &=: B_A(\psi, \psi)\,, \end{aligned} \tag{15.1}$$

where ψ is the effective wave function of the system, $U(T)$ is the unitary time evolution, $\psi \otimes \Phi$ is the initial condition, and Ψ_T is the wave function of the entire system (including possibly a piece of apparatus) at some large time T (when the experiment ends, e.g., when the result is displayed). Most importantly, the function F is a coarse-graining of configuration space. It maps microscopic configurations of the entire system to the macroscopic outcome of the experiment, e.g., to pointer positions. If $\Lambda \subset \mathbb{R}^n$ is the range of F, i.e., the set of possible outcomes, then for $A \subset \Lambda$ the probability of finding an outcome in A is given by the quadratic form $B_A(\psi, \psi)$.

According to Theorem 13.4, we can associate a bounded linear operator P_A with the sesquilinear form $B_A(\psi, \varphi)$ such that

$$B_A(\psi, \varphi) = \langle P_A \psi \,|\, \varphi \rangle\,, \quad \text{for all } \psi, \varphi \in \mathscr{H}\ (= L^2)\,.$$

With $B_A(\varphi, \varphi) \geq 0$ for all φ, we also have $\langle P_A \varphi | \varphi \rangle \geq 0$ for all φ, and such operators are said to be positive. Positive operators on complex Hilbert spaces are self-adjoint. From

$$0 \leq \langle P\varphi | \varphi \rangle = \langle \varphi | P\varphi \rangle^* = \langle \varphi | P\varphi \rangle\,,$$

D. Dürr, S. Teufel, *Bohmian Mechanics*, DOI 10.1007/978-3-540-89344-8_15,

we see that P is symmetric with respect to the diagonal, and polarization yields

$$\langle P\varphi|\psi\rangle = \langle\varphi|P\psi\rangle \quad \forall\varphi,\psi\in\mathscr{H} .$$

Now $A\subset\Lambda$, or more precisely $A\in\mathscr{B}(\Lambda)$, if F is a Borel measurable function. Hence there is a family of positive operators $(P_A)_{A\in\mathscr{B}(\Lambda)}$ which has the properties of a probability measure, but with values in the positive operators instead of $[0,1]$.

Definition 15.1. Let $\Lambda\subset\mathbb{R}^n$ be a measurable set with the corresponding σ-algebra $\mathscr{B}(\Lambda)$. A family $(P_A)_{A\in\mathscr{B}(\Lambda)}$ of bounded linear operators $P_A\in\mathscr{L}(\mathscr{H})$ is called a positive operator valued measure (POVM) if it has the following properties:

(i) Each P_A is positive.
(ii) $P_\emptyset = 0$, $P_\Lambda = \mathbb{1}_{\mathscr{H}}$.
(iii) If $(A_j)_{j\in\mathbb{N}}$ are pairwise disjoint measurable sets, i.e., $A_j\in\mathscr{B}(\Lambda)$ and also $A_i\cap A_j=\emptyset$ for $i\neq j$, then

$$P_{\bigcup A_j} = \operatorname*{s-lim}_{N\to\infty}\sum_{j=1}^{N} P_{A_j} ,$$

where s-lim is the strong limit, i.e.,

$$\lim_{N\to\infty}\left\|\left(P_{\bigcup A_j}-\sum_{j=1}^{N}P_{A_j}\right)\psi\right\| = 0 , \quad \text{for all } \psi\in\mathscr{H} .$$

Remark 15.1. Integration with Respect to a POVM
For each $\varphi\in\mathscr{H}$, it follows that

$$\begin{aligned} P^\varphi : \mathscr{B}(\Lambda) &\longrightarrow [0,\infty) , \\ A &\longmapsto P^\varphi(A) := \langle\varphi|P_A\varphi\rangle , \end{aligned}$$

defines a bounded positive Borel measure on Λ, which we can use to integrate bounded measurable functions on Λ, and we thus define

$$\left\langle\varphi\middle|\left[\int_\Lambda f(\lambda)\mathrm{d}P_\lambda\right]\varphi\right\rangle := \int_\Lambda f(\lambda)\,\mathrm{d}P^\varphi(\lambda) .$$

In order to define the operator $\int_\Lambda f(\lambda)\mathrm{d}P_\lambda$ using the representation result of Theorem 13.4, we need to introduce the complex Borel measures

$$\begin{aligned} P^{\psi,\varphi} : \mathscr{B}(\Lambda) &\longrightarrow \mathbb{C} , \\ A &\longmapsto P^{\psi,\varphi}(A) := \langle\psi|P_A\varphi\rangle , \end{aligned}$$

and define

$$\left\langle\psi\middle|\left[\int_\Lambda f(\lambda)\mathrm{d}P_\lambda\right]\varphi\right\rangle := \int_\Lambda f(\lambda)\,\mathrm{d}P^{\psi,\varphi}(\lambda) .$$

Note that according to the polarization identity, the complex measure $P^{\psi,\varphi}$ is just the following sum of positive measures:

$$P^{\psi,\varphi} = P^{\psi+\varphi} - P^{\psi-\varphi} + \mathrm{i}P^{\psi-\mathrm{i}\varphi} - \mathrm{i}P^{\psi+\mathrm{i}\varphi} ,$$

and this provides a way of defining integration with respect to such a complex measure. ■

The simplest example of (15.1) is $\Psi_T = \psi$ and $F = \mathrm{id}$, which leads us to the position POVM $(O_A)_{A\in\mathscr{B}(\mathbb{R}^n)}$ on $L^2(\mathbb{R}^n, \mathrm{d}^n x)$ [see (12.21)]. It is given by multiplication by the characteristic function,

$$O_A : \varphi \longmapsto \chi_A(x)\varphi(x) = \begin{cases} \varphi(x) , & x \in A , \\ 0 , & \text{otherwise} . \end{cases} \tag{15.2}$$

The probability for the configuration to be in A is

$$O^{\psi}(A) = \langle \psi | O_A \psi \rangle = \int |\psi(x)|^2 \chi_A(x)\, \mathrm{d}^n x = \int_A |\psi(x)|^2 \mathrm{d}^n x . \tag{15.3}$$

It is easy to see that the properties (i–iii) in Definition 15.1 hold for O_A. With the position POVM, we can use (15.3) to compute the probability distribution of the system configuration if its wave function is ψ. The position POVM has an additional property, which was mentioned in Chap. 12. In fact, it satisfies

Definition 15.1 (iv) $\qquad P_{A_1 \cap A_2} = P_{A_1} P_{A_2} ,$

since $O_{A_1 \cap A_2} = \chi_{A_1 \cap A_2} = \chi_{A_1}\chi_{A_2} = O_{A_1} O_{A_2}$, and, as a consequence, in particular

Definition 15.1 (iv)′ $\qquad P_A^2 = P_A .$

This structure plays an important role, as it is responsible for the quantum formalism. A linear operator P that satisfies $P^2 = P$ is called a *projector*. If a projector P is also self-adjoint, then P is positive:

$$\langle \varphi | P \varphi \rangle = \langle \varphi | P^2 \varphi \rangle = \langle P\varphi | P\varphi \rangle \geq 0 .$$

To see that not all projections are self-adjoint, consider the projection onto a one-dimensional subspace of $\mathbb{R}^2$ along a family of parallel lines that are not orthogonal to the subspace (see Fig. 15.1).

The self-adjoint projections are therefore called *orthogonal* projections. They satisfy $P^2 = P$ and $P^* = P$.

Definition 15.2. A family $(P_A)_{A\in\mathscr{B}(\Lambda)}$ of bounded linear operators $P_A \in \mathscr{L}(\mathscr{H})$ is called a projection-valued measure (PVM), if it satisfies Definition 15.1(i–iv), or equivalently, Definition 15.1(i–iii) and (iv′).

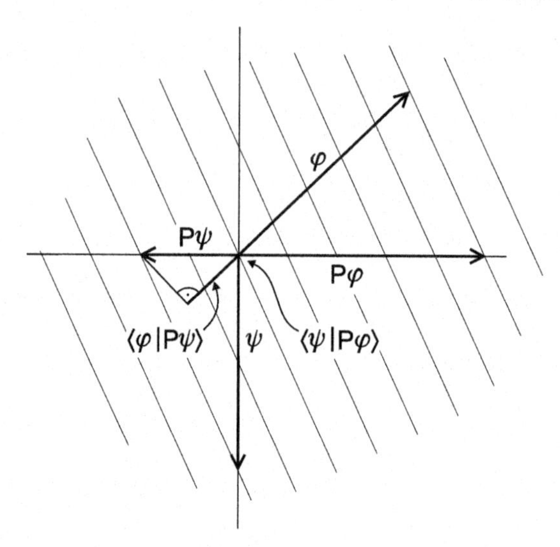

Fig. 15.1 Projection $(\|\varphi\| = 1)$

Hence a PVM is a POVM in which all operators are orthogonal projections.

Remark 15.2. Indeed (15.1)(i–iii) together with (iv′) implies (iv). This follows from $P_A P_{B\setminus A} = 0$ for $A \subset B$, which can be seen as follows. First,

$$\begin{aligned} P_B = P_B^2 = (P_A + P_{B\setminus A})^2 &= P_A + P_A P_{B\setminus A} + P_{B\setminus A} P_A + P_{B\setminus A} \\ &= P_B + P_A P_{B\setminus A} + P_{B\setminus A} P_A \end{aligned}$$

implies that

$$P_A P_{B\setminus A} + P_{B\setminus A} P_A = 0 \,.$$

Multiplying this by P_A once from the left and once from the right yields

$$P_A P_{B\setminus A} = P_{B\setminus A} P_A = -P_A P_{B\setminus A} P_A \,,$$

and therefore $P_A P_{B\setminus A} = P_{B\setminus A} P_A = 0$. ■

We will return to PVMs shortly, but first let us recall our example POVM (12.39) for a position measurement with some uncertainty. We defined

$$\tilde{O}_A = \int p(\mathbf{y}-\mathbf{x}) \chi_A(\mathbf{y}) \mathrm{d}^n y \,,$$

that is,

$$\tilde{O}_A : \varphi \longmapsto \int_A p(\mathbf{y}-\mathbf{x}) \mathrm{d}^n y \; \varphi(\mathbf{x}) \,,$$

where

$$\tilde{O}_A^2 \neq \tilde{O}_A ,$$

whenever $p(\mathbf{x}) \neq \delta(\mathbf{x})$.

After this example let us emphasize once more that the statistics of any experiment in the sense of the sequence (15.1) are described by a POVM. These are all experiments where in the end the result can be read off from the configuration of the system or some apparatus. Therefore it might be surprising to find that these POVMs play little or no role in theoretical physics. The reason is that PVMs give rise to an elegant mathematical formalism, while POVMs give rise to nothing. They are merely abstract descriptions of the statistics of some measurement, and one could just as well do without them, i.e., just stop after the first equality in (15.1). The justification for the abstraction lies in the idealized description of the measurement process, as in (12.3) to (12.9). The book-keeping operators which stem from special POVMs, namely the PVMs, give rise to an elegant and compact formalism – a textbook formalism – for computing expectation values, variances, and higher moments. We will shortly discuss three such operators which arise in the relevant measurement situations.

But before that, we should emphasize that the idea of a measurement where one can "stick in" an arbitrary wave function and always get out some measurement result in the form of a pointer position is completely unrealistic. Any experiment will only lead to sensible outcomes for a quite special class of initial wave functions. Take for example scattering experiments. Here we are only interested in wave functions where the particle (that is the system) moves towards a target and scatters. Most initial wave functions will never reach the region of the target and will never be detected. They are irrelevant. The moral is that experiments are designed and conducted for a small class of special initial data. As a consequence, the general structure of a POVM or PVM defined on the full Hilbert space is quite uninteresting and irrelevant from the point of view of the physics.

15.1 Examples of PVMs and Their Operators

We have already noted that quantum equilibrium, i.e., Born's statistical law, leads trivially to the position POVM. For a single particle, we have $(O_A)_{A\in\mathscr{B}(\mathbb{R}^3)}$ and can use O_A to compute the moments $\mathbb{E}^{\varphi}(\mathbf{X}^n)$ of the position distribution. Since the position $\mathbf{x} \in \mathbb{R}^3$ is a vector, we need to specify what we mean by $\mathbf{x}^n$. Most straightforwardly, we take $\mathbf{x}^n$ as meaning any of the possibilities

$$\mathbf{x}^n = x_1^l x_2^k x_3^m , \quad \text{with } l+k+m=n , \quad l,k,m \in \mathbb{N} ,$$

for $\mathbf{x} = (x_1, x_2, x_3)$, or any linear combination thereof.

Approximating the integral with sums over a disjoint decomposition Δ_k of $\mathbb{R}^3$, we write formally

$$\begin{aligned}
\mathbb{E}^{\varphi}(\mathbf{X}^n) &= \int \mathbf{x}^n |\varphi(\mathbf{x})|^2 \,\mathrm{d}^3x \\
&= \lim_{N\to\infty} \sum_{k=1}^{N} \mathbf{x}_k^n \int |\varphi(\mathbf{x})|^2 \chi_{\Delta_k}(\mathbf{x}) \,\mathrm{d}^3x \\
&= \lim_{N\to\infty} \sum_{k=1}^{N} \mathbf{x}_k^n \langle \varphi \,| O_{\Delta_k} \varphi \rangle \\
&= \lim_{N\to\infty} \left\langle \varphi \middle| \sum_{k=1}^{N} \mathbf{x}_k^n O_{\Delta_k} \varphi \right\rangle \\
&= \lim_{N\to\infty} \left\langle \varphi \middle| \left(\sum_{k=1}^{N} \mathbf{x}_k O_{\Delta_k} \right)^n \varphi \right\rangle ,
\end{aligned} \tag{15.4}$$

where we have used $O_{\Delta_k} O_{\Delta_j} = \delta_{kj} O_{\Delta_j}$. Hence we obtain

$$\mathbb{E}^{\varphi}(\mathbf{X}^n) = \left\langle \varphi \middle| \left(\int \mathbf{x} \,\mathrm{d}^3 O_{\mathbf{x}} \right)^n \varphi \right\rangle =: \langle \varphi \,| \widehat{\mathbf{x}}^n \varphi \rangle .$$

This motivates the definition of the self-adjoint operator $\widehat{\mathbf{x}}$, the position operator. We will discuss its domain later on:[1]

$$\widehat{\mathbf{x}} = \int \mathbf{x} \,\mathrm{d}^3 O_{\mathbf{x}} , \tag{15.5}$$

and we have

$$\widehat{\mathbf{x}}^n = \int \mathbf{x}^n \,\mathrm{d}^3 O_{\mathbf{x}} . \tag{15.6}$$

Remark 15.3. Vector Operators

When speaking of the position operator $\widehat{\mathbf{x}}$, we always have in mind the *family* $(\widehat{x}_1, \widehat{x}_2, \widehat{x}_3)$ of commuting self-adjoint operators (see Sect. 12.2.2)

$$\widehat{x}_i = \int x_i \,\mathrm{d}O_{x_i} ,$$

with

$$[\widehat{x}_i, \widehat{x}_j] = \widehat{x}_i \widehat{x}_j - \widehat{x}_j \widehat{x}_i = 0 ,$$

as follows from (15.5). The position operator $\widehat{\mathbf{x}}$ is thus a map

[1] We say this rather lightly here, although we know that, for an unbounded operator, the domain plays an important role. This is because, for multiplication operators like $\widehat{\mathbf{x}}$, the domain of self-adjointness is straightforward: it is just the maximal domain.

$$\widehat{\mathbf{x}} : L^2(\mathbb{R}^3) \longrightarrow L^2(\mathbb{R}^3) \oplus L^2(\mathbb{R}^3) \oplus L^2(\mathbb{R}^3) ,$$

$$\psi(\mathbf{x}) \longmapsto \widehat{\mathbf{x}}\psi(\mathbf{x}) = \begin{pmatrix} x_1\, \psi(\mathbf{x}) \\ x_2\, \psi(\mathbf{x}) \\ x_3\, \psi(\mathbf{x}) \end{pmatrix} .$$

This is often called the position representation of the position operator. We shall say more about that later. ■

The abstract one-to-one correspondence (15.5) between self-adjoint operators and PVMs is the content of the spectral theorem, which we will discuss in detail later on.

15.1.1 Heisenberg Operators

Let us move to the second operator. It comes out of the sequence (15.1), where again the apparatus plays no role, i.e., we just look at the system at time $T = t$, viz., $\Psi_T = \psi_t$ und $F = \mathrm{id}$. Then for $A \in \mathscr{B}(\mathbb{R}^3)$ we have

$$\begin{aligned}
\mathbb{P}^{\psi_t}(A) &= \int_A |\psi(\mathbf{x},t)|^2 \,\mathrm{d}^3x \\
&= \int_{\mathbb{R}^3} \psi^*(\mathbf{x},t)\chi_A(\mathbf{x})\psi(\mathbf{x},t)\,\mathrm{d}^3x \\
&= \langle U(t)\psi \,|\, O_A U(t)\psi\rangle \\
&= \langle \psi \,|\, U(t)^* O_A U(t)\psi\rangle ,
\end{aligned}$$

with the POVM

$$O_A(t) := U(t)^* O_A U(t) .$$

Not only O_A, but also $O_A(t)$ is actually a PVM:

$$O_A(t)^2 = U(t)^* O_A U(t) U(t)^* O_A U(t) = U(t)^* O_A O_A U(t) = U(t)^* O_A U(t) .$$

And as in (15.5), we can introduce the corresponding self-adjoint operator

$$\widehat{\mathbf{x}}(t) = \int \mathbf{x}\,\mathrm{d}^3 O_{\mathbf{x}}(t) = U^*(t)\widehat{\mathbf{x}}U(t) , \tag{15.7}$$

the so-called Heisenberg position operator at time t.

The Heisenberg position operator at time t captures the statistics of the particle position at time t. But we should not forget that this is just a way of rewriting equivariance. The (Bohmian) position $\mathbf{X}(t,\mathbf{x})$ of the particle at time t is a random variable with distribution

$$\mathbb{P}^{\psi_t}(A) = \mathbb{P}^{\psi}\big(\{\mathbf{x} \,|\, \mathbf{X}(t,\mathbf{x}) \in A\}\big) . \tag{15.8}$$

The position at time t is random because the initial position of the particle is random! We will use this again for the third operator, where we have the same situation, but with a nontrivial function F.

15.1.2 Asymptotic Velocity and the Momentum Operator

We come back now to Sect. 9.4 and ask for the asymptotic velocity of a freely moving particle, i.e., $\mathbf{X}(t,\mathbf{x})/t$. To simplify the notation, we once again set $\hbar/m = 1$ and thus look at the Schrödinger equation

$$\mathrm{i}\frac{\partial}{\partial t}\varphi(\mathbf{x},t) = -\frac{1}{2}\Delta\varphi(\mathbf{x},t) , \qquad \varphi(\mathbf{x},0) = \varphi_0(\mathbf{x}) , \qquad \mathbf{x} \in \mathbb{R}^3 . \tag{15.9}$$

We need an asymptotic expression for $\varphi(\mathbf{x},t)$ at large times, which we will justify rigorously in Remark 15.8. This is not a big issue, but the mathematics fits better there than here. The stationary phase argument from Sect. 9.4 suffices for the moment, where we found for the asymptotics of the solution of (15.9) that

$$\varphi(\mathbf{x},t) \sim \frac{\mathrm{e}^{\mathrm{i}\mathbf{x}^2/2t}}{(\mathrm{i}t)^{3/2}}\,\widehat{\varphi}_0\left(\frac{\mathbf{x}}{t}\right) , \quad \text{for } t \longrightarrow \infty . \tag{15.10}$$

The Bohmian trajectories following the asymptotic wave function are straight lines. We pick up (9.26) once more, and recall how we get the following nice and relevant example from (15.10), bringing us back to (15.1) and the PVMs.

Consider the function $F_t(\mathbf{x}) = \mathbf{x}/t$. Then, with (15.8), we find

$$\begin{aligned}
\lim_{t\longrightarrow\infty}\mathbb{P}^{\varphi_t}\left(F_t^{-1}(A)\right) &= \lim_{t\longrightarrow\infty}\mathbb{P}^{\varphi_t}\left(\frac{\mathbf{X}}{t}\in A\right)\\
&= \lim_{t\longrightarrow\infty}\mathbb{P}^{\varphi}\left(\left\{\mathbf{x}\,\middle|\,\frac{\mathbf{X}(t,\mathbf{x})}{t}\in A\right\}\right)\\
&= \lim_{t\longrightarrow\infty}\int|\varphi(\mathbf{x},t)|^2\chi_A\left(\frac{\mathbf{x}}{t}\right)\mathrm{d}^3x\\
&= \lim_{t\longrightarrow\infty}\int\frac{1}{t^3}\left|\widehat{\varphi}_0\left(\frac{\mathbf{x}}{t}\right)\right|^2\chi_A\left(\frac{\mathbf{x}}{t}\right)\mathrm{d}^3x \qquad \text{[by (15.10)]}\\
&= \lim_{t\longrightarrow\infty}\int|\widehat{\varphi}_0(\mathbf{k})|^2\chi_A(\mathbf{k})\,\mathrm{d}^3k\\
&= \langle\widehat{\varphi}\,|\,O_A\widehat{\varphi}\rangle ,
\end{aligned} \tag{15.11}$$

where we substituted $\mathbf{k} = \mathbf{x}/t$. This is now a very interesting result. If we interpret the random variable

$$\mathbf{V}_t(\mathbf{x}) := \frac{\mathbf{X}(t,\mathbf{x})}{t}$$

as an approximation to the velocity of the Bohmian particle, then we have shown that $\mathbf{V}_t(\mathbf{x})$ converges in distribution to

$$\mathbf{V}_\infty(\mathbf{x}) = \lim_{t\to\infty} \mathbf{V}_t(\mathbf{x}) ,$$

where the random variable $\mathbf{V}_\infty$ has distribution with density $|\widehat{\varphi}(\mathbf{k})|^2$. And this brings us to the momentum operator of quantum mechanics.

We now have the random variable $\mathbf{V}_\infty$, the asymptotic velocity. In order to continue as in (15.4), we need to express the probability

$$\mathbb{P}^\varphi(\mathbf{V}_\infty \in A) = \langle \widehat{\varphi} \,|\, O_A \widehat{\varphi} \rangle$$

in terms of a POVM, say $(V_A)_{A\in\mathscr{B}(\mathbb{R}^3)}$. With the Fourier transformation $\mathscr{F}$ [see (13.15)], which is unitary on L^2, we find that

$$\langle \widehat{\varphi} \,|\, O_A \,\widehat{\varphi} \rangle = \langle \varphi \,|\, \mathscr{F}^* O_A \mathscr{F} \varphi \rangle =: \langle \varphi \,|\, V_A \,\varphi \rangle . \tag{15.12}$$

Thus $V_A = V_A^*$ (since $O_A = O_A^*$) and

$$V_A^2 = \mathscr{F}^* O_A \mathscr{F} \mathscr{F}^* O_A \mathscr{F} = \mathscr{F}^* O_A \mathscr{F} \mathscr{F}^{-1} O_A \mathscr{F} = \mathscr{F}^* O_A^2 \mathscr{F} = \mathscr{F}^* O_A \mathscr{F} = V_A ,$$

whence $(V_A)_{A\in\mathscr{B}(\mathbb{R}^3)}$ is a PVM.

The asymptotic velocity is experimentally an easily accessible quantity, and it is therefore convenient to introduce the corresponding self-adjoint velocity operator $\widehat{\mathbf{v}}_\infty$ (or the momentum operator $\widehat{\mathbf{p}} = m\widehat{\mathbf{v}}_\infty$). It is constructed from $(V_A)_{A\in\mathscr{B}}$ analogously to (15.5) [see (15.3)] through

$$\widehat{\mathbf{v}}_\infty = \int \mathbf{k}\,\mathrm{d}^3 V_{\mathbf{k}} . \tag{15.13}$$

Again we delay the discussion about its domain.

Analogously to (15.4), the moments $\mathbb{E}^\varphi(\mathbf{V}_\infty^n)$ of the asymptotic velocity are given by

$$\mathbb{E}^\varphi(\mathbf{V}_\infty^n) = \langle \varphi \,|\, \widehat{\mathbf{v}}_\infty^n \varphi \rangle , \tag{15.14}$$

where

$$\widehat{\mathbf{v}}_\infty^n = \int \mathbf{k}^n \,\mathrm{d}V_{\mathbf{k}} .$$

Hence with (15.12), we have

$$V^\varphi(A) = \langle \varphi \,|\, V_A \,\varphi \rangle = \langle \widehat{\varphi} \,|\, O_A \,\widehat{\varphi} \rangle = O^{\widehat{\varphi}}(A) ,$$

and thus

$$\mathbb{E}^{\varphi}(\mathbf{V}_{\infty}^{n}) = \int \mathbf{k}^{n}\,\mathrm{d}V^{\varphi}(\mathbf{k}) = \int \mathbf{k}^{n}\,\mathrm{d}O^{\widehat{\varphi}}(\mathbf{k}) = \int \mathbf{k}^{n}|\widehat{\varphi}(\mathbf{k})|^{2}\mathrm{d}^{3}k\,.$$

What about the domains now? We emphasized in the previous chapter that an operator is only well defined if its domain is specified, which is a tricky matter in general. However, for $\widehat{\mathbf{x}}$ and $\widehat{\mathbf{v}}_{\infty}$, this is simple. Each component $\widehat{x}_i$ $(i = 1,2,3)$ is a self-adjoint operator on $\mathscr{D}(\widehat{x}_i)$, and the action of $\widehat{x}_i$ on $\psi \in \mathscr{D}(\widehat{x}_i)$ is just multiplication:

$$\widehat{x}_i : \psi \longrightarrow x_i\psi(\mathbf{x})\,.$$

It is straightforward to see that real-valued multiplication operators are self-adjoint on their maximal domain (if the latter is dense). Hence,

$$\mathscr{D}(\widehat{x}_i) = \left\{\psi \in L^2 \,\middle|\, \|x_i\psi\| < \infty\right\}, \qquad \mathscr{D}(\widehat{\mathbf{x}}) = \left\{\psi \in L^2 \,\middle|\, \big\||\mathbf{x}|\psi\big\| < \infty\right\},$$

are the proper domains for the position operator. To be precise, in (15.4), we thus have to pick φ in the domain.

For the asymptotic velocity operator, we find with (13.15) that

$$\langle\varphi|\widehat{\mathbf{v}}_{\infty}\psi\rangle = \int \widehat{\varphi}^{*}(\mathbf{k})\,\mathbf{k}\,\widehat{\psi}(\mathbf{k})\,\mathrm{d}^{3}k = \left\langle\widehat{\varphi}\middle|\widehat{(-\mathrm{i}\nabla\psi)}\right\rangle = \langle\varphi|(-\mathrm{i}\nabla)\psi\rangle\,.$$

Hence, for

$$\psi \in \mathscr{D}(\widehat{\mathbf{v}}_{\infty}) = \left\{\phi \in L^2 \,\middle|\, \big\||\nabla\phi|\big\| < \infty\right\},$$

we have

$$\widehat{\mathbf{v}}_{\infty} : \psi \mapsto \frac{1}{\mathrm{i}}\nabla\psi\,. \tag{15.15}$$

Thus $\widehat{\mathbf{x}}$ and $\widehat{\mathbf{v}}_{\infty}$ are unbounded self-adjoint operators.

We should note that there is a noteworthy correspondence between the distributions $|\varphi|^2$ and $|\widehat{\varphi}|^2$. The variances are in some sense inverse. For Gaussians, this is easy to see. The Fourier transform of a Gaussian is again a Gaussian, with width inversely proportional to the width of the original. In general, one can only say that the product of the variances is greater than or equal to 1/4. Here is an elegant proof. For the variances $\Delta\mathbf{x}$ and $\Delta\mathbf{v}_{\infty}$, and for $\varphi \in \mathscr{D}(\widehat{\mathbf{x}}) \cap \mathscr{D}(\widehat{\mathbf{v}}_{\infty})$, one has

$$(\Delta\mathbf{x})^2 = \langle\varphi|\widehat{\mathbf{x}}^2\varphi\rangle - \langle\varphi|\widehat{\mathbf{x}}\varphi\rangle^2 =: \left\langle\left(\widehat{\mathbf{x}} - \langle\widehat{\mathbf{x}}\rangle\right)^2\right\rangle =: \langle\bar{\mathbf{x}}^2\rangle\,,$$

$$(\Delta\mathbf{v}_{\infty})^2 = \langle\varphi|\widehat{\mathbf{v}}_{\infty}^2\varphi\rangle - \langle\varphi|\widehat{\mathbf{v}}_{\infty}\varphi\rangle^2 =: \left\langle\left(\widehat{\mathbf{v}}_{\infty} - \langle\widehat{\mathbf{v}}_{\infty}\rangle\right)^2\right\rangle =: \langle\bar{\mathbf{v}}_{\infty}^2\rangle\,,$$

where $\mathbf{xy}$ is the matrix $(x_iy_j)_{i,j}$, and thus

$$\Delta\mathbf{x} = \langle\bar{\mathbf{x}}^2\rangle^{1/2}\,, \qquad \Delta\mathbf{v}_{\infty} = \langle\bar{\mathbf{v}}_{\infty}^2\rangle^{1/2}\,.$$

Cauchy–Schwarz and self-adjointness imply

$$\langle|\bar{\mathbf{x}}\bar{\mathbf{v}}_\infty|\rangle \le \langle\bar{\mathbf{x}}^2\rangle^{1/2}\langle\bar{\mathbf{v}}_\infty^2\rangle^{1/2}, \qquad \langle|\bar{\mathbf{v}}_\infty\bar{\mathbf{x}}|\rangle \le \langle\bar{\mathbf{x}}^2\rangle^{1/2}\langle\bar{\mathbf{v}}_\infty^2\rangle^{1/2}.$$

Now (15.15) suggests looking at

$$\left|\langle\bar{\mathbf{x}}\bar{\mathbf{v}}_\infty - \bar{\mathbf{v}}_\infty\bar{\mathbf{x}}\rangle\right| = \left|\langle[\bar{\mathbf{x}},\bar{\mathbf{v}}_\infty]\rangle\right| = \left|\langle[\widehat{\mathbf{x}},\widehat{\mathbf{v}}_\infty]\rangle\right| \le 2\langle\bar{\mathbf{x}}^2\rangle^{1/2}\langle\bar{\mathbf{v}}_\infty^2\rangle^{1/2},$$

since, with the unit matrix E_3, we have

$$[\widehat{\mathbf{x}},\widehat{\mathbf{v}}_\infty]\varphi = \mathbf{x}\left(\frac{1}{\mathrm{i}}\nabla\varphi\right)(\mathbf{x}) - \frac{1}{\mathrm{i}}\nabla\big(\mathbf{x}\varphi(\mathbf{x})\big) = -\frac{1}{\mathrm{i}}\varphi(\mathbf{x})\mathsf{E}_3\,. \tag{15.16}$$

For $\varphi\in\mathscr{D}(\widehat{\mathbf{v}}_\infty)\cap\mathscr{D}(\widehat{\mathbf{x}})$, we thus have

$$\Delta\mathbf{x}\Delta\mathbf{v}_\infty = \langle\bar{\mathbf{x}}^2\rangle^{1/2}\langle\bar{\mathbf{v}}_\infty^2\rangle^{1/2} \ge \frac{1}{2}\mathsf{E}_3\,. \tag{15.17}$$

Recall that we put $\hbar/m = 1$, since we are focused on mathematics in this chapter. Reinstating $\hbar$ and m, we find that

$$\widehat{\mathbf{v}}_\infty = \frac{1}{m}\frac{\hbar}{\mathrm{i}}\nabla\,,$$

and with the momentum operator

$$m\widehat{\mathbf{v}}_\infty = \frac{\hbar}{\mathrm{i}}\nabla = \widehat{\mathbf{p}}\,, \tag{15.18}$$

our result (15.17) is just Heisenberg's uncertainty principle for position and momentum. The relation (15.16) is Heisenberg's famous commutation relation

$$[\widehat{\mathbf{x}},\widehat{\mathbf{p}}] = \mathrm{i}\hbar\mathsf{E}_3\,. \tag{15.19}$$

Now this is mathematically interesting, but not especially exciting, since the physical mechanism behind it is absolutely trivial: the more the support of the initial wave function φ_0 is localized, the faster it spreads under the time evolution. Hence, the better one knows the initial position of the particle (i.e., the smaller the width of $|\varphi_0|^2$), the broader will be the distribution of the asymptotic position of the freely moving particle.

Although that is really all there is to say, we remark as an aside that we can differentiate the Heisenberg position operator (15.7) with respect to time to find, using $U(t)=\exp(-\mathrm{i}tH/\hbar)$, that

$$\frac{\mathrm{d}}{\mathrm{d}t}\widehat{\mathbf{x}}(t) = \mathrm{i}\big[H,\widehat{\mathbf{x}}(t)\big] = \mathrm{i}U^*(t)\big[H,\widehat{\mathbf{x}}\big]U(t)\,.$$

With (15.18), we can write

$$H = \frac{\widehat{\mathbf{p}}^2}{2m} + V(\widehat{\mathbf{x}}) ,$$

and the commutator between $\widehat{\mathbf{x}}$ and H becomes, using (15.16),

$$\mathrm{i}[H,\widehat{\mathbf{x}}] = \frac{\widehat{\mathbf{p}}}{m} ,$$

and therefore

$$\frac{\mathrm{d}}{\mathrm{d}t}\widehat{\mathbf{x}}(t) = \frac{\widehat{\mathbf{p}}(t)}{m} .$$

Hence, with (15.19), we get for small t that

$$[\widehat{\mathbf{x}},\widehat{\mathbf{x}}(t)] \approx \left[\widehat{\mathbf{x}},\widehat{\mathbf{x}} + t\frac{\mathrm{d}}{\mathrm{d}t}\widehat{\mathbf{x}}(t)\Big|_{t=0}\right] = \frac{\mathrm{i}\hbar t}{m}\mathsf{E}_3 .$$

This is an uncertainty relation between the initial position and the position after some short time, and once again it just expresses the fact that the wave function spreads. If the wave function can evolve freely, it separates into a superposition of wave packets with locally well defined velocity, and we find the situation considered before.

Differentiating once more yields

$$\frac{\mathrm{d}}{\mathrm{d}t}\widehat{\mathbf{p}}(t) = \nabla V(\widehat{\mathbf{x}})(t) ,$$

which is another form of Ehrenfest's theorem, called Heisenberg's equation of motion. The simplicity and conciseness of this operator equation are impressive, and might prompt one to rethink the role of these operators. Maybe they are indeed more fundamental than we have been ready to admit? Why do these equations come out so nicely? Maybe the particle does have a momentum after all? Now the answer is simple. In Bohmian mechanics, the particle has no momentum, while in Newtonian or Hamiltonian mechanics it has momentum. That is all there is to it.

Remark 15.4. The Position Operator for the Harmonic Oscillator
Heisenberg's equation of motion for the harmonic oscillator is easily solved, and one finds that the position operator is periodic with period T, i.e., $\widehat{\mathbf{x}}(T) = \widehat{\mathbf{x}}(0)$. The statistics of the particle positions at time T and time 0 are identical. Therefore, according to Remark 12.2, we can "measure the position operator at time 0" by measuring the position of the particle at time T. However, for this to be true, the position of the particle at time T need not agree with the position at time 0, since only the distributions actually need to agree. One can easily see from the example in (8.20) that, in general, the trajectory is not periodic in position space.

In conclusion, we have found that a "measurement of the position operator" (a nonsensical statement anyway) is not always a measurement of the position. This is as unimportant as anything could be, but nevertheless often leads to confusion, and we should therefore take note of that. ■

15.2 The Spectral Theorem

15.2.1 The Dirac Formalism

We now return to the statistical calculus based on self-adjoint operators. Recall the book-keeping operator (12.14), viz.,

$$\widehat{A} = \sum \lambda_\alpha P_\alpha \,, \tag{15.20}$$

which contains the statistics of a measurement process with possible outcomes $\lambda_\alpha \in \Lambda$. The orthogonal projectors P_α project onto the orthogonal subspaces $\mathscr{H}_\alpha$ of initial states leading to the outcome λ_α. If the possible results λ_α are real, then A is self-adjoint. Why should measurement results be real? Because, they typically count the number of some units, and we use real numbers to count.

If we compare (15.20) with (15.5) or (15.13), and write P_{λ_α} for P_α, we see that the structure is the same. Moreover, the projectors P_{λ_α} form a PVM (at least on the range of $\sum P_{\lambda_\alpha}$), but this time we obtain a discrete measure supported on the points λ_α:

$$P^\varphi(A) = \sum_{\{\alpha \,|\, \lambda_\alpha \in A\}} \langle \varphi | P_{\lambda_\alpha} \varphi \rangle \,.$$

In the *Dirac notation*, one writes one-dimensional orthogonal projectors P_{λ_α} as

$$P_{\lambda_\alpha} = |\varphi_{\lambda_\alpha}\rangle\langle\varphi_{\lambda_\alpha}| \,,$$

where the $\varphi_{\lambda_\alpha} \in \mathrm{Ran} P_{\lambda_\alpha}$ satisfy

$$\langle \varphi_{\lambda_\alpha} | \varphi_{\lambda_\beta} \rangle = \delta_{\alpha,\beta} \,.$$

If $(\varphi_{\lambda_\alpha})_\alpha$ forms a basis, then

$$\sum_{\lambda_\alpha} |\varphi_{\lambda_\alpha}\rangle\langle\varphi_{\lambda_\alpha}| = \mathrm{id}_{\mathscr{H}} \,.$$

One immediately sees that λ_α are the eigenvalues of $\widehat{A}$, with $\mathscr{H}_\alpha$ the corresponding eigenspaces, spanned by the eigenvectors φ_{λ_α}. The advantage of this spectral representation is that one can take functions of the operator:

$$f(\widehat{A}) = \sum f(\lambda_\alpha) P_{\lambda_\alpha} \,. \tag{15.21}$$

This is one reason for diagonalizing matrices.[2] Does such a representation also exist for $\widehat{\mathbf{x}}$ or $\widehat{\mathbf{v}}_\infty$? In fact, we can use the Dirac notation as a powerful formalism for doing just that [see (9.27)]. Write

[2] This is just another way of writing the diagonal form of a self-adjoint matrix.

$$\mathrm{d}^3 O_{\mathbf{x}} = |\mathbf{x}\rangle\langle\mathbf{x}|\mathrm{d}^3 x\,, \tag{15.22}$$

$$\langle\mathbf{x}|\mathbf{x}'\rangle = \delta(\mathbf{x}-\mathbf{x}')\,,$$

$$\int_{\mathbb{R}^3} \mathrm{d}^3 O_{\mathbf{x}} = O_{\mathbb{R}^3} = \mathsf{I}_{L^2(\mathbb{R}^3)} = \int_{\mathbb{R}^3} |\mathbf{x}\rangle\langle\mathbf{x}|\mathrm{d}^3 x\,,$$

and

$$\mathrm{d}^3 V_{\mathbf{k}} = |\mathbf{k}\rangle\langle\mathbf{k}|\mathrm{d}^3 k\,,$$

$$\langle\mathbf{k}|\mathbf{k}'\rangle = \delta(\mathbf{k}-\mathbf{k}')\,,$$

$$\int_{\mathbb{R}^3} \mathrm{d}^3 V_{\mathbf{k}} = V_{\mathbb{R}^3} = \mathsf{I}_{L^2(\mathbb{R}^3)} = \int_{\mathbb{R}^3} |\mathbf{k}\rangle\langle\mathbf{k}|\mathrm{d}^3 k\,,$$

where, for $\psi, \varphi \in L^2(\mathbb{R}^3, \mathrm{d}^3 x)$, we put

$$\langle\varphi|\mathbf{x}\rangle\langle\mathbf{x}|\psi\rangle = \varphi^*(\mathbf{x})\psi(\mathbf{x})$$

and

$$\langle\varphi|\mathbf{k}\rangle\langle\mathbf{k}|\psi\rangle = \widehat{\varphi}^*(\mathbf{k})\widehat{\psi}(\mathbf{k})\,. \tag{15.23}$$

One can now easily guess what $\langle\mathbf{x}|\mathbf{k}\rangle$ should look like. From here on, everybody can find his or her own way of thinking about (15.22) to (15.23). For example, we can pretend that $|\mathbf{x}\rangle$ and $|\mathbf{k}\rangle$ are elements of $L^2(\mathbb{R}^3)$ in order to associate the usual geometric pictures with the formulas. We then write

$$\widehat{\mathbf{x}} = \int \mathbf{x}|\mathbf{x}\rangle\langle\mathbf{x}|\mathrm{d}^3 x\,, \tag{15.24}$$

instead of (15.5), and

$$\widehat{\mathbf{v}}_\infty = \int \mathbf{k}|\mathbf{k}\rangle\langle\mathbf{k}|\mathrm{d}^3 k\,, \tag{15.25}$$

instead of (15.13). For example, by showing the version of (15.6) and (15.14) in the Dirac notation, i.e.,

$$\widehat{\mathbf{x}}^n = \int \mathbf{x}^n|\mathbf{x}\rangle\langle\mathbf{x}|\mathrm{d}^3 x\,, \qquad \widehat{\mathbf{v}}^n_\infty = \int \mathbf{k}^n|\mathbf{k}\rangle\langle\mathbf{k}|\mathrm{d}^3 k\,,$$

we see how nicely and intuitively we can work with this formalism.

We now have (15.24) and (15.25), and it very much resembles (15.20). Formally, we can even call $\mathbf{x}$ (and analogously $\mathbf{k}$) the eigenvalues of the position operator $\widehat{\mathbf{x}}$

corresponding to the eigenvectors $|\mathbf{x}\rangle$, although we must not of course take this too seriously. Clearly, the "eigenfunction" $|\mathbf{x}\rangle$ is not an element of L^2, and for this reason one often calls it a generalized eigenfunction.

Remark 15.5. Almost Eigenvalues and Approximate Eigenfunctions
Now any $\lambda \in \mathbb{R}$ is almost an eigenvalue of the (one-dimensional) position operator in the following sense.[3] For $\varepsilon > 0$, consider $\psi_\lambda^\varepsilon(x) = (2\varepsilon)^{-1/2}\chi_{[\lambda-\varepsilon,\lambda+\varepsilon]}(x)$. Then $\|\psi_\lambda^\varepsilon\| = 1$ and

$$\|(\widehat{x}-\lambda)\psi_\lambda^\varepsilon\|^2 = \int_{\mathbb{R}} |x-\lambda|^2 |\psi_\lambda^\varepsilon(x)|^2 \,\mathrm{d}x = \int_{\lambda-\varepsilon}^{\lambda+\varepsilon} |x-\lambda|^2 |\psi_\lambda^\varepsilon(x)|^2 \,\mathrm{d}x \le \varepsilon^2 \|\psi_\lambda^\varepsilon\|$$

shows that ψ_λ^ε is a sequence of normalized vectors such that

$$\lim_{\varepsilon\to 0} (\widehat{x}-\lambda)\psi_\lambda^\varepsilon = 0 .$$

■

One thus has a unified language without a uniform meaning. This carries with it a danger of confusion, and for the mathematical formulation we therefore introduce new names: spectrum, spectral family (=PVM), and spectral measure. The spectrum of the operator $\widehat{A}$ is the set of its eigenvectors $\{\lambda_\alpha\}$, and the spectrum of the operators $\widehat{\mathbf{x}}$ and $\widehat{\mathbf{v}}_\infty$ is just $\mathbb{R}^3$. The PVM of $\widehat{A}$ is $(P_A = \sum_{\lambda_\alpha \in A} P_{\lambda_\alpha})_{A\in\mathscr{B}(\mathbb{R})}$, and the PVM of $\widehat{\mathbf{x}}$ or $\widehat{\mathbf{v}}_\infty$ is $(O_A)_{A\in\mathscr{B}(\mathbb{R}^3)}$ or $(V_A)_{A\in\mathscr{B}(\mathbb{R}^3)}$, respectively.

The corresponding spectral measures are $\langle\psi|P_A\psi\rangle$, $\langle\psi|O_A\psi\rangle$, and $\langle\psi|V_A\psi\rangle$, where $\langle\psi|P_A\psi\rangle = \sum_{\lambda_\alpha\in A}\langle\psi|P_{\lambda_\alpha}\psi\rangle$ is a discrete measure. Moreover, note that $\widehat{\mathbf{x}}$ is diagonal, in the sense that $\widehat{\mathbf{x}}$ acts as a multiplication operator on $L^2(\mathbb{R}^3, \mathrm{d}^3x)$ ("position representation"). In the same way $\widehat{\mathbf{v}}_\infty$ is diagonal after Fourier transformation ("momentum representation"). More precisely [see (15.12)],

$$\frac{1}{\mathrm{i}}\frac{\partial}{\partial\mathbf{x}} = \widehat{\mathbf{v}}_\infty \text{ on } L^2 \text{ corresponds to } \widehat{\mathbf{x}} = \mathscr{F}\widehat{\mathbf{v}}_\infty\mathscr{F}^* \text{ on } L^2 .$$

It is this simple action as a multiplication operator in the corresponding "representation" which lies at the heart of Dirac's formalism.

15.2.2 Mathematics of the Spectral Theorem

Now recall the Schrödinger equation and the Schrödinger operator, which must also be self-adjoint, and recall the lesson from linear algebra that it is very fruitful to look at the world, i.e., the vector space, from the point of view of the operator under consideration. This means going to the eigenbasis of the operator, i.e., its spectral

[3] Analogous statements hold for operators with a continuous spectrum, something to which we shall come soon.

representation. So can we do this for the Schrödinger operator H? Is there also a spectrum, a PVM, and the corresponding spectral measure for H? The answer is that there is indeed a corresponding calculus for general self-adjoint operators T, called a T-representation. This is the content of the spectral theorem. It comes in several different flavors which are, in a sense, all equivalent:

- The functional calculus form of the spectral theorem says that one can take functions $f(T)$ of self-adjoint operators T in a consistent way.
- The multiplication operator form says that any self-adjoint operator T becomes a multiplication operator in a suitable representation.
- The PVM form says that any self-adjoint operator T has a unique associated PVM P_A such that

$$T = \int \lambda \mathrm{d}P_\lambda \, .$$

We have already seen how all these various aspects can be very useful when working with concrete operators like $\widehat{\mathbf{x}}$ or $\widehat{\mathbf{v}}_\infty$. So in this section, we will discuss all three aspects in some detail. We start with some basic definitions. One central object is the spectrum of an operator, which is the generalization of the set of eigenvectors of a matrix. Since in general we no longer have eigenvalues, we need to generalize the eigenvalue equation in a clever way.

Definition 15.3. Let T be a closed operator with dense domain $\mathscr{D}(T)$ on a Hilbert space $\mathscr{H}$. The resolvent set $\rho(T)$ is the set of all $\lambda \in \mathbb{C}$ such that

$$(\lambda - T)^{-1} : \mathscr{H} \longrightarrow \mathscr{D}(T) \subset \mathscr{H}$$

exists as a bounded operator in $\mathscr{L}(\mathscr{H})$. $R_\lambda(T) = (\lambda - T)^{-1}$ is called the resolvent of T at λ. The complement $\sigma(T)$ of $\rho(T)$ in $\mathbb{C}$ is the spectrum of T.

Remark 15.6. On the Resolvent and the Spectrum

(i) Let T be bounded. For $\lambda > \|T\|$, we can express the resolvent as a convergent Neumann series,

$$R_\lambda(T) = \frac{1}{\lambda - T} = \frac{1}{\lambda} \frac{1}{1 - T/\lambda} = \frac{1}{\lambda} \left[1 + \sum_{n=1}^{\infty} \left(\frac{T}{\lambda} \right)^n \right] . \tag{15.26}$$

Note that with $\|T^n\| \leq \|T\|^n$, which is immediate from the definition

$$\|T\| = \sup_{\|\psi\|=1} \|T\psi\| \, ,$$

we have

$$\|R_\lambda(T)\| \leq \frac{1}{|\lambda|} \sum_{n=0}^{\infty} \frac{\|T^n\|}{|\lambda|^n} \leq \frac{1}{|\lambda|} \sum_{n=0}^{\infty} \frac{\|T\|^n}{|\lambda|^n} . \tag{15.27}$$

Hence the series converges in the operator norm and one can see, as in

$$(1-x)\sum_{n=0}^{N} x^n = 1 - x^{N+1} \longrightarrow 1\,, \quad N \to \infty\,,$$

that the limit is indeed the resolvent.

(ii) The resolvent set $\rho(T)$ is open and as a consequence the spectrum $\sigma(T)$ is closed. This can be seen as follows. Let $\lambda_0 \in \rho(T)$. The series expansion

$$\frac{1}{\lambda - t} = \frac{1}{\lambda - \lambda_0 + \lambda_0 - t} = \frac{1}{\lambda_0 - t}\,\frac{1}{1 - \dfrac{\lambda_0 - \lambda}{\lambda_0 - t}}$$
$$= \frac{1}{\lambda_0 - t}\sum_{n=0}^{\infty}\left(\frac{\lambda_0 - \lambda}{\lambda_0 - t}\right)^n, \quad \text{for } \frac{|\lambda_0 - \lambda|}{|\lambda_0 - t|} < 1\,,$$

suggests defining the operator

$$\tilde{R}_\lambda(T) = R_{\lambda_0}(T)\sum_{n=0}^{\infty}(\lambda_0 - \lambda)^n R_{\lambda_0}(T)^n\,, \tag{15.28}$$

which is well defined for $|\lambda - \lambda_0| < \|R_{\lambda_0}(T)\|^{-1}$. One checks as in (i) that $(\lambda - T)\tilde{R}_\lambda(T) = \mathsf{I} = \tilde{R}_\lambda(T)(\lambda - T)$, and therefore $\tilde{R}_\lambda(T) = R_\lambda(T)$. Hence $\lambda \in \rho(T)$ and $\rho(T)$ is open. At the same time we have shown that the map

$$\rho(T) \to \mathscr{L}(\mathscr{H}), \quad \lambda \mapsto R_\lambda(T)\,,$$

is analytic, i.e., that it can be expressed locally as the convergent power series (15.28) with coefficients in $\mathscr{L}(\mathscr{H})$.

(iii) If $\lambda \in \mathbb{C}$ is an eigenvalue of T, i.e., $T\varphi_\lambda = \lambda\varphi_\lambda$ holds for some $\varphi_\lambda \in \mathscr{H}$ with $\varphi_\lambda \neq 0$, then $\lambda - T$ is not invertible, and thus λ is in the spectrum $\sigma(T)$. The set of eigenvalues is called the point spectrum $\sigma_{\mathrm{p}}(T)$ of T, and we have $\sigma_{\mathrm{p}}(T) \subset \sigma(T)$. ■

Theorem 15.1. *Let T be self-adjoint. Then*

(i) $\sigma(T) \subset \mathbb{R}$ *and* $\|(T-z)^{-1}\| \leq |\Im(z)|^{-1}$.

(ii) $\sup_{\lambda\in\sigma(T)}|\lambda| = \|T\|$.

(iii) For $\lambda, \lambda' \in \sigma_{\mathrm{p}}(T)$ *and* $\lambda \neq \lambda'$, *one has* $\langle\varphi_\lambda|\varphi_{\lambda'}\rangle = 0$.

Proof. (i) Let $\lambda = x + \mathrm{i}y$ with $x, y \in \mathbb{R}$ and $y \neq 0$. Then like T, the operator $y^{-1}(x-T)$ is also self-adjoint. According to Theorem 14.3, $S = y^{-1}(x-T) + \mathrm{i}$ is a bijection and thus has a bounded inverse by the open mapping theorem (see, e.g., [1]). But then $yS = \lambda - T$ also has a bounded inverse and $\lambda \in \rho(T)$ follows. The bound on the norm of the resolvent follows from

$$\|(T - x - \mathrm{i}y)\psi\|^2 = \|(T-x)\psi\|^2 + \|y\psi\|^2 \geq y^2\|\psi\|^2\,, \quad \text{for all } \psi \in D(T)\,,$$

for $z = x + \mathrm{i}y$.

(ii) We showed already in the previous remark that $\bar{\lambda} := \sup_{\lambda \in \sigma(T)} |\lambda| \leq \|T\|$. The other direction is intuitively clear, when thinking of diagonal matrices with the eigenvalues on the diagonal. To see it in general, we use analyticity of the resolvent. Since the Neumann series (15.26) is the Laurent series for $R_\lambda(T)$ with expansion point $+\infty$, it converges for all $\lambda > \bar{\lambda}$. The radius of convergence is limited by the singularity at $\bar{\lambda}$. Hence

$$\sum_{n=0}^{\infty} \left(\frac{T}{\lambda}\right)^n$$

converges absolutely for $\lambda > \bar{\lambda}$. However, for self-adjoint operators, one has

$$\|T^n\| = \|T\|^n \,, \tag{15.29}$$

and therefore

$$\sum_{n=0}^{\infty} \frac{\|T^n\|}{|\lambda|^n} = \sum_{n=0}^{\infty} \frac{\|T\|^n}{|\lambda|^n} < \infty \,,$$

which implies that $\lambda > \|T\|$ and thus $\bar{\lambda} \geq \|T\|$. We still need to show (15.29), which will follow immediately once we have the spectral theorem. For the argument above, it suffices to show $\|T^{2n}\| = \|T\|^{2n}$, which can be seen as follows. For any bounded operator T, one has $\|T^2\| \leq \|T\|^2$, and also for self-adjoint T,

$$\|T\|^2 = \sup_{\|\psi\|=1} \|T\psi\|^2 = \sup_{\|\psi\|=1} \langle T\psi, T\psi\rangle = \sup_{\|\psi\|=1} \langle \psi, T^2\psi\rangle \leq \|T^2\| \,,$$

where we used Cauchy–Schwarz in the last step.

(iii) is shown as in linear algebra. ∎

For later use we note that, for self-adjoint T, we have just seen that

$$\|T^2\| = \sup_{\|\psi\|=1} |\langle T\psi|T\psi\rangle| \,,$$

which suggests that

$$\|T\| = \left(\sup_{\|\psi\|=1} |\langle \sqrt{T}\psi|\sqrt{T}\psi\rangle| \right) = \sup_{\|\psi\|=1} |\langle \psi|T\psi\rangle| \,, \tag{15.30}$$

where the bracketed part in the middle is completely formal at the moment. But the equality does indeed hold. Reality of the norm implies

$$\|T\| = \sup_{\|\psi\|=1} \Re \left\langle \frac{T\psi}{\|T\|} \Big| T\psi \right\rangle \leq \sup_{\|\varphi\|=1, \|\psi\|=1} \Re \langle \varphi|T\psi\rangle \,,$$

and with polarization, we get, for self-adjoint T and $\|\varphi\| = \|\psi\| = 1$,

$$\begin{aligned}\Re\langle\varphi|T\psi\rangle &= \frac{1}{4}\Big[\langle\varphi+\psi|T(\varphi+\psi)\rangle - \langle\varphi-\psi|T(\varphi-\psi)\rangle\Big] \\ &\leq \frac{1}{4}\big(\|\varphi+\psi\|^2 + \|\varphi-\psi\|^2\big) \sup_{\|\eta\|=1}\langle\eta|T\eta\rangle \\ &= \sup_{\|\eta\|=1}\langle\eta|T\eta\rangle \leq \|T\|\,.\end{aligned}$$

Hence (15.30) follows.

For a not too technical start, we first construct the PVM associated with a *bounded* self-adjoint operator T. The case of unbounded operators will be treated later on. The idea of the construction is quite simple. It is clear that the PVM is supported on the spectrum $\sigma(T)$, and with (15.21) in mind, we should build the characteristic function $\chi_A(T)$ of a subset $A \in \sigma(T)$, since $\chi_{\lambda_\alpha}(\widehat{A}) = P_{\lambda_\alpha}$.

The characteristic functions of all Borel subsets of the spectrum yield the PVM of T. However, this involves slightly more work than one would initially expect. One might want to start with polynomials of T – no problem with that – and then approximate characteristic functions in a suitable sense. But here comes the problem. What is a suitable sense? One probably has pointwise convergence of polynomials p_n in mind. To conclude the existence of a limiting operator $\chi(T)$, one would like to use completeness of the space $\mathscr{L}(\mathscr{H})$ of linear bounded operators on a Hilbert space $\mathscr{H}$ with respect to the operator norm. This completeness does indeed hold true and is easily shown as an exercise. This works analogously to showing that the space of continuous functions on a compact set is complete under the sup-norm.

So we know that $\mathscr{L}(\mathscr{H})$ with the operator norm $\|\cdot\|$ is complete. Now we need to show that convergence of polynomials $p_n \to f$ implies convergence of the corresponding operators $p_n(T) \to f(T)$. However, one quickly realizes that pointwise convergence is not the right notion to start with here. It is much easier to consider uniform convergence first, as the sup-norm goes well with the operator norm. Consider a polynomial $p : \sigma(T) \longrightarrow \mathbb{C}$. Then

$$\|p(T)\| = \|p\|_\infty := \sup_{\lambda\in\sigma(T)} |p(\lambda)|\,. \tag{15.31}$$

The proof is not completely trivial, since we allow p to have complex coefficients. We do not really need this now, but it will help us later on, when considering the notion of a cyclic vector.

For real polynomials $p(T)$ is self-adjoint, but for complex polynomials only $p^*(T)p(T)$ is self-adjoint. But we planned ahead for this when showing (15.30):

$$\begin{aligned}\|p(T)\|^2 &= \sup_{\|\psi\|=1} \langle p(T)\psi | p(T)\psi \rangle = \sup_{\|\psi\|=1} \langle \psi | p^*(T)p(T)\psi \rangle \\ &= \left\| p^*(T)p(T) \right\| = \sup_{\lambda \in \sigma(p^*(T)p(T))} |\lambda| = \sup_{\lambda \in \sigma(T)} \left| p^*(\lambda)p(\lambda) \right| \\ &= \|p\|_\infty^2 .\end{aligned}$$

However, we are not quite through yet, since we used the fact that, for polynomials p, one has

$$\sigma\bigl(p(T)\bigr) = p\bigl(\sigma(T)\bigr) := \bigl\{\mu \,|\, \mu = p(\lambda),\, \lambda \in \sigma(T)\bigr\} , \tag{15.32}$$

which then implies the equality

$$\sup_{\lambda \in \sigma(p(T))} |\lambda| = \sup_{\lambda \in \sigma(T)} |p(\lambda)| .$$

To see (15.32), let $\lambda \in \sigma(T)$. Note that $p(x) - p(\lambda) = (x - \lambda)q(x)$ with a polynomial factor $q(x)$, since λ is a zero of the left-hand side. This implies

$$p(T) - p(\lambda) = (T - \lambda)q(T) .$$

Since $(T - \lambda)$ is not invertible for $\lambda \in \sigma(T)$, $p(T) - p(\lambda)$ is not invertible for $\lambda \in \sigma(T)$. We thus have $p(\lambda) \in \sigma\bigl(p(T)\bigr)$, and therefore $p\bigl(\sigma(T)\bigr) \subseteq \sigma\bigl(p(T)\bigr)$.

Conversely, let $\mu \in \sigma\bigl(p(T)\bigr)$. According to the fundamental theorem of algebra, $p(x) - \mu$ factorizes and therefore

$$p(T) - \mu = \prod_{i=1}^{N} (T - \lambda_i) ,$$

with the zeros $\lambda_1, \ldots, \lambda_N$. Since $p(T) - \mu$ is not invertible, there is at least one λ_i such that $T - \lambda_i$ is not invertible, and with $p(\lambda_i) = \mu$ we conclude that $\sigma\bigl(p(T)\bigr) \subseteq p\bigl(\sigma(T)\bigr)$.

We can now make precise the idea of taking functions of operators on $\mathscr{L}(\mathscr{H})$. The following theorem usually comes under the heading of a functional calculus for self-adjoint operators.

Theorem 15.2. *Let $T \in \mathscr{L}(\mathscr{H})$ be self-adjoint. There exists a unique mapping (the functional calculus)*

$$\Phi : C\bigl(\sigma(T)\bigr) \longrightarrow \mathscr{L}(\mathscr{H}) ,$$

such that

$$\Phi(f) = \text{“}f(T)\text{”} ,$$

where the latter is defined by the following requirements. For all $f, g \in C(\sigma(T))$ and $\alpha, \lambda \in \mathbb{C}$:

(i) $\Phi(1) = \mathsf{I}_{\mathscr{H}}$,
(ii) $f(x) = x \implies \Phi(f) = T$,
(iii) Φ *is linear, i.e.,* $\Phi(f+g) = \Phi(f) + \Phi(g)$ *and* $\Phi(\alpha f) = \alpha\Phi(f)$,
(iv) Φ *is multiplicative, i.e.,* $\Phi(fg) = \Phi(f)\Phi(g)$,
(v) $\Phi(f^*) = \Phi(f)^*$,
(vi) $\sigma(\Phi(f)) = f(\sigma(T)) := \{\mu \in \mathbb{C} \,|\, \mu = f(\lambda),\ \lambda \in \sigma(T)\}$,
(vii) $\|\Phi(f)\| = \|f\|_\infty$,
(viii) $T\psi = \lambda\psi \implies \Phi(f)\psi = f(\lambda)\psi$.

Conditions (i–iv) imply that, for polynomials p, we must define $\Phi(p) := p(T)$. Since the remaining statements are evident or have been shown before for polynomials $p(T)$, we have a unique functional calculus for polynomials. The lift to continuous functions goes by density. Since $\sigma(T)$ is compact, polynomials are dense in $C\big(\sigma(T)\big)$ for the sup-norm. With (vii), this translates nicely to the level of operators.

Lemma 15.1. *There is a unique extension of Φ from the linear space of polynomials to its closure $C\big(\sigma(T)\big)$ satisfying (i–viii).*

Sketch of the Proof. Since Φ is a bounded linear map from a dense subspace (the polynomials) of the normed space $\big(C(\sigma(T)), \|\cdot\|_\infty\big)$ into the complete normed space $\big(\mathscr{L}(\mathscr{H}), \|\cdot\|\big)$, it has a unique bounded extension to all of $C\big(\sigma(T)\big)$.[4] We just need to show that the properties (i–viii) in Theorem 15.2 survive the limit.

For (iv), approximate f and g by polynomials, $p_n \to f$ and $q_n \to g$. Then $p_n q_n \to fg$ and a simple triangulation using uniform boundedness of $\Phi(p_n)$ yields the result. Straightforward approximation also yields (v), (vii), and (viii). For (vi), first let $\mu \notin \operatorname{Ran}(f)$. Then $g = (f-\mu)^{-1}$ exists as a bounded function and $\Phi(g)\Phi(f-\mu) = \Phi(1) = \mathsf{I}$ implies that

$$\Phi(g) = \big(\Phi(f-\mu)\big)^{-1} = \big(\Phi(f)-\mu\big)^{-1}$$

is bounded. Thus $\mu \notin \sigma\big(\Phi(f)\big)$. Conversely, let $\mu \in \operatorname{Ran}(f)$, i.e., $\mu = f(\lambda)$, for some $\lambda \in \sigma(T)$. How can we show that $\mu \in \sigma\big(\Phi(f)\big)$? We already know that, for any polynomial p_n in an approximating sequence $p_n \to f$, we do indeed have $p_n(\lambda) \in \sigma\big(\Phi(p_n)\big)$. Hence for any $\varepsilon > 0$ there exists an almost eigenvector ψ_n with $\|\psi_n\| = 1$ such that

$$\Big\| \big[\Phi(p_n) - p_n(\lambda)\big]\psi_n \Big\| < \varepsilon \,.$$

[4] Pick a Cauchy sequence p_n of polynomials that converges uniformly to $f \in C\big(\sigma(T)\big)$. Then by

$$\big\|\Phi(p_n) - \Phi(p_m)\big\| = \|\Phi(p_n - p_m)\| = \|p_n - p_m\|_\infty \,,$$

we also know that $\Phi(p_n)$ is a Cauchy sequence in the complete space $\mathscr{L}(\mathscr{H})$. Its limit defines the unique extension.

Choosing n large enough to ensure that $\|\Phi(p_n)-\Phi(f)\|<\varepsilon$ and $|p_n(\lambda)-\mu|<\varepsilon$, it follows that

$$\begin{aligned}\left\|\left[\Phi(f)-\mu\right]\psi_n\right\| &= \left\|\left[\Phi(f)-\Phi(p_n)+\Phi(p_n)-p_n(\lambda)+p_n(\lambda)-\mu\right]\psi_n\right\| \\ &\leq \|\Phi(f)-\Phi(p_n)\|+\left\|\left[\Phi(p_n)-p_n(\lambda)\right]\psi_n\right\|+|p_n(\lambda)-\mu| \\ &\leq 3\varepsilon\,.\end{aligned}$$

Since $\varepsilon>0$ was arbitrary, $\left(\Phi(f)-\mu\right)^{-1}$ cannot be bounded.

Now we can use continuous functions to approximate measurable functions, which is where we want to get to in the end. While this approximation can no longer be done with respect to the sup-norm, we recall that, under an integral, pointwise convergence is usually enough. Clearly, the mapping

$$\begin{aligned} l: C\big(\sigma(T)\big) &\longrightarrow \mathbb{C}\,, \\ f &\longmapsto l(f):=\langle\psi|f(T)\psi\rangle\,, \end{aligned}$$

is linear and bounded. Moreover, it is positive in the sense that it takes non-negative values $l(f)$ on non-negative functions f, since for $f\geq 0$ and $g:=\sqrt{f}$, we have

$$l(f)=\langle\psi\,|\,f(T)\psi\rangle=\langle\psi\,|\,g(T)g(T)\psi\rangle=\langle g(T)\psi\,|\,g(T)\psi\rangle=\|g(T)\psi\|^2\geq 0\,.$$

Therefore, we can quote a classical representation theorem – the Riesz–Markov theorem – which tells us that there is a unique measure μ_T^ψ on $\sigma(T)$, associated with this linear functional, such that

$$l(f)=\langle\psi|f(T)\psi\rangle=\int_{\sigma(T)}f(\lambda)\,\mathrm{d}\mu_T^\psi(\lambda)\,,\quad\text{for all } f\in C\big(\sigma(T)\big)\,. \tag{15.33}$$

Indeed, the theorem states that μ_T^ψ is a regular Borel measure.[5] We do not give the proof here, which is somewhat laborious, but not very demanding.[6]

---

[5] A Borel measure μ is a measure defined on the Borel σ-algebra $\mathscr{B}(\Omega)$ of a topological space Ω. It is regular if compact sets have finite measure and if it is compatible with approximation by compact sets from inside and open sets from outside:

$$\mu(A)=\sup\left\{\mu(C)\,\middle|\,C\subset A, C \text{ compact}\right\}=\inf\left\{\mu(O)\,\middle|\,A\subset O, A \text{ open}\right\}\,,$$

for all $A\in\mathscr{B}$.

[6] The idea is simple. One can use continuous functions to approximate characteristic functions of Borel subsets of the compact set $\sigma(T)$ in order to define an outer measure

$$\mu^*(C)=\inf\left\{l(f)\,\middle|\,f\in C\big(\sigma(T)\big)\,,\ f\geq\chi_C\right\}\,,$$

for closed $C\subset\sigma(T)$ and

$$\mu^*(A)=\sup\left\{\mu^*(C)\,\middle|\,C\subset A, C \text{ closed}\right\}\,,$$

Definition 15.4. The measure μ_T^ψ on $\sigma(T)$ defined in (15.33) is called the spectral measure of T for the vector ψ.

The rest is straightforward. By polarization, we define the complex measures

$$\mu_T^{\varphi,\psi} := \frac{1}{4}\left(\mu_T^{\varphi+\psi} - \mu_T^{\varphi-\psi} + \mathrm{i}\mu_T^{\varphi-\mathrm{i}\psi} - \mathrm{i}\mu_T^{\varphi+\mathrm{i}\psi}\right),$$

where the map $(\varphi,\psi) \mapsto \mu_T^{\varphi,\psi}$ is sesquilinear by construction. Hence, for bounded Borel-measurable functions f, we can turn (15.33) into a definition:

$$\langle\varphi\,|\,f(T)\psi\rangle := \int_{\sigma(T)} f(\lambda)\,\mathrm{d}\mu_T^{\varphi,\psi}(\lambda)\,.$$

Again this uniquely defines the operator $f(T) \in \mathscr{L}(\mathscr{H})$ according to Theorem 13.4. Since we can now work with the integral representation of $\Phi(f) = f(T)$, pointwise approximation of measurable functions by continuous functions[7] suffices to lift the properties (i–viii) of Φ to $\mathscr{M}(\sigma(T))$, the space of bounded Borel functions on the spectrum. This functional calculus on $\mathscr{M}\big(\sigma(T)\big)$ is unique if we add the following property to Theorem 15.2:

(ix) If a uniformly bounded sequence (f_n) in $\mathscr{M}\big(\sigma(T)\big)$ converges pointwise to f, then

$$\underset{n\to\infty}{\text{s-lim}}\, f_n(T) = f(T)\,.$$

In conclusion, we now have Theorem 15.2 for functions in $\mathscr{M}(\sigma(T))$ with the additional property (ix). This is all we need! It follows directly from (i–ix) that the family of characteristic functions $(\chi_A(T))_{A\in\mathscr{B}(\sigma(T))}$ defines a PVM. Each operator $\chi_A(T)$ is an orthogonal projector because of (iv) and (v), $\chi_{\sigma(T)}(T) = \mathbb{1}_{\mathscr{H}}$ is just (i), and sigma additivity follows from (iii) and (ix). Furthermore, by construction, the spectral measure is given in terms of the PVM through

$$\mu_T^\psi(A) = \int_{\sigma(T)} \chi_A(\lambda)\,\mathrm{d}\mu_T^\psi(\lambda) = \langle\psi\,|\,\chi_A(T)\psi\rangle\,.$$

Thus we have

$$f(T) = \int_{\sigma(T)} f(\lambda)\,\mathrm{d}\chi_\lambda(T)\,. \tag{15.34}$$

This allows us to formulate the spectral theorem for bounded self-adjoint operators.

---

for arbitrary subsets $A \subset \sigma(T)$. Then Carathéodory's construction yields the desired Borel measure μ_T^ψ.

[7] Lebesgue's theorem of dominated convergence is used here once again.

Theorem 15.3. *There is a one-to-one correspondence between bounded self-adjoint operators and compactly supported PVMs on* $\mathbb{R}$ *given by*

$$T \longrightarrow \big(\chi_A(T)\big)_{A\in\mathscr{B}(\sigma(T))}\,, \quad \textit{with (15.34)},$$

and

$$(P_A)_{A\in\mathscr{B}(\Lambda)} \longrightarrow T = \int \lambda\,\mathrm{d}P_\lambda\,.$$

But is it really clear that a given PVM on a compact subset of $\mathbb{R}$ defines a self-adjoint operator as claimed in the theorem? The answer is affirmative here because, by Remark 15.1, we have

$$\begin{aligned}\langle\psi|T\varphi\rangle &= \int_\Lambda \lambda\,\mathrm{d}P^{\psi,\varphi}(\lambda) = \int_\Lambda \lambda\,\mathrm{d}\overline{P^{\varphi,\psi}(\lambda)} = \overline{\int_\Lambda \lambda\,\mathrm{d}P^{\varphi,\psi}(\lambda)} = \overline{\langle\varphi|T\psi\rangle}\\ &= \langle T\psi|\varphi\rangle\,.\end{aligned}$$

15.2.3 Spectral Representations

In this short section, we make some remarks concerning spectral measures and "diagonalization" of self-adjoint operators. The idea is to write a self-adjoint operator T as a multiplication operator on $L^2\big(\sigma(T),\mathrm{d}\mu\big)$ with a suitable measure μ. The position operator, for example, is already given in this form. This is another variant of the spectral theorem and essentially an elaboration of (15.33).

Let $T\in\mathscr{L}(\mathscr{H})$ be self-adjoint. We are looking for a measure μ on $\sigma(T)$ and a unitary map $U:\mathscr{H}\to L^2\big(\sigma(T),\mathrm{d}\mu\big)$, such that UTU^{-1} on $L^2\big(\sigma(T),\mathrm{d}\mu\big)$ is diagonal, i.e., multiplication by the function $h(\lambda)=\lambda$. To get the analogy with the diagonal representation of a self-adjoint $n\times n$ matrix A, one should think of a vector $v\in\mathbb{C}^n$ as a function on the n distinct eigenvalues of A, i.e., on $\sigma(A)=\{\lambda_1,\dots,\lambda_n\}$.[8] Now multiplication by the diagonal matrix $\mathrm{diag}(\lambda_1,\dots,\lambda_n)$ translates into multiplication by the function λ on $L^2\big(\sigma(T),\sum_{i=1}^n\delta(\lambda-\lambda_i)\big)$.

How can we get an analogous result for operators? One idea would be to look, not only for eigenvectors, but also for "generalized eigenvectors", that is, for "all" solutions of $T\varphi=\lambda\varphi$ (φ not necessarily in $\mathscr{H}$). For example, the operator of asymptotic velocity can be diagonalized by Fourier transformation, which can be seen as representation with respect to the "generalized eigenfunctions" $\mathrm{e}^{\mathrm{i}kx}$. However, it clearly makes no sense, for abstract Hilbert spaces $\mathscr{H}$, to ask for solutions of $T\varphi=\lambda\varphi$ outside of $\mathscr{H}$. So we present a slightly different point of view which works in general, and come back to generalized eigenfunctions later on.

Definition 15.5. Let T be a bounded self-adjoint operator on $\mathscr{H}$. A vector $\eta\in\mathscr{H}$ is called a cyclic vector for T if $\mathrm{span}(T^n\eta)_{n=0}^\infty$ is dense in $\mathscr{H}$.

[8] The case of degenerate eigenvalues requires one more step, as explained below.

We will give an example that shows that there need not be a cyclic vector, but that will present no problem. One can always split $\mathscr{H}$ into a direct sum of orthogonal subspaces for which cyclic vectors exist. But now assume that a cyclic vector η exists. Then, for $\psi, \varphi \in \mathscr{H}$, there are bounded Borel functions g and f such that $\psi = g(T)\eta$ and $\varphi = f(T)\eta$. It follows that

$$\begin{aligned} \langle \psi | T\varphi \rangle &= \langle g(T)\eta \,|\, Tf(T)\eta \rangle = \langle \eta \,|\, g(T)^* Tf(T)\eta \rangle \\ &= \int_{\sigma(T)} g(\lambda)^* \lambda f(\lambda) \mathrm{d}\langle \eta \,|\, \chi_\lambda(T)\eta \rangle \qquad \text{[by (15.34)]} \\ &= \int_{\sigma(T)} g(\lambda)^* f(\lambda) \lambda \, \mathrm{d}\mu_T^\eta(\lambda) \,, \end{aligned} \tag{15.35}$$

where $\mathrm{d}\mu_T^\eta$ is the spectral measure corresponding to the cyclic vector η. Now we simply define

$$\begin{aligned} U : \mathscr{H} &\longrightarrow L^2\big(\sigma(T), \mathrm{d}\mu_T^\eta(\lambda)\big) \,, \\ \psi &\longmapsto g \,, \end{aligned}$$

and observe that U is unitary:

$$\begin{aligned} \langle U\psi | U\varphi \rangle_{L^2(\sigma(T), \mathrm{d}\mu_T^\eta(\lambda))} &= \int g^*(\lambda) f(\lambda) \, \mathrm{d}\mu_T^\eta(\lambda) = \langle \eta \,|\, g^*(T) f(T)\eta \rangle_{\mathscr{H}} \\ &= \langle g(T)\eta \,|\, f(T)\eta \rangle_{\mathscr{H}} \\ &= \langle \psi \,|\, \varphi \rangle_{\mathscr{H}} \,. \end{aligned} \tag{15.36}$$

As (15.35) shows, T acts as multiplication by λ on $L^2\big(\sigma(T), \mathrm{d}\mu_T^\eta\big)$:

$$\big(UTU^{-1} g\big)(\lambda) = \lambda g(\lambda) \,.$$

However, the spectral measure $\mathrm{d}\mu_T^\eta$ is not unique, since there will be many cyclic vectors. If η is cyclic and $f \in \mathscr{M}\big(\sigma(T)\big)$ is strictly positive, i.e., $f(\lambda) \geq c > 0$, then $f(T)$ is invertible and commutes with T, whence $f(T)\eta$ is also cyclic. It is not hard to see that multiplication by λ on $L^2\big(\sigma(T), \mathrm{d}\mu\big)$ and multiplication by λ on $L^2\big(\sigma(T), \mathrm{d}\nu\big)$ are unitarily equivalent if and only if the measures are equivalent, i.e., if and only if they have the same sets of measure zero.

Let us note finally that the non-existence of cyclic vectors in general, and the resulting need to split the Hilbert space into a direct sum of cyclic subspaces, is related to the existence of degenerate eigenvalues. Let us give a simple example. Consider $\mathscr{H} = \mathbb{R}^3$ and $T = P_{\mathbf{e}_1}$, the orthogonal projection onto the first canonical basis vector $\mathbf{e}_1$. Then

$$\operatorname{span}\big((P_{\mathbf{e}_1})^n \mathbf{v}\big)_{n=0}^\infty = \operatorname{span}(\mathbf{v}, P_{\mathbf{e}_1}\mathbf{v}) = \operatorname{span}(\mathbf{v}, \mathbf{e}_1)$$

is at most a two-dimensional plane, spanned by $\mathbf{v}$ and $\mathbf{e}_1$. Hence there is no cyclic vector for $P_{\mathbf{e}_1}$. This is due to the twofold degenerate eigenvalue 0, so that we need two vectors $\mathbf{v}$ and $\mathbf{v}\wedge\mathbf{e}_1$ as "cyclic" vectors. Then

$$\mathbb{R}^3 = \operatorname{span}\left(P^n_{\mathbf{e}_1}\mathbf{v}\right)^{\infty}_{n=0} \oplus \operatorname{span}\left(P^n_{\mathbf{e}_1}(\mathbf{v}\wedge\mathbf{e}_1)\right)^{\infty}_{n=0} = \operatorname{span}(\mathbf{v},\mathbf{e}_1)\oplus\operatorname{span}(\mathbf{v}\wedge\mathbf{e}_1)\,.$$

If T on $\mathbb{R}^3$ has three different eigenvalues with eigenvectors $\mathbf{v}_i$, then each $\mathbf{v}=\sum\alpha_i\mathbf{v}_i$ with $\alpha_i\neq 0$ is cyclic.

15.2.4 Unbounded Operators

Life would be easier if we could avoid this topic altogether. But the Schrödinger operator is unbounded, so we should say briefly how the spectral theorem can also be obtained for unbounded self-adjoint operators. The reader should bear in mind that we made heavy use above of the fact that $\sigma(T)$ is a compact set, which is true only for bounded operators. There are several different ways to derive the functional calculus, the spectral representation, and the spectral measures for unbounded self-adjoint operators. Instead of describing one method in detail, it is more instructive to sketch the different approaches.

One way to proceed is to make use of the spectral theorem for bounded self-adjoint operators. More precisely, one could look for a bounded self-adjoint operator in the form of a bounded function of T and use its spectral representation to get the same for T. However, this idea is somewhat circular, because only the spectral theorem for unbounded operators would allow us to talk about functions of T. However, there is one bounded function of T that gets defined independently of the functional calculus, namely the resolvent. Recall that, according to Theorem 15.1 for self-adjoint T, we know that $\sigma(T)\subset\mathbb{R}$ and therefore $(T\pm\mathrm{i})^{-1}$ are bounded operators. On the other hand, they are not self-adjoint, since with

$$\begin{aligned}\left\langle (T-\mathrm{i})^{-1}\psi\,\middle|\,\varphi\right\rangle &= \left\langle (T-\mathrm{i})^{-1}\psi\,\middle|\,(T+\mathrm{i})(T+\mathrm{i})^{-1}\varphi\right\rangle\\ &= \left\langle (T+\mathrm{i})^{*}(T-\mathrm{i})^{-1}\psi\,\middle|\,(T+\mathrm{i})^{-1}\varphi\right\rangle = \left\langle \psi\,\middle|\,(T+\mathrm{i})^{-1}\varphi\right\rangle\,,\end{aligned}$$

we have $\left[(T-\mathrm{i})^{-1}\right]^{*} = (T+\mathrm{i})^{-1}$. However, not only self-adjoint matrices, but also normal ones are diagonalizable. We can prove the spectral theorem for normal bounded operators[9] in the same way as we did for self-adjoint bounded operators. Since $(T+\mathrm{i})$ and $(T-\mathrm{i})$ commute, $(T+\mathrm{i})^{-1}$ and $(T-\mathrm{i})^{-1}$ also commute. Thus $(T-\mathrm{i})^{-1}$ is normal and one can represent it as a multiplication operator on a suit-

[9] Each bounded operator can be written as the sum of two self-adjoint operators:

$$R=\frac{R+R^*}{2}+\mathrm{i}\frac{R-R^*}{2\mathrm{i}}\,.$$

They can be diagonalized simultaneously if and only if they commute, which is the case if and only if R and R^* commute. Such operators are said to be *normal*.

able L^2-space. We skip the construction of spectral measures for normal bounded operators and also the construction of the spectral measures of T from the one for the resolvent, but instead follow an alternative route to the spectral theorem for unbounded operators.

Recall that the first step in the case of bounded self-adjoint operators was the construction of a functional calculus in Theorem 15.2. There the starting point was to consider polynomials. However, for an unbounded operator T, any polynomial $p(T)$ is itself an unbounded operator, and approximation by polynomials is not an option. The next idea would be to take polynomials of the resolvent $R_z = (T-z)^{-1}$, for z in the resolvent set of T. Such polynomials are dense in $C_\infty(\mathbb{R})$, and one could indeed use them in order to set up a functional calculus for unbounded self-adjoint operators, if one could show the analogue of (vii) in Theorem 15.2. In the following we will pursue the idea of writing $f(T)$ in terms of the resolvent $(T-z)^{-1}$, but in a more explicit way than just saying "by density". It turns out that the resulting explicit formula is very useful in applications.

To motivate the formula, consider first a compactly supported function $f: \mathbb{R}^2 \to \mathbb{C}$ which is continuously differentiable. In the following, we will identify the domain $\mathbb{R}^2$ with $\mathbb{C}$ via $z = x + \mathrm{i}y$. However, f is not assumed to be holomorphic (it cannot be, since it is compactly supported). The reader may know that the Cauchy integral formula of complex analysis is just an application of Stokes' theorem. We will now repeat the corresponding derivation, however, for our function f, which is not holomorphic. Let z_0 be a point in the support of f, Γ a compact subset of $\mathbb{C}$ with smooth boundary $\partial\Gamma$ such that $\operatorname{supp} f$ is contained in the interior Γ° of Γ, and $B_\delta(z_0)$ the ball of radius δ around z_0. Interpreting

$$\omega = f(z)(z_0 - z)^{-1}\mathrm{d}z$$

as a complex 1-form on the complement of $B_\delta(z_0)$, one finds that

$$\begin{aligned}
\mathrm{d}\omega &= \frac{\partial}{\partial z}\left[f(z)(z_0-z)^{-1}\right]\mathrm{d}z \wedge \mathrm{d}z + \frac{\partial}{\partial \bar{z}}\left[f(z)(z_0-z)^{-1}\right]\mathrm{d}\bar{z} \wedge \mathrm{d}z \\
&= \frac{\partial f}{\partial \bar{z}}(z)(z_0-z)^{-1}\mathrm{d}\bar{z} \wedge \mathrm{d}z\,,
\end{aligned}$$

where $\mathrm{d}z \wedge \mathrm{d}z$ and $\partial_{\bar{z}}(z_0-z)^{-1}$ both vanish. Note that $\mathrm{d}\bar{z} \wedge \mathrm{d}z = 2\mathrm{i}\mathrm{d}x\,\mathrm{d}y$, and take $\delta > 0$ small enough to ensure that $B_\delta(z_0) \subset \Gamma^\circ$. Then Stokes' theorem implies

$$\int_{\partial(\Gamma \setminus B_\delta(z_0))} \omega = \int_{\partial\Gamma} \omega - \int_{\partial B_\delta(z_0)} \omega = \int_{\Gamma \setminus B_\delta(z_0)} \mathrm{d}\omega\,. \tag{15.37}$$

By continuity of $\partial_{\bar{z}} f(z)$ and integrability of $(z_0-z)^{-1}$, the right-hand side of (15.37) is given in the limit $\delta \to 0$ by

$$\lim_{\delta\to 0}\int_{\Gamma\setminus B_\delta(z_0)} \mathrm{d}\omega = 2\mathrm{i}\lim_{\delta\to 0}\int_{\Gamma\setminus B_\delta(z_0)} \frac{\partial f}{\partial \bar z}(z)(z_0-z)^{-1}\mathrm{d}x\,\mathrm{d}y$$

$$= 2\mathrm{i}\int_\Gamma \frac{\partial f}{\partial \bar z}(z)(z_0-z)^{-1}\mathrm{d}x\,\mathrm{d}y\,.$$

For the left-hand side of (15.37), we get, by continuity of f,

$$\lim_{\delta\to 0}\int_{\partial B_\delta(z_0)} \omega = \lim_{\delta\to 0}\int_{\partial B_\delta(z_0)} f(z)(z_0-z)^{-1}\mathrm{d}z = -2\pi\mathrm{i}f(z_0)\,,$$

and since $f|_{\partial\Gamma} = 0$,

$$\int_{\partial\Gamma}\omega = \int_{\partial\Gamma} f(z)(z_0-z)^{-1}\mathrm{d}z = 0\,.$$

Putting everything together, we have thus shown that

$$f(z_0) = \frac{1}{\pi}\int_\Gamma \frac{\partial f}{\partial \bar z}(z)(z_0-z)^{-1}\mathrm{d}x\,\mathrm{d}y\,. \tag{15.38}$$

Note that, for holomorphic f, one has $\mathrm{d}\omega = 0$, but then the integral over $\partial\Gamma$ contributes. In that case (15.37) yields the Cauchy integral formula.

Back to operators. The idea is now to replace the variable z_0 in (15.38) by a possibly unbounded self-adjoint operator T, and to define

$$f(T) = \frac{1}{\pi}\int_\Gamma \frac{\partial f}{\partial \bar z}(z)(T-z)^{-1}\mathrm{d}x\,\mathrm{d}y\,.$$

There are two problems with this formula. One is that, while the integrand is singular at a point in (15.38) and thus integrable, $(T-z)^{-1}$ is singular on the spectrum of T, which might contain intervals of the real line. On the other hand, we aim at a functional calculus for functions on $\sigma(T)\subset\mathbb{R}$, and the second problem is that it is not clear what $\partial_{\bar z}f(z)$ should mean for a function on $\mathbb{R}$.

The clever trick in the construction is now to solve both problems together. We extend $f:\mathbb{R}\to\mathbb{C}$ to a function $\tilde f:\mathbb{C}\to\mathbb{C}$ in such a way that $\partial_{\bar z}\tilde f(z)$ vanishes as $|\Im(z)|$ when z approaches the real axis, and thus compensates the possible $|\Im(z)|^{-1}$ divergence of the resolvent [see Theorem 15.1 (i)]. Such extensions are said to be almost analytic, since $\partial_{\bar z}\tilde f(z)$ would vanish identically for an analytic extension. This also suggests a way to construct almost analytic extensions.

Let $f\in C_0^{n+1}(\mathbb{R})$. Then we can define $\tilde f(x+\mathrm{i}y)$ by Taylor expansion at x, pretending that the Cauchy–Riemann equation $\partial_y\tilde f(z) = \mathrm{i}\partial_x\tilde f(z)$ holds:

$$\tilde f(z) := \sum_{j=0}^{n}\frac{f^{(j)}(x)}{j!}(\mathrm{i}y)^j\,.$$

It is easy to check that

$$\frac{\partial}{\partial \bar{z}}\tilde{f}(z) = \frac{1}{2}\frac{f^{(n+1)}(x)}{n!}(\mathrm{i}y)^n \sim |y|^n = |\Im(z)|^n \,, \quad \text{as } |\Im(z)| \to 0\,,$$

and thus $\tilde{f}$ is an almost analytic extension. Now $\tilde{f}$ is neither compactly supported nor integrable, and therefore we need to multiply $\tilde{f}$ by a smooth cutoff function $\chi(y)$ such that $\chi|_{[-1,1]} = 1$. But then the integral

$$f(T) = \frac{1}{\pi}\int_{\Gamma}\frac{\partial \tilde{f}}{\partial \bar{z}}(z)(T-z)^{-1}\mathrm{d}x\,\mathrm{d}y \tag{15.39}$$

converges in norm and defines a bounded operator $f(T)$.

It is not difficult to show that this $f(T)$ is independent of n and the specific form of the cutoff, and that this formula defines a functional calculus for functions $f \in C_0^2\big(\sigma(T)\big)$ satisfying $\|f(T)\| \le \|f\|_\infty$. One can then pass by density to the closure

$$\overline{C_0^2\big(\sigma(T)\big)}^{\|\cdot\|_\infty} = C_\infty\big(\sigma(T)\big)\,,$$

the set of continuous functions vanishing at infinity. Using Riesz–Markov again, which yields *finite* measures μ_T^ψ on $\mathscr{B}(\sigma(T))$, one finally extends the functional calculus to bounded Borel functions $\mathscr{M}\big(\sigma(T)\big)$. In this way one obtains the following result.

Theorem 15.4. *Let T with domain $D(T)$ be a self-adjoint operator on $\mathscr{H}$. There exists a unique functional calculus*

$$\Phi : \mathscr{M}\big(\sigma(T)\big) \to \mathscr{L}(\mathscr{H})$$

with the following properties:

(i) *Φ is linear and multiplicative.*
(ii) $\Phi(f^*) = \Phi(f)^*$.
(iii) $\|\Phi(f)\| = \|f\|_\infty$.
(iv) *For $z \in \mathbb{C}\setminus\mathbb{R}$ and $r_z(x) = (x-z)^{-1}$, one has $\Phi(r_z) = (T-z)^{-1}$.*
(v) *If a uniformly bounded sequence (f_n) in $\mathscr{M}\big(\sigma(T)\big)$ converges pointwise to f, then*

$$\operatorname*{s\text{-}lim}_{n\to\infty}\Phi(f_n) = \Phi(f)\,.$$

Note that (iv) replaces requirement (ii) in the bounded case and guarantees that the functional calculus is compatible with the definition of the resolvent. The proof of (iv) is actually the most difficult part, and for this one extends formula (15.39) to a class of functions containing $r_z(x)$ (see [2]).

Given the functional calculus, we can now define the PVM associated with a self-adjoint operator T as $\big(\chi_A(T)\big)_{A\in\mathscr{B}(\sigma(T))}$, where for bounded functions $f \in \mathscr{M}\big(\sigma(T)\big)$, we once again have, by construction,

$$f(T) = \int_{\sigma(T)} f(\lambda)\,\mathrm{d}\chi_\lambda(T)\,.$$

However, for unbounded operators T, the spectrum $\sigma(T)$ is not bounded, and thus the function $f(\lambda) = \lambda$ is not bounded on $\sigma(T)$. Hence,

$$T = \int_{\sigma(T)} \lambda\,\mathrm{d}\chi_\lambda(T)$$

is not an immediate consequence of the functional calculus. To understand this precisely in the situation of unbounded operators, it is convenient to look first at the spectral representation.

Assume for simplicity that T has a cyclic vector η, i.e., every $\psi \in \mathscr{H}$ can be written as $g(T)\eta$ for some $g \in \mathscr{M}\big(\sigma(T)\big)$. Then one can construct the unitary mapping

$$\begin{aligned} U : \mathscr{H} &\longrightarrow L^2\big(\sigma(T), \mathrm{d}\mu_T^\eta(\lambda)\big)\,, \\ \psi &\longmapsto g\,, \end{aligned}$$

as in the case of bounded operators. We find once again that bounded functions of T are mapped to multiplication by the corresponding function, i.e., for $f \in \mathscr{M}\big(\sigma(T)\big)$,

$$Uf(T)U^{-1} = f(\lambda)\,. \tag{15.40}$$

The only additional point to show is that T is indeed mapped under U to multiplication by λ. To see this, note first that (15.40) applied to the resolvent implies that the range of the resolvent is mapped to the range of $r(\lambda) = (\lambda - \mathrm{i})^{-1}$, i.e.,

$$UD(T) = \Big\{\psi \in L^2\big(\sigma(T)\big)\,\big|\,\lambda\psi(\lambda) \in L^2\big(\sigma(T)\big)\Big\}\,.$$

But then, for $\psi \in UD(T)$, we have $\psi = r\varphi$ for some $\varphi \in L^2\big(\sigma(T)\big)$, and

$$UTU^{-1}\psi = UTU^{-1}r\varphi = UT(T-\mathrm{i})^{-1}U^{-1}\varphi = (\mathrm{i}r+1)\varphi = \lambda r\varphi = \lambda\psi\,.$$

At this point, let us formulate our findings as a theorem. However, before we do so, we should get rid of the assumption that a cyclic vector exists. The following lemma is not hard to show.

Lemma 15.2. *Let T with domain $D(T)$ be a self-adjoint operator on the separable Hilbert space $\mathscr{H}$. Then there exists a sequence of pairwise orthogonal cyclic subspaces $L_n \subset \mathscr{H}$ with cyclic vectors*[10] *η_n such that*

$$\mathscr{H} = \overline{\bigoplus_{n=1}^{N} L_n}\,,$$

where N can be finite or ∞. Note that one only needs to take the closure for $N = \infty$.

[10] Recall that η is a cyclic vector for $L \subset \mathscr{H}$ if $L = \mathrm{span}\big\{f(T)\eta\,\big|\,f \in \mathscr{M}\big(\sigma(T)\big)\big\}$. It is clear that a cyclic subspace is invariant for T, i.e., $f(T)L = L$ for any $f \in \mathscr{M}\big(\sigma(T)\big)$.

Using the cyclic vector η_n, we can map each cyclic subspace L_n unitarily to $L^2\big(\sigma(T), \mathrm{d}\mu^{\eta_n}\big)$. In summary, we then obtain the following multiplication operator version of the spectral theorem.

Theorem 15.5. *Let T with domain $D(T)$ be a self-adjoint operator on a separable Hilbert space $\mathscr{H}$. Let $(L_n, \eta_n)_{n=1}^N$ be a sequence of cyclic subspaces L_n with cyclic vectors η_n, as in Lemma 15.2. Then there is a unitary mapping*

$$U : \mathscr{H} = \overline{\bigoplus_{n=1}^{N} L_n} \longrightarrow \overline{\bigoplus_{n=1}^{N} L^2\big(\sigma(T), \mathrm{d}\mu_T^{\eta_n}\big)} =: L^2\big(\sigma(T) \times \{1, \dots, N\}, \mathrm{d}\mu\big) ,$$

such that, for $f \in \mathscr{M}\big(\sigma(T)\big)$ and $\psi \in L^2\big(\sigma(T) \times \{1, \dots, N\}, \mathrm{d}\mu\big) =: L^2(\mathrm{d}\mu)$,

$$\big(Uf(T)U^{-1}\psi\big)(\lambda, n) = f(\lambda)\psi(\lambda, n) .$$

Moreover,

$$UD(T) = \big\{\psi \in L^2(\mathrm{d}\mu) \,\big|\, \lambda\psi(\lambda, n) \in L^2(\mathrm{d}\mu)\big\} ,$$

and for all $\psi \in UD(T)$,

$$\big(UTU^{-1}\psi\big)(\lambda, n) = \lambda\psi(\lambda, n) .$$

We can now use this spectral representation to formulate the precise connection between unbounded self-adjoint operators and PVMs. To simplify the notation, we assume that there is a cyclic vector η for T or, equivalently, we reduce our considerations to one cyclic subspace. The key observation is that, with $g = U\psi$ and $f = U\varphi$, we have, by construction,

$$\big\langle\psi \,\big|\, \chi_A(T)\varphi\big\rangle = \big\langle\eta \,\big|\, g^*(T)\chi_A(T)f(T)\eta\big\rangle = \int_A g^*(\lambda)f(\lambda)\,\mathrm{d}\mu_T^\eta ,$$

and hence

$$\mathrm{d}\big\langle\psi \,\big|\, \chi_\lambda(T)\varphi\big\rangle = g^*(\lambda)f(\lambda)\,\mathrm{d}\mu_T^\eta . \tag{15.41}$$

Thus for $\varphi \in D(T)$ and $\psi \in \mathscr{H}$, it follows that

$$\langle\psi | T\varphi\rangle = \int_{\sigma(T)} g^*(\lambda)\lambda f(\lambda)\,\mathrm{d}\mu_T^\eta(\lambda) = \int_{\sigma(T)} \lambda\,\mathrm{d}\big\langle\psi \,\big|\, \chi_\lambda(T)\varphi\big\rangle .$$

This yields the first part of the PVM version of the spectral theorem.

Theorem 15.6. *There is a one-to-one correspondence between self-adjoint operators and PVMs on $\mathbb{R}$:*

(i) Let T be self-adjoint. Then $\big(\chi_A(T)\big)_{A \in \mathscr{B}(\sigma(T))}$ is a PVM and, for $f \in \mathscr{M}\big(\sigma(T)\big)$,

$$f(T) = \int_{\sigma(T)} f(\lambda)\, \mathrm{d}\chi_\lambda(T)\,.$$

For $\varphi \in D(T)$ and $\psi \in \mathscr{H}$, one has

$$\langle \psi | T\varphi \rangle = \int_{\sigma(T)} \lambda\, \mathrm{d}\langle \psi | \chi_\lambda(T)\varphi \rangle\,.$$

(ii) Given a PVM $(P_A)_{A\in\mathscr{B}(\Lambda)}$, let

$$D(T) := \left\{ \varphi \in \mathscr{H} \,\Big|\, \int_{\mathbb{R}} \lambda^2 \mathrm{d}\langle \varphi | P_\lambda \varphi \rangle < \infty \right\}.$$

Then

$$\langle \psi | T\varphi \rangle := \int_{\mathbb{R}} \lambda\, \mathrm{d}\langle \psi | P_\lambda \varphi \rangle\,, \quad \textit{for } \psi, \varphi \in D(T)\,,$$

defines a self-adjoint operator T with domain $D(T)$.

To understand the second part of the theorem, take a sequence of simple functions $f_n(\lambda) := \sum_{j=1}^n \lambda_j^{(n)} \chi_{\Delta_j^{(n)}}(\lambda)$ that converges monotonically to $f(\lambda) = |\lambda|$. Then, for $\psi, \varphi \in D(T)$, we have

$$\begin{aligned}
&\int_{\mathbb{R}} \lambda^2 \mathrm{d}\langle \varphi + \psi | P_\lambda(\varphi + \psi)\rangle \\
&\qquad = \lim_{n\to\infty} \int_{\mathbb{R}} f_n^2(\lambda)\, \mathrm{d}\langle \varphi + \psi | P_\lambda(\varphi + \psi)\rangle \\
&\qquad = \lim_{n\to\infty} \sum_{j=1}^n (\lambda_j^{(n)})^2 \langle \varphi + \psi | P_{\Delta_j^{(n)}}(\varphi + \psi)\rangle \\
&\qquad \le \lim_{n\to\infty} \sum_{j=1}^n (\lambda_j^{(n)})^2 \Big[\langle \varphi | P_{\Delta_j^{(n)}}\varphi\rangle + \langle \psi | P_{\Delta_j^{(n)}}\psi\rangle + 2\big|\langle \varphi | P_{\Delta_j^{(n)}}\psi\rangle\big| \Big] \\
&\qquad \le \int_{\mathbb{R}} \lambda^2 \mathrm{d}\langle \varphi | P_\lambda \varphi\rangle + \int_{\mathbb{R}} \lambda^2 \mathrm{d}\langle \psi | P_\lambda \psi\rangle \\
&\qquad\quad + \lim_{n\to\infty} \sum_{j=1}^n (\lambda_j^{(n)})^2 \langle \varphi | P_{\Delta_j^{(n)}}\varphi\rangle^{1/2} \langle \psi | P_{\Delta_j^{(n)}}\psi\rangle^{1/2} < \infty\,,
\end{aligned}$$

since $\psi, \varphi \in D(T)$, and

$$\lim_{n\to\infty}\sum_{j=1}^{n}(\lambda_j^{(n)})^2\langle\varphi\,|\,P_{\Delta_j^{(n)}}\varphi\rangle^{1/2}\langle\psi\,|\,P_{\Delta_j^{(n)}}\psi\rangle^{1/2}$$

$$\leq\lim_{n\to\infty}\left[\sum_{j=1}^{n}(\lambda_j^{(n)})^2\langle\varphi\,|\,P_{\Delta_j^{(n)}}\varphi\rangle\right]^{1/2}\left[\sum_{j=1}^{n}(\lambda_j^{(n)})^2\langle\psi\,|\,P_{\Delta_j^{(n)}}\psi\rangle\right]^{1/2}$$

$$=\left(\int_{\mathbb{R}}\lambda^2\,\mathrm{d}\langle\varphi\,|\,P_\lambda\varphi\rangle\right)^{1/2}\left(\int_{\mathbb{R}}\lambda^2\,\mathrm{d}\langle\psi\,|\,P_\lambda\psi\rangle\right)^{1/2}<\infty\,.$$

Hence, $\varphi+\psi\in D(T)$, and $D(T)$ is indeed a subspace. It is also dense, since for any $\psi\in\mathscr{H}$, we clearly have $P_{[-n,n]}\psi\in D(T)$ and, by Definition 15.1(i), and (iii),

$$\lim_{n\to\infty}P_{[-n,n]}\psi=\psi\,.$$

To see that the integral $\int_{\mathbb{R}}\lambda\,\mathrm{d}\langle\psi\,|\,P_\lambda\varphi\rangle$ exists for $\psi,\varphi\in D(T)$, note that the explicit decomposition of the complex measure $\mu_T^{\psi,\varphi}(A):=\langle\psi\,|\,P_A\varphi\rangle$ is

$$\mu_T^{\psi,\varphi}=\frac{1}{4}\left(\mu_T^{\psi+\varphi}-\mu_T^{\psi-\varphi}+\mathrm{i}\mu_T^{\psi-\mathrm{i}\varphi}-\mathrm{i}\mu_T^{\psi+\mathrm{i}\varphi}\right)\,.$$

Let $\phi\in\{\psi+\varphi,\psi-\varphi,\psi+\mathrm{i}\varphi,\psi-\mathrm{i}\varphi\}$. Then $\phi\in D(T)$ and

$$\int_{\mathbb{R}}|\lambda|\,\mathrm{d}\langle\phi\,|\,P_\lambda\phi\rangle\leq\langle\phi\,|\,P_{[-1,1]}\phi\rangle+\int_{\mathbb{R}}|\lambda|^2\,\mathrm{d}\langle\phi\,|\,P_\lambda\phi\rangle<\infty\,.$$

Thus T is a densely defined and obviously symmetric operator. To conclude that T is self-adjoint, we use the criterion of Theorem 14.3, i.e., we show that $\mathrm{Ran}(T\pm\mathrm{i})=\mathscr{H}$. So let $\psi\in\mathscr{H}$ and define

$$\varphi:=\int(\lambda+\mathrm{i})^{-1}\mathrm{d}P_\lambda\,\psi\,.$$

We will show that, for any $\phi\in\mathscr{H}$,

$$\mathrm{d}\langle\phi\,|\,P_\lambda\varphi\rangle=\frac{1}{\lambda+\mathrm{i}}\mathrm{d}\langle\phi\,|\,P_\lambda\psi\rangle\,,\qquad\mathrm{d}\langle\varphi\,|\,P_\lambda\varphi\rangle=\frac{1}{\lambda^2+1}\mathrm{d}\langle\psi\,|\,P_\lambda\psi\rangle\,.\tag{15.42}$$

From this it immediately follows that $\varphi\in D(T)$ and, for any $\phi\in\mathscr{H}$,

$$\langle\phi\,|\,(T+\mathrm{i})\varphi\rangle=\int(\lambda+\mathrm{i})\,\mathrm{d}\langle\phi\,|\,P_\lambda\varphi\rangle=\int\mathrm{d}\langle\phi\,|\,P_\lambda\psi\rangle=\langle\phi\,|\,\psi\rangle\,.$$

Hence $(T+\mathrm{i})\varphi=\psi$ and T is therefore self-adjoint.

In order to prove (15.42), it is convenient to remark that, for any bounded Borel function f, one can define the integral with respect to a PVM as a norm-convergent limit of operators. Let $f_n=\sum_{j=1}^{n}\lambda_j^{(n)}\chi_{\Delta_j^{(n)}}$ be a sequence of simple functions converging uniformly to f. Then

$$\lim_{n\to\infty}\int f_n(\lambda)\mathrm{d}P_\lambda := \lim_{n\to\infty}\sum_{j=1}^{n}\lambda_j^{(n)}P_{\Delta_j^{(n)}}$$

converges in norm, and the limit is independent of the sequence f_n and equal to the operator $\int f(\lambda)\,\mathrm{d}P_\lambda$, as defined in Remark 15.1. This follows from the easily checked fact that, for any simple function $g = \sum_{j=1}^n \lambda_j \chi_{\Delta_j}$, one has

$$\left\|\int g(\lambda)\mathrm{d}P_\lambda\right\| := \left\|\sum_{j=1}^{n}\lambda_j P_{\Delta_j}\right\| \leq \|g\|_\infty \,.$$

Now back to (15.42). Let $\sum_{j=1}^n \alpha_j^{(n)} \chi_{\Delta_j^{(n)}}$ be a sequence of simple functions converging uniformly to $(\lambda+\mathrm{i})^{-1}$. We can assume that, for each n, the sets $\Delta_j^{(n)}$ are pairwise disjoint. Then for any Borel set $A \subset \mathbb{R}$ and any $\psi \in \mathscr{H}$, we find that

$$\begin{aligned}\langle \phi \,|\, P_A \varphi\rangle &= \lim_{n\to\infty}\Big\langle \phi \,\Big|\, P_A \sum_{j=1}^{n}\alpha_j^{(n)}P_{\Delta_j^{(n)}}\psi\Big\rangle = \lim_{n\to\infty}\sum_{j=1}^{n}\alpha_j^{(n)}\langle \phi \,|\, P_{\Delta_j^{(n)}\cap A}\psi\rangle \\ &= \int_A \frac{1}{\lambda+\mathrm{i}}\mathrm{d}\langle \phi \,|\, P_\lambda \psi\rangle \,.\end{aligned}$$

This proves the first equality in (15.42). The second follows analogously by

$$\begin{aligned}\langle \varphi \,|\, P_A \varphi\rangle &= \lim_{n\to\infty}\Big\langle \sum_{k=1}^{n}\alpha_k^{(n)}P_{\Delta_k^{(n)}}\psi \,\Big|\, P_A \sum_{j=1}^{n}\alpha_j^{(n)}P_{\Delta_j^{(n)}}\psi\Big\rangle \\ &= \lim_{n\to\infty}\sum_{j=1}^{n}|\alpha_j^{(n)}|^2\langle \psi \,|\, P_{\Delta_j^{(n)}\cap A}\psi\rangle \\ &= \int_A \frac{1}{\lambda^2+1}\mathrm{d}\langle \psi \,|\, P_\lambda \psi\rangle \,.\end{aligned}$$

15.2.5 Unitary Groups

Recall that one motivation for looking at self-adjoint operators was the expectation that self-adjoint operators are the generators of unitary groups. With the spectral theorem to hand it is now straightforward to show the following theorem.

Theorem 15.7. *Let H with domain $D(H)$ be a self-adjoint operator. Then*

$$U(t) = \mathrm{e}^{-\mathrm{i}Ht}$$

defined by the functional calculus is a strongly continuous unitary group and H is its generator.

Proof. The group property and unitarity follow from properties (i) and (ii) of the functional calculus of Theorem 15.4. Strong continuity follows from (v), because $\mathrm{e}^{-\mathrm{i}xt}$ converges pointwise to 1 for $t \to 0$, so we have

$$\operatorname*{s-lim}_{t\to 0} \mathrm{e}^{-\mathrm{i}Ht} = \mathsf{I}_{\mathscr{H}} .$$

Hence $U(t)$ is indeed a strongly continuous unitary group. To show that H is indeed the generator, we need to show that $\mathrm{e}^{-\mathrm{i}Ht}\psi$ is differentiable if and only if $\psi \in D(H)$, and that the derivative is $-\mathrm{i}H\psi$. Note first that the difference quotient

$$\left\| \frac{\mathrm{e}^{-\mathrm{i}Ht} - \mathsf{I}}{t} \psi \right\|^2 = \int \underbrace{\left| \frac{\mathrm{e}^{-\mathrm{i}\lambda t} - 1}{t} \right|^2}_{\to \lambda^2 \text{ as } t \to 0} \mathrm{d}\langle \psi | P_\lambda \psi \rangle$$

remains bounded as $t \to 0$ if and only if $\psi \in D(H)$. For $\psi \in D(H)$, the limit is indeed $-\mathrm{i}H\psi$, as can be seen from

$$\lim_{t\to 0} \left\| \frac{\mathrm{e}^{-\mathrm{i}Ht} - \mathsf{I}}{t} \psi + \mathrm{i}H\psi \right\|^2 = \lim_{t\to 0} \int \underbrace{\left| \frac{\mathrm{e}^{-\mathrm{i}\lambda t} - 1}{t} + \mathrm{i}\lambda \right|^2}_{\to 0 \text{ as } t \to 0} \mathrm{d}\langle \psi | P_\lambda \psi \rangle = 0$$

and dominated convergence.

Stone's theorem now states that any strongly continuous unitary group is generated by a self-adjoint operator. This is comforting since it means that we can focus on the self-adjoint generators without taking the risk of missing some interesting unitary evolution groups.

Theorem 15.8. *(Stone's Theorem) Every strongly continuous unitary group $U(t)$ has a self-adjoint generator H, i.e., $U(t) = \mathrm{e}^{-\mathrm{i}Ht}$.*

The proof is not difficult with the machinery we have developed. But that is enough abstract mathematics for now. Let us return to quantum mechanics.

15.2.6 $H_0 = -\Delta/2$

Now we have more or less proved that there is a PVM for every self-adjoint operator, in particular also for the Schrödinger operator H, or any other operator that seems relevant to us. But what is the PVM for $H = H_0 + V$? Good question! In general we do not know, but what we are looking for are basically eigenfunctions in a suitable generalized sense. More precisely, we also look for eigenfunctions for the continuous spectrum. We shall now explain this for H_0 and discuss the extension to Schrödinger operators with a potential in the next chapter. So for the moment, we shall look for suitable solutions of

$$H_0\varphi = -\frac{1}{2}\Delta\varphi = \lambda\varphi\,, \quad \lambda \in \mathbb{R}\,,$$

in which we ignore physical constants. The suitable solutions are of course the plane waves, i.e., the Fourier "basis functions"

$$\varphi = \mathrm{e}^{\pm \mathrm{i}\mathbf{k}\cdot\mathbf{x}}\,, \qquad \lambda = \frac{1}{2}k^2\,.$$

This is nice, since we already know that the Fourier transformation is unitary on L^2 and that, in the Fourier representation, H_0 becomes the operator for multiplication by $k^2/2$. We can thus write the PVM of H_0 explicitly as

$$P^{H_0}(A) = \mathscr{F}^{-1}\chi_A(k^2)\mathscr{F}\,,$$

where A is a Borel subset of $\mathbb{R}$. We see immediately that the support of P^{H_0}, i.e., the spectrum, is $\sigma(H_0) = [0,\infty)$. More explicitly we have

$$\left(P^{H_0}(A)\psi\right)(\mathbf{x}) = \frac{1}{(2\pi)^3}\int_{\{\mathbf{k}\,|\,k^2/2\in A\}} \mathrm{e}^{+\mathrm{i}\mathbf{k}\cdot\mathbf{x}}\left[\int \mathrm{e}^{-\mathrm{i}\mathbf{k}\cdot\mathbf{y}}\psi(\mathbf{y})\,\mathrm{d}^3y\right]\mathrm{d}^3k\,.$$

Recall the Dirac notation where this is written formally as

$$P^{H_0}(A) = \int_{\{\mathbf{k}\,|\,k^2/2\in A\}} |\mathbf{k}\rangle\langle\mathbf{k}|\,\mathrm{d}^3k\,,$$

i.e., as projection onto the generalized eigenfunction $|\mathbf{k}\rangle$. To get the result in the $\mathbf{x}$ representation, we need to project onto $|\mathbf{x}\rangle$:

$$\langle\mathbf{x}|P^{H_0}(A)|\psi\rangle = \int_{\{\mathbf{k}\,|\,k^2/2\in A\}} \langle\mathbf{x}|\mathbf{k}\rangle\langle\mathbf{k}|\psi\rangle\,\mathrm{d}^3k = \int_{\{\mathbf{k}\,|\,k^2/2\in A\}} \mathrm{e}^{\mathrm{i}\mathbf{k}\cdot\mathbf{x}}\,\widehat{\psi}(\mathbf{k})\,\mathrm{d}^3k\,.$$

Finally, we can write H_0 as

$$H_0 = \int \frac{1}{2}k^2|\mathbf{k}\rangle\langle\mathbf{k}|\,\mathrm{d}^3k\,,$$

which is just another notation for the statement we started with, i.e.,

$$H_0 = \mathscr{F}^{-1}\frac{1}{2}k^2\mathscr{F}\,.$$

This is a good point to use this representation in order to compute a few functions of H_0 explicitly.

Remark 15.7. The Free Propagator
In n dimensions, the so-called free propagator is

$$\left\langle\mathbf{y}\middle|\mathrm{e}^{-\mathrm{i}tH_0}\middle|\mathbf{x}\right\rangle = \frac{1}{(2\pi\mathrm{i}t)^{n/2}}\mathrm{e}^{\mathrm{i}|\mathbf{y}-\mathbf{x}|^2/2t}\,.$$

This integral kernel yields the solutions of the free Schrödinger evolution. For an initial wave function $\varphi \in L^2(\mathbb{R}^n)$, we have

$$\begin{aligned}\varphi(\mathbf{x},t) = \left(\mathrm{e}^{-\mathrm{i}tH_0}\varphi\right)(\mathbf{x}) &= L^2\text{-lim}\frac{1}{(2\pi\mathrm{i}t)^{n/2}}\int \mathrm{e}^{\mathrm{i}|\mathbf{y}-\mathbf{x}|^2/2t}\varphi(\mathbf{y})\,\mathrm{d}^n y \\ &\text{“=”} \int \left\langle \mathbf{x}|\mathrm{e}^{-\mathrm{i}tH_0}|\mathbf{y}\right\rangle \langle \mathbf{y}|\varphi\rangle\,\mathrm{d}^n y\,.\end{aligned}$$

■

One can compute the free propagator as follows. In the Fourier representation we have

$$\mathrm{e}^{-\mathrm{i}tH_0} = \mathscr{F}^{-1}\mathrm{e}^{-\mathrm{i}tk^2/2}\mathscr{F}\,.$$

Recalling that (inverse) Fourier transformation turns multiplication into convolution, we can compute this directly for $\varphi \in \mathscr{S}$:

$$\mathscr{F}^{-1}\underbrace{\mathrm{e}^{-\mathrm{i}tk^2/2}}_{=:G(\mathbf{k})}\mathscr{F}\varphi = (2\pi)^{-n/2}(\mathscr{F}^{-1}G) * \varphi\,,$$

and the inverse Fourier transform of a Gaussian is again a Gaussian [see (5.7) and (9.19)]:

$$(\mathscr{F}^{-1}G)(\mathbf{x}) = \frac{1}{(2\pi\mathrm{i}t)^{n/2}}\mathrm{e}^{\mathrm{i}|\mathbf{x}|^2/2t}\,.$$

To see that the integral representation holds for all functions in L^2 is somewhat technical, and we refer the interested reader to [3].

From the explicit formula, one can directly read off the spreading of the wave packet:

$$\sup_{\mathbf{x}}\left|\mathrm{e}^{-\mathrm{i}tH_0}\varphi(\mathbf{x})\right| \le \frac{1}{(2\pi|t|)^{n/2}}\|\varphi\|_1\,,$$

or the sojourn probability in a domain $G \subset \mathbb{R}^n$:

$$\int_G \left|\left(\mathrm{e}^{-\mathrm{i}tH_0}\varphi\right)(\mathbf{x})\right|^2 \mathrm{d}^n x \le \frac{1}{(2\pi t)^n}\|\varphi\|_1^2 |G|\,.$$

We already anticipated the precise asymptotics of the free time evolution for $t \to \infty$ when discussing $\mathbf{v}_\infty$, using the heuristic argument in Sect. 9.4. Now the time has come to give a proof.

Remark 15.8. The "Free Asymptotics" and Stationary Phase
We will show that, for $\varphi \in L^2(\mathbb{R}^n)$,

$$\left\| e^{-itH_0}\varphi - \frac{1}{(it)^{n/2}} e^{ix^2/2t} \widehat{\varphi}\left(\frac{\mathbf{x}}{t}\right) \right\| \longrightarrow 0\,, \quad \text{for } t \to \infty\,. \tag{15.43}$$

It suffices to prove this for φ in the dense subset of Schwartz functions $\mathscr{S}$, since our claim is that the difference between two t-dependent families of unitary operators converges strongly to zero. And for $\varphi \in \mathscr{S}$ the proof is just a simple computation. Multiplying out the square in the exponent in

$$\left(e^{-iH_0 t}\varphi\right)(\mathbf{x}) = \varphi(\mathbf{x},t) = \int \frac{1}{(2\pi i t)^{n/2}} \exp\left[i\frac{(\mathbf{x}-\mathbf{y})^2}{2t}\right] \varphi(\mathbf{y})\, d^n y\,,$$

we find that

$$\begin{aligned}
\varphi(\mathbf{x},t) &= \frac{1}{(it)^{1/2}} \exp\left(i\frac{\mathbf{x}^2}{2t}\right) \int \frac{1}{(2\pi)^{1/2}} \exp\left(-i\frac{\mathbf{x}\cdot\mathbf{y}}{t}\right) \exp\left(i\frac{\mathbf{y}^2}{2t}\right) \varphi(\mathbf{y})\, d^n y \\
&= \frac{\exp\left(i\mathbf{x}^2/2t\right)}{(it)^{1/2}} 2\widehat{\varphi}\left(\frac{\mathbf{x}}{t}\right) + \underbrace{\frac{\exp\left(i\mathbf{x}^2/2t\right)}{(2\pi i t)^{1/2}} \int \exp\left(-i\frac{\mathbf{x}\cdot\mathbf{y}}{t}\right) \left[\exp\left(i\frac{\mathbf{y}^2}{2t}\right) - 1\right] \varphi(\mathbf{y})\, d^n y}_{r_t}\,.
\end{aligned}$$

Put

$$h_t(\mathbf{y}) = \left[\exp\left(i\frac{\mathbf{y}^2}{2t}\right) - 1\right] \varphi(\mathbf{y}) \in L^2\,.$$

Then

$$r_t = \frac{1}{(it)^{1/2}} \exp\left(i\frac{\mathbf{x}^2}{2t}\right) \widehat{h}_t\left(\frac{\mathbf{x}}{t}\right)$$

and

$$\|r_t\|^2 = \int \frac{1}{t^n} \widehat{h}_t^*\left(\frac{\mathbf{x}}{t}\right) \widehat{h}_t\left(\frac{\mathbf{x}}{t}\right) d^n x = \langle \widehat{h}_t | \widehat{h}_t \rangle = \langle h_t | h_t \rangle = \|h_t\|^2\,.$$

Now $h_t(\mathbf{y})$ converges pointwise to zero for $t \to \infty$ and, with $|h_t(\mathbf{y})|^2 \le 4|\varphi(\mathbf{y})|^2 \in L^1$, we can use dominated convergence to conclude that $\|h_t\|^2 \to 0$ for $t \to \infty$.

Now recall (15.11). In the fourth step there, we really get

$$\lim_{t\to\infty} \int |\varphi(\mathbf{x},t)|^2 \chi_A\left(\frac{\mathbf{x}}{t}\right) d^n x = \lim_{t\to\infty} \int \frac{1}{t^n} \left|\widehat{\varphi}\left(\frac{\mathbf{x}}{t}\right)\right|^2 \chi_A\left(\frac{\mathbf{x}}{t}\right) d^n x + R(t)\,,$$

and it remains to show that the remainder $R(t)$ goes to zero. With the above result, Cauchy–Schwarz, and Parseval, this follows directly:

$$\begin{aligned}
R(t) &= \int |r_t(\mathbf{x})|^2 \chi_A\left(\frac{\mathbf{x}}{t}\right) d^n x + 2\Re\left[\int \frac{1}{t^n} \widehat{\varphi}\left(\frac{\mathbf{x}}{t}\right) \widehat{h}_t\left(\frac{\mathbf{x}}{t}\right) \chi_A\left(\frac{\mathbf{x}}{t}\right) d^n x\right] \\
&\le \|r_t\|^2 + 2\|\varphi\|\,\|h_t\| \overset{t\to\infty}{\longrightarrow} 0\,.
\end{aligned}$$

∎

Remark 15.9. The Stationary Phase Method away from the Stationary Point
In (9.18), we discussed the stationary phase argument and just described the leading order term rigorously. We can now also study the error terms, since we know how a free wave packet moves for large times. Indeed, for large t, the wave function is supported at positions $\mathbf{x}$ in such a way that $\mathbf{x}/t$ lies in the support of the Fourier transform. Everywhere else, the wave function goes to zero. And it is easy to estimate how fast it goes to zero. Here is a precise statement.

Theorem 15.9. *Let $\varphi \in \mathscr{S}$ and $K = \mathrm{supp}(\widehat{\varphi})$ be compact. Let $\mathscr{U}$ be an open ε-neighborhood of K, i.e., $\mathrm{dist}(\mathscr{U}^c, K) = \varepsilon > 0$. Then for any $N \in \mathbb{N}$, there is a constant C_N, such that, for any pair $\mathbf{x}$, t with $\mathbf{x}/t \notin \mathscr{U}$ and $|t| \geq 1$,*

$$\left|\left(\mathrm{e}^{-\mathrm{i}tH_0}\varphi\right)(\mathbf{x})\right| \leq C_N\left(1+|t|\right)^{-N}.$$

This simple "no stationary phase" statement is just based on integration by parts. Observe that

$$\begin{aligned}\left(\mathrm{e}^{-\mathrm{i}tH_0}\varphi\right)(\mathbf{x}) &= \frac{1}{(2\pi\mathrm{i}t)^{1/2}}\int_K \exp\left[\mathrm{i}\left(\mathbf{k}\cdot\mathbf{x}-\frac{1}{2}k^2t\right)\right]\widehat{\varphi}(\mathbf{k})\,\mathrm{d}^n k\\ &= \frac{1}{(2\pi\mathrm{i}t)^{1/2}}\int_K \exp\left[\mathrm{i}(1+|t|)\left(\frac{\mathbf{k}\cdot\mathbf{x}-k^2t/2}{1+|t|}\right)\right]\widehat{\varphi}(\mathbf{k})\,\mathrm{d}^n k\,,\end{aligned}$$

and put $\alpha = 1+|t|$. Then the integrand contains the oscillating exponential $\mathrm{e}^{\mathrm{i}\alpha S}$ with the "phase function"

$$S(\mathbf{k}) = \frac{\mathbf{k}\cdot\mathbf{x}-k^2t/2}{1+|t|}\,, \qquad \text{where } \nabla_{\mathbf{k}}S = \frac{\mathbf{x}-\mathbf{k}t}{1+|t|}\,.$$

But by assumption, on the support K of $\widehat{\varphi}$, we have

$$\left|\nabla_{\mathbf{k}}S(\mathbf{k})\right| = \frac{\left|\dfrac{\mathbf{x}}{t}-\mathbf{k}\right|}{\dfrac{1}{|t|}+1} \geq \frac{1}{2}\mathrm{dist}(\mathscr{U}^c, K) = \frac{1}{2}\varepsilon > 0\,.$$

The gradient of S is therefore bounded from below by $\varepsilon/2$, and we get the identity

$$\mathrm{e}^{\mathrm{i}\alpha S(\mathbf{k})} = \left[\frac{1}{\mathrm{i}\alpha}\left|(\nabla S)(\mathbf{k})\right|^{-2}\nabla S(\mathbf{k})\cdot\nabla\right]^N \mathrm{e}^{\mathrm{i}\alpha S(\mathbf{k})}\,,$$

where we have dropped the $\mathbf{x}$ and t dependence in the notation. Integration by parts then yields ($\varphi \in \mathscr{S}$)

$$\int_K \mathrm{e}^{\mathrm{i}\alpha S(\mathbf{k})}\widehat{\varphi}(\mathbf{k})\,\mathrm{d}^n k = (-1)^N\left(\frac{1}{\mathrm{i}\alpha}\right)^N\int_K \mathrm{e}^{\mathrm{i}\alpha S(\mathbf{k})}\left\{\left[\nabla\cdot\frac{(\nabla S)(\mathbf{k})}{\left|(\nabla S)(\mathbf{k})\right|^2}\right]^N\widehat{\varphi}(\mathbf{k})\right\}\mathrm{d}^n k\,.$$

Now it is easy to see that the integrand is bounded independently of $\mathbf{x}$ and t, whence the absolute value of the integral yields exactly the constant C_N.

Note that, in Theorem 15.9, the stronger statement

$$\left|(\mathrm{e}^{-\mathrm{i}tH_0}\varphi)(\mathbf{x})\right| \leq C_N\left(1+|t|+|x|\right)^{-N}$$

actually holds. This can be shown by putting $\alpha = 1+|t|+|x|$ in the proof. However, estimating $\nabla_{\mathbf{k}}S$ and the integrand above then takes more effort. We have skipped this in order to focus on the simple structure of the argument. Moreover, for the following application, our simple result is sufficient.

Assume that $\widehat{\varphi} \in C_0^\infty(\mathbb{R}^n \setminus \{0\})$, i.e., there is $a > 0$, such that $k > a$ for all $\mathbf{k} \in \operatorname{supp}(\widehat{\varphi})$. It is now a simple consequence of the above computation that the domain from which the wave function escapes grows with time. More precisely, it holds that

$$\left\|\chi\left(|\mathbf{x}| < a|t|\right)\mathrm{e}^{-\mathrm{i}tH_0}\varphi\right\| \leq C_N\left(1+|t|\right)^{-N}. \tag{15.44}$$

The proof of this estimate comes immediately out of the constraint that $|\mathbf{x}|/t \leq a < k$ and the resulting estimate

$$\left|(\mathrm{e}^{-\mathrm{i}tH_0}\varphi)(\mathbf{x})\right| \leq C_M\left(1+|t|\right)^{-M},$$

which yields

$$\int \chi\left(|\mathbf{x}| < a|t|\right)\left(\mathrm{e}^{-\mathrm{i}tH_0}\varphi\right)(\mathbf{x})\mathrm{d}^n x \leq C'_M\left(1+|t|\right)^{-M}\left(a|t|\right)^n \leq C''_M\left(1+|t|\right)^{-N},$$

for appropriate constants C, C', C'', and M large enough. ■

As we saw from the support of its PVM, the spectrum of H_0 is $[0,\infty)$. Thus the resolvent $(H_0-\lambda)^{-1}$ exists on $\mathbb{C}\setminus\mathbb{R}^+$ and we will now compute its integral kernel.

Remark 15.10. The Resolvent of $-\Delta$

We will show that for all $\kappa \in \mathbb{C}$ with $\Re\kappa > 0$ and all $\varphi \in L^2(\mathbb{R}^3)$,

$$\left[(-\Delta+\kappa^2)^{-1}\varphi\right](\mathbf{x}) = \frac{1}{4\pi}\int \frac{\mathrm{e}^{-\kappa|\mathbf{x}-\mathbf{y}|}}{|\mathbf{x}-\mathbf{y}|}\varphi(\mathbf{y})\,\mathrm{d}^3 y. \tag{15.45}$$

Before we give the derivation of this formula, note that the integral kernel or Green's function

$$G(x) := \frac{1}{4\pi}\frac{\mathrm{e}^{-\kappa|\mathbf{x}|}}{|\mathbf{x}|}$$

is integrable and square integrable, i.e., $G \in L^1(\mathbb{R}^3)\cap L^2(\mathbb{R}^3)$. Hence by the Cauchy–Schwarz inequality, the integral on the right-hand side of (15.45) exists for all $\varphi \in L^2(\mathbb{R}^3)$ and $\mathbf{x} \in \mathbb{R}^3$. By Young's inequality, convolution with G defines a bounded operator on $L^2(\mathbb{R}^3)$. More precisely, let $\psi := G*\varphi$. Then with Fubini, we have that

$$\|\psi\|_{L^2}^2 \leq \iint G^*(\mathbf{y})G(\tilde{\mathbf{y}}) \left[\int |\varphi^*(\mathbf{x}-\mathbf{y})\varphi(\mathbf{x}-\tilde{\mathbf{y}})|\, \mathrm{d}^3x\right] \mathrm{d}^3y\, \mathrm{d}^3\tilde{y}$$
$$\leq \|G\|_{L^1}^2 \|\varphi\|_{L^2}^2 .$$

Hence both sides in (15.45) define bounded operators on L^2 and it suffices to prove the equality for φ in the dense set of Schwartz functions $\mathscr{S}$. To show this, we use the Fourier representation again:

$$(-\Delta-\lambda)^{-1} = \mathscr{F}^{-1}\frac{1}{k^2-\lambda}\mathscr{F} ,$$

where it is convenient to put $\lambda = -\kappa^2$ and to assume $\Re\,\kappa > 0$. (We thereby cover the whole resolvent set $\mathbb{C}\setminus\mathbb{R}^+$ of $-\Delta$. But we could just as well take the other square root of λ. Then in the application of the residue theorem below, we would close the path of integration through the lower half plane instead of the upper half plane.)

Let $f(\mathbf{k}) := (k^2+\kappa^2)^{-1}$ and $f_R(\mathbf{k}) = \chi_{k<R}(\mathbf{k})f(\mathbf{k})$, where $\chi_{k<R}(\mathbf{k})$ is the characteristic function of the ball B_R around the origin with radius R. Then according to Remark 13.5, we have, for $\varphi \in \mathscr{S}$ and for all $\mathbf{x}$,

$$\left((-\Delta+\kappa^2)^{-1}\varphi\right)(\mathbf{x}) = \left(\mathscr{F}^{-1}f\widehat{\varphi}\right)(\mathbf{x}) = \lim_{R\to\infty}\left(\mathscr{F}^{-1}f_R\widehat{\varphi}\right)(\mathbf{x})$$
$$= \lim_{R\to\infty}\left(\frac{1}{2\pi}\right)^{3/2}\int (\mathscr{F}^{-1}f_R)(\mathbf{x}-\mathbf{y})\varphi(\mathbf{y})\mathrm{d}^3y .$$

Here we have used the fact that $f_R\widehat{\varphi}$ converges to $\widehat{\varphi}$ in L^1, whence the inverse Fourier transform even converges uniformly.

We will now show that

$$\lim_{R\to\infty}\left(\frac{1}{2\pi}\right)^{3/2}\mathscr{F}^{-1}f_R = G ,$$

in L^2. Then, by Cauchy–Schwarz, (15.45) follows. Now,

$$\lim_{R\to\infty}\left(\frac{1}{2\pi}\right)^{3/2}\left(\mathscr{F}^{-1}f_R\right)(\mathbf{x}) = \lim_{R\to\infty}\left(\frac{1}{2\pi}\right)^{3}\int_{B_R}\frac{\mathrm{e}^{\mathrm{i}\mathbf{k}\cdot\mathbf{x}}}{k^2+\kappa^2}\mathrm{d}^3k$$
$$= \lim_{R\to\infty}\left(\frac{1}{2\pi}\right)^{3}\int_0^R\int_0^{2\pi}\int_{-1}^{1}\frac{\mathrm{e}^{\mathrm{i}kx\cos\theta}}{k^2+\kappa^2}k^2\,\mathrm{d}(\cos\theta)\,\mathrm{d}\varphi\,\mathrm{d}k$$
$$= \lim_{R\to\infty}\frac{(2\pi)^{-2}}{\mathrm{i}x}\int_{-R}^{R}\frac{\mathrm{e}^{\mathrm{i}kx}}{(k-\mathrm{i}\kappa)(k+\mathrm{i}\kappa)}k\,\mathrm{d}k .$$

Now we can apply the residue theorem. We close the integration along a rectangular path in the upper half plane, going from R to $R+\mathrm{i}\sqrt{R}$, from there to $-R+\mathrm{i}\sqrt{R}$, and finally to $-R$. For R large enough, this path encircles the pole at $k = \mathrm{i}\kappa$. Hence we find that

$$\lim_{R\to\infty}\left(\frac{1}{2\pi}\right)^{3/2}\left(\mathscr{F}^{-1}f_R\right)(\mathbf{x}) = \frac{1}{4\pi}\frac{\mathrm{e}^{-\kappa|\mathbf{x}|}}{|\mathbf{x}|},$$

granted that the contributions of the extra pieces of the path vanish in the limit $R\to\infty$. We show this for the piece $\gamma(t) = R + t\mathrm{i}\sqrt{R}$, $t\in[0,1]$, from R to $R+\mathrm{i}\sqrt{R}$. Observe that

$$\begin{aligned}\left|\frac{1}{\mathrm{i}x}\int_\gamma \frac{\mathrm{e}^{\mathrm{i}kx}}{k^2+\kappa^2}k\,\mathrm{d}k\right| &= \left|\frac{1}{\mathrm{i}x}\int_0^1 \frac{\mathrm{e}^{\mathrm{i}(R+t\mathrm{i}\sqrt{R})x}(R+t\mathrm{i}\sqrt{R})\mathrm{i}\sqrt{R}}{(R+t\mathrm{i}\sqrt{R})^2+\kappa^2}\mathrm{d}t\right| \\ &\le \frac{C_\kappa}{x}\int_0^1 \frac{\mathrm{e}^{-t\sqrt{R}x}}{\sqrt{R}}\mathrm{d}t \\ &= \frac{C_\kappa}{x^2}\frac{(1-\mathrm{e}^{-\sqrt{R}x})}{R},\end{aligned}$$

with a κ-dependent constant C_κ. In spherical coordinates, ignoring constants, we find for the L^2-norm,

$$\begin{aligned}\int_0^\infty \frac{\left(1-\mathrm{e}^{-\sqrt{R}x}\right)^2 x^2}{x^4R^2}\mathrm{d}x &= \int_0^\infty \frac{\left(1-\mathrm{e}^{-\sqrt{R}x}\right)^2}{x^2R^2}\mathrm{d}x \\ &= \int_0^{1/R} \frac{\left(1-\mathrm{e}^{-\sqrt{R}x}\right)^2}{x^2R^2}\mathrm{d}x + \int_{1/R}^\infty \frac{\left(1-\mathrm{e}^{-\sqrt{R}x}\right)^2}{x^2R^2}\mathrm{d}x \\ &\le \int_0^{1/R}\frac{Rx^2}{x^2R^2}\mathrm{d}x + \int_{1/R}^\infty \frac{1}{x^2R^2}\mathrm{d}x \\ &\le \frac{1}{R^2}+\frac{1}{R} \longrightarrow 0, \quad \text{as } R\to\infty.\end{aligned}$$

■

Remark 15.11. The Resolvent of $-c^2\Delta$
In the case of H_0, one has a prefactor in front of $-\Delta$, whence

$$\begin{aligned}\left(\left(-c^2\Delta+\kappa^2\right)^{-1}\varphi\right)(\mathbf{x}) &= c^{-2}\left(\left(-\Delta+\frac{\kappa^2}{c^2}\right)^{-1}\varphi\right)(\mathbf{x}) \\ &= \frac{1}{4\pi c^2}\int\frac{\mathrm{e}^{-|\mathbf{x}-\mathbf{y}|\kappa/c}}{|\mathbf{x}-\mathbf{y}|}\varphi(\mathbf{y})\mathrm{d}^3y.\end{aligned}$$

■

15.2.7 The Spectrum

We understand now that a self-adjoint operator can be written as a sum or an integral over a family of pairwise orthogonal projections, i.e., in terms of its associated PVM. The PVM is a measure supported on the spectrum of the operator, and we may ask what further relevance the spectrum has. This is a very natural question, given that the quick acceptance of quantum mechanics was mainly based on the successful explanation of spectral lines of atoms in terms of eigenvalues of the Schrödinger operator. The eigenvalues are part of the spectrum. But the spectrum can also be continuous, as in the case of H_0 or $\widehat{x}$. Or it can have both eigenvalues and continuous parts, like the Hamiltonian of the hydrogen atom, for example. How can one see what type of spectrum a given operator has?

The spectral measure corresponding to an eigenvector φ_0 with eigenvalue λ_0 is a point measure $\mu_H^{\varphi_0} = \delta(\lambda - \lambda_0)\|\varphi_0\|^2$. To see this, note first that $(H-z)^{-1}\varphi_0 = (\lambda_0 - z)^{-1}\varphi_0$ and the functional calculus of Theorem 15.4 imply that, for any bounded Borel function f, we have $f(H)\varphi_0 = f(\lambda_0)\varphi_0$. But then it follows for all $f \in \mathscr{M}\big(\sigma(H)\big)$ that

$$f(\lambda_0)\|\varphi_0\|^2 = \langle \varphi_0 \,|\, f(H)\varphi_0 \rangle = \int f(\lambda)\,\mathrm{d}\mu_H^{\varphi_0}(\lambda)\,,$$

and thus $\mu_H^{\varphi_0} = \delta(\lambda - \lambda_0)\|\varphi_0\|^2$.

In contrast to the spectral measure of an eigenvector, the spectral measure μ_H^{η} generated by a cyclic vector η is supported on all of $\sigma(H)$, and we can therefore think of it as "the" spectral measure of H. It will have continuous parts and parts supported on points. This suggests first looking at general measures μ on the spectrum, independently of a corresponding vector. But this is once again abstract and simple mathematics: a regular Borel measure, which in particular is finite on compact sets, can be decomposed into three parts. Let $P := \{x \in \mathbb{R} \,|\, \mu(\{x\}) \neq 0\}$ be the set of points which have nonzero measure and define, for measurable $A \subset \mathbb{R}$, the point measure part of μ as

$$\mu_{\mathrm{pp}}(A) := \sum_{x \in P \cap A} \mu(\{x\}) = \mu(P \cap A)\,.$$

If $\mu = \mu_{\mathrm{pp}}$, we say that μ is a point measure. Naturally, one defines the continuous part of μ as $\mu_{\mathrm{c}} = \mu - \mu_{\mathrm{pp}}$. Applied to spectral measures, the idea is that μ_{c} is supported on the continuous part of the spectrum. But this continuous support can be very different from what we think of as the "continuum". For example, the Cantor set has the same cardinality as the continuum. And what we have in mind is a measure like Lebesgue measure. Let us say more precisely what we mean by "like" Lebesgue measure.

Definition 15.6. We say that a measure μ on $\mathbb{R}$ is absolutely continuous with respect to Lebesgue measure λ, and write $\mu \ll \lambda$, if

$$\lambda(A) = 0 \quad \Longrightarrow \quad \mu(A) = 0\,,$$

i.e., if all Lebesgue null sets are also μ-null sets.

The Radon–Nikodym theorem states that $\mu \ll \lambda$ implies that μ has a density with respect to λ, i.e., that $\mathrm{d}\mu = \rho\, \mathrm{d}\lambda$ for some locally integrable function ρ. Let us give a short sketch of the proof. Let $(\mathbb{R}, \mathscr{B}(\mathbb{R}), \mathbb{P})$ with $\mathbb{P} = \mu + \lambda$ and $L^2(\mathbb{R}, \mathrm{d}\mathbb{P})$. With spectral measures in mind, we can assume that $\mu(\mathbb{R}) < \infty$, which implies that, for any μ-integrable f, we have

$$\int |f|\, \mathrm{d}\mu \leq \mu(\mathbb{R}) \left(\int |f|^2 \mathrm{d}\mu \right)^{1/2} \leq \mu(\mathbb{R}) \left(\int |f|^2 \mathrm{d}\mathbb{P} \right)^{1/2} .$$

Hence, $\ell(f) := \int f\, \mathrm{d}\mu$ defines a bounded linear functional on $L^2(\mathbb{R}, \mathrm{d}\mathbb{P})$, and according to Theorem 13.2, there is a unique vector $g \in L^2(\mathbb{R}, \mathrm{d}\mathbb{P})$ with

$$\ell(f) = \int f\, \mathrm{d}\mu = \int f g\, \mathrm{d}\mathbb{P} = \int f g\, \mathrm{d}\mu + \int f g\, \mathrm{d}\lambda ,$$

or, rearranged,

$$\int f(1-g)\, \mathrm{d}\mu = \int f g\, \mathrm{d}\lambda .$$

Let G_1 be the set on which $g \geq 1$. Then, by inserting $f = \chi_{G_1} \in L^2(\mathbb{R}, \mathrm{d}\mathbb{P})$ into this equation, it follows that $\lambda(G_1) = 0$. Analogously, one sees that the set G_2 with $g < 0$ is a Lebesgue null set. The assumption of absolute continuity then implies that $\mu(G_1 \cup G_2) = 0$. Let $G = G_1 \cup G_2$. Then for $A \in \mathscr{B}(\mathbb{R})$ and

$$f = \frac{1}{1-g} \chi_A \chi_{G^c} ,$$

we find that

$$\int \chi_A\, \mathrm{d}\mu = \int \chi_A \rho\, \mathrm{d}\lambda ,$$

with the non-negative density

$$\rho = \frac{g}{1-g} \chi_G .$$

The next observation is that any regular Borel measure μ can be split into a part μ_{ac} which is absolutely continuous with respect to Lebesgue measure and a part μ_{sing} which is singular and contains, in particular, its pure point part. This is easy to see. Trivially, we have $\mu \ll \mu + \lambda$, and hence there is a density ρ such that

$$\mu(A) = \int_A \rho\, \mathrm{d}\mu + \int_A \rho\, \mathrm{d}\lambda ,$$

with $\rho \leq 1$. Let $F = \{x \,|\, \rho(x) < 1\}$ and $F^c = \{x \,|\, \rho(x) = 1\}$. Then clearly $\lambda(F^c) = 0$, and therefore $\mu_{\mathrm{ac}}(A) = \mu(A \cap F)$ and $\mu_{\mathrm{sing}}(A) = \mu(A \cap F^c)$. Note that $\mu_{\mathrm{ac}} \ll \lambda$, since, for $\lambda(N) = 0$ and $\mu_{\mathrm{ac}}(N) = \mu(N \cap F) > 0$, we would get the contradiction that

$$\mu(N \cap F) = \int_{N \cap F} \rho \, \mathrm{d}\mu + \int_{N \cap F} \rho \, \mathrm{d}\lambda = \int_{N \cap F} \rho \, \mathrm{d}\mu < \mu(N \cap F) \, .$$

Hence, $\mu_{\mathrm{ac}}(N) = 0$.

Now we can subtract the pure point part from the singular part and define the singular continuous part of a regular Borel measure as $\mu_{\mathrm{sc}} = \mu_{\mathrm{sing}} - \mu_{\mathrm{pp}}$. In summary, we get the decomposition

$$\mu = \mu_{\mathrm{pp}} + \mu_{\mathrm{ac}} + \mu_{\mathrm{sc}} \, . \tag{15.46}$$

Let us return now to the spectral measure μ_H^η of a cyclic vector η. Its decomposition immediately yields a decomposition of the Hilbert space in the form

$$L^2\big(\sigma(H), \mathrm{d}\mu_H^\eta\big) = L^2\big(\sigma(H), \mathrm{d}\mu_{\mathrm{pp}}^\eta\big) \oplus L^2\big(\sigma(H), \mathrm{d}\mu_{\mathrm{ac}}^\eta\big) \oplus L^2\big(\sigma(H), \mathrm{d}\mu_{\mathrm{sc}}^\eta\big) \, ,$$

and by inverting the unitary map U from $\mathscr{H}$ to $L^2\big(\sigma(H), \mathrm{d}\mu_H^\eta\big)$, we get the decomposition

$$\mathscr{H} = \mathscr{H}_{\mathrm{pp}} \oplus \mathscr{H}_{\mathrm{ac}} \oplus \mathscr{H}_{\mathrm{sc}} \, . \tag{15.47}$$

We can also understand this in the following way. Suppose for example that $\varphi \in \mathscr{H}_{\mathrm{ac}}$. Then the corresponding spectral measure μ_H^φ is absolutely continuous. This is because $f = U\varphi \in L^2\big(\sigma(H), \mathrm{d}\mu_{\mathrm{ac}}^\eta\big)$, and therefore, with (15.41),

$$\mathrm{d}\mu_H^\varphi(\lambda) = \left|f(\lambda)\right|^2 \mathrm{d}\mu_{\mathrm{ac}}^\eta(\lambda) = \left|f(\lambda)\right|^2 \rho^\eta(\lambda) \, \mathrm{d}\lambda \, .$$

This shows in particular that the ambiguity in the spectral measure (the cyclic vector is never unique) has no influence on the decomposition (15.47). The latter is unique!

Finally, one can also decompose the spectrum into different, not necessarily disjoint components. One defines

$$\begin{aligned}
\sigma_{\mathrm{pp}}(H) &:= \sigma(H|_{\mathscr{H}_{\mathrm{pp}}}) = \mathrm{supp}(\mu_{\mathrm{pp}}^\eta) \, , \\
\sigma_{\mathrm{ac}}(H) &:= \sigma(H|_{\mathscr{H}_{\mathrm{ac}}}) = \mathrm{supp}(\mu_{\mathrm{ac}}^\eta) \, , \\
\sigma_{\mathrm{sc}}(H) &:= \sigma(H|_{\mathscr{H}_{\mathrm{sc}}}) = \mathrm{supp}(\mu_{\mathrm{sc}}^\eta) \, .
\end{aligned}$$

Here the pure point spectrum σ_{pp} is the closure of the set of eigenvalues, where the latter was previously denoted by σ_{p}.

Remark 15.12. On the Spectrum of Unitarily Equivalent Operators
Let S and T be self-adjoint, and let $T = USU^*$ with U unitary. Then, clearly, $\sigma(S) = \sigma(T)$ because $\lambda - S$ is invertible if and only if $\lambda - T = U(\lambda - S)U^*$ is invertible. Let us go quickly through an argument which shows that even more is true, viz., the spectral types agree. This follows once we show that the PVMs P^S of S and P^T of T transform according to

$$P^S = UP^T U^* , \qquad (15.48)$$

since the spectral measures then agree:

$$\mu_S^{\psi}(A) = \langle \psi | P_A^S \psi \rangle = \langle U^* \psi | P_A^T U^* \psi \rangle = \mu_T^{U^* \psi}(A) .$$

For (15.48), recall that $P_A^S = \chi_A(S)$ and $P_A^T = \chi_A(T)$. Using (15.39) and approximating as in Theorem 15.4 (v), we find that (15.48) follows from $(T-z)^{-1} = U(S-z)^{-1}U^*$. ■

Now back to physics. Think of the Schrödinger operator H. The subspace $\mathscr{H}_{\mathrm{pp}}$ contains linear combinations of eigenfunctions. These are the bound states, i.e., states which stay within bounded regions during the time evolution. In $\mathscr{H}_{\mathrm{ac}}$, there are the wave functions which spread and propagate to infinity, while in $\mathscr{H}_{\mathrm{sc}}$, there are all the wave functions which behave neither way, and which one would like to ignore altogether. As an example let us compute $\langle \varphi \,|\, \mathrm{e}^{-\mathrm{i}Ht} \psi \rangle$ for $\psi \in \mathscr{H}_{\mathrm{ac}}$ and $\varphi \in \mathscr{H}$:

$$\langle \varphi \,|\, \mathrm{e}^{-\mathrm{i}Ht} \psi \rangle = \int \mathrm{e}^{-\mathrm{i}\lambda t} \mathrm{d}\langle \varphi \,|\, P_\lambda \psi \rangle = \int \mathrm{e}^{-\mathrm{i}\lambda t} \rho^{\varphi,\psi}(\lambda)\,\mathrm{d}\lambda = \widehat{\rho}^{\varphi,\psi}(t) \stackrel{t\to\infty}{\longrightarrow} 0 ,$$

where we have used the fact that, like μ_H^{ψ}, $\mu_H^{\varphi,\psi}$ is also absolutely continuous, and we have applied the Riemann–Lebesgue lemma. So the overlap of $\psi(t) = \mathrm{e}^{-\mathrm{i}Ht}\psi$ with any fixed wave function φ goes to zero for $t \longrightarrow \infty$, or, in other words, $\psi(t)$ goes to zero weakly. While this does not yet show that $\psi(t)$ goes to spatial infinity, it gives a first idea of how spectral and dynamical properties are related.

There is an area of mathematical physics called scattering theory, which makes this picture much more precise for the case of spatially decaying interactions. In particular one tries to show that states in $\mathscr{H}_{\mathrm{ac}}$ move asymptotically according to the "free" dynamics, and such states are called scattering states. As a byproduct, one often obtains $\mathscr{H}_{\mathrm{sc}} = \{0\}$, and such Hamiltonians are said to be asymptotically complete. Among other more important things, this will be touched upon in the next and final chapter.

References

1. M. Reed, B. Simon: *Methods of Modern Mathematical Physics I: Functional Analysis*, revised and enlarged edn. (Academic Press, San Diego, 1980)
2. E.B. Davies: *Spectral Theory and Differential Operators*, Cambridge Studies in Advanced Mathematics, Vol. 42 (Cambridge University Press, Cambridge, 1995)
3. M. Reed, B. Simon: *Methods of Modern Mathematical Physics. II. Fourier Analysis, Self-Adjointness* (Academic Press [Harcourt Brace Jovanovich Publishers], New York, 1975)

Chapter 16
Bohmian Mechanics on Scattering Theory

The quantum equilibrium distribution tells us the probability for a system to be in a certain configuration at a given time t. That is the basis for the quantum formalism of POVMs, PVMs, and self-adjoint observables on a Hilbert space. In this last chapter we shall return to the beginning of it all, namely to Born's 1926 papers [1, 2], in which he applies Schrödinger's wave equation to a scattering situation. In this application, Born recognized the importance of the quantum equilibrium distribution $\rho = |\psi|^2$ as the distribution of the random position of the particle after scattering.

Curiously though, the application of quantum mechanics to scattering situations comes along with a shift of emphasis on the meaning of the quantum equilibrium distribution. In scattering theory, the *crossing probability* of spacetime surfaces becomes meaningful. This is naively clear when one pictures the scattering situation as in Fig. 16.1. There are detectors surrounding the scattering potential. The question is: What is the probability that the detector will click?

Prior to answering that one must first clarify the following question: Is the time at which a detector clicks a fixed given time, i.e., a time the experimenter can choose? Intuitively and correctly, the answer is no. The time is random. The detector clicks when the particle arrives at the detector surface and crosses it. So both the *where* and the *when* of the detection event are random.

It is immediately clear that these are questions which Bohmian mechanics is tailored-made to answer, since the notion of where and when the particle crosses a surface is a natural one when trajectories exist. It is another matter to find a closed formula for the crossing probability. A nice formula can be found when one considers the scattering regime, which is a space regime where the particles move essentially along straight lines. There are plenty of books on scattering theory, and we shall not invent scattering theory anew. But true to our intention to provide a clear ontological picture and true to our maxim expressed by Melville:

> While you take in hand to school others, and to teach them by what name a whale-fish is to be called in our tongue leaving out, through ignorance, the letter H, which almost alone maketh the signification of the word, you deliver that which is not true. – HACKLUYT
>
> Melville (1851), Moby Dick, Chap. 32 [3]

D. Dürr, S. Teufel, *Bohmian Mechanics*, DOI 10.1007/978-3-540-89344-8_16,

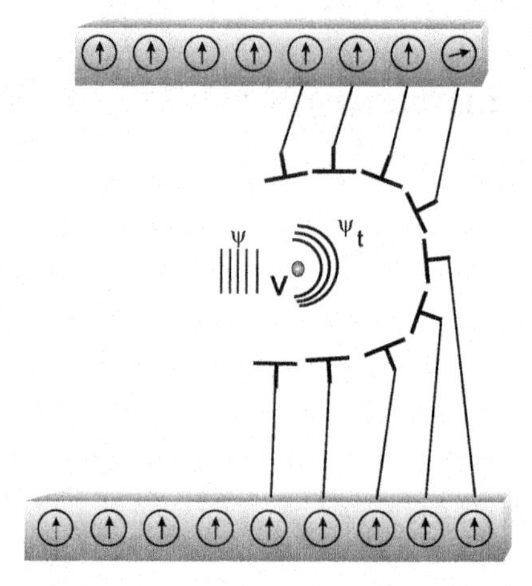

Fig. 16.1 A scattering experiment. A particle with wave function ψ is sent to a target, here a potential V. The detectors sit far away from the target and wait for the particle to arrive. The scattering experiment is in principle very much like the first-exit experiment in Fig. 16.2, with the difference that, in a scattering experiment, the particle comes into the target region from far away and the detection is far away from the target

we shall elaborate on surface-crossing probabilities for Bohmian trajectories. Then following this line of thought, we shall examine the essential elements of scattering theory, until we end with Born's formula for the scattering cross-section.

16.1 Exit Statistics

Consider the experiment sketched in Fig. 16.2. A particle is located within the region G at time $t=0$. The wave function of the particle is ψ and $\operatorname{supp}\psi \subset G$ at time $t=0$. The wave function obeys Schrödinger's equation. When and where does the particle leave G? That question is in general not well posed, because the particle can leave and reenter the region G. A good question would be: When and where does the particle leave the region for the first time. In other words, when and where does the particle cross the boundary ∂G of G for the first time?

This problem is not the common textbook problem of quantum mechanics. The reason is that time is not an observable. There are various arguments for that, but they are not important for us, since we have learnt that any experiment which ends with pointers pointing to values has POVM statistics. The same holds here. Moreover, we shall have no need to address the question as to whether a unique form of the associated POVM exists, and if it does, what it is. Why should we worry about the POVM, which by its very meaning handles all wave functions? If one is interested only in very special wave functions, as we are, for example, those which are well localized at time zero within G, we do not need an abstract formalism.

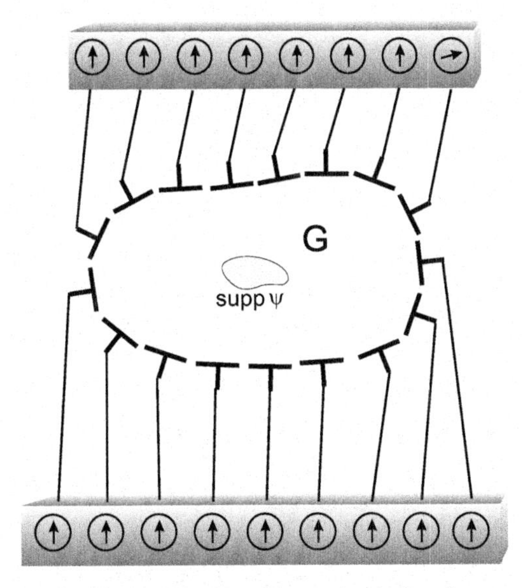

Fig. 16.2 Experiment to determine the exit position and exit time from a region G. A detector clicks at a random position and random time

Another question which might be worrisome is whether the interaction with the detectors must be taken into account to obtain the observed scattering statistics. In other words, why is the unmeasured crossing probability equal to the measured one? Detectors interact with the particle and interaction means change of wave function, which means change of trajectory, which means change of exit statistics. All this is true. So the experimenter must be careful with the choice of detectors, so that the effect of the detection on the wave function is small enough, or of such a quality, that the trajectories are not substantially altered. This may be particularly important when measuring first-exit times of the general kind we discuss below. It is presumably less important in the scattering situation which we discuss after that, because in scattering setups, the interaction energy between the particles and the detector is small compared with the energy of the particles. Having said this, we shall follow the common practice of quantum physics and henceforth not worry about the presence of detectors, simply taking it for granted that the detection is designed in such a way that it does not mess up the trajectories too much.

Let us now start with a simple argument which deals with the first-exit time of the particle from G. That time is a random variable on the initial positions within the support of the initial wave function ψ. It is defined by

$$\tau(\mathbf{x}) = \inf\left\{t \middle| \mathbf{X}(t,\mathbf{x}) \notin G,\ \mathbf{x} \in \operatorname{supp}\psi\right\} . \tag{16.1}$$

Then, using the quantum equilibrium distribution, we can attempt to compute the "distribution function"

$$
\begin{aligned}
\mathbb{P}^{\psi}\big(\{\mathbf{x}|\tau(\mathbf{x})>t\}\big) &= \mathbb{P}^{\psi}\big(\mathbf{X}(s,\mathbf{x})\in G \text{ for all } s\leq t\big) \\
&\text{“}=\text{”}\ \mathbb{P}^{\psi}\big(\mathbf{X}(t,\mathbf{x})\in G\big) \\
&= \mathbb{P}^{\psi_t}(G) \\
&= \int_G |\psi(\mathbf{x},t)|^2 \mathrm{d}^3x\,. \qquad (16.2)
\end{aligned}
$$

This is simple enough, but the second equality is only correct if the particle never returns once it has left G. The third equality is equivariance. As usual one obtains the exit-time density $\rho_{\mathrm{e}}^{\psi}(t)$ from the distribution function $\mathbb{P}^{\psi}(\tau>t)$ of τ by differentiation

$$
\begin{aligned}
\rho_{\mathrm{e}}^{\psi}(t) &= -\frac{\mathrm{d}}{\mathrm{d}t}\mathbb{P}^{\psi}(\tau>t) = -\int_G \frac{\partial|\psi(\mathbf{x},t)|^2}{\partial t}\mathrm{d}^3x \\
&= -\frac{\partial}{\partial t}\langle\psi_t|O_G\psi_t\rangle \\
&= \frac{\mathrm{i}}{\hbar}\Big(\langle H\psi_t|O_G\psi_t\rangle - \langle\psi_t|O_G H\psi_t\rangle\Big)\,. \qquad (16.3)
\end{aligned}
$$

We have introduced the position PVM O_G to make it look like advanced quantum mechanics. We have also used the notation ψ_t to highlight the time dependence of the wave function $\psi(\mathbf{x},t)$. For $[t_1,t_2]\subset\mathbb{R}^+$, we get

$$
\mathbb{P}^{\psi}\big(\tau\in[t_1,t_2]\big) = \int_{t_1}^{t_2}\rho_{\mathrm{e}}^{\psi}(t)\mathrm{d}t\,.
$$

So we have a bilinear form, but is it positive? Do we get a POVM? In fact, we do not, because of the inverted commas on one of the equals signs in (16.2).

Returning to (16.3), we introduce the quantum flux $\mathbf{j}^{\psi}=|\psi|^2\mathbf{v}^{\psi}$ in the third equality, recalling the quantum flux equation (7.17). Using Gauss's theorem with $\mathrm{d}\mathbf{S}$ as oriented surface element of ∂G,

$$
\begin{aligned}
\rho_{\mathrm{e}}^{\psi}(t) &= -\int_G \frac{\partial|\psi(\mathbf{x},t)|^2}{\partial t}\mathrm{d}^3x = \int_G \nabla\cdot\mathbf{j}^{\psi}(\mathbf{x},t)\mathrm{d}^3x \\
&= \int_{\partial G}\mathbf{j}^{\psi}(\mathbf{x},t)\cdot\mathrm{d}\mathbf{S}\,.
\end{aligned}
$$

Therefore,

$$
\int_{t_1}^{t_2}\rho_a^{\psi}(t)\mathrm{d}t = \int_{t_1}^{t_2}\int_{\partial G}\mathbf{j}^{\psi}(\mathbf{x},t)\cdot\mathrm{d}\mathbf{S}\,\mathrm{d}t
$$

is the "net flow" through the *spacetime surface* $\partial G\times[t_1,t_2]$, which can be negative. It can be interpreted as a probability if the following positivity condition holds. Suppose we define the normal vectors to ∂G as pointing outwards. Then

$$
\mathbf{j}^{\psi}(\mathbf{x},t)\cdot\mathrm{d}\mathbf{S}\geq 0\,,\quad \text{for all } x,t\in\partial G\times[t_1,t_2]\,. \qquad (16.4)
$$

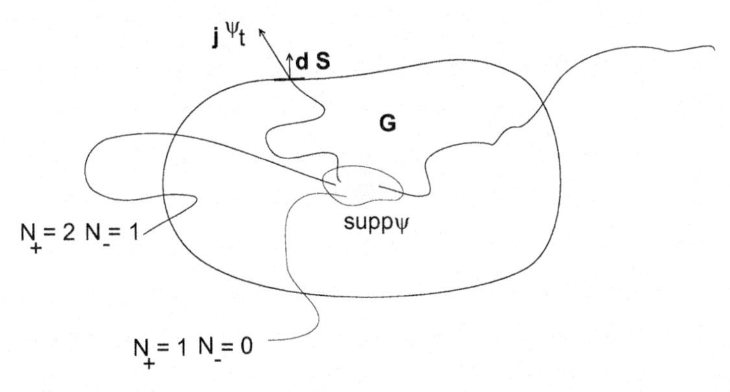

Fig. 16.3 Signed crossings of the boundary ∂G of G

If (16.4) holds, then

$$\begin{aligned}\mathbf{j}^{\psi}(\mathbf{x},t)\cdot\mathrm{d}\mathbf{S}\,\mathrm{d}t &= -\frac{i\hbar}{2m}\left[\psi^*(\mathbf{x},t)\nabla\psi(\mathbf{x},t)-\psi(\mathbf{x},t)\nabla\psi^*(\mathbf{x},t)\right]\cdot\mathrm{d}\mathbf{S}\,\mathrm{d}t\\ &= \text{"crossing probability of the surface element } \mathrm{d}\mathbf{S} \text{ in } \mathrm{d}t\text{"}. \qquad (16.5)\end{aligned}$$

Now this is news. But can we trust it? What does the flux integral mean if (16.4) does not hold?[1] Bohmian mechanics gives the answers. Let us look at the Bohmian trajectories.

The Bohmian trajectories $\mathbf{X}(t)_{t\geq 0}$ are randomly distributed according to the random distribution of initial values $\mathbf{X}(0)$, which are $|\psi(0)|^2$-distributed. The trajectories define the random number $N(\Delta\partial G,\Delta t)$ of crossings of $\mathbf{X}(t)_{t\geq 0}$ through $\Delta\partial G$ within the time interval Δt. This number is a function of the initial positions of the trajectories, and inherits its randomness from the $|\psi|^2$-distribution of the initial positions. The number is naturally decomposable into two random numbers

$$N(\Delta\partial G,\Delta t) = N_+(\Delta\partial G,\Delta t) + N_-(\Delta\partial G,\Delta t)\,,$$

with $N_+(\Delta\partial G,\Delta t)$ as outward crossings and $N_-(\Delta\partial G,\Delta t)$ as returning inward crossings of $\Delta\partial G$ within Δt, as shown in Fig. 16.3. The number of *signed* crossings is the difference

$$N_s(\Delta\partial G,\Delta t) := N_+(\Delta\partial G,\Delta t) - N_-(\Delta\partial G,\Delta t)\,.$$

We now cut the set $\Delta\partial G\times\Delta t$ into small pieces $\Delta\partial G_j\times\Delta t_i$, which are assumed to be so small that they can only be crossed once, either positively or negatively: $N_s(\Delta\partial G_j,\Delta t_i)\in\{\pm 1, 0\}$. Then the total number $N(\Delta\partial G_j,\Delta t_i)$ becomes the indicator function of Boltzmann's collision cylinder, as we shall explain now.

[1] The positivity condition on the quantum flux and the relation to mathematical scattering theory is discussed further in [4].

Boltzmann's statistical mechanics argument, which is crucial for crossing probabilities, observes that the particle can only cross $\Delta\partial G_j$ within the time interval Δt_i if it is at time t_i in the volume (the collision cylinder[2])

$$\Delta C_{i,j} = \left|\mathbf{v}^{\psi_{t_i}} \Delta t_i \cdot \Delta \mathbf{S}_j\right| .$$

Therefore

$$N\big(\Delta\partial G_j, \Delta t_i\big) = \chi_{\Delta C_{i,j}}\big(\mathbf{X}(t_i)\big) .$$

The probability for the particle to be in the cylinder is given by the quantum equilibrium (qu. eq.) distribution and thus by equivariance

$$\begin{aligned}
\mathbb{E}^{\psi_{t_i}}\Big(N\big(\Delta\partial G_j, \Delta t_i\big)\Big) &= \mathbb{E}^{\psi_{t_i}}\Big(\chi_{\Delta C_{i,j}}\big(\mathbf{X}(t_i)\big)\Big) \\
&= \mathbb{P}^{\psi_{t_i}}(\Delta C_{i,j}) \\
&\overset{\text{qu. eq.}}{=} |\psi_{t_i}|^2 \left|\mathbf{v}^{\psi_{t_i}} \Delta t_i \cdot \Delta \mathbf{S}_j\right| = \left|\mathbf{j}^{\psi_{t_i}} \cdot \Delta \mathbf{S}_j\right| \Delta t_i .
\end{aligned}$$

Using

$$N(\Delta\partial G, \Delta t) = \sum_{i,j} N\big(\Delta\partial G_j, \Delta t_i\big) ,$$

and by *linearity of the expectation value*, we can compute the expectation value of the total number of crossings through the boundary of G:

$$\mathbb{E}^{\psi}\Big(N\big(\Delta\partial G, [t_1, t_2]\big)\Big) = \int_{\Delta\partial G \times [t_1, t_2]} \left|\mathbf{j}^{\psi} \cdot \mathrm{d}\mathbf{S}\right| \mathrm{d}t .$$

Observing that we obtain a -1 from the scalar product of the velocity and the surface normal when a trajectory returns, we likewise obtain the expectation of the number of signed crossings $N_{\mathrm{s}}\big(\Delta\partial G, [t_1, t_2]\big)$:

$$\mathbb{E}^{\psi}\Big(N_{\mathrm{s}}\big(\Delta\partial G, [t_1, t_2]\big)\Big) = \int_{\Delta\partial G \times [t_1, t_2]} \mathbf{j}^{\psi} \cdot \mathrm{d}\mathbf{S}\, \mathrm{d}t . \tag{16.6}$$

The flux integrated across a surface and over some time interval is therefore in general the expected number of signed crossings of the trajectories of the surface in that time interval. It is important to understand that it is the expected value of signed crossings and not the probability. The expectation is additive, while the probability is not in general. The crossing probability is in general not additive because a trajectory may cross the surface more than once. Therefore the events where a trajectory crosses the surface $\Delta\partial G$ say at time t_1 and at time t_2 are not disjoint, as one immediately sees when one follows the trajectory back in time to its starting region. Hence

[2] The name "collision" comes from its use in the statistical mechanics of interacting particles. In our context, "collision" means "crossing".

the probabilities of these two events do not add to give the corresponding crossing probability.

Additivity holds, however, when each trajectory crosses only once, which is ensured by the positivity condition (16.4). Assuming $\mathbf{j}^{\Psi_t}\cdot d\mathbf{S} > 0$ for all times means assuming that the scalar product between the Bohmian velocity and the surface element is positive, so that the trajectory crosses from inside to outside. This in turn means that the trajectory can cross the surface only once, whence the signed number equals the total number of crossings, which is either zero or unity. Then denoting the first-exit time from G by $\tau(\mathbf{x})$ [see (16.1)], and denoting the first-exit position by $X_\tau := X\big(\tau(\mathbf{x})\big)$, where $\mathbf{x}$ is the starting point of the trajectory, we obtain

$$\begin{aligned}\mathbb{E}^{\Psi}\Big(N\big(\Delta\partial G,[t_1,t_2]\big)\Big) &= \mathbb{E}^{\Psi}\Big(N_{\mathrm{s}}\big(\Delta\partial G,[t_1,t_2]\big)\Big)\\ &= 0\cdot\mathbb{P}^{\Psi}\big(\tau\notin[t_1,t_2]\text{ or }X_\tau\notin\Delta\partial G\big)\\ &\qquad +1\cdot\mathbb{P}^{\Psi}\big(\tau\in[t_1,t_2]\text{ and }X_\tau\in\Delta\partial G\big)\,.\end{aligned}$$

Hence, by virtue of (16.6) and the positivity condition (16.4), the crossing probability is

$$\begin{aligned}\mathbb{P}^{\Psi}\big(X_\tau\in\Delta\partial G;\tau\in[t_1,t_2]\big) &= \mathbb{E}^{\Psi}\Big(N_s\big(\Delta\partial G,[t_1,t_2]\big)\Big)\\ &= \int_{t_1}^{t_2}\int_{\Delta\partial G}\mathbf{j}^{\Psi}\cdot d\mathbf{S}dt\,.\end{aligned}\tag{16.7}$$

If the positivity condition does not hold, the flux integral on the right of (16.7) is in general not positive, and the exit probability is not given by a simple expression. It is easy to see that the set of "good" wave functions which satisfy the positivity condition (16.4) is not a linear set, which means that the superposition of two "good" wave functions is not in general a "good" wave function. A first-exit statistics POVM will only be given by the flux on "good" wave functions, and it is not clear what it will look like in general. Thus we have an example of a measurement situation where we are only interested in the statistics of particular wave functions, and where we do not care at all about the general quantum formalism. Is this in any way problematic? Of course, it is not!

To help appreciate the fact that Bohmian mechanics yields in the most straightforward manner that the crossing probability is determined by the quantum flux, we contrast this with other versions of quantum theories with trajectories. For example in stochastic mechanics [5, 6], where the trajectories are like Brownian motion paths, the distribution of the first-exit time will bear no relation whatsoever with the quantum flux.

Remark 16.1. A Four-Dimensional View

The quantum flux equation (7.17) for one particle can be viewed (like any continuity equation) as the assertion that the current j^μ, $\mu = 0,1,2,3$, $j = (\rho^\Psi,\mathbf{j}^\Psi)$ is divergence-free:

$$\partial_\mu j^\mu = 0 \,. \tag{16.8}$$

In this four-dimensional view, a four-current j^μ defines the particle worldlines X^μ by requiring that these worldlines are integral curves along the four-current vector field. Choosing an appropriate parametrization, we can write

$$\frac{\mathrm{d}X^\mu}{\mathrm{d}\tau} = j^\mu(X^\mu) \,.$$

The four-current version of the quantum flux equation gives another perspective on equivariance. Let $\mathscr{F}$ denote a (three-dimensional) smooth hypersurface in spacetime, which is crossed only once by each worldline. If the worldlines do not turn backward in time, as is the case of interest here, the assumption is simply that the surface has a timelike normal at each point. Then for a subset $\Delta\mathscr{F} \in \mathscr{F}$, we read

$$\int_{\Delta\mathscr{F}} j\cdot \mathrm{d}\sigma = \mathbb{P}^\Psi\big(\Delta\mathscr{F} \text{ is crossed by a worldline}\big) \,, \tag{16.9}$$

where we take the probability as being unity on $\mathscr{F}$.

This reading of the integrated four-current as crossing probability is consistent, because the analogous statement holds for any other surface $\mathscr{F}'$ which is also crossed only once by each trajectory, and the two probabilities are connected by the current. To see this, consider the worldline cylinder C with base $\Delta\mathscr{F}$ and cap $\Delta\mathscr{F}' \in \mathscr{F}'$, which is the image of $\Delta\mathscr{F}$ under the flow map given by the four current. In other words, the lateral surface of the cylinder is made up of worldlines. The cylinder surface is taken to be oriented with outward pointing normal vectors. Then, by Gauss's theorem, and by virtue of (16.8), we obtain

$$\int_{\partial C} j\cdot \mathrm{d}\sigma = -\int_{\Delta\mathscr{F}} j\cdot \mathrm{d}\sigma + \int_{\Delta\mathscr{F}'} j\cdot \mathrm{d}\sigma = \int_C \frac{\partial j^\mu}{\partial x^\mu}\mathrm{d}^4x = 0 \,, \tag{16.10}$$

where, in the last equality, we used the fact that the current is orthogonal (by construction of the current cylinder) to the lateral surface of the cylinder. Hence the crossing probability is preserved when the current is divergence-free.

In Bohmian mechanics the natural type of surface $\mathscr{F}$ is a $x^0 = t =$ const. hyperplane with normal $(1,0,0,0)$, and for that (16.9) yields the quantum equlibrium distribution, i.e., Born's statistical law, because $j\cdot \mathrm{d}\sigma = |\psi_t|^2\mathrm{d}^3x$. Furthermore, the expression always has the same form, a property which we called equivariance.

But we can also consider skewed hypersurfaces. In the above, we considered surfaces given by the parametrization

$$\Sigma(x^0,u,v) = \big(x^0,x^1(u,v),x^2(u,v),x^3(u,v)\big) \in \mathbb{R}^4 \,,\; x^0 \in [t_1,t_2] \,,\; (u,v) \in G \subset \mathbb{R}^2 \,.$$

The scalar product $j\cdot \mathrm{d}\Sigma$ can be computed as the determinant of the enlarged four-dimensional matrix made from the Jacobi matrix of Σ and j as fourth column. [Recall that, if instead we list the canonical unit vectors $\mathbf{e}_k$, $k = 0,1,2,3$, as fourth column, we obtain the surface normal in $\Sigma(x^0,u,v)$ by expanding the determinant with respect to the fourth column.] With $\sigma(u,v) = \left(x^1(u,v), x^2(u,v), x^3(u,v)\right) \in \mathbb{R}^3$, we obtain

$$j \cdot \mathrm{d}\Sigma = \det \begin{pmatrix} 1 & 0 & 0 & j^0 \\ 0 & \partial_u x^1 & \partial_v x^1 & j^1 \\ 0 & \partial_u x^2 & \partial_v x^2 & j^2 \\ 0 & \partial_u x^3 & \partial_v x^3 & j^3 \end{pmatrix} \mathrm{d}t\, \mathrm{d}u\, \mathrm{d}v = \mathbf{j} \cdot \partial_u \sigma \wedge \partial_v \sigma \mathrm{d}t\, \mathrm{d}u\, \mathrm{d}v\,.$$

In the language of forms, j is a 3-form

$$\omega_j = j^0 \mathrm{d}x^1 \wedge \mathrm{d}x^2 \wedge \mathrm{d}x^3 - j^1 \mathrm{d}x^0 \wedge \mathrm{d}x^2 \wedge \mathrm{d}x^3 + j^2 \mathrm{d}x^0 \wedge \mathrm{d}x^1 \wedge \mathrm{d}x^3 - j^3 \mathrm{d}x^0 \wedge \mathrm{d}x^1 \wedge \mathrm{d}x^2\,.$$

The 3-form is by its very meaning an object which is to be integrated over a three-surface, and by definition,

$$\int_\Sigma \omega_j = \int \omega_j(\partial_{x^0}, \partial_u, \partial_v) \mathrm{d}t\, \mathrm{d}u\, \mathrm{d}v\,,$$

where $\mathrm{d}x^k(\partial_y)$ is defined as $\mathrm{d}x^k(\partial_y) = \partial x^k / \partial y$. Naturally, this yields the same as the determinant above.

The 3-form is to be replaced by a $3N$-form in the case of N particles. The corresponding current is $(3N+1)$-dimensional and it gets integrated over $3N$-dimensional hypersurfaces, e.g., over the $t = $ const. hyperplanes, yielding the quantum equilibrium distribution. ■

16.2 Asymptotic Exits

We shall now analyze a physical situation where we expect the positivity condition (16.4) to apply, and hence also (16.7), and we thus have a formula for the crossing probability. This is the situation in which the detectors are far away from where the wave function is initially localized, or where, in Fig. 16.3, the boundary ∂G is far away. We connect this now with the asymptotic form of the wave function which we discussed in Remark 15.8, and which says heuristically that the Bohmian trajectories move asymptotically along straight lines.

For simplicity, we take for G a ball of radius R, and consider the exit position $\mathbf{X}_\mathrm{e}$ of the particle through a piece of the spherical surface $\Sigma_R = R\Sigma$, $\Sigma \subset B_1$ (see Fig. 16.4) when R gets large, i.e., we take the large R limit of (16.7), and with t integrated over all times (since we are only interested in the position), we should have

$$\lim_{R\to\infty} \mathbb{P}^\Psi(\mathbf{X}_\mathrm{e} \in \Sigma_R) = \lim_{R\to\infty} \int_0^\infty \int_{\Sigma_R} \left|\mathbf{j}^\Psi \cdot \mathrm{d}\mathbf{S}\right| \mathrm{d}t = \lim_{R\to\infty} \int_0^\infty \int_{\Sigma_R} \mathbf{j}^\Psi \cdot \mathrm{d}\mathbf{S}\, \mathrm{d}t\,. \quad (16.11)$$

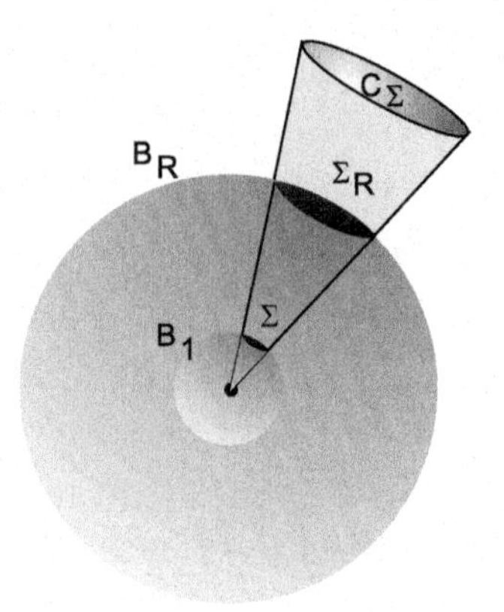

Fig. 16.4 Geometry for the computation of a flux across a surface

Note that the equality of the last two integrals ensures that the flux integral is positive, whence it does indeed give the crossing probability.

To compute this, we recall (15.10) (setting $\hbar/m = 1$ for notational convenience)

$$\psi(\mathbf{x},t) \overset{x,t\,\text{large}}{\sim} \frac{1}{(\mathrm{i}t)^{3/2}} \mathrm{e}^{\mathrm{i}\mathbf{x}^2/2t} \widehat{\psi}\left(\frac{\mathbf{x}}{t}\right) , \tag{16.12}$$

where $\widehat{\psi}$ is the Fourier transform of the wave function ψ at time zero. This asymptotic form of the wave function says that, at large times, the wave function will be at positions $\mathbf{x}$ for which $\mathbf{x}/t \in \operatorname{supp}\widehat{\psi}$.

We shall now carry out an instructive calculation which, in some sense, has already been done in (15.11), whence the result will not be particularly surprising. It is nevertheless gratifying to see how everything fits together. We replace the flux in (16.11) by the flux of the asymptotic wave function (16.12) for which we first observe that, in

$$\nabla \frac{1}{(\mathrm{i}t)^{3/2}} \mathrm{e}^{\mathrm{i}\mathbf{x}^2/2t} \widehat{\psi}\left(\frac{\mathbf{x}}{t}\right) = \frac{1}{(\mathrm{i}t)^{3/2}} \mathrm{i}\frac{\mathbf{x}}{t} \mathrm{e}^{\mathrm{i}\mathbf{x}^2/2t} \widehat{\psi}\left(\frac{\mathbf{x}}{t}\right) + \frac{1}{t}\frac{1}{(\mathrm{i}t)^{3/2}} \mathrm{e}^{\mathrm{i}\mathbf{x}^2/2t} \widehat{\psi}'\left(\frac{\mathbf{x}}{t}\right) ,$$

the second term is of smaller order since $x/t = \mathscr{O}(1)$, due to the support condition just mentioned. Therefore, for large x,t,

$$\mathbf{j}^{\psi}(\mathbf{x}) = \Im(\psi_t^* \nabla \psi_t) \approx \frac{\mathbf{x}}{t}\left(\frac{1}{t}\right)^3 \left|\widehat{\psi}\left(\frac{\mathbf{x}}{t}\right)\right|^2 . \tag{16.13}$$

This means that the flux is asymptotically radial, i.e., $\mathbf{j}^{\psi}(\mathbf{x}) \sim \mathbf{x} = \mathbf{R}$, so that on the surface,

$$\mathbf{j}^{\psi} \cdot \mathrm{d}\mathbf{S} \geq 0 .$$

This takes care of the equality of the integrated absolute flux and the integrated flux in (16.11). To compute (16.11), observe the following:

1. The surface element reads $\mathrm{d}\mathbf{S} = R\mathbf{R}\mathrm{d}^2\omega$, with $\mathrm{d}^2\omega$ as the surface element of the unit sphere.
2. Substituting $k = R/t$ for t yields $\mathrm{d}k = -k\mathrm{d}t/t$ as the new integration variable in the integral.
3. C_Σ denotes the cone with solid angle Σ, as depicted in Fig. 16.4.

Then

$$\int_0^\infty \int_{\Sigma_R} \mathbf{j}^\psi \cdot \mathrm{d}\mathbf{S}\mathrm{d}t \overset{(16.13)}{\approx} \int_0^\infty \int_{\Sigma_R} \left(\frac{1}{t}\right)^3 \left|\widehat{\psi}\left(\frac{\mathbf{R}}{t}\right)\right|^2 \frac{\mathbf{R}}{t} \cdot \mathrm{d}\mathbf{S}\,\mathrm{d}t \tag{16.14}$$

$$\overset{\text{by (1)}}{=} \int_0^\infty \mathrm{d}t \int_\Sigma \left(\frac{1}{t}\right)^3 \left|\widehat{\psi}\left(\frac{\mathbf{R}}{t}\right)\right|^2 \frac{R^3}{t} \mathrm{d}^2\omega$$

$$\overset{\text{by (2)}}{=} \int_0^\infty \mathrm{d}k\, k^2 \int_\Sigma \mathrm{d}^2\omega\, |\widehat{\psi}(\mathbf{k})|^2 = \int_{C_\Sigma} |\widehat{\psi}(\mathbf{k})|^2\, \mathrm{d}^3 k\,.$$

Hence, for (16.11), we get

$$\lim_{R\to\infty} \mathbb{P}^\psi(\mathbf{X}_\mathrm{e} \in \Sigma_R) = \lim_{R\to\infty} \int_0^\infty \int_{\Sigma_R} \mathbf{j}^\psi \cdot \mathrm{d}\mathbf{S}\mathrm{d}t = \int_{C_\Sigma} |\widehat{\psi}(\mathbf{k})|^2\, \mathrm{d}^3 k\,,$$

or a little less precisely,

$$\mathbb{P}^\psi(\mathbf{X}_\mathrm{e} \in \Sigma_R) \overset{R \text{ large}}{\approx} \int_{C_\Sigma} |\widehat{\psi}(\mathbf{k})|^2\, \mathrm{d}^3 k\,. \tag{16.15}$$

There are various mathematically rigorous assertions around (16.15). The asymptotic equality between the integrated absolute flux and the integrated flux in (16.11) is part of the so-called flux-across-surfaces theorem, of which many versions have been proven (see Remark 16.2). The above result is made rigorous in [7].

The formula (16.15) is basic to scattering theory. In scattering situations, the detectors are far away from the scattering centers, and after the scattered wave has left the scattering potential, i.e., when it is far away, it does move freely. Therefore (16.15) applies to scattering, with one caveat, however. The Fourier transform of the wave function ψ at time zero appears in (16.15). Of course, time zero could be any time, but which ψ is appropriate in formula (16.15) when the wave function also interacts with a potential, i.e., when it does not move freely all the time? One might have the idea of taking ψ_T for some large T, a time whereafter the wave function does move freely. But such a time is not sharply defined. Mathematically at least, there will always be some part of the wave function overlapping the potential. The question then is: Does there exist an asymptotic expression for a given ψ for large times, with which we can compute large time statistics? We shall address this question next. In fact, we shall compute the exit distribution now in the physically relevant situation of scattering. In doing so, we shall restrict ourselves to the simplest case of one-particle potential scattering.

16.3 Scattering Theory and Exit Distribution

We continue with our discussion of the exit distribution and consider its relevance for scattering situations. In scattering theory the wave does not evolve freely. There is a scattering potential V which influences the evolution of the wave, as shown in Fig. 16.1, i.e., the Schrödinger operator is $H = H_0 + V$. In scattering theory, we picture a wave approaching the scattering potential from far away, interacting, and then leaving as a "scattered wave". Far away from the potential, where the detectors wait for the particle to arrive, the wave should move freely (assuming that the potential falls off fast enough). Such wave functions will be called scattering states.

How can one phrase in mathematical terms the fact that a wave moves asymptotically freely? The easiest way to approach this is perhaps to think of asymptotics in time rather than spatial asymptotics, since these should be roughly equivalent points of view. The question is then: What is the asymptotic form of $\mathrm{e}^{-\mathrm{i}tH}\psi$ for large t? Tracing the time evolution of the wave function $\mathrm{e}^{-\mathrm{i}tH}\psi$ as a path in Hilbert space helps to answer this (see Fig. 16.5). We can picture free motion asymptotically in time by the path approaching an asymptote which is defined by a freely moving wave function $\left(\mathrm{e}^{-\mathrm{i}tH_0}\psi_{\mathrm{out}}\right)_{t\in\mathbb{R}}$, sketched in Fig. 16.5 as a "straight line" in Hilbert space. This suggests making the requirement

$$\lim_{t\longrightarrow\infty}\left\|\mathrm{e}^{-\mathrm{i}tH}\psi - \mathrm{e}^{-\mathrm{i}tH_0}\psi_{\mathrm{out}}\right\| = 0\,. \tag{16.16}$$

This is a requirement for the existence of the wave function ψ_{out} whose free evolution defines the asymptote for the evolution of ψ. All wave functions ψ for which such an outgoing (and ingoing, see below) asymptote exists are called scattering states of the Hamiltonian H.

Since $\psi_{\mathrm{out}}(t)$ evolves freely in time, we know of course that (16.12) holds for ψ_{out}, and (16.16) then suggests that

$$\psi(\mathbf{x},t) \overset{x,t\ \mathrm{large}}{\sim} \frac{1}{(\mathrm{i}t)^{3/2}}\mathrm{e}^{\mathrm{i}\mathbf{x}^2/2t}\,\widehat{\psi}_{\mathrm{out}}\left(\frac{\mathbf{x}}{t}\right)\,. \tag{16.17}$$

One may hope that the amount of time in which the scattering wave is influenced by the potential is in a reasonable sense limited, so that the replacement we made in (16.13) to compute the flux integral (16.14) can also be made with (16.17). Then from (16.15), replacing $\widehat{\psi}$ by $\widehat{\psi}_{\mathrm{out}}$, we obtain the formula

$$\mathbb{P}^{\psi}(\mathbf{X}_{\mathrm{e}} \in \Sigma_R) \overset{R\ \mathrm{large}}{\approx} \int_{C_\Sigma} |\widehat{\psi}_{\mathrm{out}}(\mathbf{k})|^2\mathrm{d}^3k\,. \tag{16.18}$$

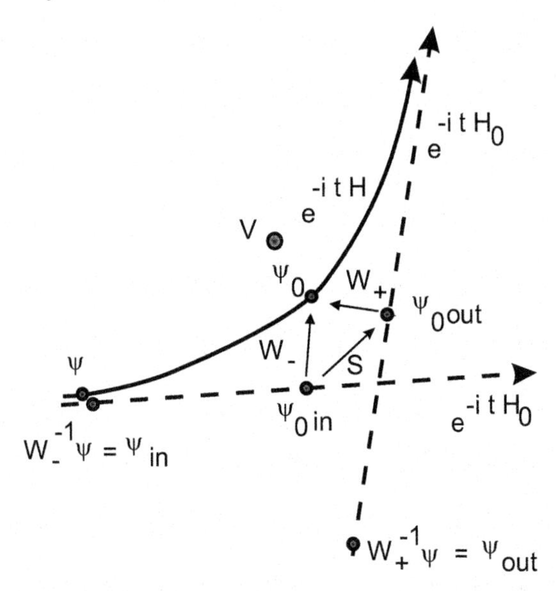

Fig. 16.5 Schematic representation of wave function evolution in Hilbert space. The time evolution of a scattering state ψ_0 is shown at some arbitrary time, e.g., $t = 0$. Associated with it are two states $\psi_{0\text{in}}$ and $\psi_{0\text{out}}$ which evolve freely. The backwards-in-time evolved $\psi_{0\text{in}}$ is the asymptote for $t \longrightarrow -\infty$ (16.43) and the forward-in-time evolved $\psi_{0\text{out}}$ is the asymptote for $t \longrightarrow \infty$. See text for explanation

Remark 16.2. On Scattering into Cones and the Flux Across Surfaces

The formula (16.18) can be turned into a mathematically rigorous assertion, usually referred to as the flux-across-surfaces theorem [7–11]. It holds for a large class of potentials. The expression (16.18) gives the probability that the particle exits through the surface piece Σ_R (i.e., gets detected by the detector which covers that solid angle). The explanation as to why the probability is given by the momentum distribution of ψ_{out} is that the particle gets scattered into the (spatial) cone C_Σ. In other words, because the particle moves more or less along a straight line – a ray, given by the direction of the momentum $\hbar\mathbf{k}$ in C_Σ which lies in that cone – it crosses the detector surface. This idea also underlies the so-called scattering-into-cones theorem [12, 13], which asserts that the probability that the particle is in the cone C_Σ at time T is given by the probability that the momentum lies in the cone, provided T is large enough. In other words this probability is also given by the right-hand side of (16.18):

$$\lim_{t\to\infty}\int_{C_\Sigma} |\psi(\mathbf{x},t)|^2 \mathrm{d}^3x = \int_{C_\Sigma} |\widehat{\psi}_{\text{out}}(\mathbf{k})|^2 \mathrm{d}^3k\,.$$

In contrast, the flux-across-surfaces theorem connects directly with the crossing of the detector surface at *random* times, and its proof supports the picture of straight line trajectories. But the rigorous assertion that the Bohmian trajectories become straight lines requires a few more technicalities, and has been established in [11].

The scattering-into-cones theorem, and in particular the flux-across-surfaces theorem, have been put forward as fundamental assertions for many-particle scattering theory [14]. There is of course no problem with writing down the N-particle quantum flux

$$\mathbf{j}^{\psi}(\mathbf{x}_1,\dots,\mathbf{x}_N,t)=\Im\big(\psi^*(\mathbf{x}_1,\dots,\mathbf{x}_N,t)\nabla\psi(\mathbf{x}_1,\dots,\mathbf{x}_N,t)\big)\,,$$

but what does it mean? For many-particle scattering there is not just one random time at which the detector clicks, but rather N random times at which the N particles arrive at the detectors. The N-particle quantum flux does not handle that. It has therefore been observed [15] that, in many-particle scattering, the quantum flux loses its meaning, and that the relevant crossing probabilities of the scattered particles through various detectors must be based on the Bohmian trajectories and the Boltzmann collision cylinder argument leading to (16.6). This argument can be straightforwardly generalized to the case of many particles. A many-particle scattering version of (16.18) based on Bohmian trajectories has accordingly been advanced in [16]. ■

16.4 More on Abstract Scattering Theory

Formula (16.16) leads directly to the definition of the operator

$$W_+ = \underset{t\to\infty}{\text{s-lim}}\, \mathrm{e}^{\mathrm{i}tH}\mathrm{e}^{-\mathrm{i}tH_0}\,, \tag{16.19}$$

where $\text{s-lim}_{t\to\infty}$ indicates the strong L^2-limit

$$\lim_{t\to\infty}\left\|\left(W_+-\mathrm{e}^{\mathrm{i}tH}\mathrm{e}^{-\mathrm{i}tH_0}\right)\varphi\right\|=0\,,\quad \text{for all } \varphi\in L^2\,.$$

In other words,

$$\psi = W_+\psi_{\text{out}}\,. \tag{16.20}$$

Figure 16.5 shows the maps $W_\pm$ (where W_- will be introduced further below) at time $t=0$, which is an arbitrary time. It will soon become clear that shifting time will only produce an irrelevant phase factor in the formulas relevant for us. That is why we omit the time index 0.

We have cheated a bit in representing the situation in such an innocent-looking way. We really want to find ψ_{out} for given ψ, so that the true aim is the inverse operator W_+^{-1}, which is defined on $\text{Ran}(W_+)$, the range of W_+. On that,

$$W_+^{-1}=W_+^*=\underset{t\to\infty}{\text{s-lim}}\,\mathrm{e}^{\mathrm{i}tH_0}\mathrm{e}^{-\mathrm{i}tH}\,. \tag{16.21}$$

To know the domain of W_+^{-1}, i.e., to know Ran(W_+) is a way of phrasing the so-called completeness problem, which we shall elaborate on below. In short, we do not know beforehand which wave functions are scattering states. In fact, this is exactly what we wish to find out. On the other hand, the operator W_+ acts naturally on L^2 (provided the limit exists), because *all* states evolve under the free dynamics into almost plane outgoing wave packets which, when evolved backwards in time with the full time evolution, become scattering states, provided the Hilbert space sketch of the asymptotic approach shown in Fig. 16.5 is correct. The latter should be the case if the potential falls off "fast enough" at infinity.

The operator is called the *wave operator*, and it was one of the earliest inventions in scattering theory. In other accounts of scattering theory the wave operator W_+ is often written as Ω_-, where the difference is mainly a sign convention, explained in footnote 3. In our notation, the + sign indicates that the time limit in the definition of W_+ is for large positive times. Later on we shall introduce W_-, in which the limit towards large negative times is considered.

How can one get a handle on the limit in the definition of the wave operator? With a simple trick, as old as the wave operator itself. Since the wave operator encodes the potential V, we need a condition on the potential which ensures the existence of the limit. So how can we make the potential visible in the wave operator W_+? The answer is, by writing

$$\begin{aligned} W_+ &= \lim_{T\to\infty} \mathrm{e}^{\mathrm{i}TH}\mathrm{e}^{-\mathrm{i}TH_0} = 1 + \lim_{T\to\infty}\int_0^T \frac{\mathrm{d}(\mathrm{e}^{\mathrm{i}tH}\mathrm{e}^{-\mathrm{i}tH_0})}{\mathrm{d}t}\mathrm{d}t \\ &= 1 + \lim_{T\to\infty}\int_0^T \mathrm{i}\mathrm{e}^{\mathrm{i}tH}(H-H_0)\mathrm{e}^{-\mathrm{i}tH_0}\mathrm{d}t \\ &= 1 + \lim_{T\to\infty}\int_0^T \mathrm{i}\mathrm{e}^{\mathrm{i}tH}V\mathrm{e}^{-\mathrm{i}tH_0}\mathrm{d}t\,. \end{aligned} \tag{16.22}$$

This gives us now the opportunity to see the stationary phase method (15.44) at work. The task of establishing that the wave operator W_+ is well defined is now reduced by (16.22) to showing the existence of the integral on L^2. Thus it is sufficient to establish that, on a dense set of wave functions,

$$\lim_{T\to\infty}\int_0^T \left\|\mathrm{e}^{\mathrm{i}tH}V\mathrm{e}^{-\mathrm{i}tH_0}\psi\right\|\mathrm{d}t = \lim_{T\to\infty}\int_0^T \left\|V\mathrm{e}^{-\mathrm{i}tH_0}\psi\right\|\mathrm{d}t \le C\|\psi\|\,.$$

In view of (15.44), this is easy if we lay down some conditions which we choose here for the sake of simplicity, while preserving the spirit of what needs to be done. The dense set we choose is the set of wave functions with support in Fourier space bounded away from zero, let us say by a distance a, so that $\mathbf{k}\in\operatorname{supp}\widehat{\psi} \Longrightarrow k > a$. Then we split the integrand, introducing the characteristic function χ,

$$\begin{aligned} \left\|V\mathrm{e}^{-\mathrm{i}tH_0}\psi\right\| &= \left\|V\big(\chi(x<at)+\chi(x\ge at)\big)\mathrm{e}^{-\mathrm{i}tH_0}\psi\right\| \\ &\le \left\|V\chi(x<at)\mathrm{e}^{-\mathrm{i}tH_0}\psi\right\| + \left\|V\chi(x\ge at)\mathrm{e}^{-\mathrm{i}tH_0}\psi\right\|\,. \end{aligned}$$

If V is bounded in the operator norm, we can pull $\|V\|$ out of the first term, and what remains is integrable in time by virtue of the stationary phase argument (15.44). For the second term, we assume that $|V(\mathbf{x})| \sim x^{-1-\varepsilon}$, so that

$$\begin{aligned}\left\|V\chi(x \geq at)\mathrm{e}^{-\mathrm{i}tH_0}\psi\right\| &\leq \left\|V\chi(x \geq at)\right\|_{\infty}\|\psi\| \\ &\sim t^{-1-\varepsilon}\,,\end{aligned}$$

which is also integrable in time.

Remark 16.3. On the Scattering Program in Mathematical Physics
Since we are talking about the behavior of wave functions, let us take the opportunity of talking about a classification of wave functions which is of interest in the mathematical physics of scattering theory. Given a Schrödinger operator there are three different spectral subspaces of the Hilbert space, as discussed at the end of Chap. 15. There is the subspace $\mathscr{H}_{\mathrm{pp}}$ belonging to the pure point spectrum spanned by eigenfunctions. These are stationary or bound states, and they do not move to infinity, whence the particles also remain in finite regions.

Then there is the subspace $\mathscr{H}_{\mathrm{cont}}$ belonging to the continuous spectrum. Can the wave functions in that subspace also be dynamically characterized? Now, the continuous spectrum splits into two spectra, the absolutely continuous spectrum and the singular one. Correspondingly, one has two more subspaces $\mathscr{H}_{\mathrm{ac}}$ and $\mathscr{H}_{\mathrm{sc}}$. What can be said about these spaces? Writing

$$\mathrm{e}^{\mathrm{i}(t+s)H}\mathrm{e}^{-\mathrm{i}(t+s)H_0} = \mathrm{e}^{\mathrm{i}sH}\mathrm{e}^{\mathrm{i}tH}\mathrm{e}^{-\mathrm{i}tH_0}\mathrm{e}^{-\mathrm{i}sH_0}\,,$$

we obtain the so called intertwining property

$$\mathrm{e}^{-\mathrm{i}sH}W_{+} = W_{+}\mathrm{e}^{-\mathrm{i}sH_0}\,, \tag{16.23}$$

from which we infer that the range $\mathrm{Ran}(W_{+})$ is invariant under the H-time evolution. Strong differentiation with respect to s of (16.23) yields

$$HW_{+} = W_{+}H_0\,, \tag{16.24}$$

and thus, since

$$W_{+}^{-1} = W_{+}^{*}$$

on $\mathrm{Ran}(W_{+})$, we have

$$W_{+}^{*}HW_{+} = H_0\,. \tag{16.25}$$

This says that the restriction of H to $\mathrm{Ran}(W_{+})$ is unitarily equivalent to H_0. And this in turn means that the restriction has absolutely continuous spectrum (see Remark 15.12), so that we get the important "a priori" result

$$\mathrm{Ran}(W_{+}) \subset \mathscr{H}_{\mathrm{ac}}\,. \tag{16.26}$$

Completeness of the scattering problem now means that $\mathrm{Ran}(W_+) = \mathscr{H}_{\mathrm{ac}}$, and the next task is to show that. For the singular part of the spectrum, one wants to show *asymptotic completeness*, namely that it is an empty set, so that one does not need to worry about what those wave functions do. In other words, $\mathscr{H}_{\mathrm{cont}} = \mathscr{H}_{\mathrm{ac}}$ (see for instance [4, 17, 18]). ■

We may now invert (16.20), i.e., we write

$$\psi_{\mathrm{out}} = W_+^{-1}\psi \,, \tag{16.27}$$

and introducing this into (16.18), we can also express the exit distribution as

$$\mathbb{P}^{\psi}(\mathbf{X}_{\mathrm{e}} \in \Sigma_R) \overset{R\,\text{large}}{\approx} \int_{C_\Sigma} |\widehat{\psi}_{\mathrm{out}}(\mathbf{k})|^2 \mathrm{d}^3k = \int_{C_\Sigma} \left|\widehat{W_+^{-1}\psi}(\mathbf{k})\right|^2 \mathrm{d}^3k \,. \tag{16.28}$$

This is the formula from which the scattering cross-section, the basic empirical import of scattering theory, arises. Note also that changing the origin of time produces a phase factor $\mathrm{e}^{-\mathrm{i}k^2t/2}$ in the Fourier transform, because ψ_{out} evolves with the free time evolution or because of the intertwining property (16.24). This factor then drops out.

Before we compute the scattering cross-section, we shall allow ourselves a short interlude at this point to introduce another technically important notion, namely the notion of generalized eigenfunctions. They can also be used to phrase the long time asymptotics of the wave function, but they do much more than that: they diagonalize the Hamiltonian on the subspace $\mathscr{H}_{\mathrm{ac}}$.

16.5 Generalized Eigenfunctions

Another important and somewhat less abstract notion, but equivalent to the idea of wave operators, is the notion of generalized eigenfunctions. They can be introduced in a straightforward manner, but since we have already talked about the wave operator, we shall introduce them using the wave operator. Let us get more familiar with the wave function in (16.28):

$$\widehat{\psi}_{\mathrm{out}}(\mathbf{k}) = \widehat{W_+^{-1}\psi}(\mathbf{k}) = \langle \mathbf{k} | W_+^{-1}\psi \rangle \,, \tag{16.29}$$

By unitarity of the wave operator $W_+^{-1} = W_+^*$ on $\mathrm{Ran}(W_+)$, this may be written as

$$\widehat{W_+^{-1}\psi}(\mathbf{k}) = \langle W_+\mathbf{k} | \psi \rangle \,. \tag{16.30}$$

Recalling that

$$-\frac{1}{2}\Delta \mathrm{e}^{\mathrm{i}\mathbf{k}\cdot\mathbf{x}} = \frac{k^2}{2}\mathrm{e}^{\mathrm{i}\mathbf{k}\cdot\mathbf{x}} \,,$$

which means that

$$H_0|\mathbf{k}\rangle = \frac{k^2}{2}|\mathbf{k}\rangle\,,$$

and in view of (16.24), we see that the vector

$$|+,\mathbf{k}\rangle := W_+|\mathbf{k}\rangle \tag{16.31}$$

is an *eigenvector* of H, albeit a generalized one:

$$H|+,\mathbf{k}\rangle = HW_+|\mathbf{k}\rangle = W_+H_0|\mathbf{k}\rangle = \frac{k^2}{2}W_+|\mathbf{k}\rangle = \frac{k^2}{2}|+,\mathbf{k}\rangle\,. \tag{16.32}$$

The "**x** representation" yields the generalized eigenfunction

$$\varphi_+(\mathbf{x},\mathbf{k}) := (2\pi)^{3/2}W_+|\mathbf{k}\rangle = (2\pi)^{3/2}\langle\mathbf{x}|+,\mathbf{k}\rangle\,, \tag{16.33}$$

which is not square integrable, and hence not an element of L^2. But this is in no way disquieting since we are already familiar with

$$\langle\mathbf{x}|\mathbf{k}\rangle = (2\pi)^{-3/2}\mathrm{e}^{\mathrm{i}\mathbf{k}\cdot\mathbf{x}}\,,$$

the generalized eigenfunctions of H_0, which are normalized to $\langle\mathbf{k}|\mathbf{k}'\rangle = \delta(\mathbf{k}-\mathbf{k}')$. This is exactly how we should think of the generalized eigenfunctions $\varphi_+(\mathbf{x},\mathbf{k})$ [likewise normalized to $\delta(\mathbf{k}-\mathbf{k}')$], namely as wave functions which diagonalize H restricted to $\mathrm{Ran}(W_+)$ in the very same sense as the Fourier transform diagonalizes H_0. The generalized eigenfunctions play the same role as the plane waves for the free Hamiltonian. They solve the eigenvalue equation

$$H\varphi(\mathbf{x},\mathbf{k}) = \frac{k^2}{2}\varphi(\mathbf{x},\mathbf{k}) \tag{16.34}$$

in the space of bounded functions, and one can take this as the starting point for scattering theory.

The problem of completeness is now expressible as the problem of establishing an isometry between L^2 and $\mathscr{H}_{\mathrm{ac}}$ via a generalized Fourier transform, taking as "basis" elements the generalized eigenfunctions. This is completely analogous to the way plane waves define the Fourier transformation. Combining (16.29), (16.30), (16.31), and (16.33), we see that the function $\widehat{\psi}_{\mathrm{out}}$, which is the Fourier transform of ψ_{out}, is also the *generalized Fourier transform* of $\psi \in \mathscr{H}_{\mathrm{ac}}$:

$$\widehat{\psi}_{\mathrm{out}}(\mathbf{k}) = (2\pi)^{-3/2}\int \varphi_+(\mathbf{x},\mathbf{k})^*\psi(\mathbf{x})\mathrm{d}^3x\,. \tag{16.35}$$

[We have ignored here the mathematical detail that the Fourier transform is not an integral transformation in general, i.e., the integral is in general defined via an approximation (see Chap. 14).] Like the ordinary Fourier transform, the generalized Fourier transform on L^2 is an isometry, represented here by the wave operators.

The eigenfunction expansion (16.35) is a powerful tool for studying the time evolution of the wave function, which, for the free Hamiltonian, is of course well known. One can get a good handle on the generalized eigenfunctions when one rewrites the eigenvalue equation (16.34) in integral form. This is called the Lippmann–Schwinger equation. We start with the eigenvalue equation (16.32),

$$H|+,\mathbf{k}\rangle = (H_0+V)|+,\mathbf{k}\rangle = \frac{k^2}{2}|+,\mathbf{k}\rangle \,,$$

and reorder the terms to get

$$\left(H_0 - \frac{k^2}{2}\right)|+,\mathbf{k}\rangle = V|+,\mathbf{k}\rangle \,.$$

We want to solve this for $|+,\mathbf{k}\rangle$, but the operator $(H_0 - k^2/2)$ is not invertible, since

$$\left(H_0 - \frac{k^2}{2}\right)|\mathbf{k}\rangle = 0 \,.$$

Observing this, we may write the inversion formally in a suggestive way as

$$|+,\mathbf{k}\rangle = \left(H_0 - \frac{k^2}{2}\right)^{-1} V|+,\mathbf{k}\rangle + |\mathbf{k}\rangle \,,$$

with the idea in mind that the generalized eigenfunctions should be asymptotically, i.e., far from the range of the potential, like plane waves, and where $(H_0 - k^2/2)^{-1}$ must be defined in an appropriate way, as spelt out below. In other words, we look for solutions of the eigenvalue equation (16.34) which satisfy the boundary condition

$$\lim_{|\mathbf{x}|\to\infty} \left[\varphi_{\pm}(\mathbf{x},\mathbf{k}) - \mathrm{e}^{\mathrm{i}\mathbf{k}\cdot\mathbf{x}}\right] = 0 \,.$$

This is the formal Lippmann–Schwinger equation, but we still need to find the proper Green's function, i.e., the kernel of $(H_0 - k^2/2)^{-1}$.

There are various ways to arrive at the Green's function. Since we introduced the eigenfunction as kernel function of the wave operator in (16.33), the straightforward way starts from a variant of (16.22). By virtue of (16.33),

$$|\mathbf{k}\rangle = W_+^*|+,\mathbf{k}\rangle \,,$$

and in view of (16.22), we write

$$W_+^* = 1 - \lim_{T\to\infty} \int_0^T \mathrm{i}\mathrm{e}^{\mathrm{i}tH_0} V \mathrm{e}^{-\mathrm{i}tH} \mathrm{d}t \,. \tag{16.36}$$

Combining these, we compute

$$\begin{aligned}|\mathbf{k}\rangle = W_+^*|+,\mathbf{k}\rangle &= |+,\mathbf{k}\rangle - \lim_{T\to\infty}\int_0^T \mathrm{i}\mathrm{e}^{\mathrm{i}tH_0}V\mathrm{e}^{-\mathrm{i}tH}|+,\mathbf{k}\rangle \mathrm{d}t \\ &= |+,\mathbf{k}\rangle - \lim_{T\to\infty}\int_0^T \mathrm{i}\mathrm{e}^{\mathrm{i}tH_0-\mathrm{i}tk^2/2}V|+,\mathbf{k}\rangle \mathrm{d}t \\ &= |+,\mathbf{k}\rangle - \lim_{T\to\infty}\int_0^T \mathrm{i}\mathrm{e}^{\mathrm{i}t\left(H_0-k^2/2\right)}V|+,\mathbf{k}\rangle \mathrm{d}t\,,\end{aligned}$$

using the intertwining property $\mathrm{e}^{-\mathrm{i}tH}|+,\mathbf{k}\rangle = \mathrm{e}^{-\mathrm{i}tk^2/2}|+,\mathbf{k}\rangle$ in the third equality. The integral is again the undefined inverse of $H_0 - k^2/2$, but we now take the *Abel limit*, which exists:

$$|\mathbf{k}\rangle = |+,\mathbf{k}\rangle - \lim_{\varepsilon\downarrow 0}\int_0^\infty \mathrm{i}\exp\left[\mathrm{i}t\left(H_0 - \frac{k^2}{2} + \mathrm{i}\varepsilon\right)\right]V|+,\mathbf{k}\rangle \mathrm{d}t\,.$$

But we have already computed this! The integral is the resolvent

$$\left(H_0 - \frac{k^2}{2} + \mathrm{i}\varepsilon\right)^{-1} = \left(-\frac{1}{2}\Delta - \frac{k^2}{2} + \mathrm{i}\varepsilon\right)^{-1},$$

which we discussed in Remark 15.10. Substituting $t/2$ for t yields a factor of 2 in front of the integral, and then we put $\kappa^2 = -k^2 + 2\mathrm{i}\varepsilon$ and $\lim_{\varepsilon\downarrow 0}\kappa = \mathrm{i}k$ in Remark 15.10, in accordance with our agreement to take the positive real part. Hence in the "**x** representation" this becomes

$$\langle \mathbf{x}|\mathbf{k}\rangle = \langle \mathbf{x}|+,\mathbf{k}\rangle + \frac{1}{2\pi}\int \frac{\mathrm{e}^{-\mathrm{i}|\mathbf{x}-\mathbf{y}|k}}{|\mathbf{x}-\mathbf{y}|}V(\mathbf{y})\langle \mathbf{y}|+,\mathbf{k}\rangle \mathrm{d}^3y\,,$$

where the integrand kernel is the Green's function of the Schrödinger equation in three dimensions. Using (16.33) and reordering yields the Lippmann–Schwinger equation

$$\varphi_+(\mathbf{x},\mathbf{k}) = \mathrm{e}^{\mathrm{i}\mathbf{k}\cdot\mathbf{x}} - \frac{1}{2\pi}\int \frac{\mathrm{e}^{-\mathrm{i}|\mathbf{x}-\mathbf{y}|k}}{|\mathbf{x}-\mathbf{y}|}V(\mathbf{y})\varphi_+(\mathbf{y},\mathbf{k})\mathrm{d}^3y\,. \tag{16.37}$$

The equation provides a handle on $\varphi_+(\mathbf{x},\mathbf{k})$, for example by using iterative procedures. One may start at zeroth order with $\mathrm{e}^{\mathrm{i}\mathbf{k}\cdot\mathbf{x}}$ replacing $\varphi_+(\mathbf{x},\mathbf{k})$ inside the integral. Iterating this gives the so-called Born series.

In view of the decay of the Green's function, (16.37) seems to suggest that, far from the scattering potential, the generalized eigenfunctions become ordinary plane waves. However, the decay of the Green's function is not strong enough to support the assertion that this also holds in the L^2-sense. For that we must appeal to the stationary phase argument of Remark 15.9, which in fact yields a much stronger decay in space and time. Expanding ψ in terms of generalized eigenfunctions [see (16.35)], we get

$$\begin{aligned}\psi(\mathbf{x},t) &= (2\pi)^{-3/2}\int \mathrm{e}^{-ik^2t/2}\varphi_+(\mathbf{x},\mathbf{k})\,\widehat{\psi}_{\mathrm{out}}(\mathbf{k})\,\mathrm{d}^3k \\ &= (2\pi)^{-3/2}\int \mathrm{e}^{-ik^2t/2}\mathrm{e}^{\mathrm{i}\mathbf{k}\cdot\mathbf{x}}\widehat{\psi}_{\mathrm{out}}(\mathbf{k})\,\mathrm{d}^3k \\ &\quad -(2\pi)^{-3/2}\int \mathrm{e}^{-ik^2t/2}\left[\frac{1}{2\pi}\int\frac{\mathrm{e}^{-\mathrm{i}|\mathbf{x}-\mathbf{y}|k}}{|\mathbf{x}-\mathbf{y}|}V(\mathbf{y})\varphi_+(\mathbf{y},\mathbf{k})\mathrm{d}^3y\right]\widehat{\psi}_{\mathrm{out}}(\mathbf{k})\,\mathrm{d}^3k\,.\end{aligned} \tag{16.38}$$

The first term contains the phase term of the free evolution, which yields, by the stationary phase argument, in accordance with (15.10), the new general asymptotics (16.17). The second term contains the phase function $-|\mathbf{x}-\mathbf{y}|k-k^2t/2$, which has no stationary point for positive times t, and hence by the stationary phase argument in Remark 15.9, the term shows arbitrary polynomial decay in time and space, provided $\widehat{\psi}_{\mathrm{out}}$ is smooth enough. Hence the long time and long distance behavior for positive times is governed by the first term, the free evolution. This is therefore another less abstract, and consequently more physical way to capture asymptotic free motion. There is no need to appeal to wave operators: simply use generalized eigenfunctions.

For reasons to be explained below, we introduce another equally important "eigenbasis" which is given by another class of solutions $\varphi_-(\mathbf{x},\mathbf{k})$ of the eigenvalue equation (16.34). They appear when we agree to take the negative real part $\kappa=-ik$ in the derivation of Remark 15.10. The difference is a change of sign in the exponent of the integrand in (16.37):

$$\varphi_-(\mathbf{x},\mathbf{k}) = \mathrm{e}^{\mathrm{i}\mathbf{k}\cdot\mathbf{x}} - \frac{1}{2\pi}\int\frac{\mathrm{e}^{\mathrm{i}|\mathbf{x}-\mathbf{y}|k}}{|\mathbf{x}-\mathbf{y}|}V(\mathbf{y})\varphi_-(\mathbf{y},\mathbf{k})\mathrm{d}^3y\,. \tag{16.39}$$

These functions define a generalized Fourier transform as well. But instead of (16.35), we now define

$$\widehat{\psi}_{\mathrm{in}}(\mathbf{k}) := (2\pi)^{-3/2}\int \varphi_-(\mathbf{x},\mathbf{k})^*\psi(\mathbf{x})\mathrm{d}^3x\,. \tag{16.40}$$

To understand the subscript "in", we reconsider the stationary phase argument we just went through in (16.38) for this class of eigenfunctions, and obtain analogously

$$\begin{aligned}\psi(\mathbf{x},t) &= (2\pi)^{-3/2}\int \mathrm{e}^{-ik^2t/2}\varphi_-(\mathbf{x},\mathbf{k})\,\widehat{\psi}_{\mathrm{in}}(\mathbf{k})\,\mathrm{d}^3k \\ &= (2\pi)^{-3/2}\int \mathrm{e}^{-ik^2t/2}\mathrm{e}^{\mathrm{i}\mathbf{k}\cdot\mathbf{x}}\widehat{\psi}_{\mathrm{in}}(\mathbf{k})\,\mathrm{d}^3k \\ &\quad -(2\pi)^{-3/2}\int \mathrm{e}^{-ik^2t/2}\left[\frac{1}{2\pi}\int\frac{\mathrm{e}^{\mathrm{i}|\mathbf{x}-\mathbf{y}|k}}{|\mathbf{x}-\mathbf{y}|}V(\mathbf{y})\varphi_-(\mathbf{y},\mathbf{k})\mathrm{d}^3y\right]\widehat{\psi}_{\mathrm{in}}(\mathbf{k})\,\mathrm{d}^3k\,.\end{aligned} \tag{16.41}$$

The sign makes all the difference. The phase in the second term has no stationary point for negative times, so it vanishes when $t\to-\infty$. The first term thus dominates the large negative time behavior and shows that, for large negative times, the wave function approaches another asymptotic free evolution, that of ψ_{in}.

This leads us to the "incoming side" of the scattering process, which has not yet been discussed. It is also important, perhaps even more important than the outgoing side, because this is after all the part of the scattering experiment which is supposed to be under the control of the experimenter. The experimenter usually prepares the incoming wave packet with great care. We shall say more on the nature of the incoming wave in the next section. At the moment it suffices to understand how one describes mathematically the fact that the incoming wave is prepared, i.e., that it is under the control of the experimenter. This means that at the time of preparation, the state is not yet influenced by the scattering potential.

Physically, the preparation is done far enough from the scattering center to ensure that, at the preparation place, the scattering potential is not felt, i.e., $\mathrm{e}^{-\mathrm{i}tH} \approx \mathrm{e}^{-\mathrm{i}tH_0}$. In other words, the wave function starts to evolve with the free time evolution. This is of course to be taken with a pinch of salt. The Schrödinger evolution will immediately produce a spread-out wave function, so – mathematically – the potential is always felt, but its influence is "physically negligible" at the time and place where the wave packet is prepared. The idea is then that, at the time and place of preparation of the scattering state ψ, the state is close (in the L^2 sense) to ψ_{in}, i.e., $\|\psi - \psi_{\text{in}}\| \approx 0$. We shall come back to this below.

Obviously, since the generalized eigenfunctions diagonalize H, they provide a powerful analytical tool to come to grips with the time evolution of wave functions for a general H. Curiously though, the generalized Fourier transform has become something of a wallflower in mathematical scattering theory. It has been revitalized in recent works on what is called time-dependent scattering theory [9, 10, 19–21], and what we present in this chapter is based on these references.

We can restate the above in terms of a wave operator W_- which, in accordance with (16.31) and (16.33), is related to φ_- according to[3]

$$\varphi_-(\mathbf{x},\mathbf{k}) =: (2\pi)^{3/2}\langle \mathbf{x}|W_-\mathbf{k}\rangle =: (2\pi)^{3/2}\langle \mathbf{x}|-,\mathbf{k}\rangle \,. \tag{16.42}$$

The definition of W_- goes like the definition of W_+, but with time reversed. Introduce an asymptotic wave function, namely ψ_{in}, evolving with the free Hamiltonian, and the scattering state ψ converges under its evolution for $t \to -\infty$ (see Fig. 16.5) to that freely evolving state. This reads in mathematical terms as

[3] The sign convention which introduces Ω_- instead of W_+ arises from the minus sign in the exponent of the integrand. Accordingly, W_- is denoted by Ω_+. One may wonder about the physical meaning of the different signs in the exponent $\mathrm{e}^{\pm\mathrm{i}|\mathbf{x}-\mathbf{y}|k}/|\mathbf{x}-\mathbf{y}|$ appearing in the Lippmann–Schwinger equations. They represent outgoing and incoming spherical waves. Their appearance is directly related to the way the wave operators $W_\pm$ are defined. For example, in W_+ [see (16.19)], the full time evolution acts backwards in time, developing the scattered state $\psi \approx \psi_{\text{out}}$ (for $t \to \infty$) backwards in time. Therefore from the scattering picture of outgoing spherical waves (see Fig. 16.6), the time-backwards evolved spherical waves must appear in the representation of W_+, whence φ_+ appears. By the same token the scattering picture of outgoing spherical waves will be captured by φ_-. Equivalently, we can understand the roles of $\varphi_\pm$ by virtue of the stationary phase arguments in (16.38) and (16.41).

$$\lim_{t\to-\infty}\left\|e^{-itH}\psi - e^{-itH_0}\psi_{\rm in}\right\| = 0\,, \tag{16.43}$$

yielding the wave operator

$$W_- = \operatorname*{s-lim}_{t\to-\infty} e^{itH}e^{-itH_0}\,. \tag{16.44}$$

Put another way,

$$\psi = W_-\psi_{\rm in}\,. \tag{16.45}$$

Inserting this into (16.29) and recalling (16.33), we get

$$\langle\mathbf{k}|W_+^{-1}\psi\rangle = \langle\mathbf{k}|W_+^*W_-\psi_{\rm in}\rangle =: \langle\mathbf{k}|S\psi_{\rm in}\rangle = \widehat{\psi}_{\rm out}(\mathbf{k})\,. \tag{16.46}$$

The operator combination $W_+^*W_- =: S$ is called the S-matrix. It maps (unitarily) $\psi_{\rm in}$ to $\psi_{\rm out}$: $\psi_{\rm out} = S\psi_{\rm in}$. Introducing this into (16.28), we obtain

$$\mathbb{P}^\psi(\mathbf{X}_e \in \Sigma_R) \overset{R\text{ large}}{\approx} \int_{C_\Sigma} |\langle\mathbf{k}|S\psi_{\rm in}\rangle|^2\, d^3k\,. \tag{16.47}$$

At every moment of time t, the map W_-^{-1} maps the scattering state ψ_t to the corresponding $\psi_{t,\rm in}$ (see Fig. 16.5).

In (16.45) above, ψ can be the scattering state at any arbitrary time and $\psi_{\rm in}$ the corresponding in state at that time. For the physical interpretation, however, it is useful to observe that the right-hand side of (16.47) does not change in states taken at different times, since the S-matrix commutes with the free time evolution $Se^{-itH_0} = e^{-itH_0}S$. This holds because the intertwining property (16.24) holds for W_- as well, i.e., $e^{-itH}W_- = W_-e^{-itH_0}$, as one readily sees by repeating the argument for W_- in place of W_+. Applying the corresponding intertwining properties twice then yields the commutation property of S. By the intertwining property, we have $\psi_{t,\rm in} = W_-^*e^{-itH}\psi = e^{-itH_0}\psi_{\rm in}$, and therefore the replacement of $\psi_{\rm in}$ by $\psi_{t,\rm in}$ produces a phase factor $e^{-ik^2t/2}$, which arises by evolving the plane wave $|\mathbf{k}\rangle$. This phase factor drops out, since one takes the absolute square in (16.47).

As already mentioned above, for the physical interpretation one should think of $\psi_{\rm in}$ as taken at a time where $\|\psi - \psi_{\rm in}\| \approx 0$, so that we may think of $\psi_{\rm in}$ as the scattering state ψ prior to the time of interaction with the scattering potential. In that case, we have approximately

$$\mathbb{P}^\psi(\mathbf{X}_e \in \Sigma_R) \overset{R\text{ large}}{\approx} \int_{C_\Sigma} \left|\langle\mathbf{k}|S\psi\rangle\right|^2 d^3k\,, \tag{16.48}$$

with ψ being the scattering state prior to the interaction with the potential. We shall come back to this in the last section.

16.6 Towards the Scattering Cross-Section

In a scattering experiment, one focuses on directions $\mathbf{k}$ different from the incoming directions $\mathbf{k}'$ which we assume to be well localized around $\mathbf{k}_0$. This helps to simplify formulas, because we can subtract the identity from S, and this is a subtraction of the unscattered part. The resulting operator is called the T-matrix. The following computation is for the main part already known to us. We compute $T := S - \mathsf{I}$, i.e.,

$$\begin{aligned}\langle \mathbf{k}|T\psi_{\text{in}}\rangle &= \big\langle \mathbf{k}\big|(S-\mathsf{I})\psi_{\text{in}}\big\rangle = \big\langle \mathbf{k}\big|(W_+^*W_- - \mathsf{I})\psi_{\text{in}}\big\rangle \\ &= \int \big\langle \mathbf{k}\big|(W_+^*W_- - \mathsf{I})\big|\mathbf{k}'\big\rangle\langle \mathbf{k}'|\psi_{\text{in}}\rangle \mathrm{d}^3k' \\ &= \int \big\langle \mathbf{k}\big|(W_+^*W_- - \mathsf{I})\big|\mathbf{k}'\big\rangle \widehat{\psi}_{\text{in}}(\mathbf{k}')\mathrm{d}^3k' \,, \end{aligned} \tag{16.49}$$

and we determine the kernel

$$\langle \mathbf{k}|T|\mathbf{k}'\rangle = \big\langle \mathbf{k}\big|(W_+^*W_- - \mathsf{I})\big|\mathbf{k}'\big\rangle \,. \tag{16.50}$$

Since $\mathsf{I} = W_-^*W_-$, we obtain

$$\big\langle \mathbf{k}\big|(W_+^*W_- - W_-^*W_-)\big|\mathbf{k}'\big\rangle = \big\langle \mathbf{k}\big|(W_+^* - W_-^*)W_-\big|\mathbf{k}'\big\rangle \,,$$

which is excellent because, in view of (16.36) and its analogue for W_-^*, we have

$$W_+^* - W_-^* = -\int_{-\infty}^{\infty} \mathrm{i}\mathrm{e}^{\mathrm{i}tH_0}V\mathrm{e}^{-\mathrm{i}tH}\mathrm{d}t \,. \tag{16.51}$$

Everything now fits beautifully together:

$$\begin{aligned}\langle \mathbf{k}|T|\mathbf{k}'\rangle &= \left\langle \mathbf{k}\right| - \int_{-\infty}^{\infty} \mathrm{i}\mathrm{e}^{\mathrm{i}tH_0}V\mathrm{e}^{-\mathrm{i}tH}\mathrm{d}t\, W_-\left|\mathbf{k}'\right\rangle = -\int_{-\infty}^{\infty} \langle \mathbf{k}|\mathrm{i}\mathrm{e}^{\mathrm{i}tH_0}V\mathrm{e}^{-\mathrm{i}tH}W_-|\mathbf{k}'\rangle \mathrm{d}t \\ &= -\int_{-\infty}^{\infty} \langle \mathbf{k}|\mathrm{i}\mathrm{e}^{\mathrm{i}tH_0}VW_-\mathrm{e}^{-\mathrm{i}tH_0}|\mathbf{k}'\rangle \mathrm{d}t = -\int_{-\infty}^{\infty} \langle \mathbf{k}|\mathrm{i}\mathrm{e}^{\mathrm{i}t(k^2/2-k'^2/2)}VW_-|\mathbf{k}'\rangle \mathrm{d}t \\ &= \int_{-\infty}^{\infty} \exp\left[\mathrm{i}t\frac{(k^2-k'^2)}{2}\right] \langle \mathbf{k}|\mathrm{i}V|-,\mathbf{k}'\rangle \mathrm{d}t \,, \end{aligned}$$

where we have introduced (16.42). Now use the fact that

$$\int_{-\infty}^{\infty} \exp\left[\mathrm{i}t\frac{(k^2-k'^2)}{2}\right] \mathrm{d}t = 2\pi\delta\left(\frac{k^2}{2} - \frac{k'^2}{2}\right) \,, \tag{16.52}$$

to get for (16.49)

$$\langle \mathbf{k}|T\psi_{\text{in}}\rangle = -\int 2\pi\delta\left(\frac{k^2}{2} - \frac{k'^2}{2}\right) \langle \mathbf{k}|\mathrm{i}V|-,\mathbf{k}'\rangle \widehat{\psi}_{\text{in}}(\mathbf{k}')\mathrm{d}^3k' \,. \tag{16.53}$$

Performing the integration in (16.53) in spherical coordinates (k', ω'), with appropriate substitution

$$\mathrm{d}^3k' = k'^2 \mathrm{d}k' \mathrm{d}^2\omega' = k' \mathrm{d}\frac{k'^2}{2} \mathrm{d}^2\omega' ,$$

yields $k = k'$, and we are left with the integration over the solid angle ω'. Writing $\mathbf{k} = (k', \omega)$, we get for (16.53)

$$\big\langle (k',\omega) \big| T\psi_{\text{in}} \big\rangle = 2\pi \mathrm{i} \int k \big\langle (k',\omega) \big| V \big| -, (k',\omega') \big\rangle \widehat{\psi}_{\text{in}}(k',\omega') \mathrm{d}^2\omega' . \tag{16.54}$$

The potential term in the integrand $\big\langle (k',\omega) \big| V \big| -, (k',\omega') \big\rangle$ is customarily written as the kernel function

$$\big\langle (k',\omega) \big| V \big| -, (k',\omega') \big\rangle =: T\big((k',\omega), \mathbf{k}'\big) = \frac{1}{(2\pi)^3} \int \mathrm{e}^{-\mathrm{i}k'\omega \cdot \mathbf{x}} V(\mathbf{x}) \varphi_-(\mathbf{x}, \mathbf{k}') \mathrm{d}^3x , \tag{16.55}$$

where we have introduced (16.42).

We can now insert this into (16.47) by restricting the integration to cones C_Σ bounded away from the incoming direction (which in a scattering experiment is rather sharply defined), so that we may replace S by $T = S - \mathrm{I}$. Collecting all terms, we obtain under this condition, for large R,

$$\mathbb{P}^\Psi(\mathbf{X}_\mathrm{e} \in \Sigma_R) \approx 4\pi^2 \int_0^\infty \int_{C_\Sigma} \left| \int T\big((k,\omega),(k,\omega')\big) \widehat{\psi}_{\text{in}}\big((k,\omega')\big) \mathrm{d}^2\omega' \right|^2 k^4 \mathrm{d}^2\omega \, \mathrm{d}k . \tag{16.56}$$

Our final goal is to deal with the realistic scattering situation, where the (normalized[4]) incoming wave packet has a well defined incoming momentum and the lateral (transverse to the incoming momentum) shape does not vary within the range of the potential. In other words, on the length scale of the target, the incoming wave packet looks almost like a plane wave $\mathrm{e}^{\mathrm{i}\mathbf{k}_0 \cdot \mathbf{x}}$. In this situation, the exit distribution yields the basic object of scattering theory, the *scattering cross-section*. Turning back now to physics, we return also to physical units.

16.7 The Scattering Cross-Section

Scattering theory is a cornerstone for verifying our understanding of the universe experimentally. The central quantity in a scattering experiment is the scattering cross-section $\sigma_{\mathbf{k}_0}(\Sigma)$. The cross-section is an area which has the following meaning. One

---

[4] Needless to say, physics demands that the wave packet be normalized, even when it becomes almost a plane wave! A formal way to think of the incoming wave is that $|\widehat{\psi}|^2(\mathbf{k}) \approx \delta(\mathbf{k} - \mathbf{k}_0)$.

prepares a beam of "identical" particles, whose wave packets are almost like plane waves with respect to the size of the target (see Fig. 16.6). This means that each wave packet is very sharply peaked around the wave vector $\mathbf{k}_0$ with lateral wave number, i.e., perpendicular to $\mathbf{k}_0$, close to zero. We shall be more specific about what this means in physical units later on, but the picture in physical space should be clear: each wave packet has almost flat wave fronts widely overlapping the target.

Now place a surface perpendicular to $\mathbf{k}_0$, centered on the beam in front of the target. Then count the number of particles crossing that surface per unit time. The scattering cross-section $\sigma_{\mathbf{k}_0}(\Sigma)$ is the size of the surface for which the number of particles crossing per unit time equals the number $\dot{N}_\Sigma$ of particles scattered per unit time into directions in Σ. By the law of large numbers, the number $\dot{N}_\Sigma$ is determined by the crossing probability (16.56), so when $\dot{N}_\mathrm{B}$ is the number of particles per unit time within the beam (i.e., the number of particles crossing the cross-section of the beam per unit time), then

$$\dot{N}_\Sigma \approx \dot{N}_\mathrm{B} \mathbb{P}^\psi(\mathbf{X}_\mathrm{e} \in \Sigma_R)\,. \tag{16.57}$$

We must now evaluate $\mathbb{P}^\psi(\mathbf{X}_\mathrm{e} \in \Sigma_R)$ under the conditions typically satisfied in scattering situations, to obtain the theoretical prediction for the cross-section $\sigma_{\mathbf{k}_0}(\Sigma)$.

Before doing this, in order to obtain a more complete perspective, we recall the famous derivation of the scattering cross-section by Born. In the following, we think of the origin as the place where the target is put. The distance R is the distance of the detectors from the origin. R is assumed to be very large compared to the target size.

16.7.1 Born's Formula

In his famous 1926 papers [1, 2], Born derived a theoretical expression for the scattering cross-section using a stationary picture. The derivation starts with the ansatz of an idealized wave picture, where a plane wave and an outgoing spherical wave with *scattering amplitude* f are superposed:

$$\Phi(\mathbf{x}) = \mathrm{e}^{\mathrm{i}\mathbf{k}_0\cdot\mathbf{x}} + f_{\mathbf{k}_0}(\omega)\frac{\mathrm{e}^{ik_0 x}}{x}\,, \tag{16.58}$$

with

$$\omega = \frac{\mathbf{x}}{x}\,. \tag{16.59}$$

From this one computes the quantum flux

$$\mathbf{j}^\varphi(\mathbf{x},t) = \frac{\hbar}{m}\Im\big(\varphi^*(\mathbf{x},t)\nabla\varphi(\mathbf{x},t)\big) \tag{16.60}$$

arising from the plane wave part $\mathrm{e}^{\mathrm{i}\mathbf{k}_0\cdot\mathbf{x}}$, yielding $\hbar\mathbf{k}_0/m$, and the quantum flux arising from the spherical wave part $f_{\mathbf{k}_0}(\omega)\mathrm{e}^{ik_0x}/x$, yielding

$$\frac{\hbar}{m}\left|f_{\mathbf{k}_0}(\omega)\right|^2\frac{k_0}{x^2}\omega\,.$$

The first flux is interpreted as the flux of an incoming beam of particles with momentum $\hbar\mathbf{k}_0$, and the second flux as the flux of the scattered particles through a detector surface covering the solid angle Σ. The quotient of the modulus of these fluxes is taken to be the cross-section for the solid angle Σ:

$$\sigma_{\mathbf{k}_0}(\Sigma)=\int_\Sigma\frac{\left|f_{\mathbf{k}_0}(\omega)\right|^2\frac{k_0}{x^2}}{k_0}x^2\mathrm{d}^2\omega=\int_\Sigma\left|f_{\mathbf{k}_0}(\omega)\right|^2\mathrm{d}^2\omega\,. \tag{16.61}$$

To determine $|f_{\mathbf{k}_0}(\omega)|^2$ in a physical situation where particles are scattered off a potential $V(\mathbf{x})$, i.e., where the wave function is governed by the one-particle Schrödinger equation with Schrödinger operator

$$H=-\frac{\hbar^2}{2m}\Delta+V\,,$$

one compares (16.58) with the generalized eigenfunctions $\varphi_\pm(\mathbf{x},\mathbf{k})$ solving the Lippmann–Schwinger equations (16.37) and (16.39):

$$\varphi_\pm(\mathbf{x},\mathbf{k})=\mathrm{e}^{\mathrm{i}\mathbf{k}\cdot\mathbf{x}}-\frac{1}{2\pi}\int\frac{\mathrm{e}^{\mp\mathrm{i}k|\mathbf{x}-\mathbf{x}p|}}{|\mathbf{x}-\mathbf{x}p|}V(\mathbf{x}p)\varphi_\pm(\mathbf{x}p,\mathbf{k})\mathrm{d}^3x'\,.$$

This can be used to read off the asymptotic (large x) behavior of the eigenfunctions. The bit of computation required for this involves estimating the absolute value

$$|\mathbf{x}-\mathbf{x}p|=x\sqrt{1+\frac{x'^2}{x^2}-2\omega\cdot\frac{\mathbf{x}p}{x}}\approx x-\omega\cdot\mathbf{x}p\,,$$

for large x, where we have introduced (16.59). Neglecting terms of order $1/x^2$, one readily sees that $\varphi_-(\mathbf{x},\mathbf{k})$ is the stationary solution which is asymptotically of the form (16.58). This is of course no surprise, because we already anticipated that in footnote 3. We have

$$\varphi_-(\mathbf{x},\mathbf{k})\approx\mathrm{e}^{\mathrm{i}\mathbf{k}\cdot\mathbf{x}}-\frac{\mathrm{e}^{\mathrm{i}kx}}{x}\frac{1}{2\pi}\int\mathrm{e}^{-\mathrm{i}k\omega\cdot\mathbf{x}'}V(\mathbf{x}')\varphi_-(\mathbf{x}',\mathbf{k})\mathrm{d}^3x'\,. \tag{16.62}$$

Comparing this asymptotic form of $\varphi_-(\mathbf{x},\mathbf{k})$ for large x with (16.58) suggests that

$$f_{\mathbf{k}_0}(\omega)\sim-\frac{1}{2\pi}\int\mathrm{e}^{-\mathrm{i}k_0\omega\cdot\mathbf{x}}V(\mathbf{x})\varphi_-(\mathbf{x},\mathbf{k}_0)\mathrm{d}^3x\,.$$

Therefore,

$$\left|f_{\mathbf{k}_0}(\omega)\right|^2 = \frac{1}{4\pi^2}\left|\int \mathrm{e}^{-\mathrm{i}k_0\omega\cdot\mathbf{x}} V(\mathbf{x})\varphi_-(\mathbf{x},\mathbf{k}_0)\mathrm{d}^3x\right|^2 . \tag{16.63}$$

Recalling the T-matrix kernel (16.55) and appropriately renaming the variables, we find by comparison that we may also express the scattering amplitude f in the form

$$\left|f_{\mathbf{k}_0}(\omega)\right| = 4\pi^2\left|T\big((k_0,\omega),\mathbf{k}_0\big)\right| . \tag{16.64}$$

Equation (16.61) then becomes

$$\sigma_{\mathbf{k}_0}(\Sigma) = 16\pi^4\int_\Sigma \left|T\big((k_0,\omega),\mathbf{k}_0\big)\right|^2 \mathrm{d}^2\omega . \tag{16.65}$$

The expression looks familiar! It is very similar to the right-hand side of the crossing probability (16.56), as indeed it should be, and yet it is not the same. The difference is that, in (16.56), the T-kernel is inside the squared $\mathrm{d}^2\omega'$ integration and integrated against $\widehat{\psi}_{\mathrm{in}}\big((k,\omega')\big)$. We now come to the evaluation of (16.56) yielding (16.65).

16.7.2 Time-Dependent Scattering

The explanation also answers a question which has been around since Born's derivation: Why is Born's stationary analysis appropriate? Scattering is a time-dependent physical process with normalized wave packets that move. Therefore a fundamental analysis must start with a normalized wave packet satisfying the conditions which a scattering experiment provides. From that we must calculate, under conditions appropriate for a scattering experiment, the probability for scattering. The general idea is of course that, in the scattering situation, the moving wave packet will look on the incoming side of the target almost like a plane wave packet. Here the word "almost" has to be taken with a pinch of salt, since the packet will always be square integrable. On the outgoing side it will almost look like a spherical wave, and a quasi-stationary picture emerges (see Fig. 16.6).

We can see this using the expansion (16.41) and introducing the approximation (16.62). With the identification of the scattering amplitude f, we have approximately

$$\psi(\mathbf{x},t) \approx (2\pi)^{-3/2}\int \exp\left(-\mathrm{i}\frac{\hbar k^2 t}{2m}\right)\mathrm{e}^{\mathrm{i}\mathbf{k}\cdot\mathbf{x}}\widehat{\psi}_{\mathrm{in}}(\mathbf{k})\,\mathrm{d}^3k$$
$$+(2\pi)^{-3/2}\int \exp\left(-\mathrm{i}\frac{\hbar k^2 t}{2m}\right)\frac{\mathrm{e}^{\mathrm{i}kx}}{x}f_{\mathbf{k}}(\omega)\widehat{\psi}_{\mathrm{in}}(\mathbf{k})\,\mathrm{d}^3k .$$

The phase in the first term has a stationary point at $\hbar\mathbf{k}/m = \mathbf{x}/t$. For an incoming wave ($\psi \approx \psi_{\mathrm{in}}$) with a sharply defined momentum $\hbar\mathbf{k}_0$, the stationary point is approximately $\mathbf{x}/t \sim \mathbf{k}_0$, so that this terms contributes to the wave packet moving in the direction $\mathbf{k}_0$. In other words, it concerns the unscattered part of the wave. The scattered part is to be computed from the second term. For example, one may use

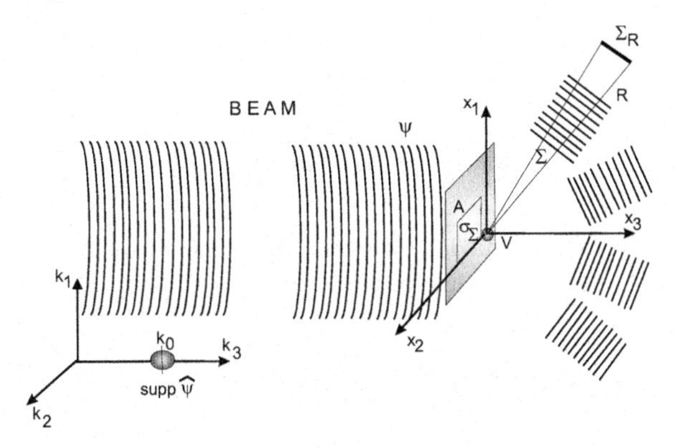

Fig. 16.6 Schematic representation of a scattering experiment. Also shown *at the bottom* is the support of the Fourier transform of the wave function ψ, which is a member of a beam of identical wave functions impinging on the target. See the text for further explanation

it to compute the probability of scattering into the cone C_Σ (see Remark 16.2). But since we already have the the crucial formula (16.56) for the crossing probability, which directly connects with the clicks in the detector, we shall now go on with that formula.

As we did with (16.48), we may replace ψ_{in} in (16.56) by the incoming scattering state ψ prior to a time where the interaction with the scattering potential becomes effective. We shall refer to the spatial location of ψ at that time by saying "in front of the target" (see Fig. 16.6). We assume that the wave packet ψ moves towards the origin (where the target sits) with a well defined momentum $\hbar\mathbf{k}_0$ such that $\mathbf{k}\cdot\mathbf{k}_0 > 0$, for all $\mathbf{k} \in \operatorname{supp}\widehat{\psi}$. We assume further that $\mathbf{k}_0$ is parallel to the $\mathbf{e}_3$ direction. To understand that the steps we take in the following are mathematically reasonable, it is perhaps best to think of $\widehat{\psi}$ as a Gaussian function. We shall formulate further conditions on ψ along the way, in particular, conditions which are typically fulfilled in scattering situations and which we shall discuss at the end.

Replacing ψ_{in} by ψ and introducing (16.64), equation (16.56) becomes

$$\mathbb{P}^{\psi}(\mathbf{X}_{\text{e}} \in \Sigma_R) \approx \frac{1}{(2\pi)^2}\int_\Sigma\int_0^\infty \left|\int f_{(k,\omega')}(\omega)\widehat{\psi}((k,\omega'))\mathrm{d}^2\omega'\right|^2 k^4\mathrm{d}k\mathrm{d}^2\omega\,.$$

We assume that the scattering amplitude varies only slowly as a function of $\mathbf{k}$ on the support of $\widehat{\psi}$ (see Fig. 16.6):

$$f_{\mathbf{k}}(\omega) \approx f_{\mathbf{k}_0}(\omega)\,, \quad \text{on } \operatorname{supp}\widehat{\psi}\,. \tag{16.66}$$

If this condition is fulfilled, the scattering amplitude may be pulled out of the integral, whence

$$\mathbb{P}^{\psi}(\mathbf{X}_{\text{e}} \in \Sigma_R) \approx \int_\Sigma \left|f_{\mathbf{k}_0}(\omega)\right|^2\mathrm{d}^2\omega\frac{1}{(2\pi)^2}\int_0^\infty\left|\int\widehat{\psi}((k,\omega'))\mathrm{d}^2\omega'\right|^2 k^4\mathrm{d}k\,. \tag{16.67}$$

Comparison with (16.61) suggests that we are nearly done. We need only understand what the factor

$$\frac{1}{(2\pi)^2}\int_0^\infty \left| \int \widehat{\psi}\big((k,\omega)\big)\mathrm{d}^2\omega \right|^2 k^4 \mathrm{d}k \tag{16.68}$$

is doing there. (We have renamed the integration variable ω' by ω.) For this purpose, introduce (k_1,k_2) as new variables for the solid angle variables (ϑ,φ). A short calculation shows that

$$k^2\mathrm{d}^2\omega = \frac{k_3(k_1,k_2)}{k}\mathrm{d}k_1\mathrm{d}k_2 \,, \qquad \text{where} \quad k_3(k_1,k_2) = \sqrt{k^2-k_1^2-k_2^2}\,.$$

This yields

$$\frac{1}{(2\pi)^2}\left| \int \widehat{\psi}\big((k,\omega)\big)k^2\mathrm{d}^2\omega \right|^2 = \frac{1}{(2\pi)^2}\left| \int \widehat{\psi}(k_1,k_2,k_3)\frac{k_3(k_1,k_2)}{k}\mathrm{d}k_1\mathrm{d}k_2 \right|^2 . \tag{16.69}$$

We now assume that

$$|k_1|,|k_2| \ll k_3 \,, \quad \text{i.e., } k \approx k_3 \,, \tag{16.70}$$

and that

$$\widehat{\psi}(k_1,k_2,k_3) \approx \widehat{\psi}(k_1,k_2,k) \,. \tag{16.71}$$

Under these conditions and using (16.69), equation (16.68) becomes

$$\frac{1}{(2\pi)^2}\int_0^\infty \left| \int \widehat{\psi}\big((k,\omega)\big)k^2\mathrm{d}^2\omega \right|^2 \mathrm{d}k = \frac{1}{(2\pi)^2}\int_0^\infty \left| \iint \widehat{\psi}(k_1,k_2,k)\mathrm{d}k_1\mathrm{d}k_2 \right|^2 \mathrm{d}k \,. \tag{16.72}$$

Denoting the partial (inverse) Fourier transform by

$$\tilde{\psi}(0,0,k) := \frac{1}{(2\pi)^2}\iint \widehat{\psi}(k_1,k_2,k)\mathrm{d}k_1\mathrm{d}k_2 \,,$$

we have

$$\frac{1}{(2\pi)^2}\int_0^\infty \left| \iint \widehat{\psi}(k_1,k_2,k)\mathrm{d}k_1\mathrm{d}k_2 \right|^2 \mathrm{d}k = \int_0^\infty |\tilde{\psi}(0,0,k)|^2 \mathrm{d}k \,.$$

Observing that $k_3 > 0$ on the support of $\widehat{\psi}$ and using the Plancherel identity, we finally obtain

$$\frac{1}{(2\pi)^2}\int_0^\infty \left| \int \widehat{\psi}\big((k,\omega)\big)k^2\mathrm{d}^2\omega \right|^2 \mathrm{d}k = \int_{-\infty}^\infty \big|\psi(0,0,x_3)\big|^2 \mathrm{d}x_3 = \int_{-\infty}^0 \big|\psi(0,0,x_3)\big|^2 \mathrm{d}x_3 \,,$$

since $\psi(x_1,x_2,x_3) = 0$ for $x_3 \geq 0$.

What is the meaning of the quantity on the right? Let A be an area perpendicular to $\mathbf{k}_0$ in front of the target, much larger than the target area but smaller than the cross-section of the beam (see Fig. 16.6), on which the wave function satisfies

$$\left|\psi(x_1,x_2,x_3)\right| \approx \left|\psi(0,0,x_3)\right| , \quad (x_1,x_2) \in A , \tag{16.73}$$

i.e., on which the lateral shape of the wave function is flat. Introduce

$$p_A := \frac{1}{|A|} \int_A \int_{-\infty}^{0} \left|\psi(x_1,x_2,x_3)\right|^2 \mathrm{d}x_1 \mathrm{d}x_2 \mathrm{d}x_3 \approx \int_{-\infty}^{0} \left|\psi(0,0,x_3)\right|^2 \mathrm{d}x_3 . \tag{16.74}$$

$p_A|A|$ is the probability that the particle is in the cylinder $A \times (-\infty, 0)$. Assuming that the wave packet does not spread too much laterally during the time the wave packet crosses the surface A, the particle will cross A with this probability, and hence we arrive at the meaning of p_A as the probability per unit area that the incoming particle crosses the surface A.

Introducing p_A into (16.67), we get

$$\mathbb{P}^{\psi}(\mathbf{X}_{\mathrm{e}} \in \Sigma_R) \approx p_A \int_{\Sigma} \left|f_{\mathbf{k}_0}(\omega)\right|^2 \mathrm{d}^2\omega . \tag{16.75}$$

Now recall (16.57). We multiply the probability density p_A accordingly by $\dot{N}_{\mathrm{B}}$, the number of particles per unit time in the beam, and get (by the law of large numbers) the number of particles per unit area per unit time crossing A, i.e., $\dot{N}_A/|A| \approx \dot{N}_{\mathrm{B}} p_A$. The scattering cross-section is by definition the area we need to multiply $\dot{N}_A/|A|$ by to obtain $\dot{N}_{\Sigma}$. We thus identify the scattering cross-section theoretically as Born's expression (16.61).

The remaining question is whether the conditions we employed along the way are consistent with the experimental scattering situation. The most demanding condition is perhaps (16.66). What does this mean in spatial terms? To see that, we approximate the generalized eigenfunction in (16.63) by a plane wave. The scattering amplitude then becomes, in this first order approximation, the so-called Born approximation, the Fourier transform of the potential, also referred to as the form factor:

$$f_{\mathbf{k}}(\omega) = \widehat{V}(\mathbf{k} - k\omega) . \tag{16.76}$$

For a rough understanding, let us use the reciprocal relation known from Gaussian shapes of the width $\Delta^F \mathbf{x}$ of a function F in space and $\Delta^{\widehat{F}} \mathbf{k}$ of the Fourier transform $\widehat{F}$, namely

$$\Delta^F \mathbf{x} \sim (\Delta^{\widehat{F}} \mathbf{k})^{-1} .$$

In terms of the form factor condition, (16.66) says that

$$\widehat{V}(\mathbf{k} - k\omega) \approx \widehat{V}(\mathbf{k}_0) , \quad \text{for } \mathbf{k} \in \operatorname{supp} \widehat{\psi} ,$$

whence $\widehat{V}$ must vary on a much larger scale than $\widehat{\psi}$, i.e., $\Delta^{\widehat{V}}\mathbf{k} \gg \Delta^{\widehat{\psi}}\mathbf{k}$. This means for the widths in space that $\Delta^{V}\mathbf{x} \ll \Delta^{\psi}\mathbf{x}$. In other words, the wave must be very much spread out compared to the spatial extent of the potential.

Now we come to conditions (16.70) and (16.71), of which (16.71) is the more demanding. As already mentioned at the beginning, in order to understand what is involved, we should think of $\widehat{\psi}$ as having a Gaussian shape (in particular it is positive), e.g.,

$$\widehat{\psi}(k_1,k_2,k_3) \sim \exp\left[-\frac{(k_1+k_2)^2}{2\sigma_\perp^2} - \frac{(k_3-k_0)^2}{2\sigma_\parallel^2}\right] .$$

Then (16.71) means roughly that, for k_1,k_2,k_3 varying within the widths $\sigma_\perp,\sigma_\parallel$,

$$\left|\widehat{\psi}(k_1,k_2,k_3) - \widehat{\psi}(k_1,k_2,k)\right| \ll \widehat{\psi}(k_1,k_2,k_3) ,$$

which leads to

$$\left|\partial_{k_3}\widehat{\psi}(k_1,k_2,k_3)(k-k_3)\right| \ll \widehat{\psi}(k_1,k_2,k_3) ,$$

that is,

$$\frac{k_3-k_0}{\sigma_\parallel^2}(k-k_3) \ll 1 .$$

Observing that

$$k-k_3 = \sqrt{k_1^2+k_2^2+k_3^2} - k_3 \approx \frac{1}{2}\frac{k_1^2+k_2^2}{k_3} \sim \frac{\sigma_\perp^2}{k_0} ,$$

one then has

$$\frac{k_3-k_0}{\sigma_\parallel^2}\frac{\sigma_\perp^2}{k_0} \sim \frac{\sigma_\parallel}{\sigma_\parallel^2}\frac{\sigma_\perp^2}{k_0} = \frac{\sigma_\perp^2}{\sigma_\parallel k_0} \ll 1 .$$

We introduce the positional spreads, i.e., the longitudinal spread $\sigma_\parallel^x = \sigma_\parallel^{-1}$ and the lateral spread $\sigma_\perp^x = \sigma_\perp{}^{-1}$, whereupon

$$\frac{\sigma_\perp}{\sigma_\perp^x}\frac{\sigma_\parallel^x}{k_0} \ll 1 . \tag{16.77}$$

This is so simple that one must wonder what it means. The answer is quite remarkable, as this condition is equivalent to another condition, which we mentioned in passing, to get the interpretation below (16.73). The point is that the wave function should not spread too much in the lateral direction when passing through the surface A. Multiplying the numerator and denominator of (16.77) by $\hbar/m$ and observing that

$$\frac{\sigma_{\parallel}^{x} m}{\hbar k_0} = T$$

is the time it takes for the wave function to cross the surface A, we set $\Delta_{\perp}^{x} = \hbar\sigma_{\perp} T/m$ for the lateral spread during that time, and we see that (16.77) amounts to the simple relation:

$$\frac{\Delta_{\perp}^{x}}{\sigma_{\perp}^{x}} \ll 1 ,$$

which is the no-spreading condition.

This insight also helps with another question one might raise, since there is a certain arbitrariness in the notion "in front of the target". Where exactly should the surface A be put? As long as the longitudinal spread of the wave function is large compared to the distance at which A is put in front of the target, the no-spreading condition ensures that the exact place where A is put does not matter.

In textbook treatments of scattering, another, much stronger no-spreading condition is often required, namely that the wave does not spread from the time of preparation to the time of detection (see, e.g., [22]). Its purpose is merely to simplify the mathematical argument, although this argument is simple enough without it, as can be seen from the above. However, one cannot do without the weaker condition we have employed here.

In real scattering experiments, the condition (16.73) is satisfied approximately for A extending over the cross-section of the beam, i.e., the decay of the wave function at the edges of the beam is concentrated in a very small region compared to the cross-section of the beam. In this case $p_A \approx 1/|A|$ [see (16.74)]. When the cross-section of the beam is very much larger than the extension of the scatterer potential, one may worry that only a very tiny fraction of the particles are being scattered. Therefore realistic scattering experiments like Rutherford scattering are done with an extended target foil, on which many scatterers are randomly distributed [23]. The incoming wave then overlaps a great many small scatterers.

Under certain conditions, which ensure no multiple scattering and no coherence effects (unlike Bragg scattering, where coherence is the key feature), the scatterers contribute independently to the numbers $\dot{N}_{\Sigma}$ of scattered particles per unit time. Therefore, if the incoming wave overlaps n_A scatterers, the relevant probability density which $\dot{N}_{\mathrm{B}}$ multiplies is no longer p_A, but

$$p_A^{\text{target foil}} \approx n_A p_A \approx \frac{n_A}{|A|} = \rho ,$$

the density of scatterers on the target. In this case, the formula for the scattering cross-section thus reads

$$\sigma_{\Sigma} = \frac{\dot{N}_{\Sigma}}{\dot{N}_{\mathrm{B}}\rho} .$$

A mathematically rigorous treatment of a scattering situation along the lines discussed in this chapter for a beam of random wave functions, which also intuitively models the random scatterer situation, can be found in [24], which includes many references on this subject.

References

1. M. Born: Z. Phys. **38**, 803 (1926)
2. M. Born: Z. Phys. **37**, 863 (1926)
3. H. Melville: *Moby Dick; or, The Whale* (William Benton, Encyclopaedia Britanica, Inc., 1952). Greatest Books of the Western World
4. D. Dürr, S. Teufel: In: *Stochastic Processes, Physics and Geometry: New Interplays, I* (Leipzig, 1999), CMS Conf. Proc., Vol. 28 (Amer. Math. Soc., Providence, RI, 2000) pp. 123–137
5. E. Nelson: *Dynamical Theories of Brownian Motion* (Princeton University Press, Princeton, N.J., 1967)
6. E. Nelson: *Quantum Fluctuations*. Princeton Series in Physics (Princeton University Press, Princeton, NJ, 1985)
7. M. Daumer, D. Dürr, S. Goldstein, N. Zanghì: J. Statist. Phys. **88** (3–4), 967 (1997)
8. W.O. Amrein, D.B. Pearson: J. Phys. A **30** (15), 5361 (1997)
9. S. Teufel, D. Dürr, K. Münch-Berndl: J. Math. Phys. **40** (4), 1901 (1999)
10. D. Dürr, T. Moser, P. Pickl: J. Phys. A **39** (1), 163 (2006)
11. S. Römer, D. Dürr, T. Moser: J. Phys. A **38**, 8421 (2005). `Math-ph/0505074`
12. J.D. Dollard: Rocky Mountain J. Math. **1** (1), 5 (1971)
13. J.D. Dollard, J. Math. Phys. **14** (6), 708 (1973). `link.aip.org/link/?JMP/14/708/1`
14. J.M. Combes, R.G. Newton, R. Shtokhamer: Physical Review D **11** (2), 366 (1975)
15. D. Dürr, S. Teufel: In: *Multiscale Methods in Quantum Mechanics: Theory and Experiment*, ed. by P. Blanchard, G.F. Dell'Antonio (Birkhäuser, Boston, 2003)
16. D. Dürr, T. Moser, S. Römer: preprint (2008)
17. M. Reed, B. Simon: *Methods of Modern Mathematical Physics III: Scattering Theory* (Academic Press, San Diego, 1979)
18. P.A. Perry: *Scattering Theory by the Enss Method*, Mathematical Reports Vol. 1, Part 1 (Harwood academic publishers, New York, 1983)
19. T. Ikebe: Archive for Rational Mechanics and Analysis **5**, 1 (1960)
20. P. Pickl: J. Math. Phys. **48** (12), 123505, 31 (2007)
21. P. Pickl, D. Dürr: Commun. Math. Phys. **282** (1), 161 (2008)
22. E. Merzbacher: *Quantum Mechanics*, 3rd edn. (John Wiley & Sons, Inc., New York, 1998)
23. M.L. Goldberger, K.M. Watson: *Collision Theory* (John Wiley & Sons, Inc., New York, 1964)
24. D. Dürr, S. Goldstein, T. Moser, N. Zanghì: Commun. Math. Phys. **266** (3), 665 (2006)

Chapter 17
Epilogue

> Finally: It was stated at the outset, that this system would not be here, and at once, perfected. You cannot but plainly see that I have kept my word. But I now leave my cetological System standing thus unfinished, even as the great Cathedral of Cologne was left, with the cranes still standing upon the top of the uncompleted tower. For small erections may be finished by their first architects; grand ones, true ones, ever leave the copestone to posterity. God keep me from ever completing anything. This whole book is but a draught – nay, but the draught of a draught. Oh, Time, Strength, Cash, and Patience!
>
> Melville (1851), Moby Dick, Chap. 32 [1]

What is a Bohmian quantum theory? A quantum theory that spells out what it is about. In other words a theory with a clear (primitive) ontology. A theory which has no place for mysticism, paradoxes, and superstition, and in which observation does not play an irreducible fundamental role. This does not mean that observation has no effect, or only a slight effect on an observed system. It does not mean that at all, as Bohmian mechanics proves. This chapter only draws one moral and its role is to encourage young scientists not to give up reason and rationality in their quest to understand how the physical world functions.

The moral is this. It has been claimed that quantum mechanics proves that a reasonable (some call that classical) understanding of the world is impossible, and that it is impossible to write down the laws of nature in terms of a clear ontology. Bohmian mechanics proves that such claims are false, even embarrassingly false, because Bohmian mechanics is an utterly straightforward completion of quantum mechanics. Indeed, it is quantum mechanics.

A satisfactory Lorentz invariant Bohmian theory still needs to be found. The fact that it has not been found yet does not prove that it is impossible to do so. It may prove, if anything, that we need to think harder, and work harder, and that we need to scrutinize our established modes of thinking and remain open to reasonable ideas. Here, reasonable means simply something like this: change our idea about space-time, change the structure of Schrödinger's equation, change the ontology, make it points, fields, strings, or make it whatever seems right for describing nature, but *never give up ontology*!

D. Dürr, S. Teufel, *Bohmian Mechanics*, DOI 10.1007/978-3-540-89344-8_17,

References

1. H. Melville: *Moby Dick; or, The Whale* (William Benton, Encyclopaedia Britanica, Inc., 1952). Greatest Books of the Western World

Bibliography

1. V. Allori, D. Dürr, S. Goldstein, and N. Zanghì: Seven steps towards the classical world, Journal of Optics B **4**, 482–488 (2002)
2. V. Allori, S. Goldstein, R. Tumulka, and N. Zanghí: On the common structure of Bohmian mechanics and the Ghirardi–Rimini–Weber theory, British Journal for the Philosophy of Science.
3. W.O. Amrein and D.B. Pearson: Flux and scattering into cones for long range and singular potentials, J. Phys. A **30**, no. 15, 5361–5379 (1997)
4. W. Appel and M.K.-H. Kiessling: Mass and spin renormalization in Lorentz electrodynamics, Ann. Physics **289**, no. 1, 24–83 (2001)
5. A. Aspect, J. Dalibard, and G. Roger: Experimental test of Bell's inequalities using time-varying analyzers, Phys. Rev. Lett. **49**, no. 25, 1804–1807 (1982)
6. G. Bacciagaluppi and A. Valentini: *Quantum Theory at the Crossroads: Reconsidering the 1927 Solvay Conference*, to be published by Cambridge University Press 2008 (2006) p. 553
7. A.O. Barut: *Electrodynamics and Classical Theory of Fields & Particles*, Dover Publications Inc., New York, 1980, corrected reprint of the 1964 original
8. A. Bassi and G. Ghirardi: Dynamical reduction models, Phys. Rep., **379**, no. 5–6, 257–426 (2003)
9. G. Bauer and D. Dürr: The Maxwell–Lorentz system of a rigid charge, Ann. Henri Poincaré **2**, no. 1, 179–196 (2001)
10. J.S. Bell: *Speakable and Unspeakable in Quantum Mechanics*, Cambridge University Press, Cambridge, 1987
11. M. Bell, K. Gottfried, and M. Veltman (Eds.): *John S. Bell on the Foundations of Quantum Mechanics*, World Scientific Publishing Co. Inc., River Edge, NJ, 2001
12. K. Berndl: *Zur Existenz der Dynamik in Bohmschen Systemen*, Ph.D. thesis, Ludwig-Maximilians-Universität München, 1994
13. K. Berndl, D. Dürr, S. Goldstein, G. Peruzzi, and N. Zanghì: On the global existence of Bohmian mechanics, Commun. Math. Phys. **173**, no. 3, 647–673 (1995)
14. P. Bocchieri and A. Loinger: Quantum recurrence theorem, Phys. Rev. **107** (2), 337–338 (1957)
15. D. Bohm: *Quantum Theory*, Prentice Hall, New York, 1951
16. D. Bohm: A suggested interpretation of the quantum theory in terms of "hidden" variables I, II, Physical Review **85**, 166–179, 180–193 (1952)
17. D. Bohm and B. J. Hiley: *The Undivided Universe. An Ontological Interpretation of Quantum Theory*, Routledge, London, 1995
18. H. Bohr: *Fastperiodische Funktionen*, Springer, 1932
19. I. Born: *The Born–Einstein Letters*, Walker and Company, New York, 1971
20. M. Born: Quantenmechanik der Stoßvorgänge, Z. Phys. **38**, 803–827 (1926)
21. M. Born: Zur Quantenmechanik der Stoßvorgänge, Z. Phys. **37**, 863–867 (1926)
22. M. Born and L. Infeld: *Foundation of the New Field Theory*, Commun. Math. Phys. A **144**, 425–451 (1934)

23. J. Bricmont: Science of chaos or chaos in science? In: *The Flight from Science and Reason* (New York, 1995), Vol. 775 of Ann. New York Acad. Sci., New York, 1996, pp. 131–175
24. J. Burnet: *Early Greek Philosophy*, A. and C. Black, London and Edinburgh, 1930
25. L. Carroll: *The Complete, Fully Illustrated Works*, Gramercy Books, New York, 1995
26. J.-M. Combes, R.G. Newton, and R. Shtokhamer: Scattering into cones and flux across surfaces, Physical Review D **11**, no. 2, 366–372 (1975)
27. J.T. Cushing: *Quantum Mechanics*. Science and Its Conceptual Foundations, University of Chicago Press, Chicago, IL, 1994. Historical contingency and the Copenhagen hegemony
28. M. Daumer, D. Dürr, S. Goldstein, and N. Zanghì: On the quantum probability flux through surfaces, J. Stat. Phys. **88**, nos. 3–4, 967–977 (1997)
29. E.B. Davies: *Spectral Theory and Differential Operators*, Vol. 42 of Cambridge Studies in Advanced Mathematics, Cambridge University Press, Cambridge, 1995
30. E. Deotto and G.C. Ghirardi: Bohmian mechanics revisited, Found. Phys. **28**, no. 1, 1–30 (1998)
31. P. Dirac: Classical theory of radiating electrons, Proc. Roy. Soc. A **178**, 148 (1938)
32. J.D. Dollard: Quantum-mechanical scattering theory for short-range and Coulomb interactions, Rocky Mountain J. Math. **1**, no. 1, 5–88 (1971)
33. J.D. Dollard: Scattering into cones. II. N-body problems, Journal of Mathematical Physics **14**, no. 6, 708–718 (1973)
34. D. Dürr, S. Goldstein, T. Moser, and N. Zanghì: A microscopic derivation of the quantum mechanical formal scattering cross section, Commun. Math. Phys. **266**, no. 3, 665–697 (2006)
35. D. Dürr, S. Goldstein, K. Münch-Berndl, and N. Zanghì: Hypersurface Bohm–Dirac models, Physical Review A **60**, 2729–2736 (1999)
36. D. Dürr, S. Goldstein, J. Taylor, R. Tumulka, and N. Zanghì: Topological factors derived from Bohmian mechanics, Ann. Henri Poincaré **7**, no. 4, 791–807 (2006)
37. D. Dürr, S. Goldstein, R. Tumulka, and N. Zanghì: Bohmian mechanics and quantum field theory, Phys. Rev. Lett. **93**, no. 9, 090402–090404 (2004)
38. D. Dürr, S. Goldstein, R. Tumulka, and N. Zanghì: Bell-type quantum field theories, J. Phys. A **38**, no. 4, R1–R43 (2005)
39. D. Dürr, S. Goldstein, R. Tumulka, and N. Zanghì: John Bell and Bell's theorem. In: *Encyclopedia of Philosophy*, ed. by D.M. Borchert, Macmillan Reference, USA, 2005
40. D. Dürr, S. Goldstein, and N. Zanghì: Quantum equilibrium and the origin of absolute uncertainty, Journal of Statistical Physics **67**, 843–907 (1992)
41. D. Dürr, S. Goldstein, and N. Zanghì: Bohmian mechanics and the meaning of the wave function. In: *Experimental Metaphysics* (Boston, MA, 1994), Vol. 193 of Boston Stud. Philos. Sci., Kluwer Acad. Publ., Dordrecht, 1997, pp. 25–38
42. D. Dürr, S. Goldstein, and N. Zanghi: Quantum equilibrium and the role of operators as observables in quantum theory, J. Stat. Phys., **116**, nos. 1–4, 959–1055 (2004)
43. D. Dürr, S. Goldstein, and N. Zanghì: *On the Weak Measurement of Velocity in Bohmian Mechanics*, J. Stat. Phys., online first
44. D. Dürr, T. Moser, and P. Pickl: The flux-across-surfaces theorem under conditions on the scattering state, J. Phys. A **39**, no. 1, 163–183 (2006)
45. D. Dürr, T. Moser, and S. Römer: *On the Exit Statistics Theorem of N-Particle Quantum Scattering*, preprint
46. D. Dürr and S. Teufel: On the role of the flux in scattering theory. In: *Stochastic Processes, Physics and Geometry: New Interplays, I* (Leipzig, 1999), Vol. 28 of CMS Conf. Proc., Amer. Math. Soc., Providence, RI, 2000, pp. 123–137
47. D. Dürr and S. Teufel: On the exit statistics theorem of many particle quantum scattering. In: *Multiscale Methods in Quantum Mechanics: Theory and experiment*, ed. by P. Blanchard and G. F. Dell'Antonio, Birkhäuser, Boston, 2003
48. H. Dym and H.P. McKean: *Fourier Series and Integrals*, Probability and Mathematical Statistics No. 14, Academic Press, New York, 1972
49. P. Ehrenfest and T. Ehrenfest: *The Conceptual Foundations of the Statistical Approach in Mechanics*, Dover Publications Inc., New York, English edn., 1990, translated from the German by Michael J. Moravcsik, with a foreword by M. Kac and G.E. Uhlenbeck

50. A. Einstein: *Investigations on the Theory of the Brownian Movement*, Dover Publications Inc., New York, 1956, edited with notes by R. Fürth, translated by A.D. Cowper.
51. A. Einstein, B. Podolsky, and N. Rosen: Can quantum-mechanical description of physical reality be considered complete? Physical Review **41**, 777 (1935)
52. A. Einstein and M.V. Smoluchowski: *Untersuchungen über die Theorie der Brownschen Bewegung/ Abhandlung über die Brownsche Bewegung und verwandte Erscheinungen*, Verlag Harry Deutsch, New York, 1997, reihe Ostwalds Klassiker 199
53. R. Feynman: *The Character of Physical Law*, MIT Press, Cambridge, 1992
54. R.P. Feynman, F.B. Morinigo, and W.G. Wagner, *Feynman Lectures on Gravitation*, ed. by B. Hatfield, Addison-Wesley Publishing Company, 1959
55. A.D. Fokker: Z. Physik **58**, 386 (1929)
56. C. Gauss: A letter to Weber on 19 March 1845
57. G.C. Ghirardi, P. Pearle, and A. Rimini: Markov processes in Hilbert space and continuous spontaneous localization of systems of identical particles, Phys. Rev. A (3) **42**, no. 1, 78–89 (1990)
58. D. Giulini, E. Joos, C. Kiefer, J. Kumpsch, I.-O. Stamatescu, and H. Zeh: *Decoherence and the Appearance of a Classical World in Quantum Theory*, Springer-Verlag, Berlin, 1996
59. M.L. Goldberger and K.M. Watson: *Collision Theory*, John Wiley & Sons, Inc., New York, 1964
60. S. Goldstein: Boltzmann's approach to statistical mechanics. In: *Chance in Physics: Foundations and Perspectives*, ed. by J. Bricmont, D. Dürr, M.C. Galavotti, G. Ghirardi, F. Petruccione, and N. Zanghí
61. S. Goldstein, J.L. Lebowitz, R. Tumulka, and N. Zanghì: Canonical typicality, Phys. Rev. Lett. **96**, no. 5, 050403 (2006)
62. S. Goldstein and W. Struyve: On the uniqueness of quantum equilibrium in Bohmian mechanics, J. Stat. Phys. **128**, no. 5, 1197–1209 (2007)
63. W. Heisenberg: *Quantentheorie und Philosophie*, Vol. 9948 of *Universal-Bibliothek [Universal Library]*, Reclam-Verlag, Stuttgart, 1987, vorlesungen und Aufsätze. Lectures and essays, with an afterword by Jürgen Busche
64. P.R. Holland: *The Quantum Theory of Motion*, Cambridge University Press, Cambridge, 1995
65. T. Ikebe: Eigenfunction expansion associated with the Schrödinger operators and their applications to scattering theory, Archive for Rational Mechanics and Analysis **5**, 1–34 (1960)
66. M. Kac: *Statistical Independence in Probability, Analysis and Number Theory*, The Carus Mathematical Monographs, No. 12, published by the Mathematical Association of America. Distributed by John Wiley and Sons, Inc., New York, 1959
67. M. Kac, G.-C. Rota, and J.T. Schwartz: *Discrete Thoughts. Scientists of Our Time*, Birkhäuser Boston Inc., Boston, MA, 1986, essays on mathematics, science, and philosophy
68. E. Kappler: Versuche zur Messung der Avogadro-Loschmidtschen Zahl aus der Brownschen Bewegung einer Drehwaage, Annalen der Physik **11**, 233–256 (1931)
69. M.K.-H. Kiessling: Renormalization in radiation reaction: New developments in classical electron theory. In: *Nonlinear Dynamics and Renormalization Group* (Montreal, QC, 1999), Vol. 27 of CRM Proc. Lecture Notes, Amer. Math. Soc., Providence, RI, 2001, pp. 87–96
70. O.E. Lanford, III: Time evolution of large classical systems. In: *Dynamical Systems, Theory and Applications* (Recontres, Battelle Res. Inst., Seattle, Wash., 1974), Lecture Notes in Physics, Vol. 38, Springer, Berlin, 1975, pp. 1–111
71. E. Madelung: Quantum theory in hydrodynamic form, Zeitschrift für Physik **40**, no. 3/4, 322–326 (1926)
72. T. Maudlin: *Quantum Non-Locality and Relativity: Metaphysical Intimations of Modern Physics*, Vol. 13 of Aristotelian Society Series, Oxford UK and Cambridge, Blackwell, 1994
73. H. Melville: *Moby Dick; or, The Whale*, William Benton, Encyclopaedia Britanica, Inc., 1952, Greatest Books of the Western World
74. E. Merzbacher: *Quantum Mechanics*, 3rd edn., John Wiley & Sons, Inc., New York, 1998
75. E. Nelson: *Dynamical Theories of Brownian Motion*, Princeton University Press, Princeton, N.J., 1967

76. E. Nelson: *Quantum Fluctuations*, Princeton Series in Physics, Princeton University Press, Princeton, NJ, 1985
77. J. Novik: A covariant formulation of classical electrodynamics for charges of finite extension, Ann. Phys. **28**, 225–319 (1964)
78. R. Penrose: *The Emperor's New Mind*, The Clarendon Press, Oxford University Press, New York, 1989. Concerning computers, minds, and the laws of physics, with a foreword by Martin Gardner
79. R. Penrose: *The Road to Reality*, Alfred A. Knopf Inc., New York, 2005. A complete guide to the laws of the universe
80. P.A. Perry: *Scattering Theory by the ENSS Method*, Mathematical Reports Vol.1, Part 1, Harwood Academic Publishers, New York, 1983
81. P. Pickl: Generalized eigenfunctions for Dirac operators near criticality, J. Math. Phys. **48**, no. 12, 123505–123531 (2007)
82. P. Pickl and D. Dürr: Adiabatic pair creation, Commun. Math. Phys. **282**, no. 1, 161–198 (2008)
83. H. Poincaré: *Wissenschaft und Hypothese*, Teubner Verlag, Leipzig, 1914
84. M. Reed and B. Simon: *Methods of Modern Mathematical Physics. II. Fourier Analysis, Self-Adjointness*, Academic Press [Harcourt Brace Jovanovich Publishers], New York, 1975
85. M. Reed and B. Simon: *Methods of Modern Mathematical Physics III: Scattering Theory*, Academic Press, San Diego, 1979
86. M. Reed and B. Simon: *Methods of Modern Mathematical Physics I: Functional Analysis*, revised and enlarged edn., Academic Press, San Diego, 1980
87. F. Rohrlich: *Classical Charged Particles*, World Scientific Publishing Co. Pte. Ltd., 3rd edn., Hackensack, NJ, 2007
88. S. Römer, D. Dürr, and T. Moser: Asymptotic behavior of Bohmian trajectories in scattering situations, J. Phys. A **38**, 8421–8443 (2005); `math-ph/0505074`
89. W. Rudin: *Principles of Mathematical Analysis*, 3rd edn., International Series in Pure and Applied Mathematics, McGraw-Hill Book Co., New York, 1976
90. F. Scheck: *Mechanics. From Newton's Laws to Deterministic Chaos*, 3rd edn., Springer-Verlag, Berlin, 1999, translated from the German
91. A. Schild: Electromagnetic two-body problem, Phys. Rev. **131**, 2762 (1962)
92. E. Schmidt: Math. Annalen **64**
93. E. Schrödinger: Die gegenwärtige Situation in der Quantenmechanik, Naturwissenschaften **23**, 807–812, 823–828, 844–849 (1935)
94. E. Schrödinger: *Unsere Vorstellung von der Materie*. CD-ROM: Was ist Materie?, Original Tone Recording, 1952, Suppose Verlag, Köln, 2007
95. K. Schwarzschild: Nachr. Ges. Wis. Gottingen **128**, 132 (1903)
96. M. Smoluchowski: Über den Begriff des Zufalls und den Ursprung der Wahrscheinlichkeitsgesetze in der Physik, Die Naturwissenschaften **17**, 253–263 (1918)
97. H. Spohn: *Dynamics of Interacting Particles*, Springer Verlag, Heidelberg, 1991
98. H. Spohn: *Dynamics of Charged Particles and the Radiation Field*, Cambridge University Press, 2004
99. H. Tetrode: Z. Physik **10**, 317 (1922)
100. S. Teufel, D. Dürr, and K. Münch-Berndl: The flux-across-surfaces theorem for short range potentials and wave functions without energy cutoffs, J. Math. Phys. **40**, no. 4, 1901–1922 (1999)
101. S. Teufel and R. Tumulka: A simple proof for global existence of Bohmian trajectories, Commun. Math. Phys. **258**, no. 2, 349–365 (2005)
102. A. Tonomura, J. Endo, T. Matsuda, T. Kawasaki, and H. Ezawa: Demonstration of single-electron build-up of an interference pattern, American Journal of Physics **57**, 117–120 (1989)
103. R. Tumulka: On spontaneous wave function collapse and quantum field theory, Proc. R. Soc. Lond. Ser. A Math. Phys. Eng. Sci. **462**, no. 2070, 1897–1908 (2006)
104. R. Tumulka: A relativistic version of the Ghirardi–Rimini–Weber model, J. Stat. Phys. **125**, no. 4, 825–844 (2006)
105. A. Valentini: Astrophysical and cosmological tests of quantum theory, J. Phys. A **40**, no. 12, 3285–3303 (2007)

106. J. Wheeler and R. Feynman: Rev. Mod. Phys. **17**, 157 (1945)
107. J.A. Wheeler and R.P. Feynman: Classical electrodynamics in terms of direct interparticle action, Rev. Modern Physics **21**, 425–433 (1949)
108. J.A. Wheeler and W.H. Zurek (Eds.): *Quantum Theory and Measurement*, Princeton Series in Physics, Princeton University Press, Princeton, NJ, 1983
109. E. Wigner: On hidden variables and quantum mechanical probabilities, American Journal of Physics **38**, 1005–1009 (1970)

Index

CPSIA information can be obtained
at www.ICGtesting.com
Printed in the USA
LVOW13*2034250717
542596LV00009B/182/P